Lecture Notes in Mechanical Engineering

Lecture Notes in Mechanical Engineering (LNME) publishes the latest developments in Mechanical Engineering—quickly, informally and with high quality. Original research reported in proceedings and post-proceedings represents the core of LNME. Volumes published in LNME embrace all aspects, subfields and new challenges of mechanical engineering. Topics in the series include:

- Engineering Design
- Machinery and Machine Elements
- Mechanical Structures and Stress Analysis
- Automotive Engineering
- Engine Technology
- Aerospace Technology and Astronautics
- Nanotechnology and Microengineering
- Control, Robotics, Mechatronics
- MEMS
- Theoretical and Applied Mechanics
- Dynamical Systems, Control
- Fluid Mechanics
- Engineering Thermodynamics, Heat and Mass Transfer
- Manufacturing
- Precision Engineering, Instrumentation, Measurement
- Materials Engineering
- Tribology and Surface Technology

To submit a proposal or request further information, please contact the Springer Editor of your location:

China: Dr. Mengchu Huang at mengchu.huang@springer.com
India: Priya Vyas at priya.vyas@springer.com
Rest of Asia, Australia, New Zealand: Swati Meherishi at swati.meherishi@springer.com
All other countries: Dr. Leontina Di Cecco at Leontina.dicecco@springer.com

To submit a proposal for a monograph, please check our Springer Tracts in Mechanical Engineering at http://www.springer.com/series/11693 or contact Leontina.dicecco@springer.com

Indexed by SCOPUS. All books published in the series are submitted for consideration in Web of Science.

More information about this series at http://www.springer.com/series/11236

L. Venkatakrishnan · Sekhar Majumdar ·
Ganesh Subramanian · G. S. Bhat ·
Ratul Dasgupta · Jaywant Arakeri
Editors

Proceedings of 16th Asian Congress of Fluid Mechanics

Springer

Editors
L. Venkatakrishnan
Experimental Aerodynamics Division
CSIR-National Aerospace Laboratories
Bengaluru, Karnataka, India

Sekhar Majumdar
Department of Mechanical Engineering
NMIT Bangalore
Bengaluru, Karnataka, India

Ganesh Subramanian
Engineering Mechanics Unit
Jawaharlal Nehru Centre for Advanced Scientific Research
Bengaluru, Karnataka, India

G. S. Bhat
Centre for Atmospheric and Oceanic Sciences
Indian Institute of Science
Bengaluru, Karnataka, India

Ratul Dasgupta
Department of Chemical Engineering
Indian Institute of Technology Bombay
Mumbai, Maharashtra, India

Jaywant Arakeri
Department of Mechanical Engineering
Indian Institute of Science
Bengaluru, Karnataka, India

ISSN 2195-4356 ISSN 2195-4364 (electronic)
Lecture Notes in Mechanical Engineering
ISBN 978-981-15-5182-6 ISBN 978-981-15-5183-3 (eBook)
https://doi.org/10.1007/978-981-15-5183-3

This Springer imprint is published by the registered company Springer Nature Singapore Pte Ltd.
The registered company address is: 152 Beach Road, #21-01/04 Gateway East, Singapore 189721, Singapore

Foreword[1]

This Sixteenth Asian Congress of Fluid Mechanics (ACFM16) takes my mind back in history to ACFM1 held in December 1980, at Bangalore, almost forty years ago. ACFM1 was then organized at the Indian Institute of Science (IISc); ACFM16 met at the Jawaharlal Nehru Centre for Advanced Scientific Research (JNCASR). So I am sure that all the fluid dynamicists here in Bangalore, and I believe all those who came from elsewhere in India as well, were happy to welcome all the participants in this Congress, mostly from Asia but quite a few from elsewhere in the world as well.

It might be appropriate to first describe how the Congress took birth, as there are probably very few attending ACFM16 that were also present in ACFM1.

The origin for the Congress was actually the idea of the late Prof. Hiroshi Sato and myself when we were both in Pasadena, California—I was visiting the Graduate Aeronautical Laboratories of the California Institute of Technology (GALCIT), and Prof. Sato was a visiting scientist at the Jet Propulsion Laboratories (JPL, managed by Caltech). We began by meeting often at the Caltech seminars and quickly got to know each other. After a few meetings, I once asked him, 'Isn't it strange that we meet each other in Pasadena, so far away for me and quite far away for you too, rather than meeting on our home grounds in Asia?'. Sato said 'A good question' and asked me what we should do. I proposed that we should have an Asian Congress of Fluid Mechanics so that scientists from all over Asia could get together and discuss this interesting and important subject of fluid mechanics once every three or four years. Sato clearly liked the idea very much, and we started planning how we might do it. So it was decided to write to leading fluid dynamicists in any Asian country where there was some activity in fluid mechanics going on. A joint statement was drafted and distributed to known fluid dynamicists in all such Asian countries. The response was overwhelming—the idea was welcomed, and every reply looked forward to the first congress. It was agreed that IISc would be the host for the meeting, and in fact, people from many other countries were also happy to come to Bangalore. At that time, there was more fluid mechanics in Bangalore than in any other city in the country, partly

[1]This Foreword is a summary of the inaugural address delivered at ACFM16 on 'Memories of the First Asian Congress of Fluid Mechanics'.

because it was the centre for aerospace research and development. Apart from IISc which had a strong teaching and research programme in the subject, there was also the National Aeronautical Laboratory (NAL) which had several wind tunnels (including a 4 × 4 ft. test section supersonic tunnel and several smaller ones at both high and low speeds) and became a partner of IISc in hosting the Congress. Furthermore, Bangalore was also the headquarters of Hindustan Aeronautics Limited (the only aircraft manufacturing company in the country), a Gas Turbine Research Establishment, and the headquarters of the Indian Space Research Organization (ISRO) and its Satellite Centre. And there were a few other laboratories and smaller industries (Figs. 1 and 2).

Professor Satish Dhawan, a graduate of GALCIT, who was the father (so to speak) of the development of modern fluid dynamics in India and was both Director of IISc and Chairman of ISRO, was very enthusiastic about the idea and helped us to get

Fig. 1 Sato and RN trying to sort out a last-minute problem at the IISc Faculty Hall where ACFM1 was held

Fig. 2 The three supporters of the Congress. (Left to right) Mr. Raj Mahindra (Chief, Design Bureau, HAL); Prof. S. Dhawan; Dr. S. R. Valluri (Director NAL)

the required resources from various institutions. Sato and I made a list of invited speakers from different parts of Asia. The only problem at the time was that we had not heard from China, although the invitation had gone there too. However, when we had almost come to the conclusion that Chinese scientists may not be able to come, we were happy to see a delegation of four leading fluid dynamicists led by Prof. Chou Pei-Yuan, also a Ph.D. from Caltech and the leader of aeronautical science in China.

The Congress assembled on 8 December 1980 at the Faculty Hall at IISc. Sato (then from the Institute of Space and Aeronautical Science of Japan) started the meeting by speaking on the background to the organization of ACFM. He then urged Asian scientists to exchange information on the science of fluids in motion, with the final aim of creating what he called 'an Asian style system of doing fluid mechanics' (which he had also said in a press interview). This was followed by the scientific programme, the first speaker being Satish Dhawan, whose subject was 'A glimpse of Fluid Mechanics research in Bangalore 25 years ago'. This was an account of the history of how modern fluid dynamics grew in the Aeronautical Engineering Department at IISc. He had joined the institute in 1951, was on the faculty in 1952 and quickly set up a High-Speed Aerodynamics Laboratory as well as a low-speed boundary layer tunnel in the department. These facilities grew rapidly, and the first step of research was about studies in transition from laminar to turbulent flow in the boundary layer tunnel. This led to some interesting results, including the successful description of the intermittency distribution in the transition zone, for which a new expression assuming concentrated breakdown in the boundary layer at a specific distance from the leading edge where, to a good first approximation, all turbulent spots were born. This was different from the proposal in the earlier literature and was followed by other studies on transition as well as relaminarization, at both low and high speeds. Dhawan went on to become a professor in the department a few years later and became director of the institute at the young age of 42. The fluid mechanics enterprise grew rapidly at both IISc and NAL in the following years.

The Congress programme continued and had a total of 16 invited lectures. These appeared in 1982 in a book called *Surveys in Fluid Mechanics* published by the Indian Academy of Sciences [1]. All invited speakers submitted their papers on their talks so that we could make a collection of interesting accounts of what was happening in Asian fluid mechanics. I would like to make a few brief comments about a selection of the invited addresses (Fig. 3).

From Japan, we had four invited speakers, beginning with the internationally known scientist Itiro Tani, who spoke on 'three-dimensional aspects of boundary layer transition'. This was a review of how transition in a 2D boundary layer could have instability of a 3D character, leading eventually to 3D turbulent spots. He also considered the 3D effects of roughness elements and 3D boundary layers on rotating discs. The lecture was a comprehensive review of the status of our understanding of boundary layer transition 40 years ago (Fig. 4).

Chou Pei-Yuan gave a lecture on solutions of the Navier–Stokes equations on the theory of homogeneous isotropic turbulence. He had done some work in turbulence even while he was at Caltech, and his interest had continued; he had also become the

Fig. 3 Cover of *Surveys in Fluid Mechanics*

leader of fluid dynamics in China. He developed an expansion method to obtain an axisymmetric solution of the Navier–Stokes equations in the form of an infinite set of nonlinear partial differential equations of the second order. He solved the zeroth-order approximation numerically and obtained results for the energy spectrum and showed that his calculations generally agreed with experiments except in a few cases (Fig. 5).

Among the invited speakers from outside Asia was Gerald B. Whitham, who headed the Department of Applied Mathematics at Caltech. He gave a lecture on finite amplitude water waves which he studied using the variational technique. The talk he gave concentrated on waves in deep water, for which he found numerical solutions that showed how the results could be extended to the case of finite depth (Fig. 6).

Another guest from Caltech, Donald Coles at GALCIT, spoke on prospects for useful research on coherent structures in turbulent shear flows. As is well known, the work of Brown and Roshko at Caltech, published in 1974 [2], had demonstrated how the turbulent mixing layer consisted of a series of nearly well-organized (turbulent) flow structures. This had a major impact on the problem of coherent structures in what apparently was a basically random process that was turbulent. He reviewed

Fig. 4 Itiro Tani (National Aerospace Laboratory and University of Tokyo)

Fig. 5 The three Asian leaders. (Left to right) Dhawan, Chou and Sato, with the Governor of Karnataka H. E. Sri Govind Narayan at a reception in Raj Bhavan

such structures in several flows, from puffs in a pipe through spiral turbulence in the flow in the gap between two rotating circular cylinders, the vortex street behind a cylinder, a vortex ring and of course the mixing layer. He made a survey of how the presence of such structures may affect the way that we think of turbulence and of how the idea may be productive when applied to a single structure. There was also

Fig. 6 G. B. Whitham (Caltech) and RN

Fig. 7 Donald Coles (GALCIT, Caltech), in a thoughtful conversation with another participant, behind whom Chuck Van Atta (a student of Coles) may be seen

the possibility that at least partial control of important flows might make exploitation of coherent structures feasible (Fig. 7).

Coherent structures were also the subject of Fazle Hussain's lecture. His written contribution was an extensive and detailed review of coherent structures in turbulent shear flows. His research at the University of Houston led to the possibility that coherent structures were dominant in transport of the early stages of their formation but not in the self-preserving regions (Fig. 8).

M. Ohji, speaking on modelling of turbulent shear flows, identified three kinds: integral, differential and numerical and discussed their individual characteristics. T. Matsui presented interesting flow visualizations of a variety of vortices: the Karman vortex street, the flow around a rotating cylinder as well as an impulsively started circular cylinder, and vortex rings.

Fig. 8 Fazle Hussain (University of Houston), with Dhawan

There were two talks of aeronautical interest on transonic flows. T. C. Lin and Jia Zhenxue (China) described a new method of analysing inviscid transonic flow in which the observation that the streamline angle is close to that in incompressible flow is exploited. P Niyogi (India) presented an analysis of inviscid transonic flow past a thin wing in a subsonic free-stream using the integral method. He applied the method to derive a solution for the shock-free supercritical flow on a thin wing at zero incidence.

All these presentations, whether experimental or theoretical, had the potential of providing greater understanding of the true structure of turbulent shear flows as well as their applications in technology. Also, in the same category, there were presentations by Koshi Oshima (Japan) on thermal and fluid mechanical problems in space, Vijay Arakeri (India) on real fluid effects on flow cavitation and Sivasegaram (from Sri Lanka) on wind tunnel tests for vertical axis wind rotors.

There were three interesting lectures of geophysical interest. Sulochana Gadgil (IISc) gave a comprehensive lecture on a meteorological problem that is still not completely solved, namely the fluid dynamics of the Indian monsoons. The talk started with a brief historical survey of the idea of the monsoons, going back to the first century BC and then through the Hadley circulation and seasonal variation in the atmosphere and oceans. She gave a description of the discovery that had been made (with D. R. Sikka) of the intertropical convergence zone in India and its cycles of motion from near the equator to near the Himalayas during the monsoon season.

Fig. 9 Sulochana Gadgil (IISc, on the right) with Neelima Narasimha

This is of considerable significance for understanding the Asian monsoon rainfall. She continued then to a discussion of other issues connected with the monsoons, involving the Hadley Cell, the conditional instabilities of the first and second kind in cumulus dynamics and interactions between cavity and synoptic scales, concluding with transition, fluctuations and climatic changes. Ruby Krishnamurti (Florida State University) first described how large-scale circulation develops on the earth's surface as the Rayleigh number increases and presented a simple mathematical model whose solution exhibits qualitative similarity to the observed flow. P. K. Das (India Meteorological Department) presented an analysis of the shallow-water equations and obtained numerical solutions which show that, in the initial stages, the divergence of the wind stress is more important than its curl, which in turn becomes more important than the divergence in later stages (Fig. 9).

At the conclusion of the sessions, we had made arrangements for exposing the delegates to things Indian. I hope the three pictures put together here, about the art of Indian music and dancing, Dhawan cutting jokes with the younger generation and the chatting ladies who helped us greatly in running the Congress, will give the reader some idea of the Indian evenings that the participants experienced during the Congress (Figs. 10, 11 and 12).

Forty years later, there has been considerable progress in some of the issues that were raised at the first Congress in Bangalore, and some things have changed and offered greater understanding of fluid mechanics in the subjects that were selected for ACFM1. Although difficulties remain (certainly in turbulent flows), there are

Fig. 10 For the evening: Dancing and music

Fig. 11 For the evening: Dhawan, with the younger generation

Fig. 12 The ladies who worked hard for the Congress

many new problems now, as I believe will become clear in the rest of this book of the Proceedings of this Sixteenth Congress.

July 2020

Roddam Narasimha
Engineering Mechanics Unit
Jawaharlal Nehru Centre for Advanced Scientific Research (JNCASR)
Bengaluru, India

References

1. Narasimha R, Deshpande SM (1982) Surveys in fluid mechanics. Indian Academy of Sciences, Bengaluru
2. Brown G L, Roshko A (1974) On density effects and large structure in turbulent mixing layers. J Fluid Mech 64:775–816

Contents

About the Editors

Dr. L. Venkatakrishnan completed his Master's in Aeronautical Engineering from the Indian Institute of Technology Bombay in 1991 and obtained his Ph.D. in Aerospace Engineering at the Indian Institute of Science in 1997, before joining the Fluid Mechanics Research Laboratory (FMRL) at Florida State University as a Postdoctoral Fellow. Dr. Venkatakrishnan joined the National Aerospace Laboratories in June 2000 and is currently Chief Scientist and Head of the Experimental Aerodynamics Division and concurrently Professor, Academy of Scientific and Innovative Research (AcSIR). His main interests are in flow diagnostics, fluid mechanics and aeroacoustics, and he has developed new measurement methodologies as well as techniques for launch vehicle noise suppression. He has about 80 publications in international refereed journals and conferences in addition to numerous invited lectures in international conferences. He is a Indian National Academy of Engineering Fellow, Royal Aeronautical Society, UK Fellow, Associate Fellow of the American Institute of Aeronautics and Astronautics (Elected 2010) and has been awarded the Prof. Satish Dhawan Award for Young Engineers in 2011 in addition to the Best Innovation Award from CSIR-NAL in 2013.

Dr. Sekhar Majumdar served the National Aerospace Laboratories (CSIR) as a Senior Scientist from 1989 till January 2008 and retired as the Head of the Computational and Theoretical Fluid Dynamics Division at NAL. He is now a Senior Professor in the department of Mechanical Engineering and the Head of CFD Centre of Excellence, NMIT Bangalore. Prof. Majumdar received his B.E. and M.E. both in Mechanical Engineering from Bengal Engineering College, Calcutta University in 1968 and 1970 respectively. He worked at the Central Mechanical Engineering Research Institute (CSIR) Durgapur with the Fluid Mechanics and the Heat Power Division of CMERI where he received his Ph.D. degree in Mechanical Engineering in 1981. From 1983 to 1989, he was a Senior Research Scientist at the Institute for Hydromechanics, University Karlsruhe, Germany and worked with Professor W.Rodi in the area of Turbulent Flow Simulation. He was invited as a Senior Visiting Fellow by the Center for Turbulence Research, Stanford University, USA, and worked with Prof. Paul Durbin on Adaptive Cartesian RANS solvers. He has published more than 100 technical papers in peer-reviewed journals and conference proceedings. He is a

Fellow of Indian National Academy of Engineers, Fellow of Aeronautical Society of India and Fellow of Institution of Engineers India, Member, International Committee of Asian Fluid Mechanics Congress and life member of the National Society of Fluid Mechanics & Fluid Power, India, and the Indian Society of Technical Education.

Prof. Ganesh Subramanian completed his B.Tech. in Chemical Engineering from the Institute of Chemical Technology (ICT) in 1996. He went on to receive his doctoral degree in Chemical Engineering, from California Institute of Technology (Caltech) in 2002, and thereafter, completed his postdoctoral research from the Department of Chemical and Biomolecular Engineering, at Cornell University, in 2005. He joined the Engineering Mechanics Unit, at the Jawaharlal Nehru Centre for Advanced Scientific Research (JNCASR) in November 2005, as a Faculty Fellow (Assistant Professor), and is currently a Professor at the same institute. His research interests extend over a wide range of topics in fluid mechanics and transport phenomena. These include the dynamics and rheology of complex fluids such as suspensions, emulsions and polymer solutions, the dynamics and statistical mechanics of active matter, vortex dynamics and stability, atmospheric (the Ramdas layer) and oceanic (biogenic mixing) sciences. His research is regularly published in flagship fluid mechanics journals including the Journal of Fluid Mechanics. He has advised many Ph.D. and M.S. students, some of whom have gone on to become faculty members at the Indian Institutes of Technology.

Prof. G. S. Bhat obtained his B.Tech. in Mechanical Engineering from IIT Bombay, Master's of Engineering from IISc and Ph.D. from the Department of Aerospace Engineering, IISc. His fields of interest include turbulent flows, tropical convection, cloud physics and dynamics, atmospheric boundary layer, laboratory modelling of atmospheric phenomena and ocean-atmosphere coupling. He has carried out laboratory simulation of cloud-like flows and field experiments over land and ocean. He was the Principal Investigator for the atmospheric component in the three Indian national observational monsoon experiments under the Indian Climate Research Program (ICRP), viz. the Bay of Bengal Monsoon Experiment (BOBMEX) carried out in 1999, the Arabian Sea Monsoon Experiment (ARMEX, 2002–05), the Continental Tropical Convergence Zone (CTCZ) experiment (2009–2015) and presently the Indian Principal Investigator of Indo-UK collaborative research program titled INCOMPASS.

Prof. Ratul Dasgupta is a faculty member at the Department of Chemical Engineering, IIT Bombay. His research interests are in the mechanics and thermodynamics of two-phase flows (interfacial waves) and amorphous materials. He received his undergraduate degree in Mechanical Engineering from NIT Rourkela (formerly REC Rourkela) in 2001 and worked with the industry for a few years. He subsequently completed his Ph.D. from the Jawaharlal Nehru Centre for Advanced Scientific Research at Bengaluru working on the mechanics of hydraulic jumps on thin film flows, in 2011. He was a Postdoctoral Fellow at the Department of Chemical Physics, Weizmann Institute of Science, Israel, from 2010 to 2013. Ratul has been on the Faculty of IIT Bombay since December 2013 where he supervises a small group of Master's & doctoral students working on computational and analytical aspects of interfacial waves. He enjoys teaching courses related to mathematics, fluid dynamics and thermodynamics to graduate and undergraduate students at IIT Bombay. His group also collaborates with researchers at TCIS Hyderabad, ICTS Bengaluru and the IITs at Chennai, Hyderabad and Ropar.

Prof. Jaywant Arakeri is in the Faculty of the Mechanical Engineering department and the Centre for Product Design and Manufacture at the Indian Institute of Science, Bangalore. All his education has been in Aeronautical Engineering, B.Tech. (IIT, Madras), M.E. (IISc) and Ph.D. (Caltech). His research is primarily focused on the fundamental understanding of various phenomena in fluid mechanics and heat transfer, in particular related to turbulence, transition to turbulence, unsteady flows, bio-fluid mechanics and evaporation from porous media. Some of the current questions being addressed in his lab relate to the role of turbulence in condensation and droplet growth in clouds; flows around flexible surfaces like fish tails and heart valves; instability of unsteady flows, including those with curvature, like those found in arteries; effect of gusts on insect flight; heat and moisture loss from soils and leaves and precision agriculture. He has written several popular science articles in Resonance. He is the Fellow of the Indian National Academy of Engineering and of the Indian Academy of Sciences.

Computational Study on Dynamic Stall and Flow Control in a Pitching Airfoil

Patibandla B. L. V. Ramana, Akshoy Ranjan Paul, Anuj Jain, and Kazuo Matsuura

Abstract Dynamic stall phenomenon on a Boeing Vetrol-VR-7 airfoil, oscillating at quarter chord with reduced frequency of 0.1, is investigated at three different slot configurations—leading edge, trailing edge and combination of this two edges at a Reynolds number of 2.5×10^6. It is shown that the use of a leading-edge slot can eliminate the dynamic stall vortex (DSV) and increase the lift coefficient by 20%, and a decrease in the drag and moment coefficient by more than 70%. It is computed that the performance at low angles of attack can be improved with the use of a non-drooped leading edge in the airfoil. Furthermore, the combination of a leading and trailing-edge slots further improves the lift characteristics.

Keywords Pitching airfoil · Leading-edge slot · Reduced frequency · Dynamic stall · Leading-edge vortex (LEV) · Dynamic stall vortex (DSV)

1 Introduction

A phenomenon that is frequently encountered in airfoils whose angle of attack (AoA) oscillates with respect to free stream velocity, like in advancing rotorcrafts, jet compressors, wind turbines, rapid maneuvering aircraft, birds and insects, is called dynamic stall. It can be characterized by an increase of lift beyond static stall angle and a sudden loss of moment coefficient followed by lift coefficient after a particular dynamic stall angle. In static stall, however, lift and moment stall occur simultaneously. This sudden drop in moment and lift coefficients causes fluctuating loads on airfoils, which leads to fatigue and flutter, which could be detrimental to structural integrity and life of rotor blades, wind turbines, etc. [1, 2].

P. B. L. V. Ramana · A. R. Paul (✉) · A. Jain
Department of Applied Mechanics, Motilal Nehru National Institute of Technology Allahabad, Prayagraj, India
e-mail: arpaul@mnnit.ac.in

K. Matsuura
Graduate School of Science and Engineering, Ehime University, Matsuyama, Ehime, Japan

L. Venkatakrishnan et al. (eds.), *Proceedings of 16th Asian Congress of Fluid Mechanics*, Lecture Notes in Mechanical Engineering,
https://doi.org/10.1007/978-981-15-5183-3_1

Dynamic stall occurs by formation and shedding of a dynamic stall vortex (DSV) which results in a sudden increase and an immediate drastic decrease in lift. This DSV forms when the vorticity flux level exceeds a limit threshold. It can be understood by the following equation:

$$\nu \frac{\partial \omega}{\partial n} = \frac{\partial U_s}{\partial t} + \frac{1}{\rho} \frac{\partial P}{\partial n} + \nu \omega \tag{1}$$

Here, the LHS is the vorticity flux, where ν is kinematic viscosity, n is the direction normal to the surface of the airfoil, and ω is spanwise vorticity. The first term in the RHS is the surface acceleration term, where U_s is the surface tangential speed, while the second term is the pressure gradient term, where s is the direction along the surface, ρ is density of the fluid and the third term is surface transpiration term. During pitching motion, the vorticity flux increases rapidly, this vorticity coalesces after a threshold and forms a dynamic stall vortex (DSV). Therefore, DSV manipulation involves not letting the vorticity to coalesce by shedding smaller pieces of this vorticity into the wake. Then the consequences of vortex induced fluctuating loads can be reduced. This can be done either by surface acceleration or by injecting or extracting fluid from the flow field as per the equation above [3].

One of the most efficient passive methods of injecting and extracting fluid from the flow field is with the help of slots on the airfoil. Slotted airfoils are independently developed by Lachman [4] to reduce flow separation and increase stall angle. These slots divide the airfoil into parts. In a three-part wing, there are two slots—one at leading edge and the other at trailing edge. The leading-edge part is called a 'slat,' the middle part is called 'main element' or 'main airfoil' and the trailing-edge part is called a 'flap.'

The major mechanisms responsible for the effectiveness and lift increasing capability of these instruments are (a) Slat effect: Reduction in the circulation of the main element because of the circulation of the slat. (b) Circulation effect: The circulation of the slat is increased by the effect of the main airfoil. (c) Dumping effect: Slat boundary is dumped at high velocity, which will reduce separation and increase lift. (d) Off the surface pressure recovery: Deceleration of the flow and pressure recovery occurs in the free stream. (e) Fresh boundary layer effect: As every element experience a new thin boundary layer, it can withstand higher adverse pressure gradient. A detailed description of all these effects is given by Smith [5].

The first use of a leading-edge slat in a dynamic stall situation is in 1983 research paper of Carr and McAlister [6]. They have used a VR-7 airfoil with a NACA 15320 slat pitching around the quarter chord. Different parameters like Reynolds number (Re), mean AoA, reduced frequency (k) were varied. The dynamic stall reappeared after mean AoA of 20° even in slatted airfoil. In basic airfoil at Mach number = 0.185, Re = 2.5×10^6, $k = 0.1$ and for AoA varied as $15° + 10° \sin(\omega t)$, reversed flow reached leading edge at AoA of 22°. However, in slatted case, it barely reached quarter chord at all. Later, McAlister and Tung [7], Tung et al. [8] and Tuncer and Sankar [9] carried out further investigation using Boeing Vetrol-VR-7 airfoil.

Van Dam [10] produced a comprehensive review of multi-element airfoils used for transport airplane application. It was observed that even though these aircrafts operate at high Reynolds numbers, because of the relatively small size of slats, the Reynolds number on the slat is still laminar. Therefore, transition modeling is an important aspect for multi-element airfoil modeling.

It is observed from literature review that very few slat geometries are tested for dynamic stall. Most of these slats, being drooped, produced high drag and less lift during low angle of attacks because of the increase in zero-lift angle of attack. Moreover, only a few researchers have implemented a non-drooped slat and conducted optimization study regarding the slats. Hence, a non-drooped slat (using a leading-edge slot) is tested in the present study along with a trailing-edge slot and a combination of both to test the efficiency of slots at two different locations. The following section describes the computational techniques, which includes geometry, grid generation and the solver settings.

2 Computational Methodology

If equations are present, they should be numbered in the order of appearing, and all symbols and variables described.

2.1 Geometry

The airfoil used in this work is a Boeing Vetrol-VR-7 airfoil with a trailing-edge 'tab.' Tabs are instruments connected at the end of an airfoil used to decrease camber and reduce pitching moment. The chord length of the airfoil is 0.61 m and the airfoil is placed at an initial AoA of 15°. The outlet angle, cutoff length and chord length of leading-edge and trailing-edge slot positions are 20°, 6.4% chord, 12% chord and 36°, 70% chord, 8% chord, respectively, and are as depicted in Fig. 1a. 6.4% cord is the position of the stagnation point at 20° AoA (2° before stall). Most of the other parameters are chosen from Greenman and Roth [11]. Only slot gap of 1% chord is used to position the slat and flap away from the main element along the chord line.

An O-type computational domain is divided into inner and outer regions. The inner region is a circular zone around the airfoil at a radius of 5 chord length. A sliding mesh technique is used to give pitching motion for the inner zone. The outer region is a stationary zone of radius 20 chord length.

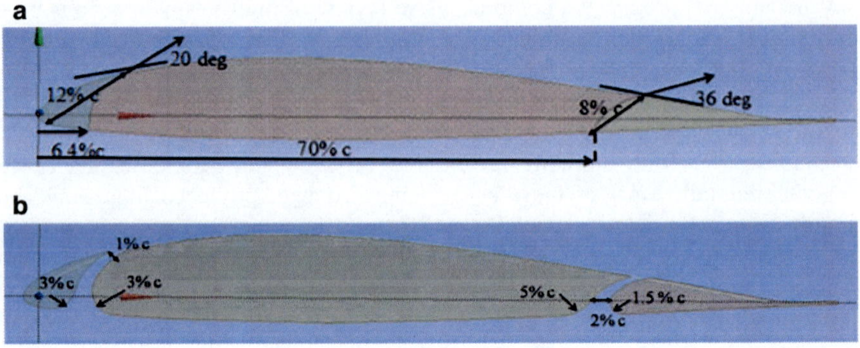

Fig. 1 **a** Slot location and geometry. **b** Slot position and geometry

2.2 *Grid Generation*

A hybrid mesh is generated, using Ansys-ICEM 16.0, for all the cases tested with quadrilateral, near wall cells and triangular cells away from the wall. First cell height of 5×10^{-6} is used with a y^{+} value less than 1. In the near wall boundary layer region, 70 nodes perpendicular to the airfoil are used to resolve the flow reversal and reattachment. This inflation layer is as shown in Fig. 2a. The equi-angle skewness of all the meshes is above 0.3 indicating good quality. The equi-angle skewness of all the meshes is above 0.3 indicating good quality.

Grid independence test is carried out on a non-slotted airfoil, as it is computationally prohibitive to carry the same test on all the other conditions. As there is sufficient literature [12, 13] on the grid independence tests, the element sizes on different zones, considered optimum by those works, are chosen. Three different time-step size (t/T = 0.01, 0.005, 0.0025) are chosen for the time-step independence study.

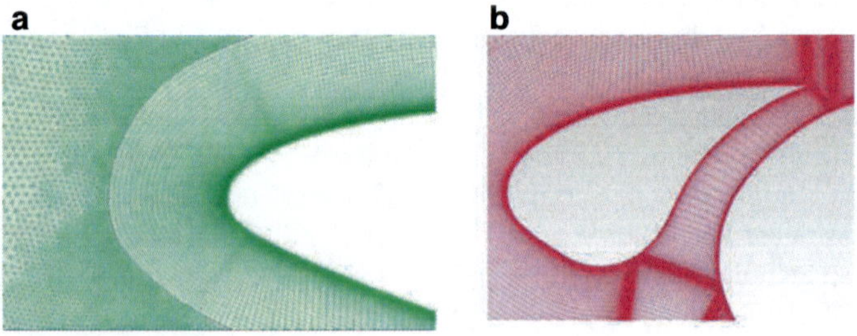

Fig. 2 **a** Inflation layer. **b** Meshing in the slot

2.3 Governing Equations and Boundary Conditions

The governing differential equations of this unsteady flow phenomenon are unsteady Navier–Stokes equations along with continuity and energy equations. Shear stress transport (SST) k-ω turbulence model is used in the present study because this turbulence model was widely used for dynamic stall prediction, for modeling adverse pressure gradients [10, 12, 13] and for grid and time convergence studies [13, 14].

An airfoil, oscillating from 5° to 25° with a reduced frequency of 0.1, freestream Mach number of 0.185 and $\mathrm{Re} = 2.5 \times 10^6$ is used to study the slot flow control of dynamic stall. Pressure far field along with inlet Mach number is used as the inlet boundary condition. Pressure outlet option is used as outlet boundary condition and the border between inner and outer zones is specified as an interface. No-slip boundary condition is also applied at the airfoil surface. The airfoil AoA varies in the following way:

$$\alpha = 15^\circ - 10^\circ \sin(\omega t)$$

Here, t is time and ω is angular frequency calculated from reduced frequency. A uniform initial flow corresponding to $\mathrm{Re} = 2.5 \times 10^6$ everywhere in the domain is chosen.

The maximum Mach number is more than 0.3 for most of the flow around airfoil. it is observed from preliminary computational results that the area of Mach number more than 0.3 is confined to the leading edge and is very small. It is also seen that no dynamic stall vortex is visible at maximum AoA. Hence, a compressible flow solver is used from henceforth to account for and examine the effect of Mach number.

2.4 CFD Solver Settings

Coupled algorithm option available in finite-volume method-based Ansys-Fluent CFD solver is used for pressure–velocity coupling as it ensures faster convergence. Green-Gauss Node-Based methods are used for gradient evaluation as this method is more accurate for unstructured meshes. Second-order upwind spatial discretization scheme is used for all the parameters like momentum, turbulent kinetic energy, rate of turbulent dissipation and energy. The convergence criterion, for all the variables, is taken as $<\sim 10^{-5}$.

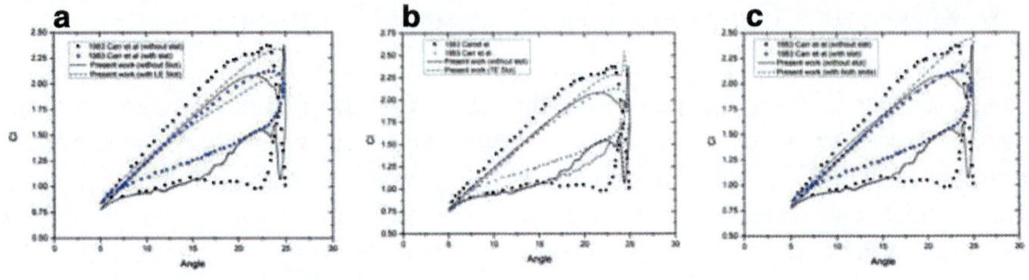

Fig. 3 **a** With LE slot. **b** With TE slot. **c** With both slots

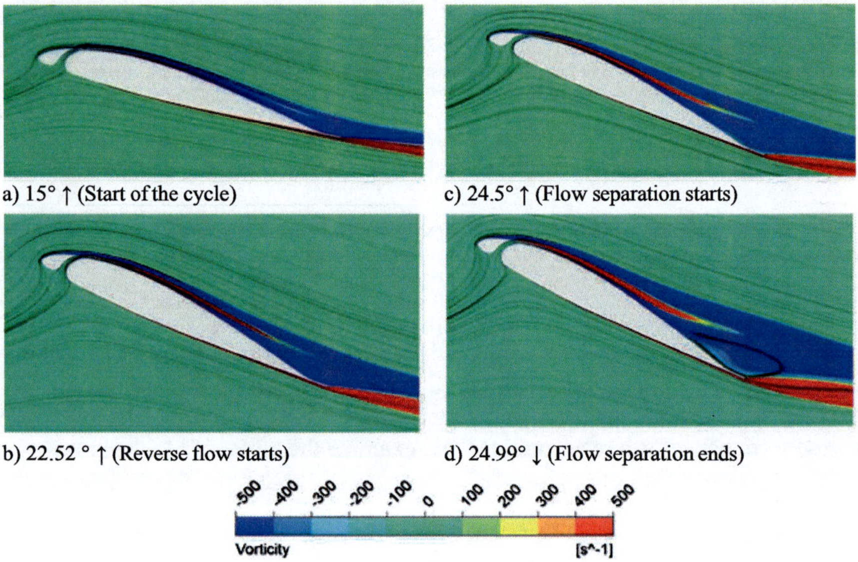

Fig. 4 Instantaneous streamline and vorticity contour on a LE slotted airfoil

3 Results and Discussion

The work of Carr and McAlister [6] on the implementation of leading-edge slat is used for validating the computational results in the present study. The CFD results of Liggett and Smith [15] are also compared with the present results to test the efficacy of the computational method adopted here.

3.1 Leading-Edge Slot

The results of the leading-edge slot are as given in Figs. 3a and 4. In Fig. 3a, variation of lift coefficient (C_L) with AoA of VR-7 airfoil with leading-edge slot is compared with those of VR-7 airfoil with NACA 15320 leading-edge slat used by Carr and McAlister [6]. It is seen that the hysteresis loop of VR-7 airfoil with leading-edge slot is much less than the hysteresis loop of the same with a NACA 15320 slat. A small increase in lift coefficient is observed even at low AoA during the upstroke. At high AoA, because of the absence of the DSV, significant lift coefficient increase is obtained. An overall increase of 20% is observed in cycle-averaged lift coefficient with the slotted airfoil. Further, a significant decrease of 72.9 and 87% in drag coefficient and negative moment coefficient are also observed with the slotted airfoil.

However, there is separated flow on the slotted airfoil too, at higher AoA. It is observed from the vorticity contours of Fig. 4b that the boundary layer of the first element mixes with the boundary layer of the main airfoil near the trailing edge and this could be the reason for elimination of the dynamic stall. It is noticed from the instantaneous streamline diagrams (Fig. 4) that the stagnation point on the main airfoil indicates a lesser circulation around it and whereas on the first element indicates a higher circulation around it. These results agree with the mechanisms of lift increase due to slotted configuration given by Smith [5].

3.2 Trailing-Edge Slot

A trailing-edge slot is also provided in the present airfoil to check its effect on the alleviation of dynamic stall. It is observed from Fig. 3b that even though the stall is postponed to higher angles of attack with the trailing-edge slot, the stall is not eliminated. It is also seen that the lift coefficient peaks of the airfoil with trailing-edge slot are greater than the lift coefficient peaks of the basic airfoil. A 2.1% increase in the cycle-averaged lift coefficient is obtained where as an increase of 2.8 and 0.41% in drag and moment coefficients is observed. This marginal change in the aerodynamic coefficients indicates that the trailing-edge slot itself has very less impact on the dynamic stall. However, because of the slot, flows around the trailing-edge element are always attached and separate from the end of the main element.

A 4° increase in the start of the reverse flow and the reattachment time from the non-slotted airfoil is noticed From Fig. 5. However, the reverse flow reached the leading edge around the same time. The DSV and secondary vortices have shed at a larger AoA as seen in Fig. 5b. The major difference between non-slotted and trailing-edge slotted airfoil is the travel of the trailing-edge vortex. In the non-slotted case, the trailing-edge vortex travels toward the leading edge before the formation DSV, whereas in trailing-edge slotted airfoil, the trailing-edge vortex forms at the end of the main element, so it travels toward the trailing edge of the second element and then to the leading edge of the first element. This decreased the effect of separation caused

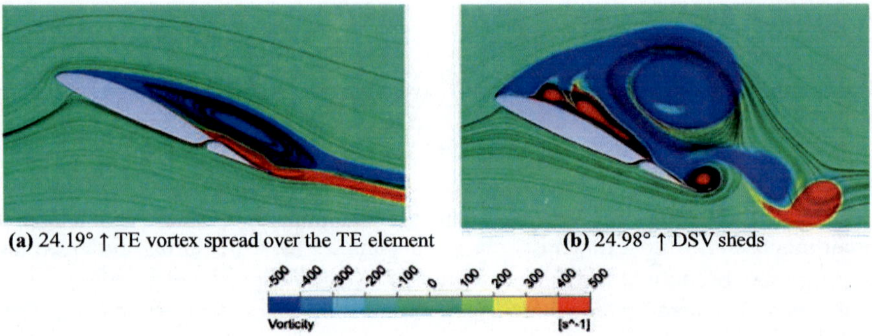

(a) 24.19° ↑ TE vortex spread over the TE element (b) 24.98° ↑ DSV sheds

Fig. 5 Vorticity contour and instantaneous streamline plots of the TE slotted airfoil

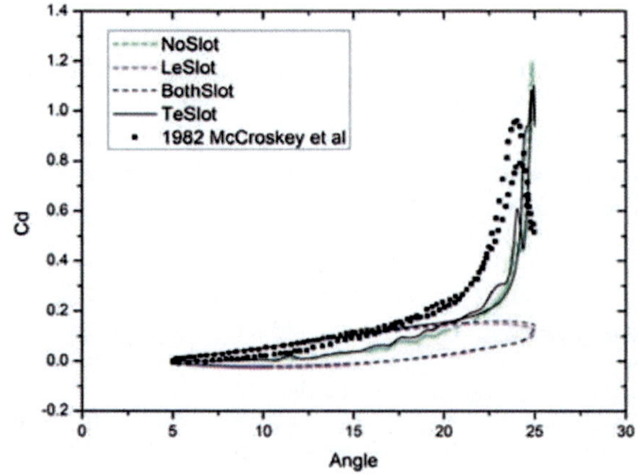

Fig. 6 Drag coefficient versus AoA of different slot configurations

by the trailing-edge vortex significantly and the first reduction in lift around 20° observed in non-slotted case does not appear in the trailing-edge slot case (Fig. 5a).

3.3 Combination of Leading and Trailing-Edge Slots

The leading-edge slot eliminated dynamic stall vortex but a small region of flow separation exists at high AoA. The trailing-edge slot improved the lift performance as a result of the movement of trailing-edge vortex but could not eliminate the dynamic stall. Based on these results, the effectiveness of the combination of both leading-edge and trailing-edge slots is investigated. It can be observed from Fig. 3c that providing both the slots further enhanced the lift profile. The impending trailing-edge flow separation clearly visible in the vorticity contours of the airfoil with leading-edge

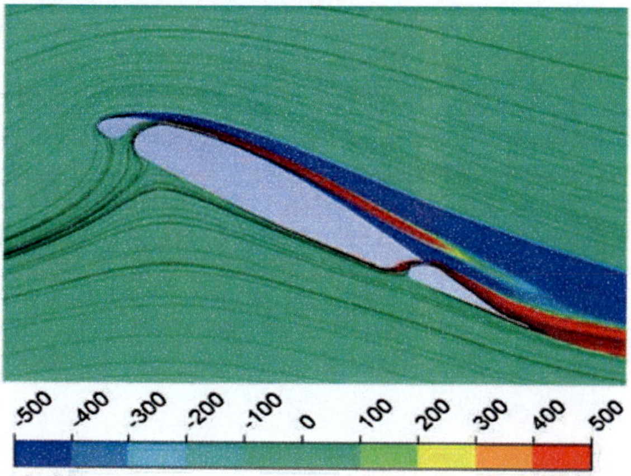

Fig. 7 At 25° ↑ maximum AoA (no separation)

slot is completely absent with the addition of trailing-edge slot (Fig. 7). No stall can be observed in the figure and the cycle-averaged lift coefficient increased by 22% compared to the basic VR-7 airfoil without slots. Additionally, the cycle-averaged drag and moment coefficients reduced by 71.7% and 80%, respectively (Fig. 6).

Although the cycle-averaged lift coefficient is highest in the airfoil with both slots, it is still marginal when compared to the cycle-averaged lift coefficient of the airfoil with single leading-edge slot. Similarly, the drag coefficient versus AoA graph in Fig. 7 shows that the drag coefficient change between the airfoil with leading-edge slot and the airfoil with both slots is also marginal. It is also noticed that the influence of trailing-edge slot is minimal in reducing drag coefficient. Hence, for understanding the influence of these different slot configurations on the lift, drag and moment coefficients, the slotted airfoils need to be further tested at wider parameters and spaces.

4 Conclusions

The present work attempts at controlling the stall behavior of the VR-7 airfoil with leading and trailing-edge slots. Three slot configurations (leading edge, trailing edge and combined) on the VR-7 airfoil have been tested at 2.5 million Reynolds number and a reduced frequency of 0.1 and compared with a non-slotted VR-7 airfoil. It has been observed that the implementation of a leading-edge slot completely eliminates the dynamic stall vortex and the airfoil with leading-edge slot has a 20% higher cycle-average lift coefficient than non-slotted airfoil. The decrease in the cycle-averaged drag and moment coefficient due to leading-edge slot is more than 70% of the non-slotted values.

In the trailing-edge slotted airfoil, the stall did occur but at a higher AoA. The lift coefficient increases by 2.1% from the non-slotted case, but the drag and moment coefficients also increase by 2.8% and 0.4%, respectively. This behavior can be attributed to the distinctive feature of the trailing-edge slotted airfoil: the reverse travel of the trailing-edge vortex. The combination of both slots further increases the lift coefficient (22%), more than leading-edge slotted case, but slightly higher values of drag and negative moment are observed. The flow separation visible at a higher AoA in the airfoil with leading-edge slot is successfully eliminated by the combination of both the slots.

References

1. McCroskey WJ (1982) Unsteady airfoils. Annu Rev Fluid Mech 14:285–311
2. Carr LW (1988) Progress in analysis and prediction of dynamic stall. AIAA J Aircr 25:6–17
3. Chandrasekhara MS, Wilder MC, Carr LW (1999) Compressible dynamic stall control: a comparison of different approaches. In: 17th Applied aerodynamics conference, AIAA-99-3122
4. Lachman G (1921) Slotted wing sections. Technical notes, NACA TM-71
5. Smith AMO (1975) High lift aerodynamics. 37th Wright Brothers lecture. AIAA J Aircr 12
6. Carr LW, McAlister KW (1983) The effect of leading edge slat on dynamic stall of an oscillating airfoil. In: AIAA conference, AIAA-83-2533. https://doi.org/10.2514/6.1983-2533
7. McAlister KW, Tung C (1993) Suppression of dynamic stall with a leading edge slat on a VR-7 airfoil. NASA Technical papers, NASA-TM-3357
8. Tung C, McAlister KW, Wang CM (1993) Unsteady aerodynamic behavior of an airfoil with and without a slat. Comput Fluids 22:529–547. https://doi.org/10.1016/0045-7930(93)90024-4
9. Tuncer IH, Sankar LN (1994) Unsteady aerodynamic characteristics of a dual element airfoil. AIAA J Aircr 31:531–537. https://doi.org/10.2514/3.46526
10. Van Dam CP (2002) The aerodynamic design of multi-element high-lift systems for transport airplanes. Prog Aerosp Sci 38:101–144
11. Greenman RM, Roth KR (1999) Minimizing computational data requirements for multi-element airfoils using neural networks. AIAA J Aircr 36
12. Szydlowski J, Costes M (2004) Simulation of flow around a static and oscillating in pitch NACA 0015 airfoil using URANS and DES. In: Proceedings of HT-FED04, ASME heat transfer/fluids engineering summer conference, Charlotte, North Carolina, USA, July 2004
13. Costes M, Gleize V, Szydlowski J, Sankar LN, Guzel G, Rhee M (2005) Grid sensitivity study for the turbulent viscous flow around a NACA 0015 airfoil at stall. https://www.researchgate.net/publication/260391102
14. Richter K, Le Pape A, Knopp T, Costes M, Gleize V, Gardner AD (2009) Improved two dimensional dynamic stall prediction with structured and hybrid numerical methods. Presented at American helicopter society 65th annual forum, Grapevine, Texas, 27–29 May 2009
15. Liggett ND, Smith MJ (2012) Temporal convergence criteria for time accurate viscous simulations of separated flows. Comput Fluids, Elsevier, 06:140–156

Computational Studies of Aerosolized Drug Deposition in Human Respiratory Tract

Vivek Kumar Srivastav, Anuj Jain, and Akshoy Ranjan Paul

Abstract The administration of drug and vaccine using pMDI through mouth in a human respiratory tract (HRT) is important for treatment of local and systemic diseases. Computational fluid dynamics (CFD) is used to gain an insight on the effect of various parameters on aerosol drug deposition in a CT-scan-based three-dimensional model of human respiratory tract (HRT). In the present study, the aerosol drug deposition is estimated using discrete phase model (DPM) and LRN k-ω turbulence model in CFD. The effect of variation of inhalation rates, particle size, aerosol injection velocity, nozzle spray angle on aerosol deposition efficiency in oral cavity and up to sixth-generation bronchi model was studied. The results reveal that optimum nozzle spray angle is around 45–50°, the particle size should be below 5 μm, and inhalation rate above 40 L/min. The outcome of the present study will be helpful for efficient design of pMDI inhaler and for the effective treatment of respiratory and other diseases.

Keywords Human respiratory tract (HRT) · CT-scan model · Pressurized metered-dose inhaler (pMDI) · Aerosol transport · Drug deposition

1 Introduction

A pressurized metered-dose inhaler (pMDI) is a medical device that on operating delivers a specific amount of medication in the form of a short burst of aerosolized medicine [1]. The medication is usually self-administered by the patient into the body through the oral cavity into the human respiratory tract via inhalation. It is the most widely used drug delivery system for treating asthma, chronic obstructive pulmonary disease (COPD), and other respiratory and systemic diseases.

V. K. Srivastav
Motihari College of Engineering, Motihari, Bihar, India

A. Jain · A. R. Paul (✉)
Motilal Nehru National Institute of Technology Allahabad, Prayagraj, Uttar Pradesh, India
e-mail: arpaul@mnnit.ac.in

L. Venkatakrishnan et al. (eds.), *Proceedings of 16th Asian Congress of Fluid Mechanics*, Lecture Notes in Mechanical Engineering,
https://doi.org/10.1007/978-981-15-5183-3_2

The efficacy of the inhaler depends on the propellant [2]. Recently, chlorofluorocarbon (CFC) propellants were replaced by hydrofluoroalkane (HFA) as the previous is linked to the depletion of the ozone layer. However, HFA-based pMDIs with good inhaler technique deposit only 10–20% of the dose in the lower respiratory tract, while most of the dose being deposited in the oral cavity [3]. High oral drug deposition can cause localized and systemic adverse effects. Hence, higher drug deposition in deep lungs is a requirement for pMDIs. The efficacy of pMDI is also being explored nowadays for administering medicines for treatment of different other diseases. Such targeted therapies that have narrow therapeutic windows, such as inhaled insulin, morphine, etc., for systemic action, where a very precise and reproducible dose may be needed [4]. Hence achieving targeted drug delivery (i.e., effective healing with low dose of medicines) for pMDIs is another criteria.

In developing effective pMDI systems, two areas are important: the device hardware and the drug formulation. The hardware consists of the vial, metering valve, actuator, and for newer pMDIs usually a dose counter [5]. The formulation comprises primarily of the propellant, drug, and often other excipients. Aerosolized drug particle size, nozzle design, injection velocity, aerosol characteristics, aerosol size distribution, spray temperature, and spray impact force, as well as the mass of drug available per shot play vital roles in contributing better drug deposition efficiency and targeted drug delivery of pMDIs [2].

The computational studies on deposition of aerosolized drug in human respiratory tract (HRT) are mainly based on two types of geometric models namely: (i) simplified model, which is based on regular pipes given by Weibel [6] and Horsfield et al. [7] and (ii) CT-scan-based model, which offers realistic geometry of HRT. The dimensions and angle between branches of simplified model are provided up to twenty-third generation. The present study is focused on drug delivery through mouth. Here, a sixth-generation HRT model having 23 bifurcation junction is generated from CT-scan slices using minimum possible smoothening factor so that the model exhibits finer geometric intricacies of HRT like cartilaginous rings, etc. [8]. The effects of various parameters (injection velocity, aerosol particle size, cone angle, and inhalation flow rates) on the aerosol deposition are also studied. It is envisaged that aerosol particle size, air inhalation flow rate, aerosol injection velocity, and aerosol injection cone angle governs the deposition efficiency of the aerosol drugs in HRT.

2 Computational Methodology

2.1 Geometry and Grid Generation

A three-dimensional geometrical model of human respiratory tract (Figs. 1 and 2) from trachea to sixth-generation bronchi was constructed using MIMICS software from 591 computed tomography (CT) scan images (DICOM files) of human airways

Fig. 1 Cartilaginous rings in trachea

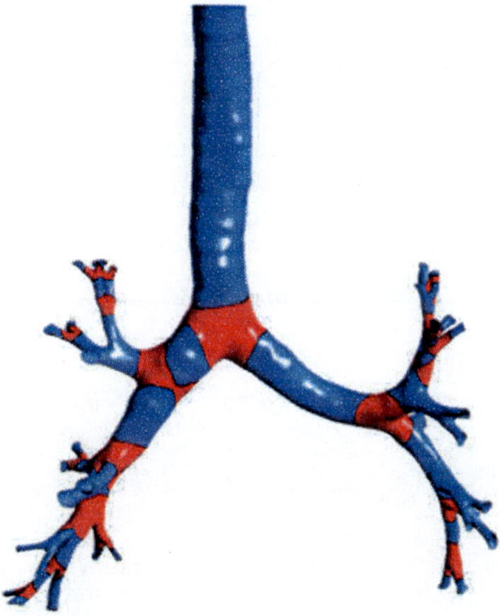

Fig. 2 Sixth-generation model with bifurcation junctions (in red color)

having pixel resolution of 300 × 203 taken at the interval of 0.625 mm. The images were taken from National Biomedical Imaging Archives (NBIA), NIH, USA.

An unstructured grid of tetrahedral elements was generated using patch independent method in Ansys-Meshing platform. The velocity at bifurcation junction was computed for different grid elements to perform grid independency test. The relative error of velocity magnitude between 2.85×10^6 and 3.7×10^6 elements is found approximately 0.49%. Thus, further results are reported corresponding to the grid of 2.85×10^6 elements. Mesh quality is considered satisfactory as only two computational element having skewness more than 0.9 exist in the flow domain.

2.2 Governing Equations

Considering the steady and incompressible air flow, the governing equations used in the study are presented below.

$$\frac{\partial u_j}{\partial x_j} = 0 \tag{1}$$

$$\rho \overline{u}_j \frac{\partial \overline{u}_i}{\partial x_j} = \rho \overline{f}_i + \frac{\partial}{\partial x_j}\left[-\overline{p}\delta_{ij} + \mu\left(\frac{\partial \overline{u}_i}{\partial x_j} + \frac{\partial \overline{u}_j}{\partial x_i}\right) - \rho\overline{u_i' u_j'}\right] \tag{2}$$

Reynolds number at the inlet of trachea corresponding to inhalation flow rate 60 L/min is estimated to be 5180. Therefore, low Reynolds number (LRN) *k*-*ω* turbulence model is used in the present study as given in Eqs. (3)–(6).

$$\overline{u_j} \frac{\partial k}{\partial x_j} = \tau_{ij} \frac{\partial \overline{u_i}}{\partial x_j} - \beta^* k\omega + \frac{\partial}{\partial x_j}\left[(\nu + \sigma_k \nu_T)\frac{\partial k}{\partial x_j}\right] \tag{3}$$

$$\overline{u_j} \frac{\partial \omega}{\partial x_j} = \alpha \frac{\omega}{k} \tau_{ij} \frac{\partial \overline{u_i}}{\partial x_j} - \beta\omega^2 + \frac{\partial}{\partial x_j}\left[(\nu + \sigma_\omega \nu_T)\frac{\partial \omega}{\partial x_j}\right] \tag{4}$$

where $\nu_T = C_\mu f_\mu k/\omega k$ and $f_\mu = \exp\left[-3.4/(I + R_T/50)^2\right]$.

Model constants are

$$C_\mu = 0.09, \alpha = 0.555, \beta = 0.8333, \beta^* = 1 \text{ and } \sigma_k = \sigma_w = 0.5$$

Discrete phase model (DPM) is also applied as the problem belongs to the multiphase flow of aerosol. The force balance equation acting on a given particle gives its trajectory. The aerosol particle velocity in the Lagrange formulation for a Cartesian coordinate system is obtained from force balance equation which is given in Eq. (5)

$$\frac{\partial u_p}{\partial t} = F_{\mathrm{D}}(u - u_p) + \frac{g_x(\rho_p - \rho)}{\rho_p} \tag{5}$$

First term on the right-hand side represent drag force per unit particle mass and second term is gravitational force. F_{D} in the drag force is defined as:

$$F_{\mathrm{D}} = \frac{18\mu}{\rho_p d_p^2}\frac{C_{\mathrm{D}}\mathrm{Re}}{24} \tag{6}$$

Re is the relative Reynolds number, which is defined as

$$\mathrm{Re} = \frac{\rho d_p \left| u_p - u \right|}{\mu} \tag{7}$$

The drag coefficient (C_{D}) can be computed from

$$C_{\mathrm{D}} = a_1 + \frac{a_1}{\mathrm{Re}} + \frac{a_3}{\mathrm{Re}^2} \tag{8}$$

where a_1, a_2 and a_3 are constants that apply to smooth spherical particles over several ranges of Reynolds number given by Morsi and Alexander [9].

The density of aerosol was considered as equivalent to that of water (1000 kg/m^3). Nozzle tip diameter is 2 mm with total 5000 aerosol particles are injected at the inlet for each inhalation flow rate to study the drug deposition in respiratory tract. All the aerosol particles are considered to be released from a conical nozzle with a given tip diameter at a particular instant. The cone is placed at the center of inlet of oral cavity.

2.3 *Boundary Conditions*

The boundary conditions for air flow used in the present computational study are: velocity at the inlet, gauge pressure at the outlets, and no slip-wall conditions at the wall. 'Escape' option for aerosol particles in DPM is defined at the inlet and outlets, because aerosol crosses these planes. Since inside wall of human airway is covered with mucous and saliva, the aerosol particles once coming in contact with wall are considered to have been stick to it and not detached or reflected from the wall. Thus, 'trap' boundary condition is enabled at the wall to estimate the deposition of aerosol particles by inertial impaction. Different aerosol particle sizes (0.1, 1, 2.5, 5, and 10 μm) were used for different inhalation conditions, namely normal (20 L/min), moderate (40 L/min), and heavy (60 L/min). To estimate the effect of spray angle, the aerosol is considered to have been ejected at different cone angle (25°, 45°, 60°, and 75°) as shown in Fig. 3.

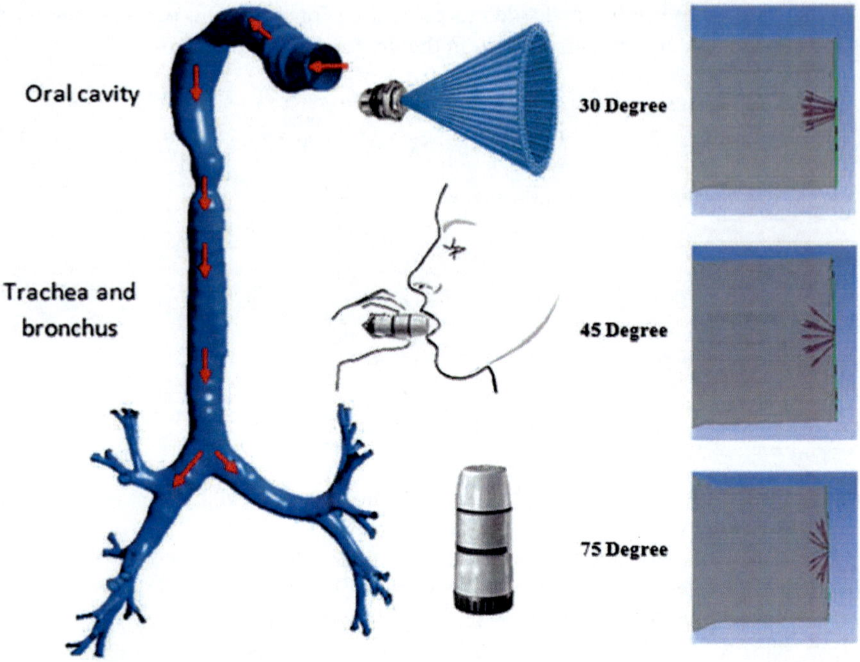

Fig. 3 Aerosol ejection cone angle [10]

2.4 CFD Solver Settings

Semi-Implicit Method for Pressure-Linked Equations Consistent (SIMPLEC) pressure–velocity coupling scheme was employed. The pressure was discretized with second-order upwind scheme while momentum, turbulent kinetic energy (k), and specific dissipation rate (ω) terms by Quadratic Upstream Interpolation for Convective Kinematics (QUICK) scheme. Numerical simulation is carried out in the finite-volume method-based CFD solver Ansys-Fluent with a convergence criteria of 10^{-5}.

2.5 Computational Validation

The validation of the present work is done in two parts: oral cavity and lower respiratory tract. The deposition efficiency of the particles (6.5, 10 and 17 μm) determined computationally in the oral cavity model was compared with the experimental results of Cheng et al. [11] and is shown in Fig. 4.

In the lower respiratory tract, the validation of present simulation was performed with the computational results reported by Nowak et al. [12]. The simulation was

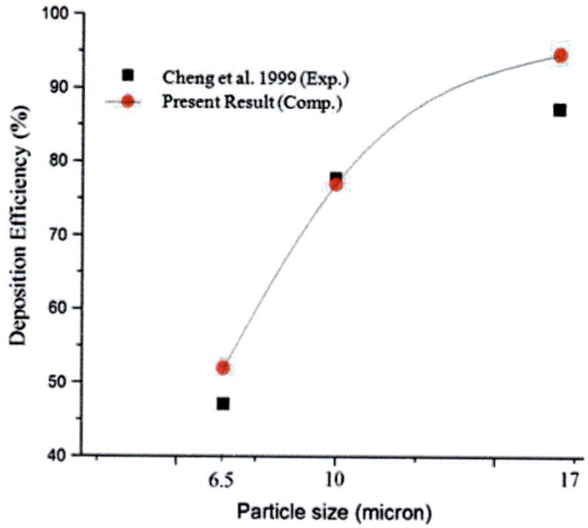

Fig. 4 Validation of computational results with the experimental results of Cheng et al. [11]

run for the inspiratory flow rate of 28.3 L/min on the same respiratory model and boundary conditions as used by Nowak et al. [12]. The validation with the velocity profile at the mid plane of fourth-generation is shown in Fig. 8 which is within the acceptable limit. After validation, same boundary conditions are considered for sixth-generation CT-scan model (Fig. 5).

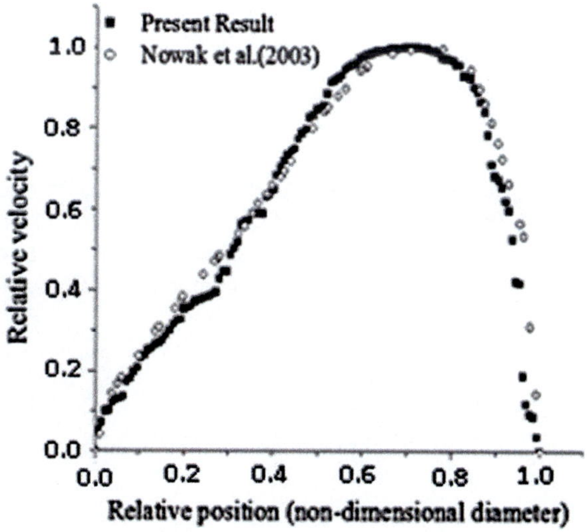

Fig. 5 Relative velocity at mid-section of bronchus-4

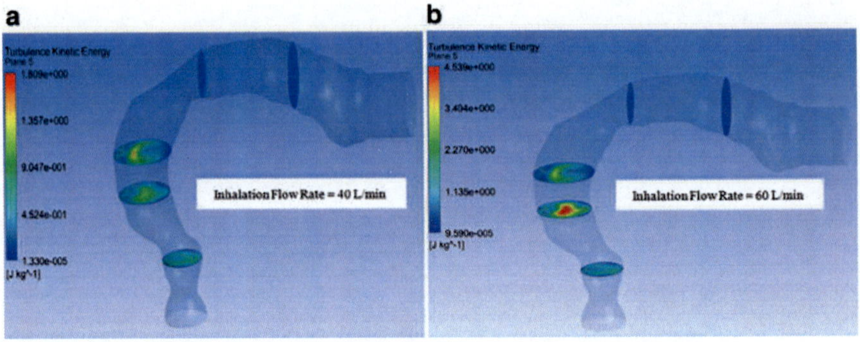

Fig. 6 TKE contours **a** inhalation flow rate = 40 L/min **b** inhalation flow rate = 60 L/min

3 Results and Discussion

In the present study, aerosol flow is simulated in HRT model using CFD. In the following sub-section, turbulence and effect of various parameters on aerosol deposition are presented to understand the internal flow physics and aerosol drug deposition. Aerosol deposition is affected by several parameters like turbulence kinetic energy, nozzle cone angles, aerosol particle size, inhalation flow rate, injection velocity, etc., which are also discussed in this section.

3.1 Turbulence Kinetic Energy (TKE)

Turbulence is an important parameter for the air flow study. The turbulence kinetic energy for inhalation flow rate of 40 and 60 L/min is shown in Fig. 6. It is seen that turbulence is formed in oral cavity because of the presence of laryngeal jet. As the inhalation flow rate increases, turbulence is also increased in the upper human airways. High turbulence is observed near the curved surface of oral cavity for higher inhalation rate (60 L/min). Hence, turbulence model is useful at moderate to higher inhalation rate. CFD simulation for normal inhalation rate may be carried out considering the flow as laminar which saves computational cost and time.

3.2 Overall Deposition Efficiency of Aerosol in Normal, Moderate and Heavy Breathing Conditions

Deposition efficiency is calculated as: Deposition efficiency = (Number of aerosol particle deposited at a particular zone/total number of aerosol particles injected at

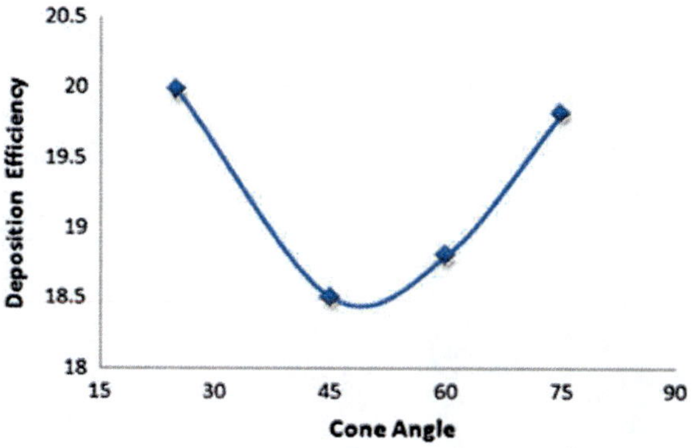

Fig. 7 Effect of cone angle (flow rate = 60 L/min, aerosol particle size = 1 μm) in oral cavity

the inlet of oral cavity) × 100. Aerosol depositions based on location and inhalation conditions are shown and discussed in the following subsections.

3.2.1 Effects of Cone Angle

To increase the reach of drug to the lower part of respiratory tract, aerosol deposition efficiency in the oral cavity must be minimum possible. Thus, the deposition efficiency of aerosol drug for different cone angle (25°, 45°, 60° and 75°) in oral cavity is studied. The aerosol deposition efficiency at different cone angle for 1 μm size particle at 60 L/min inhalation rate is shown in Fig. 7. It is found that minimum drug deposit in oral cavity occurs is at around 45° injection angle which leads corresponding to higher drug flow to lower bronchi. Similar results are obtained for other particle sizes and inhalation rates. Therefore, remaining computational simulations are performed for 45° injection cone angle.

3.2.2 Effect of Particle Size on Aerosol Deposition Efficiency in Oral Cavity

The aerosol deposition efficiency in the oral cavity for different particle sizes and different aerosol injection velocities at inspiratory air flow rate of (20, 40 and 60 L/min) and 45° nozzle angle of inhaler is studied. The simulated results reveal that the minimum deposition efficiency in the oral cavity is approximately 20% for the aerosol particles in the size range from 1 to 2.5 μm. The deposition efficiency is however increased marginally for the finer particles (0.1 μm). For the particles of 10 μm size, aerosol deposition efficiency in the oral cavity becomes high (59–66%) depending upon the aerosol injection velocity. It is also found that the deposition efficiency is higher for higher injection velocity. Further, it is observed that the effect of aerosol injection velocity is negligible as compared to effect of aerosol particle

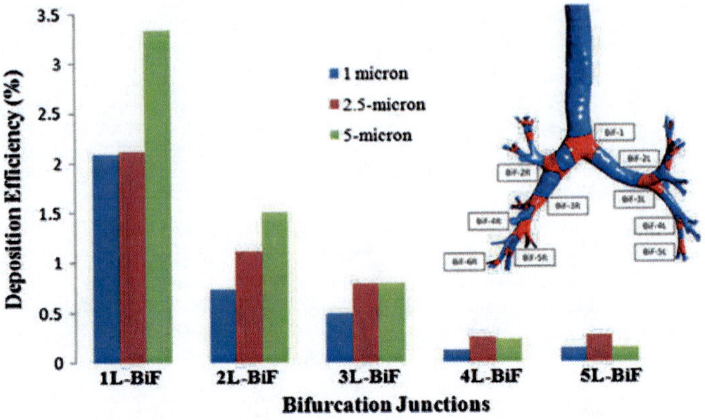

Fig. 8 Aerosol deposition efficiency in left bronchus at 40 L/min inhalation rate

size on the deposition efficiency especially near the minimum deposition efficiency zone.

3.3 Study of Sixth-Generation Bronchus

The aerosol deposition efficiency in bronchi is obtained for 40 L/min inhalation rate as this being the inhalation rate at which the aerosol deposition efficiency is minimum in the oral cavity. The efficiency for three different particle sizes at various bifurcation junctions of the left and right bronchi are presented in subsequent sections.

3.3.1 Aerosol Deposition Efficiency in Left Bronchus

The aerosol deposition efficiency at the bifurcation junctions of the left bronchus is furnished in Fig. 8. It is seen that deposition efficiency decreases from BiF-1 to BiF-5L. Maximum deposition efficiency is obtained at the first bifurcation junction while reducing deposition efficiency is observed in subsequent bifurcation junctions. This has happened because of lesser number of particles are available in subsequent junction due to deposition of particles in the previous generation bronchi. The larger size particles deposit more in initial bronchi and junctions, because of higher inertia. As its number (concentration) starts reducing in subsequent bronchi and junctions, its deposition reduces as compared to smaller size particles.

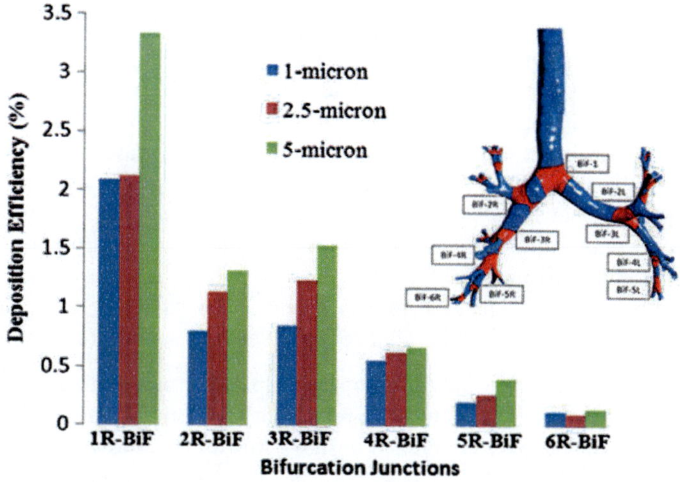

Fig. 9 Aerosol deposition efficiency in right bronchus at 40 L/min inhalation rate

3.3.2 Aerosol Deposition Efficiency in Right Bronchus

The deposition efficiency at the bifurcation junctions of the right bronchus is furnished in Fig. 9. The deposition pattern follows the similar trend as observed for the left bronchus, i.e., deposition efficiency decreases continuously in the subsequent junctions, except that there is increase in the deposition efficiency in the third bifurcation (BiF-3R) junction as compared to second. This is attributed to the irregular shape of the respiratory tract of the CT-scan-based model. Moreover, asymmetry in the deposition pattern in left and right side of bronchus is also because of asymmetry in the respiratory tract structure.

4 Conclusions

The present CFD study is carried out on realistic HRT model constructed from oral cavity to sixth-generation bronchi to understand the aerosol deposition in upper and lower respiratory tracts. The major findings are furnished below:

1. Effects of turbulence are found only in upper airways; therefore, turbulence model can be useful only in upper respiratory tract. Moreover, lower respiratory tract can be modeled using laminar flow solver.
2. It is found that minimum aerosol deposition in oral cavity occurs at around 45° injection angle which leads to higher flow of aerosol to lower bronchi.
3. The injection velocity of aerosol from the inhaler is found to have insignificant effect on the aerosol deposition efficiency.

4. The air flow and aerosol deposition in left bronchus are less as compared to that in right bronchus due to asymmetric features of HRT.
5. Aerosol particle size and inhalation rate are found to be most important factors that play key role for the drug delivery through inhaler.
6. As most of the deposition of 10 μm particle is in the oral cavity, it is not suitable for drug delivery in the HRT. For better (aerosolized) drug deposition, the particle size should be between 1 and 5 μm as it has the lower deposition efficiency in oral cavity and higher deposition efficiency in lower respiratory tract. This was also supported in Hickey et al. [13].
7. Inhalation rate of about 40 L/min is found to be sufficient to achieve least deposition in oral cavity and higher reach of aerosol in lower bronchi.

The outcome of the present study will be helpful for efficient design of pMDI inhaler and for the effective treatment of respiratory and other diseases. The present investigation is part of study targeted to design a inhaler that allows deep deposition of the drug. The present study is carried out when the neck of the patient is at upright position. Moreover, the delivery of drug depends on patient skill and head position. Further studies may be carried out for different head positions to maximize drug delivery in deep bronchi.

Acknowledgements This research work reported in the paper was supported by a R&D project entitled *Prototype Development of Dry Powder Inhaler (DPI) for Pulmonary Drug delivery* sanctioned by the Council of Science and Technology, Government of Uttar Pradesh (India) vide sanction letter no. CST/D-8249 dated 7th March 2018.

References

1. Newman SP (2005) Principles of metered-dose inhaler design. Respir Care 50(9):1177–1190
2. Myrdal PB, Sheth P, Stein SW (2014) Advances in metered dose inhaler technology: formulation development. AAPS PharmSciTech 15(2):434–455. https://doi.org/10.1208/s12249-013-0063-x
3. Lavorini F (2013) The challenge of delivering therapeutic aerosols to asthma patients. ISRN Allergy. 2013(102418). https://doi.org/10.1155/2013/102418
4. Stein SW, Thiel CG (2017) The history of therapeutic aerosols: a chronological review. J Aerosol Med Pulm Drug Deliv 30(1):20–41. https://doi.org/10.1089/jamp.2016.1297
5. Stein SW, Sheth P, Hodson PD, Myrdal PB (2014) Advances in metered dose inhaler technology: hardware development. AAPS PharmSciTech. 15(2):326–338. https://doi.org/10.1208/s12249-013-0062-y
6. Weibel ER (1963) Morphometry of the human lung. Academic, New York
7. Horsfield K, Gladys D, Olson DE, Finlay GF, Cumming G (1971) Models of the human bronchial tree. J Appl Phys 31:207–217
8. Srivastav VK, Kumar A, Shukla SK, Paul AR, Bhatt AD, Jain A (2014) Airflow and aerosol-drug delivery in a CT scan based human respiratory tract with tumor using CFD. J Appl Fluid Mech 7(2):245–256
9. Morsi SA, Alexander AJ (1972) An investigation of particle trajectories in two-phase flow systems. J Fluid Mech 55:193–208

10. Collins TP, Tabor GR, Young EPG (2007) A computational fluid dynamics study of inspiratory flow in orotracheal geometries. Med Biol Eng Comput 45:829–836
11. Cheng YS, Zhou Y, Chen BT (1999) Particle deposition in a cast of human oral airways. Aerosol Sci Technol 12:286–300
12. Nowak N, Kakade PP, Annapragada AV (2003) Computational fluid dynamics simulation of airflow and aerosol deposition in human lungs. Ann Biomed Eng 31:374–390
13. Hickey AJ, Martonen TD, Yang YD (1996) Theoretical relationship of lung deposition to the fine particle fraction of inhalation aerosols. Pharm Acta Helv 71(3):185–190

Vorticity Generation Due to Surface Tension-Driven Spreading

Anurag Pant and Baburaj A. Puthenveettil

Abstract We study the generation of vorticity at a water–air interface due to the spreading of ethanol–water droplets for ethanol concentrations in the drop ranging from $20\% \leq C_e \leq 100\%$ on the surface of a 50 mm deep water layer. On deposition onto the water layer, the lighter, miscible ethanol droplet spreads as a film on the water surface due to the ethanol–water surface tension difference. Simultaneous to the film spreading, an expanding vortex ring is found below the tip of the spreading film front. The phenomenon in the water layer is visualized using particles, and 2D PIV is used to obtain the velocity field. Vortex regions are identified using the λ_2 method for various time instants. The average vorticity generation in between the time instants is calculated. A scaling law is proposed for the dimensionless vorticity at a given instant based on the experimental observations.

Keywords Drops · Spreading · Vortices · PIV · Surface tension

1 Introduction

A lighter drop of ethanol when gently deposited at the surface of a deep water layer leads to an interesting display of dynamics driven by surface tension. The surface tension difference between ethanol and water leads to a positive spreading coefficient, which causes a thin circular film to spread out and away from the drop, while the drop itself floats at the surface. The film spreads radially outward where the film spreading radius displays a power law dependence on time. The film continues to spread, two counter rotating vortices are observed below the spreading film in particle-based side-view visualizations. The film retracts after reaching a maximum radius, and the drop completely mixes in the water by this time. The spatial evolution of droplets spreading due to surface tension difference has been a subject of extensive studies such as spreading of surfactants on organic solutions [2], spreading of continuously

A. Pant (✉) · B. A. Puthenveettil
Department of Applied Mechanics, Indian Institute of Technology Madras, Chennai, Tamil Nadu 600036, India
e-mail: akkupant@gmail.com

L. Venkatakrishnan et al. (eds.), *Proceedings of 16th Asian Congress of Fluid Mechanics*, Lecture Notes in Mechanical Engineering,
https://doi.org/10.1007/978-981-15-5183-3_3

supplied IPA (Isopropyl alcohol) droplets on a thin aqueous layer reminiscing a hydraulic jump [5] and spreading of surfactants imitating drug delivery in lungs [6]. In the case of an immiscible volatile droplet spreading on a deep water layer, Dussaud and Troian [4] found that the radius varies as $t^{\frac{1}{2}}$. However, for a miscible, volatile droplet of ethanol, Dandekar et al. [3] have shown the spreading radius to vary as $t^{\frac{1}{4}}$ and the spreading velocity as $t^{-3/4}$. The scaling was obtained by neglecting the role of vortices that travel along with the edge of the spreading film, and the associated mixing of the film with the bulk fluid. Such vortices have been observed earlier. Dussaud and Troian [4] also noticed the presence of an advancing convective roll below a film of toluene spreading on water. These rolls were attributed by them to Rayleigh Benard convection where it was assumed that surface evaporation leads to a vertical temperature gradient. They also observed the stretching of the rolls due to a shear force on the surface due to the surface tension difference. These observations were made for volatile films only.

Kim et al. [8] studied the generation of circulation when isopropyl alcohol (IPA) drops were deposited on 2 cm deep layers of water ethanol mixtures. The experiments were conducted in Hele-Shaw flow cells of thickness 160 μm and 240 μm, respectively. Steady vorticity generation was observed in the initial stages followed by a decay. They calculated the amount of circulation generated by balancing the kinetic energy of the vortical flow and the surface energy provided to obtain the scaling for the vortex strength as $\Gamma \sim \left(h^2 \Delta\sigma^{\frac{3}{2}}\right) / \left(\rho_2 R \sigma_1^{\frac{1}{2}}\right)$. where $\Delta\sigma$ is the surface tension difference, R is the drop size, σ_1 is the drop surface tension, ρ_2 is the density of the ethanol–water mixture, and h is the flow cell thickness. However, even though in Hele-Shaw cells, viscous effects were neglected in their analysis.

The origin of vorticity below spreading films above liquid substrates is still not clear. In this paper, we study the vortical flow beneath the spreading film of a miscible, volatile liquid. We present PIV measurements of the vortical flow, from which we identify the axisymmetric vortex ring manifesting itself in the form of two counter rotating vortices. The vortex region is identified, and the average vorticity generation is calculated in between two time instants. Further, a scaling is proposed for the dimensionless vorticity using the experimental data.

2 Experiments

PIV experiments were conducted in a rectangular glass tank measuring 100 mm × 100 mm × 50 mm. Water contained seeding particles (Polyamide, 20 μm mean diameter, 1.03 g/cm^3 density, Lavision GmbH) and was filled up to the brim of the tank to avoid meniscus effects during visualization. Needles of various sizes were connected to a syringe pump positioned at the center of the glass tank and the flow rates of water–alcohol solutions were adjusted to a low value (2.4 ml/h) so as to obtain periodic dripping of alcohol drops of specific sizes. The drop radius was calculated by counting the number of drops that fill a known volume. The error in the drop radius measurements was found to be ±0.3 mm. In some cases, the drop also contained 0.1 ppm of Rhodamine 6G dye for visualization purposes.

A vertical laser sheet of 1 mm thickness with 80 mJ/pulse (532 nm, Nd:YAG laser) was passed parallel to the tank in such a way that it illuminated a plane passing through the drop center. The drop was brought very close to the surface and then allowed to detach, so that inertial effects due to impingement could be avoided. The particle images in the water layer were then captured using a Imager Pro HS CCD (Lavision GmbH) camera which was positioned perpendicular to the plane illuminated by the laser sheet (Fig. 1). It was made sure that the depth of field of the camera exceeded the laser sheet thickness [1]. The separation time between the two laser pulses was set such that the out of plane displacement of particles was less than one-fourth of the sheet thickness. The images were recorded at 50 Hz. The above steps were performed for drops with C_e ranging from 20 to 100% (Table 1). The particle images were processed using adaptive multi-pass cross-correlation where a window size of 16 × 16 pixel and 50% overlap were used in DAVIS (Lavision GmbH) in

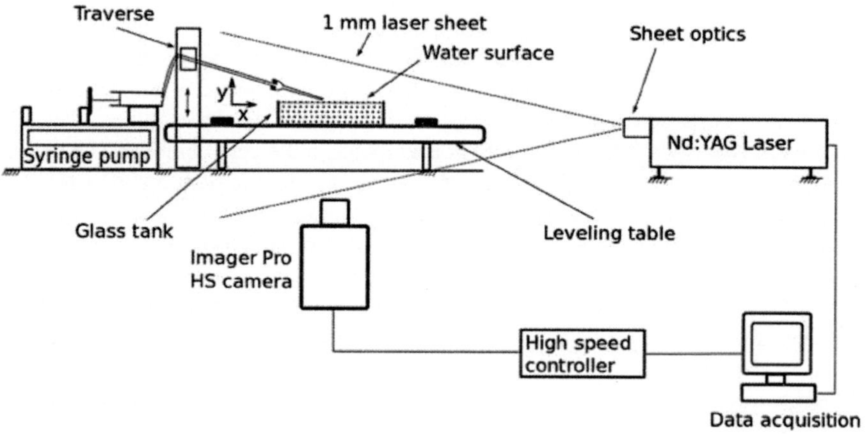

Fig. 1 Schematic of the experimental setup

Table 1 Properties of various binary solutions used in the drop

Ethanol concentration (C_e) (%)	Surface tension (σ) (N m)	Surface tension difference ($\Delta\sigma$) (N m)	Kinematic viscosity (ν) (m^2/s)	Density (ρ) (kg/m^3)
20	37.97	34.03	1.76	970.36
40	31.00	41.00	2.34	948.47
60	26.00	46.00	2.24	908.72
80	23.80	48.20	1.66	859.58
100	22.00	50.00	1.10	790.74

The properties were obtained at 25 °C

order to obtain the velocity field. Image correction was performed before processing the images to reduce any inhomogeneity in illumination of particles before, so that spurious vectors could be reduced.

3 Results

3.1 *Identification of the Vortex Region and Calculating Vorticity*

Figure 2 shows the velocity field at three time instants for a 1 mm drop deposited on a 50 mm deep water layer. On observing the flow field, we notice the presence of two counter rotating vortices on either side of the point of drop deposition. These vortices are found below the spreading film in the water surface and are found to travel outwards and away from the point of drop deposition. In order to understand the phenomenon of vortex ring generation as a result of film spreading on the surface, we needed to identify the vortex core from the vector field obtained from 2D PIV. The vortex core was identified using the λ_2 method [7] implemented using a MATLAB code. The velocity field data was processed using the PIVMat 4.10 toolbox to calculate the 2D velocity gradient tensor J for each point in the velocity field. The tensor J was then decomposed into the symmetric and antisymmetric parts S and ω, respectively. The eigenvalues were then calculated for the tensor $S^2 + \omega^2$ to obtain two eigenvalues λ_1 and λ_2. A point in the velocity field was identified as a part of the vortex core only if both the eigenvalues were negative with the condition, $\lambda_1 < \lambda_2$ being satisfied. The contour plots of λ_2 would then represent the iso-vorticity regions and hence the vortex core. Figure 3 shows the contour plot for λ_2 values, that is 80 ms after a drop with $r_d = 1$ mm and $C_e = 80\%$ is deposited on a 50 mm deep water layer. Once the vortex core region is identified for a time instant, the average vorticity in the given instant is calculated by averaging the total vorticity over a region which is defined by setting a threshold of 10% of the minimum value of λ_2. The area average vorticity at a given time instant is then estimated by summing up the vorticity in the region above the threshold value and averaging it over the area. It is normalized using a characteristic vorticity (ω_c) given by $\omega_c = (\Delta\sigma/\mu_d r_d)$.

3.2 *Scaling*

Dimensionally, the amount of vorticity generated due to a surface tension gradient $\Delta\sigma$ across a length $c_1 r_f$ can be scaled as

$$\omega_g = \frac{\Delta\sigma}{c_1 \mu_f \tau_f} \tag{1}$$

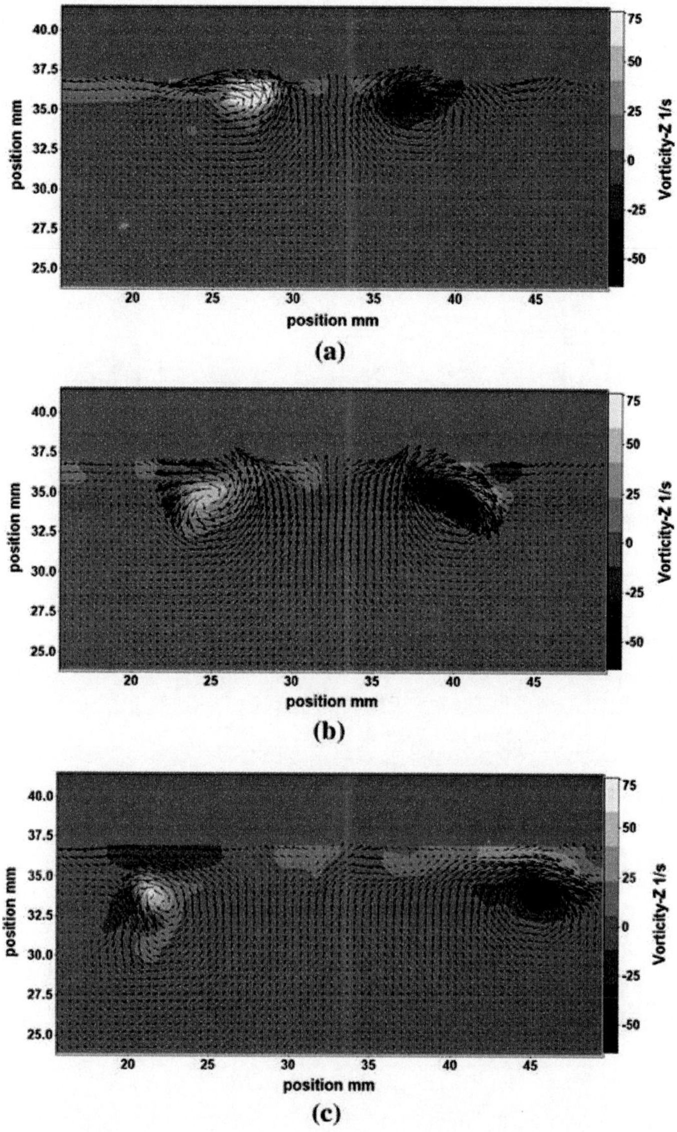

Fig. 2 Velocity fields at various times for a 1 mm droplet of $C_e = 80\%$ deposited on a 50 mm deep water layer at **a** $t = 80$ ms; **b** $t = 240$ ms, and **c** 420 ms

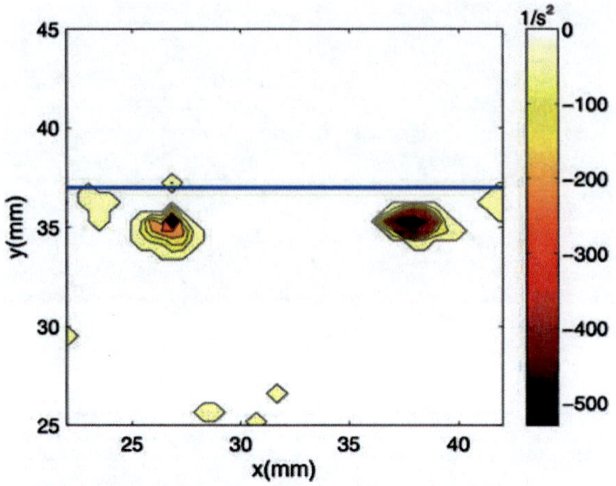

Fig. 3 Contour plot representing λ_2 values for the case of $r_d = 1$ mm and $C_e = 80\%$ at $t = 80$ ms. The solid line represents the free surface

Here, μ_f is the dynamic viscosity of the substrate liquid, and r_f is the radius of the spreading film given by Dandekar et al. [3] as

$$r_f = \frac{c_2}{f}\left(\frac{\Delta\sigma r_d^3}{\mu_d}\right)^{\frac{1}{4}} t^{\frac{1}{4}}, \tag{2}$$

where $c_2 = 0.83$, r_d is the drop radius, and f is a function of the bond number written as

$$f(\mathrm{Bo_d}) = \frac{3^{\frac{1}{4}}}{2}(1 + \mathrm{Bo_d})^{\frac{3}{4}}, \tag{3}$$

with $\mathrm{Bo_d} = \rho_d g r_d^2/\Delta\sigma$ being the drop bond number. Since we assume no mixing, μ_f can be written as μ_d.

On substituting (2) in (1), we obtain

$$\omega_g = c_3 f \left(\frac{1}{t_c^3 t} \right)^{\frac{1}{4}}, \tag{4}$$

where $t_c = \mu_d r_d / \Delta\sigma$ and $c_3 = c_1/c_2$.

The dimensionless vorticity (ω^*) can now be written as,

$$\omega^* = \int_0^t \omega_g \mathrm{d}t = c_4 \left(\frac{t}{t_c} \right)^{\frac{3}{4}}, \tag{5}$$

$c_4 = (4/3)c_3$. Figure 4 shows the variation of the dimensionless vorticity with dimensionless time for experiments with different C_e. The experimental ω^* is obtained by calculating the average vorticity at each time instant using PIV data and normalizing it using the characteristic vorticity ω_c, where $\omega_c = (1/t_c)$. Also the fit corresponding to the scaling proposed in (5) using a solid line is shown in the figure. The experimental data follows the $t^{\frac{3}{4}}$ trend shown by the fit reasonably well. Thus, we can attribute the vorticity in the substrate fluid to the generation of vorticity by the surface tension gradient across the spreading film of ethanol at the free surface.

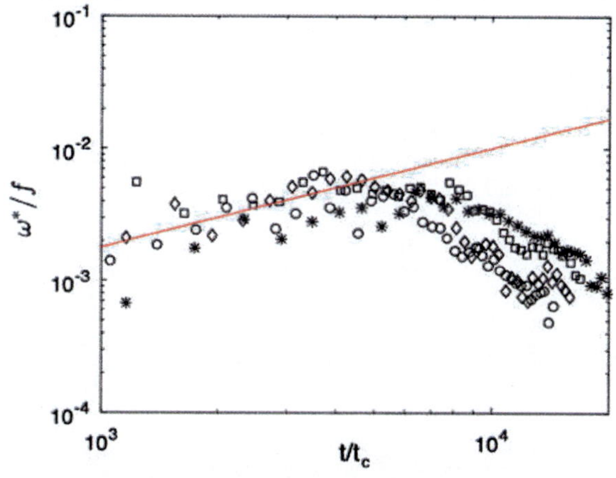

Fig. 4 Variation of the dimensionless vorticity (ω^*) with dimensionless time for a drop with $r_d =$ 1 mm. White triangle—$C_e = 100\%$; asterisk—$C_e = 80\%$; white square—$C_e = 60\%$, and white diamond—$C_e = 40\%$. The solid line shows the fit $10^{-4}(t/t_c)^{\frac{3}{4}}$

4 Conclusions

The primary result in this paper is the relation proposed for the dimensionless vorticity present at a given time instant in a water layer when a film of ethanol–water spreads on the surface of the water layer. It is shown that the vorticity at an instant shows a $t^{\frac{3}{4}}$ dependence on time. However, the vorticity generation ω_g between two time instants decreases with time, possibly due to the eventual mixing of the film in the water layer. The results are important in order to understand the vorticity generation at an interface due to a surface tension gradient.

Acknowledgements The equipment used in this study was funded by the grants SR/FST/ETII-017/2003 and SR/FST/ETII-064/2015 from DST, Government of India.

References

1. Adrian R, Westerweel J (2011) Particle image velocimetry. Cambridge University Press, Cambridge
2. Bacri L, Debrgeas G, Brochard-Wyart F (1996) Experimental study of the spreading of a viscous droplet on a non viscous liquid. Langmuir 12(26):6708–6711
3. Dandekar R, Pant A, Puthenveettil BA (2017) Film spreading from a miscible drop on a deep liquid layer. J Fluid Mech 829:304–327
4. Dussaud AD, Troian SM (1998) Dynamics of spontaneous spreading with evaporation on a deep fluid layer. Phys Fluids 10(1):23–38
5. Hernandez-Sanchez JF, Eddi A, Snoeijer JH (2015) Marangoni spreading due to a localized alcohol supply on a thin water film. Phys Fluids 27(3):032003
6. Jensen OE, Grotberg JB (1992) Insoluble surfactant spreading on a thin viscous film: shock evolution and film rupture. J Fluid Mech 240:259288
7. Jeong J, Hussain F (1995) On the identification of a vortex. J Fluid Mech 285:6994
8. Kim H, Lee J, Kim TH, Kim HY (2015) Spontaneous Marangoni mixing of miscible liquids at a liquid air contact line. Langmuir 31(31):8726–8731. pMID: 26185919

Bubble Capture, Breakup, and Coalescence in Vortex–Bubble Interaction

Subhajit Biswas and Raghuraman N. Govardhan

Abstract Driven by the motivation to understand breakup and coalescence of bubbles in turbulent flows, we experimentally study an idealization of this, namely interaction of an air bubble with a vortex ring in water. This study is carried out at different ring Reynolds numbers (Re_Γ) for a fixed ratio of bubble's volume to ring's core volume (V_R) of about 0.1. The interaction dynamics is governed by a Weber number (We), which is defined here as the ratio of the vortex core's strength to the Laplace pressure, and has a broad range of about 12–763 in the present study. Several stages of bubble dynamics starting from bubble capture by the ring, to bubble breakup and coalescence are explored using high-speed shadowgraphy technique. Changes in several bubble dynamics aspects such as bubble's equilibrium position inside a vortex, azimuthal elongation, and breakup pattern(s) are witnessed among low and high Weber number cases, and these are described in the present work.

Keywords Vortex ring · Bubble · Bubble breakup · Bubble coalescence

1 Introduction

Turbulent dispersed multiphase flows with bubbles occur in many places such as ocean waves, chemical reactors, and ship hydrodynamics, where bubble breakage and coalescence play a major role. This breakup of bubbles by turbulence is important as its usually decides the size and distribution of the bubble phase in these different flows, with the bubble size in turn being an important factor in deciding mass, momentum, and energy transport between the primary and the dispersed phase. The phenomena of breakup and coalescence of bubbles has attracted considerable attention over the past few decades, with numerous theoretical, numerical, and experimental studies [5–7, 13–17, 20, 23]. A turbulent flow is populated with vortical structures [1], and these structures are highly coupled in nature. The presence of bubbles results in a highly complex multiscale coupled interaction between vortical structures and bubbles. An

S. Biswas (✉) · R. N. Govardhan
Indian Institute of Science, Bangalore, India
e-mail: sbiswas.iisc@gmail.com

L. Venkatakrishnan et al. (eds.), *Proceedings of 16th Asian Congress of Fluid Mechanics*, Lecture Notes in Mechanical Engineering,
https://doi.org/10.1007/978-981-15-5183-3_4

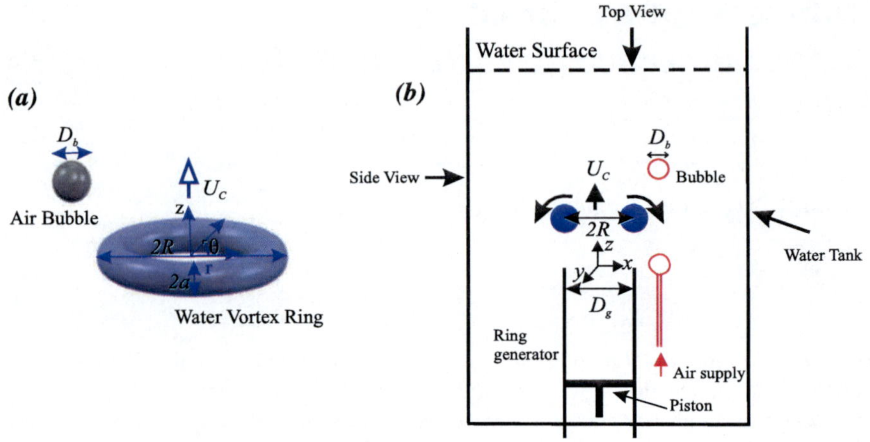

Fig. 1 **a** A schematic of an air bubble of diameter D_b and a vortex ring of ring diameter $2R$ and core diameter of $2a$, traveling vertically upward in z direction with velocity U_C. **b** Schematic of the experimental setup for a single bubble interacting with a vortex ring. A piston cylinder arrangement was used to generate the vortex ring, and the tube was connected to an air supply to generate the air bubble

idealized study of such a complex interaction can be thought to be the interaction of a vortex ring with a bubble. In the present study, we take a single air bubble and study its breakup as it interacts with a vortex ring (see Fig. 1a) generated in the water medium.

There have been very few investigations on the interaction of a bubble with vortex structures as stated by Magnaudet and Eames [18]. The numerical work by Ferrante and Elghobashi [9] on the interaction of bubbles with a Taylor–Green vortex showed that the presence of a bubble can reduce vorticity and enstrophy of the vortex. Oweis et al. [22] adopted one-way coupling method to predict the capture of a bubble in a line vortex. Rastello et al. [24], Van Nierop et al. [27], Bluemink et al. (2009), and Sridhar and Katz [26] experimentally determined correlations for drag and lift coefficients of bubbles in vortex structures. Mazzitelli et al. [19] numerically explored the two-way coupled interaction between bubbles and homogeneous isotropic turbulence where they stated lift force to be responsible for the clustering of bubbles on the vortex on the side having downward velocity. Deng et al. [8] experimentally investigated bubble's behavior in a Taylor vortex. Finn et al. [10] explored the equilibrium position of few microbubbles in a traveling vortex-tube. Higuera [11] and Revuelta [25] analyzed the interaction of a bubble and a vortex ring where they observed distinctive modes of bubble breakup depending on the Weber number and the initial vortex-to-bubble size ratio. Recently Jha and Govardhan [12] studied the interaction of a large bubble on a thin vortex ring at volume ratio of about 0.1 and observed significant permanent fragmentation of the vortex core at low Weber number. They have also presented different stages of bubble–vortex interaction along with some interesting aspects of bubble dynamics which includes bubble's capture time in the vortex core, final number of broken bubbles, and their mean size at different Weber numbers.

As an extension of Jha and Govardhan [12], some more intriguing bubble dynamics aspects related to bubble–vortex interaction, which are bubble's equilibrium position inside a vortex core, the critical aspect ratio of an elongated bubble and simultaneous breakup and coalescence of broken bubbles inside a vortex ring are explored in the present study. There are several parameters that govern the dynamics of interaction between a vortex and a bubble as given in Jha and Govardhan [12] and Biswas and Govardhan [3, 4], which include parameters on both the vortex ring side and the bubble side. Parameters from ring's side includes the vortex ring radius (R), vortex core radius (a), non-dimensional core radius (a/R), circulation strength of the ring (Γ), and ring Reynolds number ($\mathrm{Re}_\Gamma = \Gamma/\nu$, ν is kinematic viscosity of water). On the bubble side, these are the bubble diameter (D_b) and surface tension (σ) at air–water interface. Two more parameters associated with the coupled interaction of the bubble and the ring are the ratio of the bubble's volume to the ring's core volume ($V_R = (\pi/6)D^3/(2\pi^2 Ra^2)$), and Weber number (We) defined here as the ratio of the pressure difference ($\Delta P = 0.87\rho(\Gamma/2\pi a)^2$) of ring's core and outside far-field, to the Laplace pressure (σ/D_b).

The construction of the paper is as follows. In Sect. 2, we discuss about the methods and experimental approaches used in the present work. In Sect. 3, we present bubble dynamics during the interaction with the ring obtained for a wide regime of Weber numbers (We ~ 12–763) keeping V_R fixed at about 0.1. Finally, in Sect. 4, we present a brief overview of the main results.

2 Data/Methodology/Experimental Set Up

A vortex ring was generated using a piston cylinder arrangement in a water tank with transparent side walls providing optical access from all sides. The radius of the vortex ring (R), its core radius (a), and the circulation of the core (Γ) were measured from particle image velocimetry (PIV) measurements after the complete formation of the vortex ring. A large range of vortex ring circulations (Γ) was obtained by changing the impulse of the piston. In the present study, results are shown for a wide range of ring Reynolds numbers (6003–67,376); this covers laminar, transitional, and turbulent vortex rings. A large bubble of diameter (D_b) about 5.7 mm was generated next to the vortex ring generator by a small tube connected to an external air supply.

The uncertainties in the measurement of the vortex ring radius (R), core radius (a), bubble diameter, (D_b) and vortex ring strength (Γ) were estimated to be ±0.2 mm, ±0.15 mm, ±0.1 mm, and ±5%, respectively. The vortex ring used for this study was relatively thin with the ratio of the core radius of the vortex ring (a) to the radius of the ring (R) being around 0.2 (a/R ~ 0.2) [21]. Micron size bubbles were generated at the tip of the vortex ring nozzle using electrolysis and were used to trace the ring. We also traced the vortex ring using dye, with both these measurements being within experimental uncertainties. Due to separate requirement of illumination, both front and top view were captured separately using a Photron SA5 FASTCAM high-speed camera at a frame rate of 3000. The origin of the coordinate was taken at the center of the face of the vortex ring generator nozzle (see Fig. 1b), which is the exit plane of the vortex ring, x and y represent the axis in the horizontal plane, and z represents the vertical direction.

3 Results

In this section, we present bubble dynamics of a 5.7 mm bubble interacting with a thin vortex ring for a range of Weber numbers (see Table 1) keeping the volume ratio fixed at about 0.1. The circulation strength (Γ) of the vortex ring is varied by changing the impulse of the piston, and we achieved different ring Reynolds numbers and their corresponding Weber numbers (We) (see Table 1). In all cases, the vertical position of the vortex ring (z) was traced with time (t) from the side view visualization to obtain information about the convection speed of the ring. The top view visualization was also performed, and this was useful to see interesting aspects of bubble capture and breakup process within the ring. The dimensional position (z) of the ring was normalized with the ring radius (R), and dimensional time (t) was normalized with the ring radius (R) and the convection speed (U_C) of the ring, with both R and U_C being taken just before capture of the bubble. We present here both front view (Fig. 2a) and top view (Fig. 2b) of the time sequence images of visualization at We = 160. As discussed by Jha and Govardhan [12], there are broadly four stages of interaction: (i) Before capture of the bubble, (ii) capture due to the low pressure inside the vortex core followed by elongation and breakup of the bubble by the azimuthal pressure gradient within the ring, (iii) bubble breakup is complete but the vortex core can fragment, and (iv) the last stage where the bubble(s) leave the ring as the ring becomes weaker. Further detailed information of these stages can be found in Jha and Govardhan [12] and Biswas and Govardhan [4].

Now, we are going to discuss some interesting bubble dynamics aspects gleaned from the top view visualizations. We start with the bubble's radial equilibrium location inside the core of the vortex, which is the radial position of the bubble after which only azimuthal elongation occurs. In the experiment, this is measured as the radial location of the bubble within the core after bubble capture and just before the bubble starts elongating azimuthally. Once bubble's entrainment into the ring's core is completed, it tends to settle around a mean radial location within the core,

Table 1 Parameters in our study

Parameters	Range
a/R	~0.2
Re_Γ	6003–67,376
We	12–763
V_R	~0.1

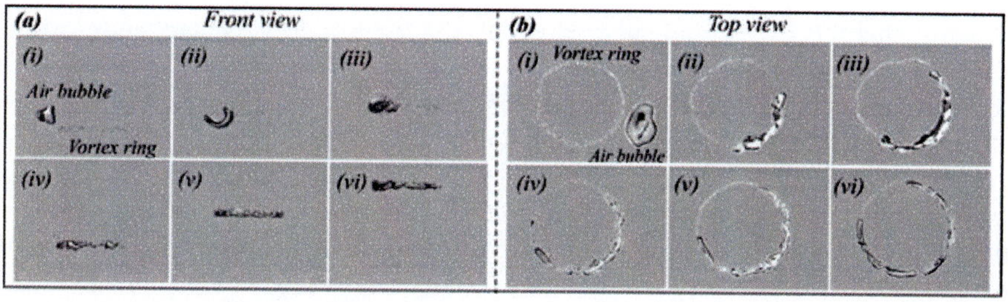

Fig. 2 **a** Front view of bubble–vortex interaction: time sequence of front view visualization images of the interaction of single bubble with a vortex ring at $Re_\Gamma = 29{,}395$ and We = 160. (i) Before capture, (ii) capture, (iii) azimuthal elongation, (iv) and (v) and (vi) breakup. Time instances (tU_C/R) are: (i) 3.32, (ii) 3.87, (iii) 4.19, (iv) 4.62, (v) 5.68, (vi) 6.73. **b** Top view of bubble–vortex interaction: time sequence of top view visualization images of the interaction of single bubble with a vortex ring at $Re_\Gamma = 29{,}395$ and We = 160. (i) Before capture, (ii) and (iii) elongation, (iv) and (v) and (vi) breakup. Time instance (tUC/R): (i) 3.17, (ii) 3.77, (iii) 3.98, (iv) 6.13, (v) 6.48, (vi) 8.72

where there is radial force equilibrium between the radially inward pressure-gradient-induced force, and the radially outward lift force (Sridhar and Katz 1999) [12]. As the bubble entrains and gets settled, its presence results in an observable distortion of the vortex core, which is a function of the bubble's settling location. In the present study, the mean radial location (r/a) of the bubble from the vortex core's center ($r/a = 0$) is seen to be dependent on Weber number, as shown in two top view visualizations shown in Fig. 3. At We of 12 (and $Re_\Gamma = 6003$), the center of the bubble is offset by a distance of about $r/a \sim 0.9$ from the vortex core's center, and as We is increased to 763, the bubble sits much closer to the core's center. Also, the figure shown is a plot of this radial location as a function of the We taken from many such visualizations,

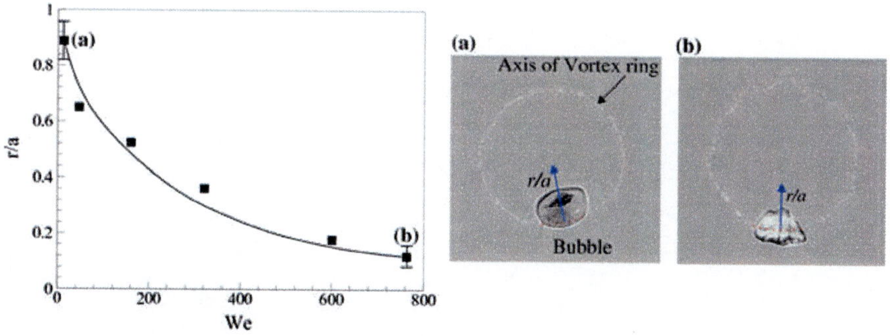

Fig. 3 Bubble's radial equilibrium position (r/a) inside the vortex core after capture at different Weber numbers. **a, b** Represent the top view of the vortex ring with the bubble being seated inside the ring at We = 12 and 763, respectively. At We = 12, bubble's radial location is offset from the ring's axis (or center of the core) by a distance of about $r/a \sim 0.9$. For the other case at We = 763, bubble sits very close to the vortex ring's axis (or center of the core). Here, $r/a = 0$ indicates the axis (or core's center) of the vortex ring

showing that this offset gradually reduces as the We is increased, and tends toward zero (center of the core) at large We.

After the bubble reaches a mean settling location inside the vortex, the bubble is subjected to azimuthal pressure gradient due to the presence of the bubble inside the vortex. This azimuthal pressure gradient stretches the bubble azimuthally which can be seen in Fig. 2b. At the same time, the bubble interface is subjected to interfacial instabilities/perturbations seen as roughness of the bubble's surface (Fig. 2b (ii) and (iii)). These perturbations grow over time and cause the elongated bubble to break at several azimuthal locations.

As discussed by Jha and Govardhan [12], the breakup of the azimuthally elongated bubble inside the vortex core could be due to Kelvin waves and Rayleigh–Plateau instability. In the present study, the aspect ratio (L_b/W) of the elongated bubble which is calculated as the ratio of the elongated length (Lb) of the bubble to the azimuthally averaged width (W) of the stretched bubble has a critical value beyond which the extended bubble starts breaking. We call this the critical aspect ratio ($(L_b/W)_{\text{critical}}$), and this seems to increase with Weber number till We of about 160 and then starts decreasing, and eventually it appears to become independent of We as seen in Fig. 4. As We increases, the bubble elongates more due to higher azimuthal pressure gradient forcing larger extension of the bubble. The critical L_b/W at which breakup begins increases till certain We (~160) as seen in the figure. Increasing the We further, we observe more number unstable waves on the bubble's surface which occurs due to the turbulent nature (large Re_Γ) of the vortex ring. At very large We (or large Re_Γ) these surface waves are seen to grow faster and penetrating the elongated bubble causing early breakup of the bubble by means of multiple fragmentation happening at several azimuthal locations of the extended bubble. Due to this early breakup of the elongated bubble at larger We (or Re_Γ), the critical L_b/W is seen to reduce beyond certain We (~160) and then gradually becoming independent of We (or Re_Γ).

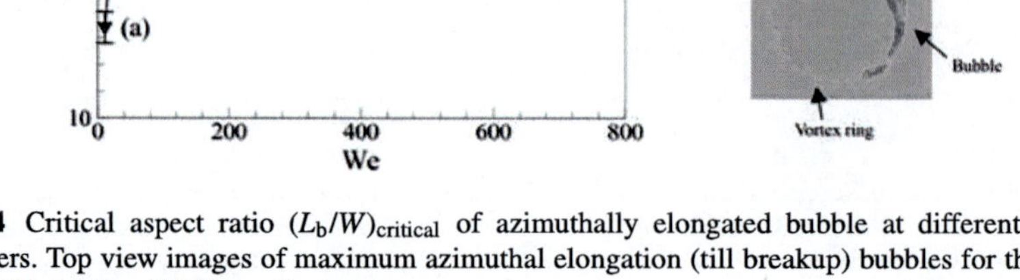

Fig. 4 Critical aspect ratio $(L_b/W)_{\text{critical}}$ of azimuthally elongated bubble at different Weber numbers. Top view images of maximum azimuthal elongation (till breakup) bubbles for three We cases (12, 160, and 763) are shown here

Fig. 5 Number of broken bubbles (N_b) with time (tU_C/R)

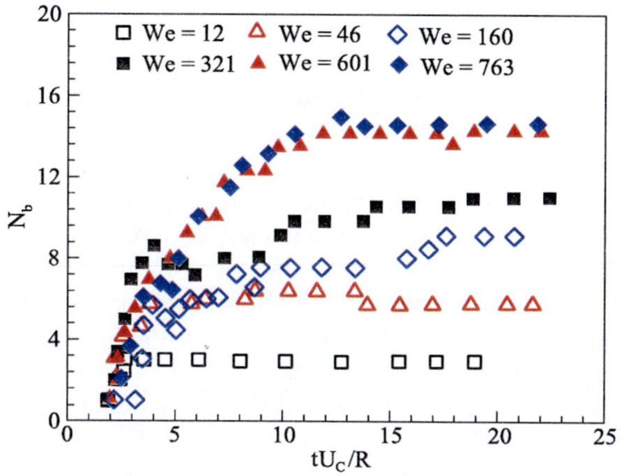

Once reaching the critical aspect ratio, the elongated bubble starts breaking and produces several broken bubbles which can further undergo breakup depending on Weber number and their size. Now we focus on the transient changes in the number of broken bubbles inside the vortex presented in Fig. 5. In the present study, from top view visualizations, we observe that the mother/main bubble which elongates azimuthally can undergo binary/tertiary breakup till We of about 321, and beyond this, rupture/fragmentation of the mother bubble is seen to happen producing relatively larger number of broken bubbles. Revuelta [25] in their numerical study of bubble breakup outside a vortex ring due to induced velocity of the vortex ring observed binary bubble breakup at relatively low We and fragmentation at large We, which is in line with our observations. In the present work, we observe breakup and coalescence of broken bubbles and depending on the We, we can classify this broadly into three categories. The first one comes under low We regime which is of about We ~ 12 in the present study, where the main bubble can undergo binary or tertiary breakup and those broken bubbles can further undergo breakup but no chances of coalescence as they are situated azimuthally sufficiently far from each other. In the second category for which We is broadly of the range of ~46–321, after binary/tertiary breakup of the main bubble, broken bubbles stay inside the ring's core relatively closer to each other and can undergo collisions which can lead to coalescence and reduction of Nb as seen in Fig. 5. Finally in the third category which has much higher We (>321), although there are relatively larger number of broken bubbles situated more closely to each other inside the vortex ring and have higher chances of collisions, these broken bubbles are less likely to coalesce as they offer more resistance to coalescence due to relatively higher surface tension force.

4 Conclusions

Motivated by the complex mechanism of bubble breakup in a turbulent environment, we have experimentally investigated an idealization of this, which is the study of bubble dynamics in a bubble–vortex interaction. The focus of the work has been to explore some aspects of bubble dynamics in this interaction as a function of the many parameters of the interaction. We have presented some important bubble dynamics aspects which are the variation of the mean equilibrium location of the bubble inside vortex, bubble's critical aspect ratio where breakup initiates, and several distinctive bubble breakup patterns starting from binary and tertiary breakup at low and intermediate We and fragmentation at large We, which are comparable with several observations on bubble breakup in bubbly turbulent flows. This idealized study can help us advance our understanding of the breakup mechanism of bubbles interacting with vortical structures in turbulent environments.

References

1. Adrian RJ, Meinhart CD, Tomkins CD (2000) Vortex organization in the outer region of the turbulent boundary layer. J Fluid Mech 422(54):1–54
2. Alopaeus V et al (2002) Gas–liquid stirred tank reactor modelling with CFD and user subroutines. AIChE Annual meeting, Indianapolis, IN, 3–8 Nov
3. Biswas S, Govardhan RN (2019) Interaction of a vortex ring with single and multiple air bubbles. APS G28-004
4. Biswas S, Govardhan RN (2020) Effect of single and multiple bubbles on a thin vortex ring. J Flow Vis Image Process 27(1)
5. Chatzi E (1983) Analysis of interactions in fluid-fluid dispersion systems in agitated vessels. Cleveland State University Press, Cleveland
6. Chatzi E et al (1989) Generalized model for prediction of the steady-state drop size distributions in batch stirred vessels. Ind Eng Chem Res 28:1704–1711
7. Coulaloglou CA, Tavlarides LL (1977) Description of interaction processes in agitated liquid–liquid dispersions. Chem Eng Sci 32:1289–1297
8. Deng R, Wang C-H, Smith K (2006) Bubble behavior in a Taylor vortex. Phys Rev E 73(3)
9. Ferrante A, Elghobashi SE (2007) On the effects of microbubbles on Taylor-Green vortex flow. J Fluid Mech 572:145–177
10. Finn J, Shams E, Apte SV (2011) Modeling and simulation of multiple bubble entrainment and interactions with two dimensional vortical flows. Phys Fluids 23(2):023301
11. Higuera FJ (2004) Axisymmetric inviscid interaction of a bubble and a vortex ring. Phys Fluids 16:1156
12. Jha NK, Govardhan R (2015) Interaction of a vortex ring with a single bubble: bubble and vorticity dynamics. J Fluid Mech 773:460–497
13. Lee CH et al (1987) Bubble breakup and coalescence in turbulent gas–liquid dispersions. Chem Eng Commun 59:65–84
14. Lee CH et al (1987) Dynamics of bubble size distribution in turbulent gas–liquid dispersions. Chem Eng Commun 61:181–195
15. Lehr F et al (2002) Bubble-size distributions and flow fields in bubble columns. AIChE J 48:2426–2443
16. Martinez-Bazan C et al (1999) On the breakup of an air bubble injected into fully developed turbulent flow. Part 1. Breakup frequency. J Fluid Mech 401:157–182

17. Martinez-Bazan C et al (1999) On the breakup of an air bubble injected into fully developed turbulent flow. Part 2. Size PDF of the resulting daughter bubbles. J Fluid Mech 401:183–207
18. Magnaudet J, Eames I (2000) The motion of high-Reynolds-number bubbles in inhomogeneous flows. Annu Rev Fluid Mech 32(1):659–708
19. Mazzitelli IM, Lohse D, Toschi F (2003) The effect of microbubbles on developed turbulence. Phys Fluids 15(1):L5
20. Narsimhan G, Gupta JP (1979) A model for transitional breakage probability of droplets in agitated lean liquid–liquid dispersions. Chem Eng Sci 34:257–265
21. Norbury J (1973) A family of steady vortex rings. J Fluid Mech 57(03):417–431
22. Oweis GF, van der Hout IE, Iyer C, Tryggvason G, Ceccio SL (2005) Capture and inception of bubbles near line vortices. Phys Fluids 17(2):022105
23. Prince MJ, Blanch HW (1990) Bubble coalescence and break-up in air-sparged bubble columns. AIChE J 36:1485–1499
24. Rastello M, Marie JL, Lance M (2011) Drag and lift forces on clean spherical and ellipsoidal bubbles in a solid-body rotating flow. J Fluid Mech 682:434–459
25. Revuelta A (2010) On the interaction of a bubble and a vortex ring at high Reynolds numbers. Eur J Mech (B/Fluids) 29:119–126
26. Sridhar G, Katz J (1995) Drag and lift forces on microscopic bubbles entrained by a vortex. Phys Fluids (1994–Present) 7(2):389–399
27. Van Nierop EA, Luther S, Bluemink JJ, Magnaudet J, Prosperetti A, Lohse D (2007) Drag and lift forces on bubbles in a rotating flow. J Fluid Mech 571:439

Comparison of Flow Features Inside a Plain Orifice Injector for Different Length-to-Diameter Ratios

Vaisakh S. Nair, Vikash Kumar, T. John Tharakan, and Abhilash Suryan

Abstract Cavitation in the flow passages of plain orifice injector affects both internal flow and jet characteristics. Cavitation occurs in flow passages of liquid engine injectors when static pressure falls below the vapor pressure at that temperature. Injector elements having narrow flow passages are more susceptible to cavitation. The flow rate through the injector is reduced significantly due to cavitation when bubbles partially block the flow. It is important to understand cavitation phenomena occurring inside the injector flow passage as it can lead to fluctuation in propellant flow rate to the combustion chamber. Cavitation structures inside the flow path of the injector are obtained by numerical simulations. The numerical results show agreement within validation uncertainty against the previously observed experimental data.

Keywords Cavitation · Injector · Internal flow · CFD

1 Introduction

The effect of cavitation on liquid jet atomization has been studied since 1959, and its effects on flow regimes have been understood. Early works dealt mainly with qualitative visualization studies highlighting the effect of cavitation on the development of the spray. Many features of cavitation are understood reasonably well that sheds light onto the dynamical behavior of the cavitation structures, to understand the key physical phenomena leading to bubble collapse and atomization enhancement. The cavitation phenomena occurring inside the injector nozzle with narrow flow passages has been considered as an important factor in fuel atomization. Injection pressure, liquid properties, and geometric features affect this phenomenon. Thus, it plays a critical role in primary breakup and atomization, in turn having a pivotal role in the

V. S. Nair (✉) · A. Suryan
Department of Mechanical Engineering, College of Engineering Trivandrum, Thiruvananthapuram, Kerala 695016, India
e-mail: vaisakh.nair28@gmail.com

V. Kumar · T. J. Tharakan
Liquid Propulsion Systems Centre, Valiamala, Trivandrum, Kerala, India

L. Venkatakrishnan et al. (eds.), *Proceedings of 16th Asian Congress of Fluid Mechanics*, Lecture Notes in Mechanical Engineering,
https://doi.org/10.1007/978-981-15-5183-3_5

combustion process. The cavitation bubbles affect the internal flow and consequent spray formation outside the injector.

The first thorough study of flow separation was conducted in 1959 by Spikes and Pennington [1] that brought out the effects of cavitation on discharge coefficient in turbulent flow regime. Further tests were carried out with inlet chamfer to determine the optimum angle and depth of chamfer for cavitation suppression. Numachi et al. [2] conducted a detailed study on the effect of cavitation occurrence on the discharge coefficient of sharp-edged orifice plates. The experimental results indicate that cavitation can exist to a minimum cavitation number of 0.2. An investigation into the different flow structures which exist in the holes of direct injection nozzles and their corresponding large-scale acrylic models is described by Soteriou et al. [3]. They explain the role of cavitation on spray abnormality, hydraulic flip, and atomization. Chaves et al. [4] found that discharge coefficient and the spray angle level off at a value that is almost independent of any further increase of injection pressure. Dabiri et al. [5] looked into the cavitation collapse in the year 2010 and proposed that the collapse of traveling bubbles increases the disturbances inside the liquid flow. In this paper, effects of cavitation on increasing the disturbances in the flow through the injection orifice were studied. In 2013, Payri et al. [6] investigated the behavior of the internal flow under cavitating conditions, and the influence of using four different fuels is studied using a transparent nozzle with a cylindrical orifice. The influence of upstream and downstream pressure conditions as well as liquid viscosity was determined. De Giorgi et al. [7] conducted experiments on a cavitating orifice to determine the influence of pressure and temperature on flow regime transition due to cavitation. It was observed that cavitating structures showed different characteristics with variations in operating conditions such as pressure inside the orifice and the flow temperature. The cavitation intensity and the incipient cavitation number is found to vary with liquid temperature. Tharakan et al. [8] conducted comprehensive modeling of spray formation in liquid injectors involving (i) internal hydrodynamics of injector (ii) breakup of liquid sheet leading to primary and secondary atomization and (iii) prediction of size and velocity distributions of droplets in the spray. In this work, progress and current trends in the individual modules have been extensively reviewed, and their implications on development of comprehensive models have been discussed. The unresolved issues and future research directions are also indicated. Altimira and Fuchs [9] studied the effect of liquid viscosity and density on the cavitation sensitivity. The flow through a throttle geometry, which has been previously investigated experimentally, is simulated under conditions corresponding to the onset of cavitation and critical cavitation number adopting the large eddy simulation approach. Results show that the onset and growth of cavitation bubbles is enhanced by a decrease in the density of the liquid.

Lee et al. [10] investigated the impact of fuel atomization in effective combustion of propulsion devices and provided an understanding of how integrated the fuel system is in accordance with the other engine components. He et al. [11] conducted an experimental study under different pressures to analyze the evolution of cavitation inside diesel nozzle and found that higher injection pressure leads to earlier cavitation inception. It was found that the occurrence of “string cavitation” has a strong relationship with the location of needle and injection pressure. Abderrezzak and Huang [12] studied the fundamental effect of cavitation bubbles on the internal nozzle flow, which is the vital link between the liquid emerging from the nozzle and the fully developed spray. The influence of the length-to-diameter ratio on cavitation and subsequently on the spray structure is also addressed. Cui et al. [13] experimentally studied the effect of diameter error and conicality that embody common deviations in nozzles geometry using transparent nozzles.

Numerical models have been developed by earlier investigators to simulate cavitation phenomena inside fuel injector nozzles. The present study involves numerical simulation of flow through a single plain orifice injector nozzle for different length-to-diameter ratios. The numerical model is developed, and meshing is done using ANSYS ICEM CFD. The simulation is done using commercial software ANSYS Fluent 15 [14]. A two-dimensional, transient, problem is solved using mixture multiphase model, implementing Schnerr–Sauer model [15], along with the k–ϵ (two-equation) turbulence model. The CFD simulation is compared with experimental study of Abderrezzak and Huang [12].

2 Numerical Setup and Methodology

A single orifice injector is modeled which is chosen as the base model for the simulation from the literature survey conducted, to understand the different flow regimes produced as the flow occurs, its interaction with the flow, and in turn the effect on flow features. The computational domain differs with respect to the orifice length.

The geometrical specifications of the nozzle and the boundary conditions implemented for the simulation are depicted in Fig. 1. Pressure-based solver was used, and analysis was performed under transient conditions. Standard k–ϵ turbulence model was used, and the convective terms were solved using second-order upwind interpolation scheme. Hybrid initialization has been performed to obtain the initial solution, and the reference values were provided from the inlet. Table 1 elucidates the general working conditions employed for the present study.

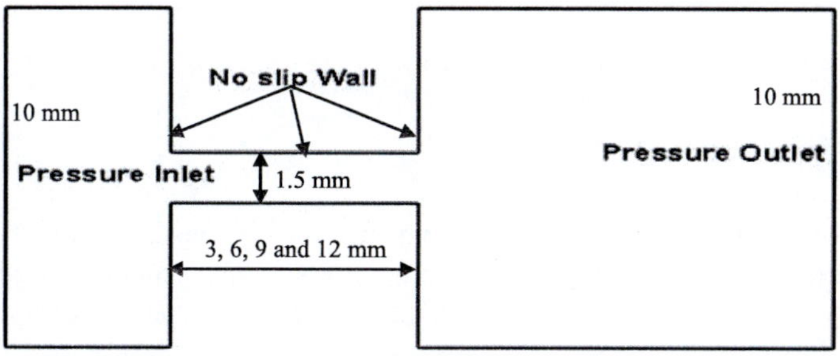

Fig. 1 Computational domain and boundary conditions

Table 1 Working conditions

Working fluid	Water
Back pressure (kPa)	101.325
Temperature (°C)	30
Density (kg/m^3)	988.25
Vapor pressure (kPa)	2.850
Dynamic viscosity (Ns/m^2)	1.002e03

In this study, two-phase flow is modeled with single fluid model (homogenous mixture). Continuity and momentum equations used for the numerical simulation are as follows:

$$\frac{\partial \rho_{\rm m}}{\partial t}+\frac{\partial \rho_{\rm m} u_i}{\partial x_i}=0 \tag{1}$$

$$\rho_{\rm m}\left(\frac{\partial u_i}{\partial t}+u_j\frac{\partial u_i}{\partial x_j}\right)=-\frac{\partial \rho}{\partial x_i}+\frac{\partial}{\partial x_j}\left[\mu_{\rm m}\left(\frac{\partial u_i}{\partial x_j}+\frac{\partial u_j}{\partial x_i}\right)\right]+\rho_{\rm m} g \tag{2}$$

where $\rho_{\rm m}$ is density of mixture and $\mu_{\rm m}$ is viscosity of mixture which is as following:

$$\rho_{\rm m}=\alpha_{\rm l}\rho_{\rm l}+\alpha_{\rm v}\rho_{\rm v} \tag{3}$$

$$\mu_{\rm m}=\alpha_{\rm l}\mu_{\rm l}+\alpha_{\rm v}\mu_{\rm v} \tag{4}$$

α is vapor volume fraction for each phase, $\alpha_{\rm v}$ is vapor volume fraction for vapor phase and $\alpha_{\rm l}$ is vapor volume fraction for liquid phase. If we assume radius of bubble as $R_{\rm b}$, number of bubbles in a unit of volume as $n_{\rm b}$, $\alpha_{\rm v}$ can be calculated as follows:

$$\alpha_{\rm v}=\frac{\frac{4}{3}\pi n_{\rm b} R_{\rm b}^3}{1+\frac{4}{3}\pi n_{\rm b} R_{\rm b}^3} \tag{5}$$

A grid independence study was carried out for the computational fluid dynamic model.

3 Results

The simulation of the flow through single orifice injector has been done. The cavitation regimes corresponding to each injection pressure ($P_{\rm inj}$) for each configuration (fixed back pressure) has been compared with the corresponding experimental values. The vapor fraction contours shows good agreement, which can be also seen evidently from the relationship between the flow rate and injection pressure as shown in Fig. 2a,

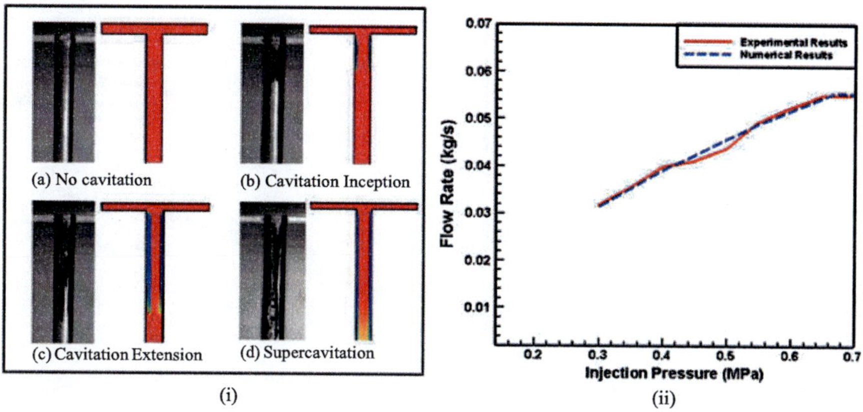

Fig. 2 Comparison of experimental and computational results (i) instantaneous flow characteristics, (ii) relationship between flow rate and injection pressure for ($L/D = 8$)

b. The numerical results show good agreement with the experimental findings of Abderrezzak and Huang [12].

From analyzing the results, with respect to the changes in geometric and flow parameters like length-to- diameter ratio (*L*/*D* ratio) and injection pressure, the following data regarding the system are plotted in Fig. 3.

Figure 3a, b depicts the pressure and velocity contours for injector with length-to-diameter ratio of 8. These contours indicate various regimes of cavitation and the related flow features. The regimes of cavitation inception, its extension, and supercavitation can be seen in these contours. Both the pressure and velocity contours elucidate the fact that at the cavitation zone, the pressure falls near to the vapor pressure and the exit velocity decreases as the regime turns to supercavitation. It is seen from Fig. 4 for all *L*/*D* ratios, the flow rate suddenly diminishes when injection

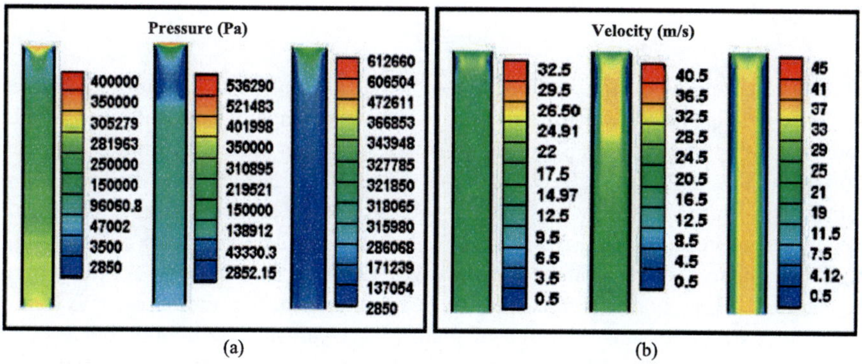

Fig. 3 Contours of **a** pressure and **b** velocity for $L/D = 8$ for different flow regimes

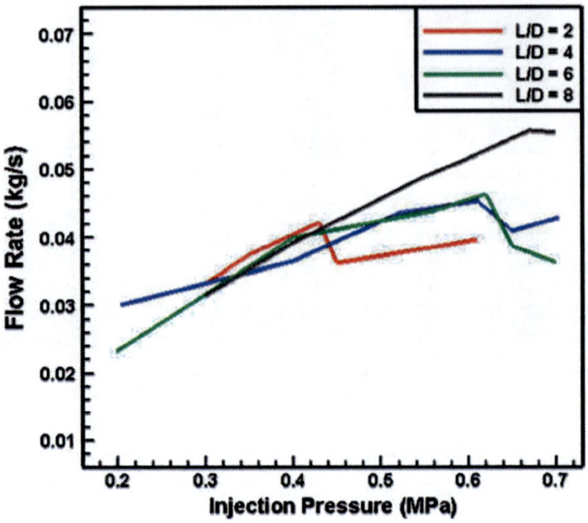

Fig. 4 Relationship between the flow rate and injection pressure for different *L*/*D* ratios

pressure is increased. This in turn sheds light on the dependence of geometrical and flow conditions on the inception and development of cavitation in the injector, which further affects flow through the injector. The flow rate increases up to a particular injection pressure which produces supercavitation, causing the flow path to reduce and in turn lowers the flow rate. Further increase in the injection pressure causes the flow rate to increase, which may be due to the increase in flow path, caused by retracting cavitation regime from the exit toward the inlet, allowing the fluid to occupy this portion of the injector passage. The pattern similar to the above is observed for the pressure and velocity contour but at varying inlet conditions, with different *L*/*D* ratios in the range of 2–8.

4 Conclusions

Numerical analysis of single orifice injector nozzle is carried out at various geometric and flow conditions. The result emphasizes the importance of flow and geometrical parameters on the cavitation phenomenon inside the plain orifice injector. A comparison of the flow field for different *L*/*D* ratios was made and its effect on formation, and growth for various flow regimes was brought out. The fluctuation of flow with the injection pressure for different *L*/*D* ratios was studied. Analysis of cavitation regimes confirms the dependence of flow with parametric variations. As the length increases, the injection pressure required to initiate cavitation increases. The future works include the effective design of plain orifice injector for conducting experimental investigation to understand these features in detail and also to visualize both internal and external flows.

Acknowledgements The authors would like to express their deepest thanks to Liquid Propulsion Systems Centre and College of Engineering Trivandrum for providing the facilities and guidance during the course of the work.

References

1. Spikes RH, Pennington GA (1959) Discharge coefficient of small submerged orifices. SAE paper, I73, No 25
2. Numachi F, Yamabe M, Oba R (1960) Cavitation effect on the discharge coefficient of the sharp-edged orifice plate. J Basic Eng
3. Soteriou C, Andrews R, Smith M (1995) Direct injection diesel sprays and the effect of cavitation and hydraulic flip on atomization. SAE technical paper series, 950080
4. Chaves H, Knapp M, Kubitzek A, Obermeier F, Schneider T (1995) Experimental study of cavitation in the nozzle hole of diesel injectors using transparent nozzles. SAE technical paper series, 950290
5. Dabiri S, Sirignano WA, Joseph DD (2010) A numerical study on the effects of cavitation on orifice flow. Phys Fluids 22
6. Payri R, Salvador FJ, Gimeno J, Venegas O (2013) Study of cavitation phenomenon using different fuels in a transparent nozzle by hydraulic characterization and visualization. Exp Thermal Fluid Sci
7. De Giorgi MG, Ficarella A, Tarantino M (2013) Evaluating cavitation regimes in an internal orifice at different temperatures using frequency analysis and visualization. Int J Heat Fluid Flow 39:160–172
8. Tharakan TJ, Nath S, Datta A, Mukhopadhyay A, Sen S, Jog MA (2013) Trends in comprehensive modeling of spray formation. Int J Spray Combust Dyn 5:123–180
9. Altimira M, Fuchs L (2014) Effect of fuel flexibility on cavitation in injector-like flows. In: 26th Annual conference on liquid atomization and spray systems
10. Lee J, Cheng K, Eklund D (2015) Challenges in fuel injection for high-speed propulsion systems. AIAA J 53(6)
11. He Z, Zhang Z, Guo G, Wang Q, Leng X, Sun S (2016) Visual experiment of transient cavitating flow characteristics in the real-size diesel injector nozzle. Int Commun Heat Mass Transfer 78:13–20
12. Abderrezzak B, Huang Y (2017) Investigation of the effect of cavitation in nozzles with different length to diameter ratios on atomization of a liquid jet. J Therm Sci Eng Appl
13. Cui J, Lai H, Feng K, Ma Y (2018) Quantitative analysis of the minor deviations in nozzle internal geometry effect on the cavitating flow. Exp Therm Fluid Sci
14. ANSYS Inc., Fluent user's guide, USA
15. Schnerr GH, Sauer J (2001) Physical and numerical modeling of unsteady cavitation dynamics. In: 4th International conference on multiphase flow, vol 1, USA

Aerodynamics and Design of Spinning Samara Model

G. Yogeshwaran, M. V. Srisha Rao, and Jagadeesh Gopalan

Abstract A particular form of winged seed (samara) dispersal technique adopted by nature uses autorotative (unpowered rotation of the wing generating thrust force against gravity) descent; for e.g. in Maple and Mahagony trees. This technique provides the lowest descent velocities among various seed dispersal techniques found in nature ensuring the safety of the delicate seeds. Bio-mimicked solutions to important engineering problems in aerospace as well as in disaster management—air dropping of life-saving packages during floods can be inspired from the samara. The samara is a complex structure having a bluff root containing the seed attached to a three-dimensional wing. The dynamics of the samara from the instant of release is entirely unsteady, involving an initial transition phase where the samara tumbles until it achieves autorotation leading to a steady descent velocity. The distribution of mass and aerodynamic forces in this single structure ensures its stability during descent. Studies to comprehensively understand the physics of the samaras are limited. Recently, leading edge vortex (LEV) has been found to be responsible for the high thrust forces achieved during autorotation. The dependence of LEV on the morphology of the seed needs to be understood to design optimal devices for engineering applications. The principal aim of this study is to understand the effect of morphology on the aerodynamics of the samara with a particular focus on the characteristics of the LEV. The flow field around the autorotating samara is experimentally obtained using particle image velocimetry (PIV) in a specially designed vertical wind tunnel. However, natural samara has limited utility for parametric studies; therefore, 3D printed model is developed that closely mimic the functions of the natural samara. Drop tests of the natural samara and the 3D printed model show that the dynamics of the model and the samara are similar. For the first time, a complete characterization of the spanwise distribution of LEV has been carried out on the samara model which enables us to understand the inter-relationship between observed flow physics, morphology and performance parameters.

G. Yogeshwaran (✉) · M. V. S. Rao · J. Gopalan
Department of Aerospace Engineering, IISc Bangalore, Bengaluru 560012, India
e-mail: yogeshwarang@iisc.ac.in

L. Venkatakrishnan et al. (eds.), *Proceedings of 16th Asian Congress of Fluid Mechanics*, Lecture Notes in Mechanical Engineering,
https://doi.org/10.1007/978-981-15-5183-3_6

Keywords Unsteady aerodynamics · Leading edge vortex · Spinning samaras · PIV

1 Introduction

Plants have limited mobility and so it adopts various seed dispersal techniques to establish its embryo bearing seed at a suitable place away from the parental plants [1]. Spinning samaras are winged seeds [2] where the seed is enclosed in a protective cover and attached to the wing structure as shown in Fig. 1. Spinning samaras undergo initial random motion for a period of time (transition phase) when it is released from the parental plant before starting to rotate about a point near to its centre of gravity (steady state) without any externally induced power. This rotation aids the samara to generate high thrust force, thereby making its descent velocity approximately 15 times lesser than the impact velocity on the ground without rotation. The low rate of descent of these samaras combined with gust of wind makes it to disperse from its parental plant. This is one of the seed dispersal techniques adopted by the plants to disperse through wind [3]. This unpowered spinning motion is called autorotation [4], similar to autorotation of helicopter under engine failed condition. The low rate of descent of the spinning samaras without use of external power promotes it to be possibly mimicked in real-life scenarios where the object has to be brought to ground safely from a height. Few existing applications are in the field of missiles, samara inspired delivery system, unmanned ariel vehicle (UAV) and designing of re-entry probes for Mars and Venus.

Theoretical studies showed that spinning samaras under stable autorotation generate an unexpected high thrust force, despite their small size and low descent velocity [3, 5]. The secret behind this high thrust force was found to be the presence of strong stable leading edge vortex (LEV) on the leeward side of their wing [5]. The angle of attack at the wing cross sections is much higher than in a conventional aircraft (post-stall region) and there is a separated flow at the leading edge [5]. This

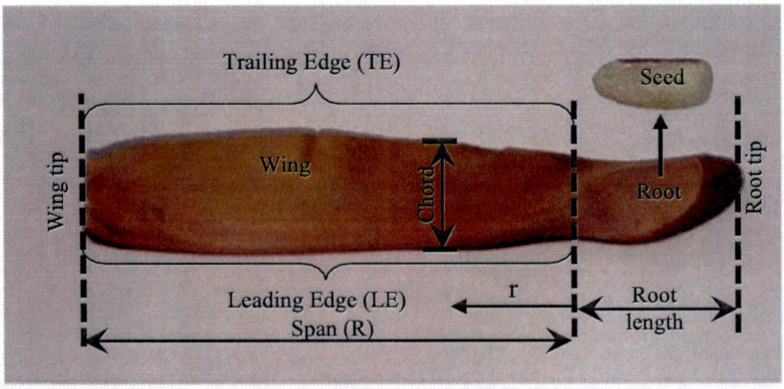

Fig. 1 Spinning samaras

separated flow combined with the high angular acceleration of the spinning samara makes it reattach and generate a strong LEV [5–7]. However, the physics behind this autorotation involves balance of inertia, centrifugal and aerodynamics forces acting on it throughout its descent. Owing to the complex interaction of these forces, especially the unsteady aerodynamic force makes it difficult to understand the physics behind this autorotation completely [5]. The uniqueness of these samaras is its rigid shape and mass distribution which governs the nature of these forces [2]. Whereas, in case of animal flight, the neuromuscular control allows the deformation of wing structure for its performance and stability [2]. Further, the cross-sectional shapes are observed to non-standard aerofoils and the wings structures resemble the helicopter blade [8]. The objective of the present work is to find out the relationship between morphological parameters of the spinning samaras and local lift force due to the presence of LEV. Since parametric studies on real samaras are not possible, samara models are designed and fabricated to understand the flow physics of real samaras.

2 Methodology

In the absence of comprehensive design methodology for engineering spinning samaras [4, 6, 9], we begin by considering a close mimicking of natural samaras. Initially, few samara models are designed and analysed to understand the fundamental required morphology for autorotation phenomena to occur. For this design purpose, depron sheet is used to create the wing planform and elastomers, wood and china clay for mass distribution. A systematic comparative study between these designed samaras models and real samaras are established. These results show that asymmetric mass and asymmetric area distribution play an important role in autorotation. Further, for maximum performance, chord length is observed to be maximum at the mid-span and minimum at the edges with minimum thickness in the wing region. Based on these observations, a 3D printed samara model is designed and developed as shown in Fig. 2. The total mass of the model is 2.58 g which is five times of the average mass of the real samaras and the planform area is 27.48 cm^2. The chord and mass distributions are maintained as per the previous analysis. VeroBlue [6] of density 1.17 g/cm^3 is chosen as resin material for its high strength and smooth surface finish with SUP705 as the support material for printing the model. Stiffeners

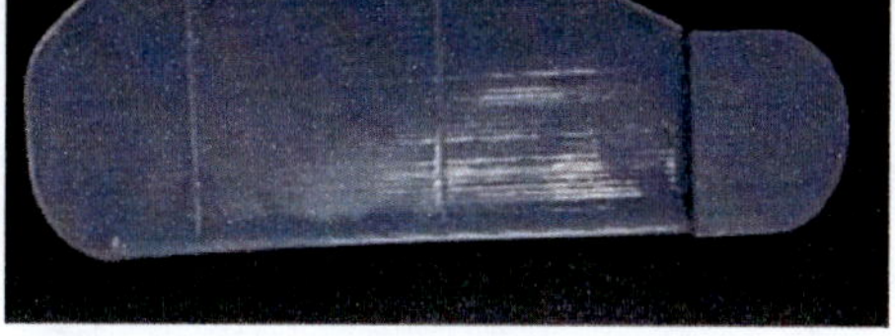

Fig. 2 Designed 3D printed samara model

are added at the sharp edges and in the wing region to provide rigidity to the model. The span of the model is chosen to be 8 cm with root length of 2 cm and maximum chord to be 3.57 cm at spanwise location of 43.20 cm from root.

3 Experiments in Vertical Wind Tunnel

An updraft of magnitude equal to the descent velocity is incident upon the model to make it autorotate at a stationary location in the test section of a vertical wind tunnel. A vertical wind tunnel [8] facility in the Laboratory of Hypersonic and Shockwave Research (LHSR) in Indian Institute of Science is used to carry out the flow visualization experiment over the designed model. The wind tunnel consists of five sections: an inlet, a diffuser, a honeycomb structure, a fan and a test section as shown in Fig. 3. The detailed description of the vertical wind tunnel facility is provided in the thesis of Sonal [8]. The tunnel was basically designed to perform flow visualization around real samaras. The descent velocity of the designed samara model is higher that the real samaras, therefore, the existing wind tunnel is improvised in terms its

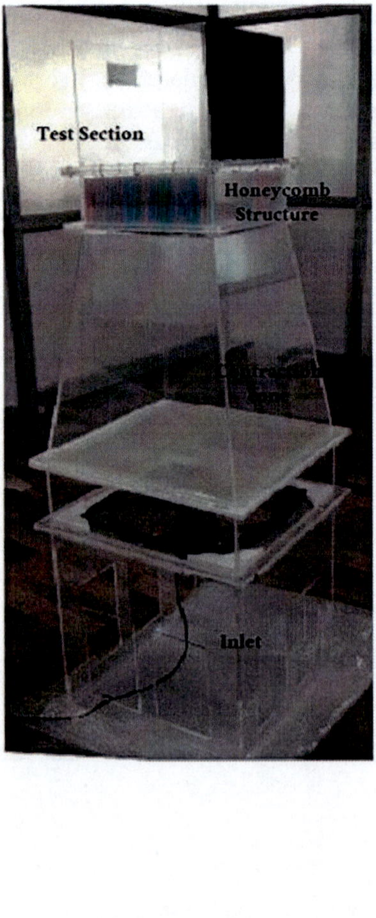

Fig. 3 Vertical wind tunnel [8]

test sectional velocity. On observing the free fall flight of the real samaras using high speed camera, it is evident that its centre of gravity moves in helical path around its axis of rotation. This additional motion prevents us to observe the flow field around the samara or model at a particular cross section on the test section of the wind tunnel, especially the LEV formed on its leeward side. To overcome this, fishing rope of diameter 0.05 cm is made to pass through the centre of gravity of the model and its two ends are attached to test section of the tunnel. The variation in performance parameters of the model or samara due to the presence of fishing rope is found to be less than 5% [10]. The test sectional velocity of the wind tunnel is maintained at descent velocity of the model by adjusting the fan speed through a voltage regulator (VARIAC). Experiments to characterize the LEV are conducted using planar particle image velocimetry (PIV) techniques. A high repetition rate double pulsed Nd:YLF laser with 20 mJ per pulse at a repetition rate of 0.844 kHz is passed through the seeded flow in the test section. Images are captured by high speed Phantom Miro 110 camera, with a pixel resolution of 1280×800 pixels, which is synchronized to the laser using a high speed controller. The time delay between two successive images is set to 100 μs and the cross-correlation algorithm is performed to obtain velocity field.

4 Results and Discussions

In order to understand the free fall behaviour, the designed 3D printed model is dropped from the height of 6 m with a suitable experimental setup in the LHSR, IISc. It is observed that designed model initially undergoes a random motion for a small period of time and starts to autorotate about its centre of gravity similar to real samaras. The descending motion of the model is captured using the high speed camera and post-processed to calculate its performance parameters. Like real samaras, the length of the transition phase is dependent on its initial orientation before it drops. The 3D printed model is dropped with different orientations and holding points to analyse the effect of orientation on the length of the transition phase. The model has minimum transition phase when its holding point is held at trailing edge with wing surface making an actuate angle with respect to vertical plane.

Similar to the free fall behaviour, the autorotation phenomena is observed on the designed model placed in the test section of vertical wind tunnel. The test sectional velocity is maintained similar to the descent velocity of the model by adjusting the fan speed. High speed camera is used to capture autorotation phenomena inside the test section and post-processed to calculate the RPM and coning angle (angle made by the model with respect to the horizontal plane in steady state). The test sectional velocity of the tunnel is calculated using fan type anemometer. The descent velocity, RPM and conning angle are found to be 1.7 ± 0.1 m/s, 538 ± 3 and $9.75° \pm 0.23°$, respectively, and it is in accordance with performance parameter calculated using free fall test. In the case of real samaras, their corresponding ranges are 0.82–1.45 m/s, 420–700, 15°–27° [3, 8, 10]. Thus, the shape and the performance parameters of the

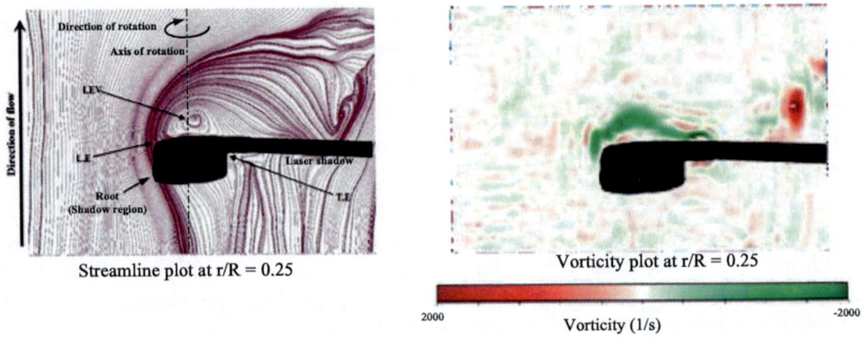

Fig. 4 Streamline and vorticity plots for the designed model at $r/R = 0.25$

designed model are similar to real samaras. The deviating morphological parameter is the mass of the model which is due to the density of the VeroBlue that cannot be altered.

Planar PIV experiment is carried across seven different spanwise locations (at $r/R = 0, 0.25, 0.38, 0.48, 0.5, 0.75, 1$) in the model when it is autorotating in the tunnel. The results show the presence of strong stable LEV from $r/R = 0.25$ to 0.50 similar to the real samaras [5, 10]. In the region between $r/R = 0$ to 0.25 and 0.5 to 1, the flow topology is similar to the bluff body which is in accordance with the result obtained from [11]. The size of the vortex is increases from 0.25 to 0.48 and then decreases. The maximum size of the vortex is found to appear around the region of $r/R = 0.48$. In general, placing the planar laser sheet at the exact location of r/R under rotating condition is a difficult task and for any case it lies in the range of ±0.01 around that location [11]. A streamline and vorticity plots around the model at spanwise location of $r/R = 0.25$ is shown in Fig. 4. The circulation strength is calculated across the span by considering the anticlockwise vorticity which corresponds to LEV. The distribution of circulation strength across the span is similar to its chord distribution.

5 Conclusion

In the present work, a 3D printed samara model is designed and fabricated to understand the relationship between the local lift force and the morphology of the model. The 3D printed model is developed based on the close mimicking of real samaras. The free fall behaviour of the model from point of release is similar to the real samaras. The decent velocity, RPM and coning angle of the model is 1.7 ± 0.1 m/s, 538 ± 3 and 9.75° ± 0.23°, respectively. Planar PIV experiments show the presence of strong stable LEV for the model in the spanwise location of $r/R = 0.25$ to 0.50 similar to real samaras. The local lift force is directly proportional to circulation strength of the LEV is calculated across the span. Thus, for zero camber, zero twist across the span and zero angle of incidence the variation of local lift force is directly proportional to its chord distribution.

Acknowledgements This research is undertaken with the assistance of resources provided Indian Institute of Science, Bangalore. Authors would like to express their sincere thanks to members of the LHSR for their assistance in conducting the experiments.

References

1. https://en.wikipedia.org/wiki/Seed_dispersal
2. Norberg RA (1973) Autorotation, self-stability, and structure of single-winged fruits and seeds (samaras) with comparative remarks on animal flight. Biol Rev 48(4):561–596
3. Azuma A, Yasuda K (1989) Flight performance of rotary seeds. J Theor Biol 138:23–53
4. Andreas K (2007) The guided samara: design and development of a controllable single-bladed autorotating vehicle. Master's thesis, The Massachusetts Institute of Technology (MIT)
5. Lentink D, Dickson WB, Van Leeuwen JL, Dickinson MH (2009) Leading-edge vortices elevate lift of autorotating plant seeds. Science 324:1438–1440
6. Ulrich ER, Pines DJ (2012) Planform geometric variation and its effect on the autorotation efficiency of a mechanical samara. J Am Helicopter Soc 57(012003):1–9
7. Lentink D, Dickinson MH (2009) Rotational accelerations stabilize leading edge vortices on revolving fly wings. J Theor Biol 212:2705–2719
8. Sonal L (2014) Aerodynamics of mahogany seed. Master's thesis, Aerospace Engineering, Indian Institute of Science
9. Minami S, Azuma A (2003) Various flying modes of wind-dispersal seeds. J Theor Biol 225:1–14
10. Salcedo E, Trevino C, Vargas RO, Martinez Suastegui L (2013) Stereoscopic particle image velocimetry measurements of the three-dimensional flow field of a descending autorotating mahogany seed (*Swietenia macrophylla*). J Exp Biol 216:2017–2030
11. Limacher E, Rival DE (2015) On the distribution of leading-edge vortex circulation in samara-like flight. J Fluid Mech 776:316–333

Recirculation Zones and Its Implications in a Taylor Bubble Flow in a Square Mini/Microchannel at Low Capillary Number

Prafulla P. Shevkar and Manoj Kumar Moharana

Abstract A 3D numerical study is carried out for the formation of (i) isolated Taylor bubble and (ii) a train of Taylor bubbles, in a square channel of side 1.0 mm for the range of a capillary number 0.005 < Ca < 0.007. The fluid channel is carved in the rectangular solid substrate of dimensions 3 mm × 2 mm × 30 mm achieves a conjugate heat transfer condition. Constant heat flux is applied at the bottom wall of the substrate of area 3 × 30 mm^2, while all the other surfaces were insulated. In the fluid channel, the bubble flows through the inlet section, the heating zone and the end section. We observed the recirculation zones all around the bubble body, the one at the back of the bubble is found to be strongest, which results in the maximum heat flux in the nearby region. Taylor bubble flow enhanced the heat transfer multifold times compared to the single-phase flow. To study the axial wall conduction in conjugate heating conditions, we carried out the simulations for the range of substrate wall to fluid thermal conductivity ratio $10 < k_{sf} < 646$ and the ratio of substrate thickness to channel depth $1 < \delta_{sf} < 5$.

Keywords Taylor bubble flow · Mini/microchannel · Axial wall conduction · Two-phase flow · Recirculation flow

1 Introduction

Taylor bubble forms when liquid and gas flow in a channel in the particular range of gas–liquid velocity ratio. Many experimental [1–3] and numerical studies [4, 5] show a Taylor bubble formation at different gas and liquid velocities in a channel of different sizes and shapes. Gupta et al. [5] give the modelling methodology to solve the Taylor bubble flow problem numerically in a microchannel based on the different

P. P. Shevkar (✉)
Indian Institute of Technology Madras, Chennai 600036, India
e-mail: prafulla145@gmail.com

M. K. Moharana
National Institute of Technology Rourkela, Rourkela, Odisha 769008, India

L. Venkatakrishnan et al. (eds.), *Proceedings of 16th Asian Congress of Fluid Mechanics*, Lecture Notes in Mechanical Engineering,
https://doi.org/10.1007/978-981-15-5183-3_7

experimental correlations and modelling studies. Moharana et al. [6] numerically studied the axial wall conduction in conjugate heating conditions and found the optimum Nusselt number for the substrate wall to fluid thermal conductivity ratio. For a given flow rate and thickness ratio, the axial wall heat conduction effect increases with an increase in conductivity ratio.

Taylor bubble finds applications in the measurement of flow velocity, cooling of electronic devices such as printed circuit board, lab-on-chip devices and microreactors. Taylor bubble is promising in cooling devices as the bubble creates recirculation in the liquid slug, which increases the heat transfer many times in a microchannel compared to a single-phase flow. The strength and the location of occurrence of recirculation zones related mainly to the shape of the liquid film around the bubble body. The bubble and the liquid film take different forms in a triangular, rectangular and elliptical channels [7]. At low capillary number, the bubble takes cylindrical shape in the middle and a hemispherical one at the tail end while at a high capillary number the tail end becomes concave. The infrared (IR) thermograms [1] and microparticle image velocimetry [2] studies confirmed the recirculating flow at the back of the bubble in the liquid slug. Many researchers studied the recirculating flow in a circular channel in the plane parallel to the primary flow direction. Due to an unevenly distributed film thickness in a non-circular channel, more in the corners than at the flatter section where the bubble body is closer to a wall; the recirculating flow generated in the corners is more complex and needs more attention. In this paper, we study the effect of recirculation zones present at the back of the bubble on the different fluid properties, flow and heat transfer characteristics. Also, the temperature contours plotted for the different substrate thicknesses show the reduced axial wall conduction in the Taylor bubble flow in a microchannel in a conjugate heating condition for the range of capillary number $0.005 < \mathrm{Ca} < 0.007$.

2 Numerical Simulations

The square microchannel of cross-sectional area 1 mm^2 is carved in the solid substrate of $3 \times 2 \times 30\ \mathrm{mm}^3$, as shown in Fig. 1. Air flows annulus to water to form a bubble through the air and water inlets, respectively, labelled in Fig. 1. The void fraction at the inlet is equal to a homogeneous void fraction $\beta = U_g/U_{TP} = 0.5$. Boundary conditions at the inlet are velocity inlet. The length of the inlet section (L_i), heating zone (L_h) and end section (L_e) of the fluid channel was 15 mm, 30 mm and 9 mm, respectively. The thermo-physical properties of the air and water were calculated at 300 K. A constant heat flux was applied at the bottom wall of the substrate while all other walls are kept adiabatic. For a temperature rise of 25 °C, a heat flux of 232,332.6 $\mathrm{W/m}^2$ was applied at the bottom wall for the value of mixture Reynolds number $\mathrm{Re} = U_{TP} \times d_h/\upsilon = 200$, where U_g the gas velocity, U_{TP} the mixture velocity ($U_{GS} + U_{LS}$), U_{LS} the superficial liquid velocity, U_{GS} the superficial gas velocity, d_h the hydraulic diameter and υ the kinematic viscosity of the liquid. The width of the substrate (ω_s) and fluid channel (ω_f) are 1 mm. The height of the fluid channel (δ_f)

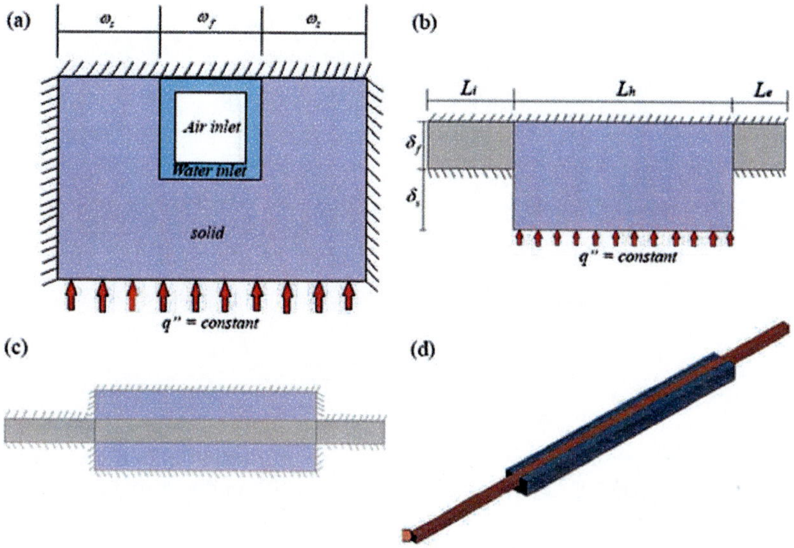

Fig. 1 **a** Left-hand side view, **b** front view, **c** top view and **d** 3D view of the microchannel

was kept constant while that of the substrate below the fluid channel (δ_s) is changed. At Ca = 0.0055, different cases are studied for the values of $k_{sf} = k_s/k_f = 12.6, 27.77, 152.9, 337.33, 495.62$ and 646 and $\delta_{sf} = \delta_s/\delta_f = 1, 2, 3$ and 5. For post-processing of the data, the non-dimensional numbers are

$$Z^* = \frac{z}{L_h}; \delta_{sf} = \frac{\delta_s}{\delta_f}; \Phi = \frac{q_w}{q''}; \mathrm{Nu}_z = \frac{h_z d_h}{k_f}; \theta = \frac{T - T_i}{T_o - T_i}$$

$$\theta_f = \frac{T_{b|z} - T_{b|i}}{T_{b|o} - T_{bli}}; T_b = \frac{\int_A \rho \times C_p \times U \times T\mathrm{d}A}{\int_A \rho \times C_p \times U\mathrm{d}A}; \mathrm{Nu}_{avg} = \int_0^L \mathrm{Nu}_z \mathrm{d}z \quad (1)$$

where T is the temperature, ρ is the density, k is the thermal conductivity, q is the heat flux, Nu is the Nusselt number and the subscripts w is the channel wall, b is the bulk fluid, z is the distance from the start of the heating zone, f is the fluid, s is the solid substrate, o is the exit and i is the inlet. Entry to the heated zone is considered as zero. The local wall temperature and the local heat flux are the average temperatures and the average heat flux, respectively, of three side walls of the solid substrate. All the fluid properties in the present study are calculated for the mixture unless specified.

The continuity, momentum and energy equations are solved for the mixture properties using the volume of fluid (VOF) model from Ansys Fluent®. Also, the colour function is solved to capture the interface between two phases. In the VOF method, a volume fraction α of one of the phases is used as a colour function. Surface tension force ($\vec{F}_{SV}$) at the interface is expressed as a body force, which is added in the source term of the momentum equation. The effect of gravity is neglected.

$$\text{Continuity: } \frac{\partial \rho}{\partial t} + \nabla \cdot \left(\rho \vec{v}\right) = 0 \tag{2}$$

$$\text{Momentum: } \frac{\partial \left(\rho \vec{v}\right)}{\partial t} + \nabla \cdot \left(\rho \vec{v} \otimes \vec{v}\right) = -\nabla P + \nabla \cdot \left(\mu\left(\nabla \vec{v} + \nabla \vec{v}^{T}\right)\right) + \vec{F}_{\text{SV}} \tag{3}$$

$$\text{Energy: } \frac{\partial (\rho e)}{\partial t} + \nabla \cdot \left(\rho \vec{v} h\right) = \nabla \cdot (k \nabla T) \tag{4}$$

$$\text{Color Function: } \frac{\partial C}{\partial t} + \vec{v} \cdot \nabla C = 0 \tag{5}$$

The transient formulations were carried out using first-order non-iterative time marching scheme using a constant time step of $5 \times e^{-6}$ s for the pressure, momentum and energy equations for all the cases. The convergence criteria were satisfied in all the simulations, and the absolute values of all the residuals were well below O ($1 \times e^{-6}$) for continuity and O ($1 \times e^{-6}$) for velocities and O ($1 \times e^{-9}$) for energy. The sufficiently fine grid at the fluid boundary captures the liquid film between the liquid–fluid interface and the wall while relatively coarser cubic grid in the core of the fluid channel reduces the errors in surface tension calculation. The liquid film thickness at the boundary was maintained while the bubble flows through the channel. For the train of bubbles, the equal bubble length, the equal slug length and the symmetric bubble breakup at the inlet are the critical parameters for a grid-independent test, which is satisfied for all the simulations. The bubble length and the liquid film thickness for different grid sizes are shown in Table 1. Grid 4 was used in all the simulations. The validation of the numerical method is carried out for the experimental study of Salman et al. [3]. The numerical simulations are carried out for the nozzle diameter of 0.34 mm and the superficial gas velocity of 0.012 m/s (Fig. 2).

Table 1 Bubble length and liquid film thickness for different grid sizes in a square channel of a side 1 mm

S. No.	Grid size	Average bubble length (mm)	Deviation in bubble length (mm)	Liquid film thickness (μm)
1	166.67 × 166.67 × 125	1.9	0.09	12.5
2	125 × 125 × 125	1.86	0.03	10
3	125 × 125 × 100	1.9	0.01	10
4	100 × 100 × 100	1.9	0.02	9.5

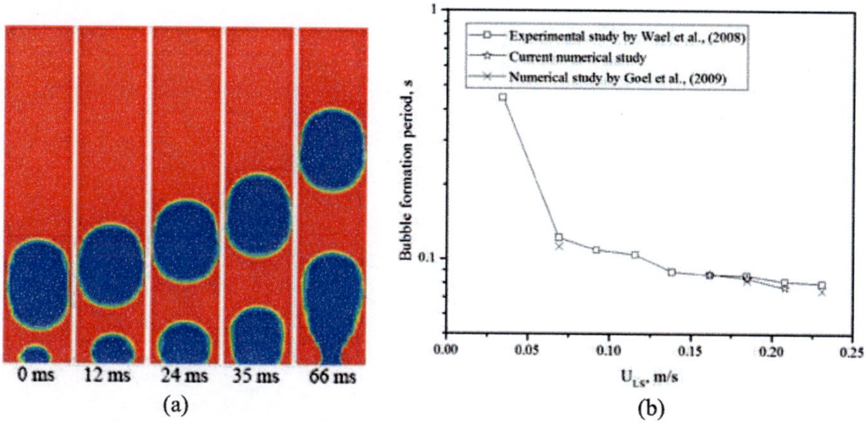

Fig. 2 3D numerical solutions showing **a** bubble shape at different time instants and **b** bubble formation period for the air-octane system agrees with the experimental results by Salman et al. [3]

3 Results and Discussion

The effect of disturbance zones around the bubble body on flow field and heat transfer characteristics of a Taylor bubble flow in a conjugate heating condition are studied in detail.

In square channels, the bubble shape is more complicated due to four corners. The film thickness around the bubble body is more at the diagonal plane than the axial plane at all Ca, as shown in Fig. 3a–e. Due to the bullet-like shape of a bubble, a liquid film shape and thickness are different at different cross sections; the film shape is more circular towards the nose while non-circular and non-symmetric at the tail as shown in Fig. 3p–s. Also, the film thickness in the corners increases towards the nose of the bubble. We found a similar type of bubble shape and film thickness forms for the range of Ca. We observe that the disturbances in the flow field are closely associated with the shape of the bubble.

At the back of bubble, the vector field in the mid-plane perpendicular to the flow shows the occurrence of the recirculating flow (shown in LHS of Fig. 4b) while the vortex occurs in the corner within the liquid film (shown in RHS of Fig. 4b). The axial pressure profile in the unit cell indicates that there is a constant pressure in

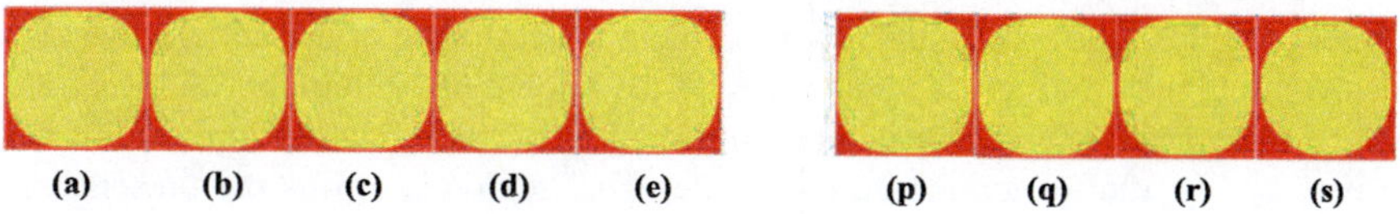

Fig. 3 Liquid film shape at centre of the bubble for β of, **a** 0.47, **b** 0.5, **c** 0.53, **d** 0.56 and **e** 0.6. Cross-sectional view of the Taylor bubble from the tail region towards the nose (**p–s**)

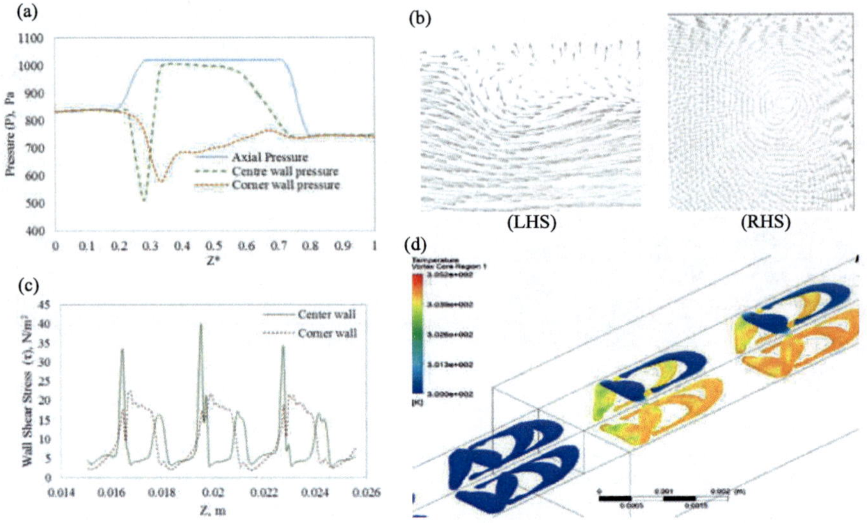

Fig. 4 **a** Static pressure variation along the length of the bubble body in a unit cell for the train of Taylor bubble, **b** recirculation zones at the back of the bubble, in the central plane (LHS) and the vortex in the corner (RHS), **c** variation of wall shear stress for three bubbles and **d** temperature plot in the vortex core region at the entrance of the heating zone

the bubble, and the pressure drops across the interfaces at the head and the tail as shown in Fig. 4a. The points at which the magnitude of the axial pressure in the liquid slug and that in the bubble equals are the extreme ends of the bubble. For the top wall, the local pressure was measured laterally at two points, one in the corner and second at centre or mid-point of the wall. By this way, the top wall pressure is recorded at two locations in multiple planes along the length of the bubble in the unit cell, plotted in Fig. 4a. The mid-point wall pressure profile shows the monotonically decreasing pressure at the head of the bubble due to disturbances occurring around the bullet-shaped head while the low-pressure zone at the back of the bubble was due to the recirculating flow. The pressure values in the corner (red dashed line) around the bubble are smaller than that of the liquid slug on both sides of the bubble. Also, the pressure profile shows another sign of low-pressure zone at the back of the bubble at the location where the vortex occurred. The lower pressure values in the corner shows that the disturbance zones were strong in the corner all around the bubble and the strongest at the location where the vortex has occurred. Both the mid-wall and corner pressure values show the low-pressure zone at the tail of the bubble. We also record the higher wall stress values around the bubble body than that in the liquid slug; the peak wall stress value was recorded nearby the recirculation zone, in Fig. 4c. In Fig. 4d, the plot of temperature in vortex core region confirms the presence of stronger disturbance zones around the bubble body.

The values of a dimensionless fluid temperature Θ_f and heat flux ϕ show the footprint of the Taylor bubble, shown in Fig. 5. The bubble is shown flowing through

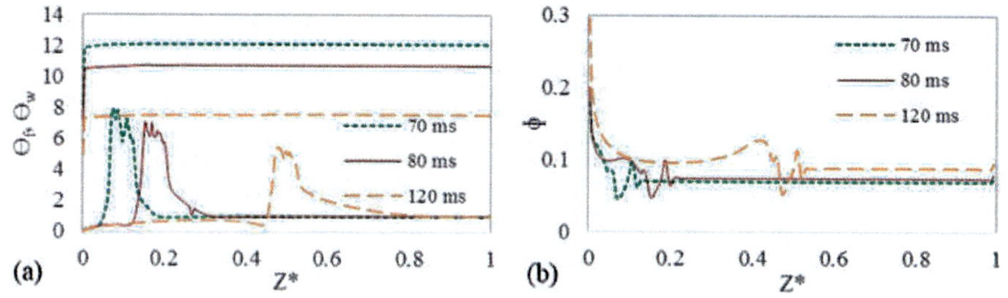

Fig. 5 Axial variation of dimensionless **a** fluid temperature Θ_f, **b** wall temperature Θ_w and **c** heat flux Φ plotted for isolated Taylor bubble at three-time instants

the heating zone of the channel at three different time instants. We observed that as the bubble passes, the surrounding fluid rushes to the low-pressure zone generated at the back of the bubble. Due to this fluid flow to the low-pressure zone, the value of heat flux attains the peak at the tail of the bubble, seen in Fig. 5b at time 120 ms. The isotherms plotted by Moharana et al. [6] at Re = 100 at $\delta_{sf} = 1$ and 16 at higher k_{sf} in the substrate show that there is a positive gradient of temperature in the direction from the channel inlet towards the end, implying the condition of the maximum possible axial wall conduction. However, in the Taylor bubble flow, shown in Fig. 6, the temperature contours plotted in the vertical plane of symmetry show that there is the gradient of temperature from the bottom substrate wall towards the fluid channel, indicating that the heat transfer was one-dimensional, perpendicular to the flow. So, it can be concluded that the Taylor bubble flow minimizes the condition of the axial wall condition in a conjugate heating condition.

4 Conclusions

The Taylor bubble flow in a square channel of diameter 1 mm is studied numerically to understand the effect of the disturbance zones around the bubble body on different flow and thermal parameters in a conjugate heating condition for the range of a Ca = 0.005–0.007. The recirculating flows at the back of the bubble create a low-pressure zone and result in increased heat flux to that region, thereby enhancing the overall heat transfer. The average Nusselt number in a Taylor bubble flow is multifold times than that for the single-phase flow, and Nu_{avg} increases with the increase in the Ca number, as shown in Fig. 7. At Ca = 0.055, the optimum Nusselt number is clearly seen to appear at $\delta_{sf} = 3$ for the range of k_{sf}. In a Taylor bubble flow, the developed temperature gradient from the bottom wall of the substrate towards the channel wall shows that the effects of axial wall conduction in a microchannel are minimized, except at higher k_{sf}. Also, the axial wall conduction effect increases with an increase in k_{sf}, seen from the inclined temperature gradients towards the exit of the channel, developed in the substrate, in Fig. 6d–g at $k_{sf} = 337$ and 646, respectively. Further,

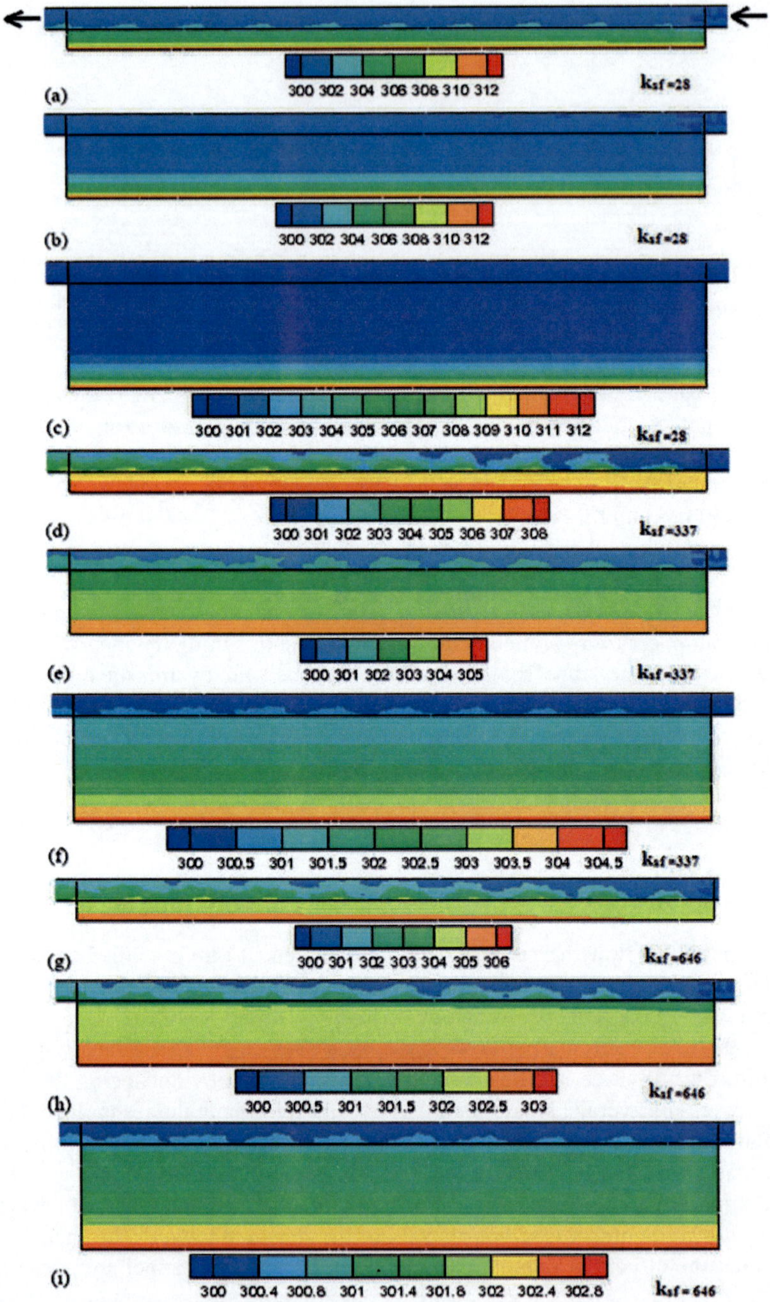

Fig. 6 Axial temperature distribution across the vertical plane of symmetry in a fluid and a solid sections at Ca = 0.0055 for $\delta_{sf} = 1$ (**a**, **d**, **g**), $\delta_{sf} = 3$ (**b**, **e**, **h**) and $\delta_{sf} = 5$ (**c**, **e**, **i**), the corresponding k_{sf} values are indicated in figure

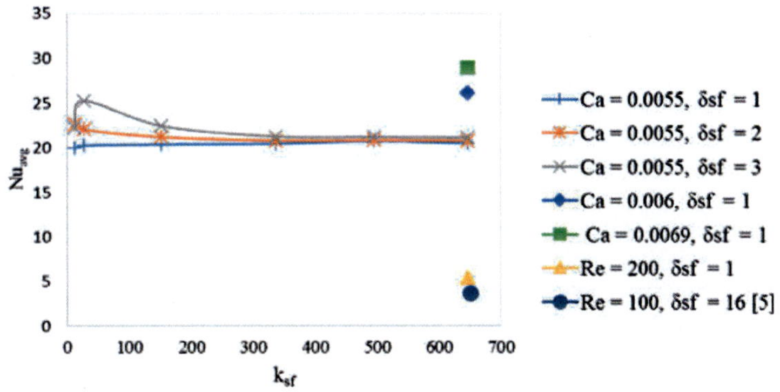

Fig. 7 Variation of Nu_{avg}, (i) for steady, single-phase (water) flow at Re = 100 and Re = 200 and (ii) for the train of Taylor bubble flow at $t = 130$ ms, as a function k_{sf} and δ_{sf} at Ca = 0.0055, 0.006 and 0.0069

in Taylor bubble flow, at lower k_{sf}, the relatively higher values of Nu_{avg} at larger δ_{sf} compared to that at lower δ_{sf} are need to be studied.

References

1. Balkrishna M, Sameer K (2014) Taylor bubble-train flows and heat transfer in the context of pulsating heat pipes. Int J Heat Mass Transf 79:279–290
2. Thulasidas TC, Abraham MA, Cerro RL (1997) Flow patterns in liquid slugs during bubble-train flow inside capillaries. Chem Eng Sci 52(17):2947
3. Salman W, Gavriilidis A, Angeli P (2006) On the formation of Taylor bubbles in small tubes. Chem Eng Sci 61:6653–6666
4. Deepak G, Buwa VV (2009) Numerical simulations of bubble formation and rise in microchannels. Ind Eng Chem Res 48:8109–8120
5. Gupta R, Fletcher DF, Haynes BS (2009) On the CFD modelling of Taylor flow in microchannels. Chem Eng Sci 64(12):2941–2950
6. Moharana MK, Singh PK, Khandekar S (2012) Optimum Nusselt number for simultaneously developing internal flow under conjugate conditions in a square microchannel. ASME J Heat Transf 134:071703 (1–10)
7. Kolb WB, Cerro RL (1991) Coating the inside of a capillary of square cross-section. Chem Eng Sci 46(9):2181–2195

Effect of Large-Scale Flow on the Boundary Layer Velocity Field in Turbulent Convection

Prafulla P. Shevkar, Sanal K. Mohanan, and Baburaj A. Puthenveettil

Abstract We study the effect of shear by the inherent large-scale flow in the bulk on the boundary layers on the hot plate in Rayleigh–Bénard convection for a range of Prandtl numbers $4.69 \leq \text{Pr} \leq 5.88$ and Rayleigh numbers $10^5 \leq \text{Ra} \leq 10^9$. We observe that, at each Ra, at an instant, the distribution of horizontal velocities within the boundary layer in a horizontal plane is either of a unimodal nature or of a bimodal nature. Unimodal distributions occur either at low Ra or at high Ra while the bimodal distributions occur more at intermediate Ra. The peak of the unimodal distribution at low Ra occurs at values at around V_{bl}, the natural convection boundary layer velocity while the peak of the unimodal PDFs at high Ra occur at values greater than or equal to V_{sh}, the large-scale flow velocity. In the case of bimodal distributions, the first peak occurs in between V_{bl} and V_{sh} while the second peak occurs after V_{sh}. We show that the second peak of the bimodal distribution and the unimodal peak which occur at $\geq V_{\text{sh}}$ scales as V_{sh} scales with Ra. The first peak of bimodal distribution and the single peak occurring at around V_{bl} scales with Ra as the natural convection boundary layer velocity forced by shear.

Keywords Rayleigh–Bénard convection · Shear effect · Boundary layer velocity · PIV · Perturbation solutions

1 Introduction

In turbulent Rayleigh–Bénard Convection (RBC), near the hot plate, sheet like plumes form, merge and rise [1]. Due to the confined geometry in RBC, these rising plumes form a large-scale flow (LSF). The Reynolds number based on this large-scale flow velocity V_{sh} [1, 2] is,

$$\text{Re}_{\text{sh}} = V_{\text{sh}} H/\upsilon = 0.55 \text{Ra}_w^{4/9} \text{Pr}^{-2/3} \tag{1}$$

P. P. Shevkar (✉) · S. K. Mohanan · B. A. Puthenveettil
Department of Applied Mechanics, Indian Institute of Technology Madras, Chennai 600036, India
e-mail: prafulla145@gmail.com

L. Venkatakrishnan et al. (eds.), *Proceedings of 16th Asian Congress of Fluid Mechanics*, Lecture Notes in Mechanical Engineering,
https://doi.org/10.1007/978-981-15-5183-3_8

where $\mathrm{Ra_w} = g\beta\Delta T_w H^3/\upsilon\alpha$ is the near wall Rayleigh number, $\Delta T_w = T_w - T_B$ is the near wall temperature drop, g is the acceleration due to gravity, β is the coefficient of thermal expansion, T_w is the bottom wall temperature, T_B is the bulk fluid temperature, and H the height of fluid layer. The Rayleigh number based on ΔT with $\Delta T = 2\Delta T_w$ is $\mathrm{Ra} = 2\mathrm{Ra_w}$ and Prandtl number $\mathrm{Pr} = \upsilon/\alpha$. The LSF creates a shear near the bottom plate which interacts with the boundary layer and changes the velocity field within the boundary layers as well as their stability. In the absence of a shear, the local boundary layers near the hot plate are expected to be natural convection type. The Reynolds number based on the characteristic velocity in the natural convection boundary layer V_{bl} [1, 3] is,

$$\mathrm{Re_{bl}} = V_{bl}H/\upsilon = (47.5/2)^{1/5}\mathrm{Ra_w^{1/3}}\mathrm{Pr}^{-0.98}. \tag{2}$$

In the absence of predominant shear, the plumes originate as the outcome of the gravitational instability of these local natural convection boundary layers [4] with the spacings between these plumes being log-normally distributed at any instant [2]. The mean plume spacing in absence of shear λ_0, scales as,

$$\lambda_0 = 47.5 Z_w \mathrm{Pr}^{0.1}, \text{ with} \tag{3}$$

$$Z_w = \left(\frac{\upsilon\alpha}{g\beta\Delta T_w}\right)^{1/3} = \frac{H}{\mathrm{Ra_w^{1/3}}} \tag{4}$$

being a length scale near the plate [5]. Recent experiments shown that at locations with a dominant presence of shear, the gravitational instability of natural convection boundary layers [6] gets modified [7] and the plume spacing λ_s is increased from its no-shear value λ_0 (3). The difference of the plume spacing from its no-shear value, $\lambda_s - \lambda_0$ normalized by the height of fluid layer H as given by [8] is,

$$\frac{\lambda_s - \lambda_0}{H} = \frac{1}{D}\frac{\mathrm{Re}^2}{\mathrm{Ra}_w}, \text{ with } D = 0.0004\mathrm{Pr}^3 \text{ for } \mathrm{Pr} > 5. \tag{5}$$

The boundary layer model to study the combined effects of pressure gradient and buoyancy in boundary layer of turbulent RBC is proposed by [9]. However, the effect of shear on boundary layer velocity field is not understood in turbulent convection. In this paper, we measure the spatial distribution of horizontal velocities in water within the boundary layers in turbulent Rayleigh–Bénard convection for the range of $10^5 \leq \mathrm{Ra} \leq 10^9$ and $4.69 \leq \mathrm{Pr} \leq 5.88$ by stereo PIV. We then propose an explanation for the different types of distributions of planar horizontal velocities within the boundary layers and the scaling for the peak of such distribution of boundary layer velocities forced by the shear with Ra.

2 Experimental Details

Steady turbulent Rayleigh–Bénard convection experiments were carried out in a square glass tank of cross section 30 cm × 30 cm, in a water column confined between a hot copper plate at the bottom and a water cooled glass plate at the top in the set-up shown in Fig. 1. The side walls were insulated and the bottom copper plate was maintained at a constant heat flux by a heater plate assembly connected to a variac. The heat flux was measured by measuring the temperature drop across the glass plate in the heater assembly using T-type thermocouples at three different locations. The temperatures of the hot Cu plate (T_h) and the cold glass plate (T_c) were measured at two different locations by T-type thermocouples. The average T_h and T_c was used to calculate the constant temperature difference of $\Delta T = T_h - T_c$ across the fluid layer. Experiments were conducted over the range of Ra_w and Pr given in Table 1, obtained by changing the variac voltage, the water column height H, and the water flow rate over the top glass plate.

The velocity field in a horizontal plane (x–y) close to the hot plate at a height h_m was obtained by stereo PIV at all Ra. The height h_m was within the Prandtl–Blasius boundary layer thickness, (δ_{pb}) [10] and the natural convection velocity boundary layer thickness, (δ_{nc}) [11]. The flow was seeded with neutrally buoyant polyamide particles of mean diameter 50 μm which were illuminated by a laser sheet of thickness 1 mm from an Nd:YAG laser (Litron, 100 mJ/pulse). Two Imager Pro HS (LaVision GmbH) cameras (1024 × 1280 pixels) oriented at 32.5° with the vertical were used to capture particle images of area A_i at the center of the plate from above. The stereo cross-correlation method from Davis (LaVision GmbH) was used to calculate the 2D3C spatial vector fields, with interrogation window sizes of 32 × 32 pixels and an overlap of 50%, in the horizontal plane at a height of h_m from bottom plate at

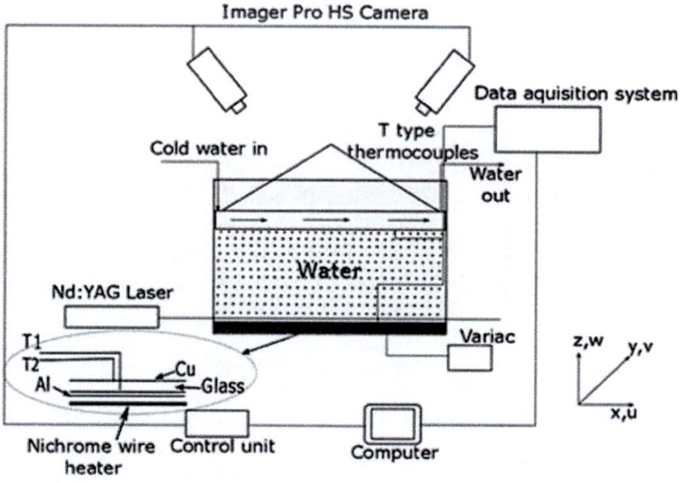

Fig. 1 Schematic of the experimental set-up

Table 1 Values of experimental parameters used in the experiments

Ra × 10^8	Pr	ΔT (°C)	T_B (°C)	H (cm)	h_m (mm)	A_i (mm^2)	δ_{pb} (mm)	δ_{nc} (mm)	V_{bl} (mm/s)	V_{sh} (mm/s)
0.0055	5.88	0.99	26.52	3	2.7	83.18 × 72.02	43.88	4.26	0.62	1.27
0.2368	5.69	3.17	27.97	7	1.8	83.36 × 73.04	18.71	2.79	0.93	2.87
0.2624	5.59	3.44	28.66	7	1.6	84.66 × 73.30	18.17	2.67	0.97	2.99
2.1848	5.24	5.17	31.42	12	1.5	83.08 × 71.94	11.06	2.18	1.15	4.42
5.5875	5.18	6.54	31.88	15	1.0	82.28 × 70.38	08.93	1.98	1.26	5.34
8.0804	4.69	8.24	36.30	15	1.0	82.48 × 68.56	07.94	1.66	1.44	6.18
12.058	5.09	9.16	32.60	17.54	1.3	84.10 × 73.76	07.48	1.77	1.40	6.43

Properties were estimated at T_B, the bulk fluid temperature frequencies of 10–15 Hz for a time period of at least ten large-scale flow circulation times. The spatial resolution for all the vector fields was approximately 1 mm.

3 Results

The probability density function (PDF) of the magnitude of horizontal velocities $U = (u^2 + v^2)^{1/2}$, where u and v are the fluid velocities in x and y directions, were obtained from the spatial distribution of velocity data at various instants at all Ra. The bin sizes for calculating each PDF were progressively decreased till we obtained a bimodal or a unimodal type of PDF after neglecting the low-frequency small variations occurring at the tail of the PDFs.

At each Ra, three types of PDFs were observed as shown in Fig. 2. The PDFs obtained at the lowest Ra were predominantly of a unimodal type with a peak at or around $U/V_{bl} \sim 1$ as shown in Fig. 2a; the maximum occurring horizontal velocities at the lowest Ra were of the order of natural convection boundary layer velocity. In contrast, the distributions obtained at the highest Ra mostly peaks at a value greater than or equal to V_{sh}/V_{bl} implying that the velocities were of the order of large-scale flow velocities (1). For intermediate Ra shown in Fig. 2b, we obtain bimodal PDFs with the first peak in between $U/V_{bl} \sim 1$ and V_{sh}/V_{bl} and the second peak at value greater than or equal to V_{sh}/V_{bl}. So, the PDFs of the boundary layer velocity field show

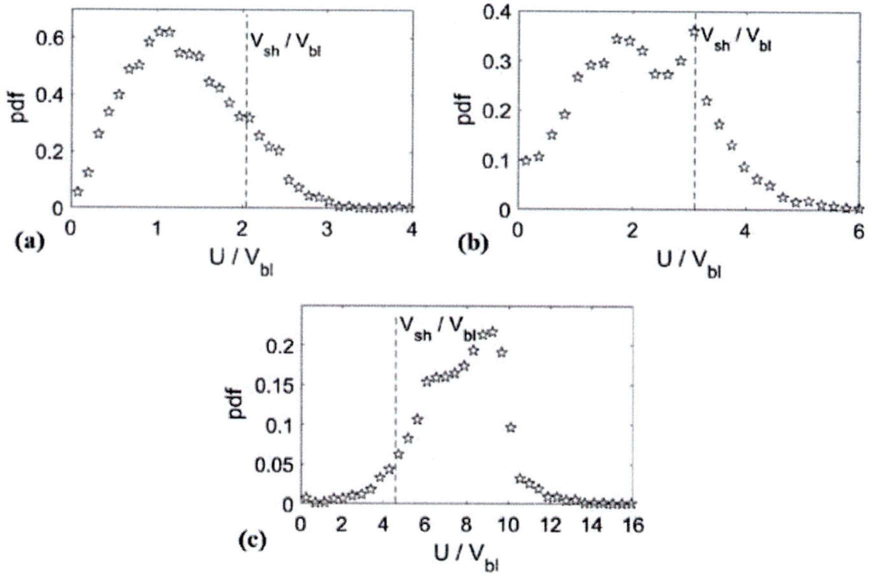

Fig. 2 PDFs of U/V_{bl} at an instant at **a** Ra = 5.5240×10^5, **b** Ra = 2.3680×10^7 and **c** Ra = 1.2058×10^9

peaks at velocities of different magnitudes based on the value of Ra. It seems like, at intermediate Ra, the first peak of the PDFs corresponds to the natural convection boundary layer velocities forced by shear, while the second peak corresponds to the large-scale flow velocities (1).

At each Ra, we calculate the mean of velocities from all the instances corresponding to the first peak of the bimodal PDFs and those corresponding to the peaks of the unimodal PDFs whichever occur at each instant. Here we neglect the instants when PDFs peak at values greater than or equal to V_{sh}/V_{bl}, similar to that at Ra = 1.2058×10^9 in Fig. 2c; we denote this mean velocity as V_1 at each Ra. Similarly, the mean of velocities corresponding to the second peak of the bimodal PDFs and those corresponding to the unimodal peak similar to that at Ra = 1.2058×10^9 in Fig. 2c are calculated and denoted as V_2; PDFs similar to that at Ra = 5.5240×10^5 in Fig. 2a are skipped in this calculation. We denote the Reynolds number calculated based on V_1, H and υ as Re_1 hereafter and that corresponding to V_2 as Re_2. In Fig. 3, we observe that the variation of Re_2 with Ra satisfactorily matches with the scaling of the large-scale flow velocity (1). The variation of Re_1 with Ra however does not follow the $\mathrm{Ra_w}^{1/3}$ scaling suggested by the natural convection boundary layer velocity relation (2). Such a deviation of Re_1 from (2) could be because the shear generated by the large-scale flow could be changing the characteristic velocity within the boundary layer. We hence now look at how the characteristic velocity within the natural convection boundary layer is changed when the boundary layers are forced by shear due to V_{sh}.

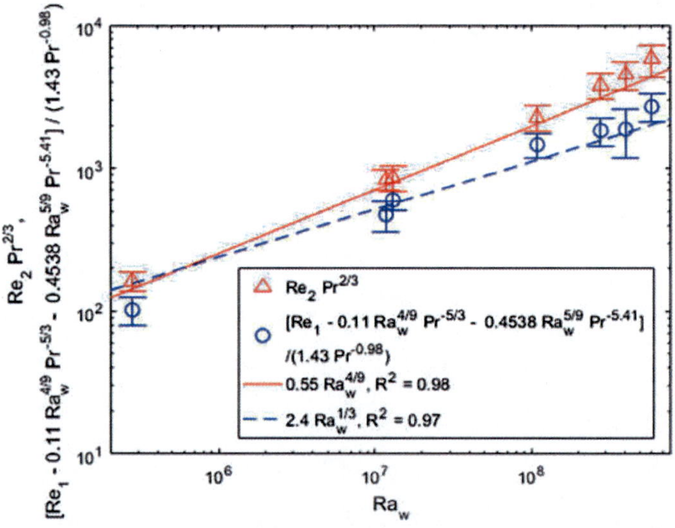

Fig. 3 Δ indicates the variation of $\mathrm{Re}_2 = V_2 H/\upsilon$ and o indicates that the modified equation of (18) with $\mathrm{Re}_b^s = \mathrm{Re}_1 = V_1 H/\upsilon$, shown to be proportional to $\mathrm{Ra_w}^{1/3}$

The 2D laminar natural convection boundary layer equations at Pr = 1, forced by a shear velocity V_{sh}, when integrated across the boundary layer thickness, δ using Boussinesq approximation are,

$$\frac{\partial}{\partial x}\int_0^{\delta} u\mathrm{d}z - V_{sh}\frac{\mathrm{d}\delta}{\mathrm{d}x} + v_{\delta} = 0, \quad (6)$$

$$\frac{\partial}{\partial x}\int_0^{\delta} u^2\mathrm{d}z - V_{sh}\frac{\partial}{\partial x}\int_0^{\delta} u\mathrm{d}z + \frac{1}{\rho}\int_0^{\delta}\frac{\partial p}{\partial x}\mathrm{d}z + \upsilon\left(\frac{\partial u}{\partial z}\right)_{z=0} = 0, \quad (7)$$

$$p = \rho g\beta\int_z^{\delta}(T - T_B)\mathrm{d}z, \text{ and} \quad (8)$$

$$\frac{\partial}{\partial x}\int_0^{\delta} u(T - T_B)\mathrm{d}z = -\alpha\left(\frac{\partial T}{\partial z}\right)_{z=0}, \quad (9)$$

where x is the horizontal direction parallel to the bottom plate, y is the vertical direction perpendicular to the bottom plate, v_{δ} is the velocity in the z-direction at $z = \delta$, p is the pressure of a fluid and $T(x, z)$ the temperature of a fluid. Using the scales,

$$u \sim U_c,\ z \sim \delta,\ x \sim \lambda_s/2, \quad (10)$$

in (6) to (9), where, U_c is the characteristic velocity inside the boundary layer, λ_s is the distance between an adjacent plumes in presence of shear, we obtain the scaling equations,

$$\frac{U_c^2\delta}{\lambda_s/2} - \frac{V_{sh}U_c\delta}{\lambda_s/2} - \frac{g\beta\Delta T_w\delta^2}{\lambda_s/2} + \frac{\upsilon U_c}{\delta} \sim 0, \text{ and} \quad (11)$$

$$\frac{U_c\Delta T_w\delta}{\lambda_s/2} \sim \frac{\alpha\Delta T_w}{\delta}. \quad (12)$$

Substituting the value of δ from (12) into (11), we obtain,

$$16\left[\frac{U_c}{2\alpha/\lambda_s}\right]^{5/2} - 4\mathrm{Re}_{\lambda}\left[\frac{U_c}{2\alpha/\lambda_s}\right]^{3/2} - \mathrm{Ra}_{\lambda} \sim 0, \quad (13)$$

where $\mathrm{Re}_{\lambda} = V_{sh}\lambda_s/\upsilon$, $\mathrm{Ra}_{\lambda} = g\beta\Delta T_w{\lambda_s}^3/\upsilon\alpha$. Using $b = [U_c/(2\alpha/\lambda_s)]$ and $\varepsilon = \mathrm{Re}_{\lambda}/\mathrm{Ra}_{\lambda} \ll 1$ for the range of the present study, we obtain from (13),

$$(16/\mathrm{Ra}_{\lambda})b^{5/2} - 4\varepsilon b^{3/2} - 1 \sim 0. \quad (14)$$

Let the solution for (14) be a power series in ε,

$$b = b_0 + \varepsilon b_1 + \varepsilon^2 b_2 + \cdots. \tag{15}$$

Substituting (15) in (14) and comparing the coefficients of different powers of ε, we get,

$$b_0 = \left(\frac{\mathrm{Ra}_\lambda}{16}\right)^{2/5}, b_1 = \frac{\mathrm{Ra}_\lambda}{10} \text{ and } b_2 = 0.49\mathrm{Ra}_\lambda^{3/5} - 0.36\mathrm{Ra}_\lambda^{8/5} \tag{16}$$

Thus, the perturbation solution of Eq. (13) is,

$$\left[\frac{U_c}{2\alpha/\lambda_s}\right] \sim \left(\frac{\mathrm{Ra}_\lambda}{16}\right)^{2/5} + \frac{\mathrm{Re}_\lambda}{10} + (0.49\mathrm{Ra}_\lambda^{3/5} - 0.36\mathrm{Ra}_\lambda^{8/5})\left(\frac{\mathrm{Re}_\lambda}{\mathrm{Ra}_\lambda}\right)^2 + \cdots \tag{17}$$

Substituting for λ_s in (17) from (5) and neglecting higher-order terms, we get,

$$\mathrm{Re}_b^s = \frac{U_c H}{\upsilon} \sim 1.43\mathrm{Ra}_w^{1/3}\mathrm{Pr}^{-0.98} + 0.11\mathrm{Ra}_w^{4/9}\mathrm{Pr}^{-2/3} + 0.45\mathrm{Ra}_w^{5/9}\mathrm{Pr}^{-5.41} + \cdots \tag{18}$$

Equation (18) gives the variation of Reynolds number based on the boundary layer velocity with Ra and Pr when the boundary layers are perturbed by the shear due to the large-scale flow. In Fig. 3, (18) is shown as the dashed line which matches the variation of Re_1. Equation (18) predicts then the effect of shear due to the large-scale flow on the velocity inside the boundary layers that form on the hot surface in turbulent natural convection.

4 Conclusions

In turbulent RBC for $10^5 \leq \mathrm{Ra} \leq 10^9$ and $4.69 \leq \mathrm{Pr} \leq 5.88$, we observed unimodal and bimodal PDFs of horizontal velocities close to the hot plate, within the boundary layer thickness. We showed that the unimodal distributions and the first mode in the bimodal distributions, with peaks less than the large-scale velocity, occurred due to the effect of the external shear on the velocities in the boundary layers. We then obtained a scaling for the variation of the peak of these distributions with Ra. We also showed that the peak of the unimodal distributions, with peak velocities larger than the large-scale velocity, as well as the second peak of the PDFs in the case of bimodal distributions, to scale with Ra as the large-scale velocities themselves. In conclusion, the velocity field in turbulent Rayleigh–Bénard convection within the boundary layers on the hot plate consists of a mix of regions having natural convection boundary layers forced by shear due to the large-scale flow and the remaining region with predominant shear due to the large-scale flow.

References

1. Gunasegarane GS, Puthenveettil BA (2014) Dynamics of line plumes on horizontal surfaces in turbulent convection. J Fluid Mech 749:37–78
2. Puthenveettil BA, Arakeri JH (2005) Plume structure in high Rayleigh number convection. J Fluid Mech 542:217–249
3. Koothur V (2013) Boundary layer velocity field in turbulent convection. MS thesis, Department of Applied Mechanics
4. Pera L, Gebhart B (1973) On the stability of natural convection boundary layer flow over horizontal and slightly inclined surfaces. Int J Heat Mass Transf 16:1147–1163
5. Theerthan SA, Arakeri JH (1998) A model for near wall dynamics in turbulent Rayleigh Bénard convection. J Fluid Mech 373:221–254
6. Gebhart B, Pera L, Schorr AW (1970) Steady laminar natural convection plume above a horizontal line heat source. Int J Heat Mass Transf 13:161–171
7. Castaing B, Gunaratne G, Heslot F, Kadanoff L, Libchaber A, Thomas S, Wu X, Zaleski S, Zanetti G (1989) Scaling of hard thermal turbulence in Rayleigh-Bénard convection. J Fluid Mech 204:1–30
8. Shevkar PP, Gunasegarane GS, Mohanan SK, Puthenveettil BA (2019) Effect of shear on coherent structures in turbulent convection. Phys Rev Fluids 4:043502
9. Ovsyannikov M, Krasnov D, Mohammad SE, Schumacher J (2016) Combined effects of prescribed pressure gradient and buoyancy in boundary layer of turbulent Rayleigh Bénard convection. Euro J Mech B/Fluids 57:64–74
10. Ahlers G, Grossmann S, Lohse D (2009) Heat transfer and large scale dynamics in turbulent Rayleigh-Bénard convection. Rev Mod Phys 81:503–537
11. Puthenveettil BA, Gunasegarane GS, Agarwal YK, Schmeling D, Bosbach J, Arakeri JH (2011) Length of near-wall plumes in turbulent convection. J Fluid Mech 685:335–364

Drag Crisis on a Smooth Sphere Exposed to Free Stream Turbulence

A. Desai, R. Shakya, and S. Mittal

Abstract Flow past a smooth sphere exposed to free stream turbulence intensity (T_u) of 0.4% and 1% in the regime $5 \times 10^4 \leq \text{Re} \leq 5 \times 10^5$ is studied via measurement of unsteady forces. The measurements span three flow regimes, namely subcritical, critical and supercritical. As proposed by Deshpande et al. (J Fluid Mech 812:815–840, 2017 [4]), subregimes: I, II and III are also observed in the critical regime. With increase in turbulence intensity, the drag crisis shifts to a lower Re and the maximum drag force experienced by the sphere reduces drastically. With increase in T_u, the drop in $\overline{C}_D$ becomes more gradual to the extent that subregime-II is not observed for $T_u = 1\%$. Probability density functions (PDF) of low-pass filtered time histories of the three force coefficients are used to investigate the various flow states in the critical regime. Some of these states are associated with asymmetric lateral forces, indicating non-axisymmetric transition of the boundary layer. It is proposed that this is due to non-axisymmetric formation of the laminar separation bubble (LSB). Towards the end of the critical regime, near-zero lateral forces and very low $\overline{C}_D$ are observed. These indicate that the LSB is axisymmetric at all time instants at this Re.

Keywords Boundary layer · Boundary layer transition · Free stream turbulence

1 Introduction

Turbulent flow over a bluff body is encountered in many practical situations. The governing parameter for this flow is Reynolds number, defined as $\text{Re} = \frac{UD}{\vartheta}$, where U is the free stream speed, D is the diameter of the sphere and ϑ is the kinematic viscosity of the fluid. Turbulent intensity is defined as $T_u = \frac{u'}{U}$, where u' is the root

A. Desai (✉) · R. Shakya · S. Mittal
Department of Aerospace Engineering, Indian Institute of Technology Kanpur, Kanpur, Uttar Pradesh 208016, India
e-mail: adityad@iitk.ac.in

R. Shakya
National Wind Tunnel Facility, Indian Institute of Technology Kanpur, Kanpur, Uttar Pradesh 208016, India

L. Venkatakrishnan et al. (eds.), *Proceedings of 16th Asian Congress of Fluid Mechanics*, Lecture Notes in Mechanical Engineering,
https://doi.org/10.1007/978-981-15-5183-3_9

mean square of the velocity fluctuations and $\overline{U}$ is the mean flow speed. The variation of time-averaged drag coefficient ($\overline{C}_D$) of the sphere with Re shows a sharp drop when boundary layer transitions to turbulent from laminar. This phenomenon is popularly referred to as drag crisis. Based on the variation of mean drag coefficient ($\overline{C}_D$) with Re, Achenbach [1] proposed four regimes of the flow—subcritical, critical, supercritical and transcritical. In the subcritical regime, the boundary layer over the surface of the sphere undergoes a laminar separation around $\varphi = 80^{\text{o}}$, leading to $\overline{C}_D$ of about 0.5. The drag crisis is observed in the critical regime, where boundary layer transitions form laminar to turbulent via formation of laminar separation bubble (LSB). In the supercritical regime, the $\overline{C}_D$ gradually increases with increase in Re. In the transcritical regime, the point of transition from laminar to turbulent moves upstream, causing $\overline{C}_D$ to increase with increase in Re. Norman and McKeon [7] studied the unsteady forces on the sphere and observed that in the critical regime, the instantaneous C_D intermittently switched between values corresponding to subcritical and supercritical regimes. Recently, Deshpande et al. [4], investigated the dynamics of formation of LSB on a sphere. They extended the definitions of different flow regimes proposed by Schewe [9] for a cylinder to the flow over a sphere. They also proposed division of critical regime into three subregimes: I, II and III. In subregime-I, the $\overline{C}_D$ decreases due to increase in suction in the region upstream of the shoulder and increase in the base pressure, $\overline{C}_{p,b}$. In subregime-II, intermittent appearance/disappearance of an axisymmetric LSB is observed. The frequency and duration of the stay of the LSB increases with increase in Re, leading to sharp drop in $\overline{C}_D$. In subregime-III, the LSB exists at all time instants and the drag coefficient decreases due to increase in base pressure.

Bearman and Morel [2] reviewed results from different researchers on turbulent flow over different bluff bodies and proposed three mechanisms by which the free stream turbulence (FST) may interact with the flow: promotion of transition to turbulence, enhanced mixing and entrainment and the distortion of the FST itself by the mean flow. Raithby and Eckert [8] conducted flow visualization on a sphere exposed to different turbulent intensities and found that the Reynolds number for the onset of critical regime decreased with increase in turbulence intensity. Son et al. [10] studied flow over a sphere exposed to turbulent intensities of 4%, 6% and 8%. Through flow visualization in the supercritical regime, they found that with increase in Re, the point of laminar separation moved downstream. However, the point of turbulent reattachment and final turbulent separation remained fixed, resulting in constant $\overline{C}_D$.

In the present study, we investigate the effect of FST on the dynamics of LSB formation and the nature of the intermediate flow states. Keeping in mind the sensitivity of boundary layer transition to small changes, we restrict ourselves to low turbulence intensities of 0.4% and 1%.

2 Experimental Setup

The experiments were conducted at National Wind Tunnel Facility (NWTF), IIT Kanpur. This is an atmospheric, closed-circuit tunnel. The cross section of the test section is 2.25 m × 3 m. Maximum speed that can be achieved in the tunnel is 80 m/s. The spatial inhomogeneity in the incoming flow was measured to be about 0.05% at a speed of 20 m/s. More details about the characterization of the tunnel can be found in Cadot et al. [3].

Figure 1 shows a schematic of the experimental setup and the coordinate system. The diameter of the sphere is 120 mm, leading to a blockage ratio of 0.16%. The model is manufactured using stereo-lithography technique with Acura 60 material. After fabrication and post-processing, the model is polished to obtained glossy smooth finish. The sphere is mounted on a horizontal sting of diameter 25 mm and length 480 mm. The horizontal sting is mounted on a vertical support which is grounded to the tunnel floor. The model and setup are similar to those used by Deshpande et al. [4].

Unsteady forces on the sphere are measured using a six component strain gauge-based force balance. The balance is installed inside the horizontal sting in such a way that the downstream end of the sphere is directly connected to it. The calibration curve of the balance is linear. The output from the balance is amplified and acquired at 500 Hz. At least 60 s of data is acquired at each Reynolds number in subcritical and supercritical regime. In the critical regime, 300 s of data is acquired at each Reynolds number to capture sufficient instances of the long-period fluctuations associated with formation of LSB. Each experiment is repeated at least twice to ensure repeatability. The non-dimensionalized force coefficients in the three directions are indicated by C_D, C_Y and C_Z, respectively. The maximum uncertainty in measurement of C_D is 1.5% at Re $= 1.5 \times 10^5$ and decreases progressively with increase in speed.

Turbulence was measured using constant temperature anemometry and free stream turbulence was found to be $T_u \leq 0.06\%$ for the entire speed range. To create additional turbulence, a square grid spanning the entire cross section of the tunnel is mounted upstream of the model. The mesh is made of round wires with diameter $d = 1.54$ mm and the mesh size is $M = 15.4$ mm. A hot wire probe made of tungsten wire of diameter 5 μm and length 1.25 mm is used for constant temperature

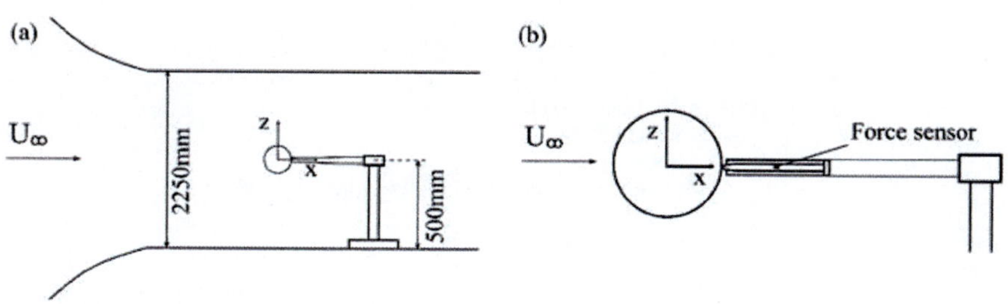

Fig. 1 Schematic **a** model mounted in the test section and **b** force measurement setup

anemometry. The overheat ratio was set at 0.8 and the cutoff frequency was 10 kHz. The probe was calibrated for free stream speed using a two-port pitot probe and digital manometer. To vary the turbulence intensity (T_u) between the experiments, two different locations of the grid were used while keeping the location of the sphere fixed. The resultant turbulence intensities, measured at the location of the centre of the sphere, were 0.4% and 1%, respectively. In the following section, experiments carried on in absence of turbulence grid are referred to as 'clean flow'.

3 Results and Discussion

The variation of time-averaged drag coefficient with Reynolds number is plotted in Fig. 2, along with data from earlier studies. The values of coefficient as well as the trend show good match with those reported by Deshpande et al. [4] and Norman et al. [6]. Particularly, values of coefficient in subcritical and supercritical regime show a good match. Some difference is visible in the critical regime which is attributed to the sensitivity of the flow in critical regime to minute manufacturing imperfections. In the present study as well as in the study by Deshpande et al. [4], the $\overline{C}_D$ in the supercritical regime is slightly lower than those reported by earlier studies. This difference is attributed to higher sting to sphere diameter ratio for the present study as compared to that of others. Hoerner [5] studied the effect of sting to sphere diameter ratio (d_{sting}/d) on the drag coefficient and found that this ratio did not have a significant effect on $\overline{C}_D$ during the subcritical regime. However, $\overline{C}_D$ the supercritical regime was seen to decrease with an increase in this ratio.

The flow regimes and subregimes, as proposed by Deshpande et al. [4], are marked in Fig. 3. Of particular interest is the critical regime, characterized by the drop in $\overline{C}_D$. The subregimes of the critical regime are also marked in colour. A gradual decrease in $\overline{C}_D$ is seen in subregime-I and III while in subregime-II, there is a sharp drop in $\overline{C}_D$. Subregime-II spans a very small range of Re over which an intermittent

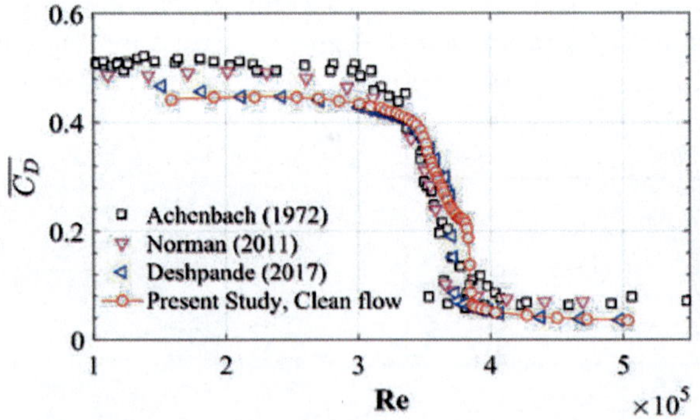

Fig. 2 Variation of mean drag coefficient with Re for present study and results from earlier studies

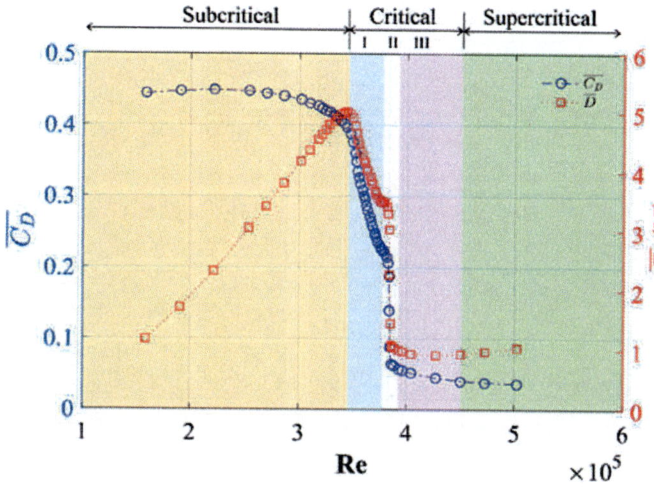

Fig. 3 Variation of mean drag force ($\overline{D}$) and ($\overline{C}_D$) with Re for clean flow. Various flow regimes are marked

laminar separation bubble (LSB) exists on the surface and leads to the flow exploring a high-drag state and a low-drag state intermittently. The intermittent behaviour of these time histories serves as an indicator for presence of an intermittent LSB in this regime. Towards the end of subregime-II, the LSB is found to exist at all time instants.

The effect of free stream turbulence intensity (T_u) on the drag crisis can be seen from Fig. 4a, b. The maximum drag force experienced by the sphere reduces drastically with increase in T_u. The critical regime is seen to shift to a lower Re with increase in T_u. It is also observed that unlike in case of cylinder, supercritical $\overline{C}_D$ is nearly identical for different T_u, which is consistent with the observations by Son

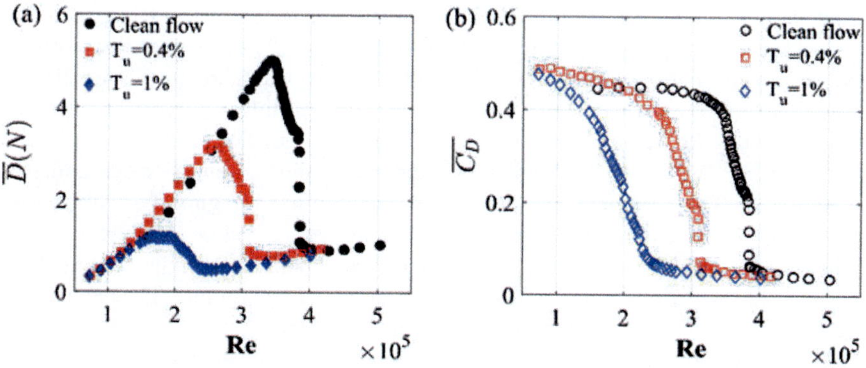

Fig. 4 Variation of **a** $\overline{D}$ and **b** $\overline{C}_D$ with Re for different levels of T_u

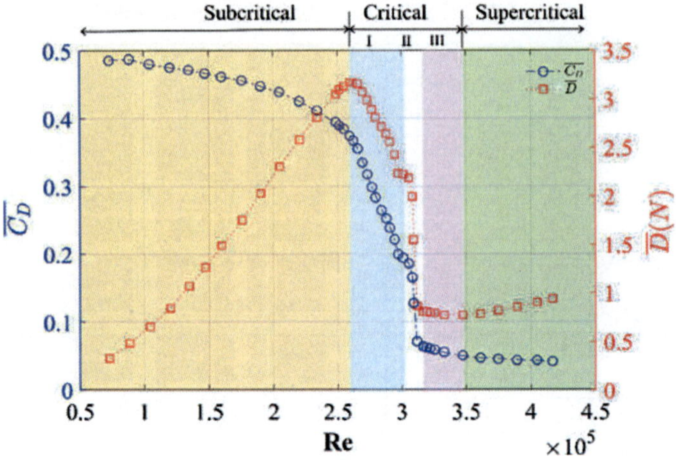

Fig. 5 Variation of mean drag force ($\overline{D}$) and ($\overline{C}_D$) with Re for $T_u = 0.4\%$. Various flow regimes are marked

et al. [10]. Among the various flow subregimes defined earlier, subregime-II is seen to be affected most prominently by the change in T_u. Upon increasing T_u to 0.4% from the clean flow, the sharp drop in $\overline{C}_D$ becomes slightly gradual. Increasing T_u form 0.4% from 1% leads to the drop becoming gradual to the extent that subregime-II is not observed.

Figure 5 shows the variation of force ($\overline{D}$) and ($\overline{C}_D$) with Reynolds number for $T_u = 0.4\%$. The flow regimes and subregimes are marked with shaded regions. The critical regime is identified as $2.58 \times 10^5 < \mathrm{Re} < 3.28 \times 10^5$.

The unsteady forces experienced by the sphere in a turbulent flow include significant buffeting which tends to overshadow other fluctuations. To get rid of this, the force time histories are filtered using low-pass filter with cut off at 10 Hz. We note that this cut off is well above the frequency associated with intermittent dynamics associated with the LSB. The filtered time histories are thus considered adequate to represent the behaviour in time of the intermittent LSB. Low-pass filtered time history of C_D at $\mathrm{Re} = 2.94 \times 10^5$ is plotted in Fig. 6a. Though this Re lies in subregime-I, the time history exhibits long-period intermittent fluctuations, similar to those seen at Re belonging to subregime-II (Fig. 6b–d). The amplitude of fluctuations at $\mathrm{Re} = 2.94 \times 10^5$ is smaller as compared to those seen at Re belonging to subregime-II. This indicates that the fluctuations seen in subregime-I are likely due to intermittent transition of boundary layer in a smaller region on the sphere.

Probability density function (PDF) of the force coefficients is used to analyse and separate the different flow states that exist in the critical regime. The parameters chosen for calculating the PDF are presented in Table 1. In order to discern any non-axisymmetric flow states that may be present, we consider PDF of all three force coefficients. For the sake of convenience, a state of the flow is represented by $[\overline{C}_{D,}\overline{C}_Y, \overline{C}_{Z,}]$ in the following discussion.

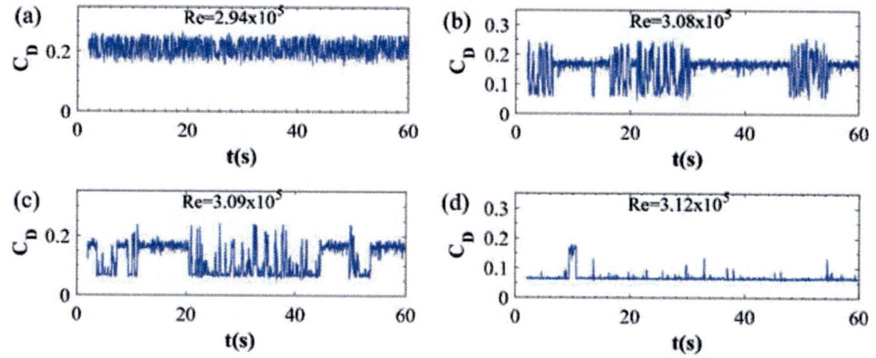

Fig. 6 Low-pass filtered time history of C_D at four different Re in the critical regime for $T_u = 0.4\%$

Table 1 Parameters used for calculation of probability density function (PDF) of the three force coefficients

Coefficient	Range chosen	No. of bins
C_D	−0.1 to 0.6	100
C_Y	−0.3 to 0.3	100
C_Z	−0.3 to 0.3	100

PDF of low-pass filtered coefficients C_D, C_Y, and C_Z for $T_u = 0.4\%$ are plotted against Re in Fig. 7a–c. For $2.58 \times 10^5 < \text{Re} < 2.77 \times 10^5$, $\overline{C}_D$ gradually decreases while $\overline{C}_Z$ gradually increases, resulting in the flow state changing from $[0.37, 0, -0.06]$ to $[0.28, 0, 0.05]$. For $2.77 \times 10^5 < \text{Re} < 2.88 \times 10^5$, $\overline{C}_D$ gradually decreases from 0.28 to 0.25. During $2.91 \times 10^5 < \text{Re} < 2.97 \times 10^5$, the flow intermittently switches between two states. The two states can be described as $[0.24, 0, 0]$ and $[0.18, 0.11, -0.17]$ respectively. The intermittent behaviour can also be seen from the time history of $\overline{C}_D$ at $\text{Re} = 2.94 \times 10^5$ plotted in Fig. 6a.

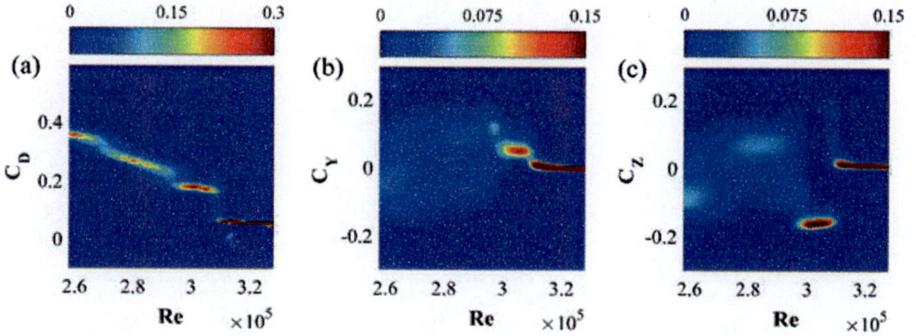

Fig. 7 Variation with Re of PDF of low-pass filtered time histories of **a** C_D, **b** C_Y and **c** C_Z; $T_u = 0.4\%$

The asymmetric lateral forces indicate that the boundary layer over only a small part of the sphere has undergone transition. The direction of the resultant lateral forces indicates that the small pocket of LSB must be located in the quadrant formed by $-y$ and $-z$ axes.

Only small fluctuations are seen for $2.97 \times 10^5 < \text{Re} < 3.01 \times 10^5$ and the flow remains in state $[0.18, 0.05, -1.17]$. Beyond $\text{Re} = 3.01 \times 10^5$, the time histories again become intermittent. The flow intermittently switches between initial state $[0.18, 0.05, -0.17]$ and final state $[0.06, 0, 0]$. This phenomenon is observed upto $\text{Re} = 3.20 \times 10^5$ and at the end of this regime, the state $[0.06, 0, 0]$ exists at all times. The symmetric nature of this final state and the low value of $\overline{C}_D$ indicates that this state corresponds to a fully formed, axisymmetric LSB. We note that as per the definition proposed by Deshpande et al. [4], the regime $3.01 \times 10^5 < \text{Re} < 3.20 \times 10^5$ corresponds to subregime-II for $T_u = 0.4\%$. The regime $3.20 \times 10^5 < \text{Re} < 3.28 \times 10^5$, corresponds to subregime-III and $\overline{C}_D$ gradually decreases from 0.06 to 0.05 in this regime.

For $T_u = 1\%$, the critical regime extends from $\text{Re} = 1.65 \times 10^5$ to $\text{Re} = 2.56 \times 10^5$. PDF of low-pass filtered coefficients C_D, C_Y, and C_Z from the critical regime for $T_u = 1\%$ are plotted against Re in Fig. 8a–c. For $1.61 \times 10^5 < \text{Re} < 1.94 \times 10^5$, there is only a gradual decrease in $\overline{C}_D$, while $\overline{C}_Z$ gradually increases. The flow goes from state $[0.34, 0, 0]$ at $\text{Re} = 1.61 \times 10^5$ to state $[0.23, 0, 0.1]$ by $\text{Re} = 1.94 \times 10^5$. Upon increasing Re, intermittent time histories of forces are observed in the regime $1.98 \times 10^5 < \text{Re} < 2.06 \times 10^5$. The two coexisting states are $[0.22, 0, 0.1]$ and $[0.17, -0.07, 0]$.. The asymmetric forces indicate that these states are accompanied formation of LSB in small pockets near $+z$ and $-z$ axes, respectively. At $\text{Re} = 2.06 \times 10^5$, the flow remains in the state $[0.17, -0.07, 0]$ at all times.

Further increase in Re to $2.06 \times 10^5 < \text{Re} < 2.14 \times 10^5$ leads to $\overline{C}_D$ decreasing gradually from 0.17 to 0.14. Increasing Re further to 2.17×10^5 leads to intermittency in C_Z while $\overline{C}_D$ and $\overline{C}_Y$ remain nearly constant. The two coexisting flow states are identified as $[0.14, -0.1, -0.05]$ and $[0.14, -0.1, 0.08]$. The asymmetry of mean lateral forces in these states and the low $\overline{C}_D$ value indicate except for a small region along the $-y$ axis, the boundary layer has undergone transition everywhere else,

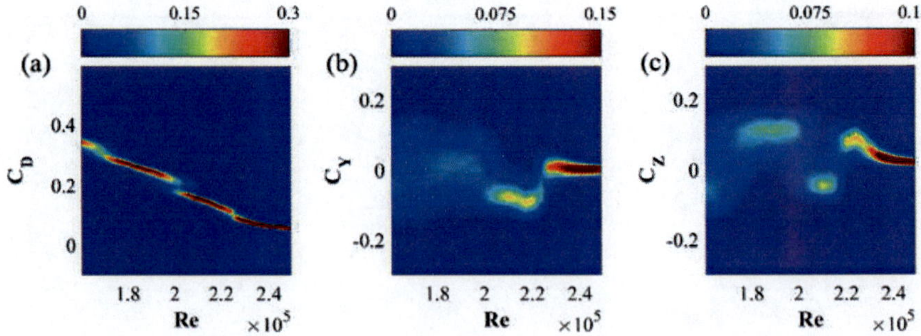

Fig. 8 Variation with Re of PDF of low-pass filtered time histories of **a** C_D, **b** C_Y and **c** C_Z, $T_u = 1\%$

forming a non-axisymmetric LSB. The regime $2.2 \times 10^5 < \text{Re} < 2.49 \times 10^5$: shows gradual decrease in all coefficients. The flow goes from initial state [0.14, −0.05, 0.1] to the final state [0.06, 0, 0]. The symmetry in lateral forces and low value of $\overline{C}_D$ indicates that this state is associated with an axisymmetric LSB at all time instants.

The different states exhibited by the flow indicate that LSB formation in presence of low FST is an inherently non-axisymmetric phenomenon. The flow is seen to explore a variety of states in which LSB appears to be formed in small spatial regions. Depending upon its location, the formation of such LSB results in intermittent, long-period fluctuations in one or more of the force coefficients. We observe that the intermittency in time histories of forces is also observed in subregime-I and is not restricted to subregime-II. Lastly, we note that since these boundary layer transition phenomenon are highly sensitive to experimental conditions, formation of non-axisymmetric LSB may come about differently for different models.

4 Conclusions

Flow past a smooth sphere exposed to free stream turbulence is investigated via measurement of unsteady forces exerted on the sphere. With increase in turbulent intensity, the drag crisis is seen to shift to a lower Re. PDF of low-pass filtered time histories of the three force coefficients is used to investigate the flow states in the critical regime. For turbulent intensity of 0.4% as well as 1%, some of the states explored by the flow are accompanied by asymmetric lateral forces, indicating non-axisymmetric transition of the boundary layer. Onset of some of these non-axisymmetric flow states is accompanied by long-period, intermittent fluctuations in force time histories, indicating the formation of a non-axisymmetric LSB. Towards the end of the critical regime, near-zero lateral forces and very low value of $\overline{C}_D$ indicate the presence of an axisymmetric LSB on the surface of the sphere at all time instants.

Acknowledgements The authors wish to thank Mr. Sharad Saxena and the team at NWTF, IIT Kanpur, for their support in conducting the experiments.

References

1. Achenbach E (1972) Experiments on the flow past spheres at very high Reynolds numbers. J Fluid Mech 54:565–575
2. Bearman PW, Morel T (1983) Effect of free stream turbulence on the flow around bluff bodies. Prog Aerosp Sci 20:97–123
3. Cadot O, Desai A, Mittal S, Saxena S, Chandra B (2015) Statistics and dynamics of the boundary layer reattachments during the drag crisis transitions of a circular cylinder. Phys Fluids 27 (2015)

4. Deshpande R, Kanti V, Desai A, Mittal S (2017) Intermittency of laminar separation bubble on a sphere during drag crisis. J Fluid Mech 812:815–840
5. Hoerner S (1935) Tests of spheres with reference to Reynolds number, turbulence, and surface roughness
6. Norman AK, Kerrigan EC, McKeon BJ (2011) The effect of small-amplitude time-dependent changes to the surface morphology of a sphere. J Fluid Mech 675:268–296
7. Norman AK, McKeon BJ (2011) Unsteady force measurements in sphere flow from subcritical to supercritical Reynolds numbers. Exp Fluids 51:1439–1453
8. Raithby GD, Eckert ERG (1968) The effect of turbulence parameters and support position on the heat transfer from spheres. Int J Heat Mass Transf 11:1233–1252
9. Schewe G (1983) On the force fluctuations acting on a circular cylinder in crossflow from subcritical up to transcritical Reynolds numbers. J Fluid Mech 133:265–285
10. Son K, Choi J, Jeon WP, Choi H (2010) Effect of free-stream turbulence on the flow over a sphere. Phys Fluids 22:1–7

Coupled Fluid–Structure Study of Inspiratory Flow Through Human Airways with Double Bifurcation

Mayank Verma, Karmanya, Achhardeep Verma, and Ashoke De

Abstract In our respiratory system, trachea–bronchi geometry is very complex in nature, which gives rise to the intricate airflow patterns. Since along with air, many other pollutants and dust particles go inside our respiratory system resulting in various lung problems like asthma, bronchiole inflammation, chronic bronchitis, and other infectious diseases like influenza and pneumonia. Thus, it is essential to understand fluid flow patterns, related stresses, and their effects inside the human airways. In most of the research till now, the role of the elasticity of the walls of human airways is neglected due to the computational difficulties. Here, we performed a two-way coupled fluid–structure interaction analysis on a double bifurcation geometry (G_3 to G_5) using finite volume approach for Re = 500 and Re = 2000, keeping the mass flow rate constant. The solid walls of human airways are considered to be of isotropic material. With including the elasticity, we were able to predict the more physical values of the stresses. We observed that maximum deformation in the geometry occurs at first bifurcation for low Reynolds number and the branch G_{5M} for high Reynolds number. High values of Von-Mises stress are also observed at the point of bifurcations.

Keywords Human airways · FSI · Bio-fluid mechanics · Double bifurcation · CFD

1 Introduction

Characteristics of flow in the human airway are of great importance because they have an impact on particle deposition and transportation in human lungs. They find their application in aerosol drug transport in human airways. Along with the flow characteristics, a proper understanding of stresses and deformation of airway walls

M. Verma (✉) · Karmanya · A. Verma · A. De
Department of Aerospace Engineering, Indian Institute of Technology Kanpur, Kanpur, Uttar Pradesh 208016, India
e-mail: vmayank@iitk.ac.in

Karmanya · A. Verma
Department of Aerospace Engineering, Punjab Engineering College, Chandigarh, India

L. Venkatakrishnan et al. (eds.), *Proceedings of 16th Asian Congress of Fluid Mechanics*, Lecture Notes in Mechanical Engineering,
https://doi.org/10.1007/978-981-15-5183-3_10

can help the medical surgeons to analyze the influence of tracheal stent implantation in a patient.

Initial studies related to human airways trace back to 1969 when Schroter and Sudlow discussed the flow patterns in human bronchial airways [1]. They addressed the velocity profiles at different locations in the airways. Many studies have been conducted since then. Experimental results of Zhao and Lieber [2] also discussed the flow patterns in the bifurcation tube at different locations for single bifurcation geometry. Many more experiments were carried out. But since it is challenging to experimentally measure characteristics of flow (such as velocity, stresses, deformations) with original conditions as that present in human airways, many computational models were also used for properly analyzing the flow. Comer et al. [3] described the flow characteristics in double bifurcation human airways using computational methods. Augusto et al. [4] also discussed the flow pattern results obtained from a computational model.

However, in most of the computational studies, wall elasticity is not taken into effect. The walls of the airways were considered to be rigid. But in reality, walls of human airways are elastic. They do get deformed and develop stresses when we inhale or exhale. So, in this study, we have tried to find the effect of the wall elasticity of human airways on the related stresses and flow patterns. A three-dimensional two-way-coupled fluid–structure interaction study is performed for two different Reynolds numbers (based on the diameter of the mother tube, $d = 6$ mm, Re = 500, and 2000) for examining the related Von-Mises stresses and corresponding deformation of the airway. The maximum value of the Von-Mises stress is observed to be at first bifurcation.

2 Numerical Details

2.1 Geometry and Meshing Details

The double bifurcation geometry used in the study is based on the one used by Comer et al. [3]. Their second bifurcation generations are based on Weibel's lung model [5] and are named as G3, G4, and G5 based on Weibel's classification scheme. Bifurcation geometry is depicted in Fig. 1.

The 3D model for the above geometry is generated in the Solidworks. The mesh for the fluid domain is multiblock-structured mesh with 0.71 million cells and generated using ICEM-CFD, while the structure mesh with 66,680 cells is made using ANSYS-Workbench with thickness as 0.41 mm.

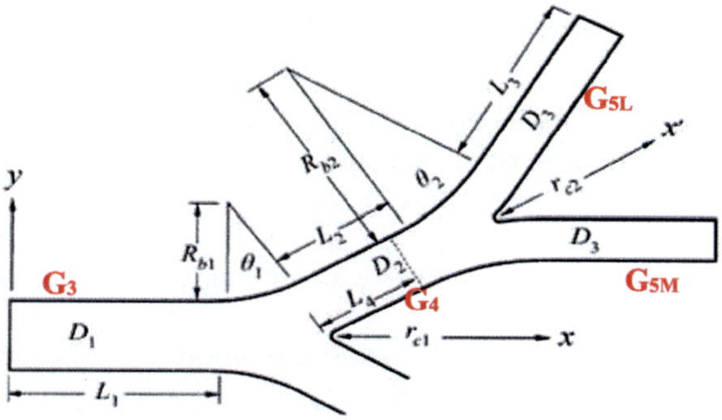

Fig. 1 Schematic of the double bifurcation from G-3 to G-5 [3]

2.2 *Material Properties and Boundary Conditions*

Simulation is done using 'air' as fluid with a density of 1.225 kg/m^3 and viscosity as 1.789 × 10^{-5} kg/ms. The walls are assumed to be of homogenous material with a density of 1365.6 kg/m^3 [6]. The isotropic material model is considered with Young's modulus of elasticity of 130.89 kPa [7] and Poisson's ratio of 0.45 [8].

The flow inside the double bifurcation is generated through the parabolic velocity profile at the inlet. No slip condition is imposed on the bifurcation walls for fluid domain. The pressure outlet condition with one atmospheric pressure is given on the outlets of the daughter tubes. For structural domain, the fixed support boundary condition is provided at both inlet and outlet. All the inner surfaces of the airways are defined as the fluid–structure interface, where the exchange of data between both the solvers takes place.

2.3 *Numerical Schemes and FSI Coupling*

The fluid domain is discretized using a finite volume approach, and Navier–Stokes equations are solved.

$$\nabla.u = 0 \tag{1}$$

$$\frac{\partial u}{\partial t} = -(u.\nabla)u + \vartheta\nabla^2 u - \frac{1}{\rho}\nabla p + f \tag{2}$$

where u is velocity, υ is the kinematic viscosity of the fluid, p is pressure, and f is the external force.

Spatial flow domain is discretized using second-order central schemes. A SIMPLE algorithm is used to solve the pressure–velocity coupling. Solid domain is discretized using the finite element approach.

2.4 *Fluid–Structure Coupling*

There are two approaches for fluid–structure coupling, (a) Monolithic approach where the governing equation for both the fluid and solid domains are in terms of same primitive variables (p and U) and discretizes the entire domain using the same scheme and (b) Partitioned approach which retains separate domains for fluid and structure and separate solvers with independent discretization are used in modeling of each domain.

The partitioned approach is used for the current simulation. The coupling is one-way if the motion of a fluid flow influences a solid structure, but the reaction of a solid upon a fluid is negligible. The other way around is also possible. This different type of coupling is applied to the problem where the motion of fluid influences a solid structure, and at the same time, the flow of fluid is influenced by the reaction of the solid structure. This is known as two-way coupling. During the first-time step, converged solutions of the fluid calculation provide the forces acting on the solid body. Then, the forces are interpolated to the structural mesh like in one-way coupling, and the solution from the structural solver is obtained with those fluid forces as boundary conditions. As a consequence, the mesh is deformed according to the response of the structure. These displacement values are interpolated to the fluid mesh, which results in the deformation of the fluid domain. This process is repeated until both force and displacement values are converged below the pre-determined limit.

3 Results and Discussions

3.1 *Validation Study*

The numerical diffusion for the fluid solver is tested by developing a laminar flow inside a cylindrical pipe, having L/D ratio as 10, and comparing the velocity profiles at the outlet and the inlet of the pipe. The results are shown in Fig. 2. The velocity profiles at the inlet and the outlet overlap confirm non-diffusive nature of the fluid solver.

3.2 *Fluid–Structure Interaction Model Validation*

Validation study for the computational fluid structure interaction model is performed on the geometry shown in Fig. 3a, b and extensively studied by Turek and Hron [9]. The incompressible flow over a fixed cylinder with a flexible plate is simulated. The domain length, $L = 2.5$ m, domain height $H = 0.41$ m, cylinder center, $C = (0.2,$

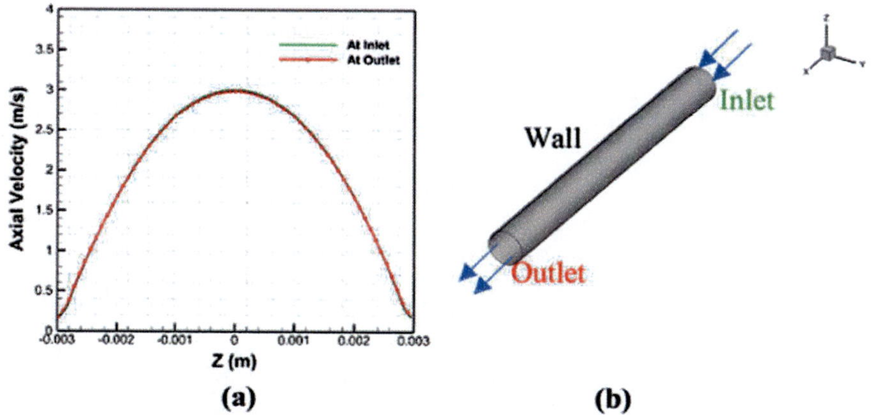

Fig. 2 Testing of fluid solver for numerical diffusion for laminar pipe flow. **a** Axial velocity profile. **b** Cylindrical pipe geometry

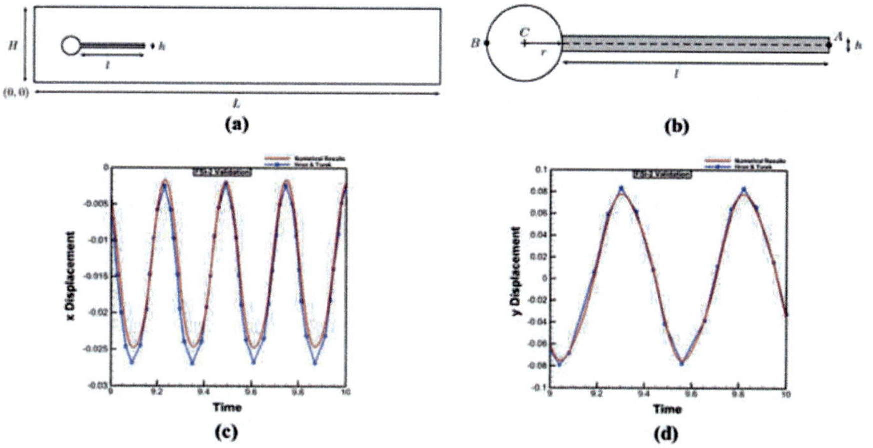

Fig. 3 **a** Computational domain. **b** Structural domain FSI model validation. **c** x-displacement of point A and **d** y-displacement of point A

0.2), cylinder radius $r = 0.05$ m, length of the elastic bar $l = 0.35$ m, height of the elastic bar $h = 0.02$ m, control point $A = (0.6, 0.2)$, and $B = (0.15, 0.2)$.

For the validation case, the structure's material properties are: elastic bar density, $\rho_s = 10{,}000\,\text{kg/m}^3$, Poisson's ratio, $\upsilon_s = 0.4$, Young's modulus of elasticity, $E = 1.4 \times 10^6\,\text{N/m}^2$. The fluid properties are: density of fluid, $\rho_f = 1000\,\text{kg/m}^3$, kinematic viscosity, $\upsilon_f = 0.001\,\text{m}^2/\text{s}$, average velocity, $U_{\text{avg}} = 1\,\text{m/s}$, and Reynolds number, $\text{Re}_f = 100$. The displacement of the point A (fixed on the elastic bar) in

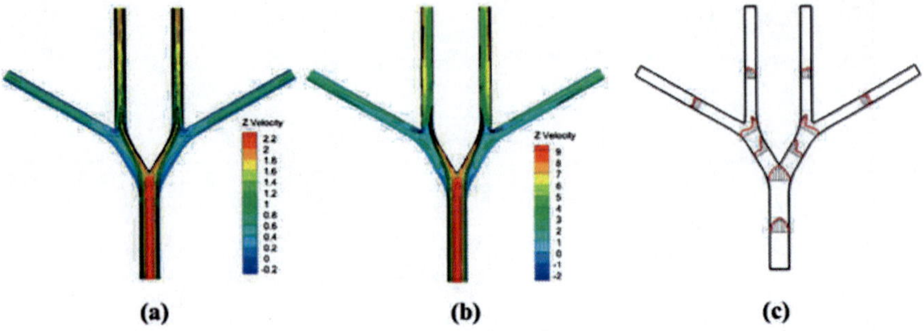

Fig. 4 Axial velocity distribution through airway at **a** Re = 500, **b** Re = 2000, **c** vectors and velocity profiles at different axial locations through the airway (for Re = 2000)

x-direction and *y*-direction is depicted with time and compared with the numerical benchmark results. The numerical results of our setup are found to be in good agreement as shown in Fig. 3.

3.3 Coupled Fluid–Structure Study for Human Airway with Double Bifurcation

The distribution of airflow velocity, airway pressure, related stresses viz. Von-Mises stress and wall shear stress for the flexible human airway walls are discussed below.

3.3.1 Air Flow Development Inside the Human Airway

Figure 4a, b depicts the velocity distribution of the airflow in the airways at Re = 500 and Re = 2000, respectively. The values of maximum airflow velocity are observed to be 2.4 and 10 m/s for Re = 500 and 2000, respectively. Figure 4c represents the vectors and velocity profiles at various axial locations. From Fig. 4c, we can observe that in G_4, the magnitude of the velocity in the inner side is more than the outer side due to development of secondary flow from the first bifurcation. However, velocity profiles in the G_5 are asymmetric, and the airflow velocity in the G_{5L} is lower than the velocity in the G_{5M}. This can be attributed due to the high velocity in the inner side of G_4.

3.3.2 Development of Airflow Pressure and Wall Shear Stresses

Figure 5a, b represents the airflow pressure distribution and the wall shear stress for Re = 500 and 2000, respectively. The maximum pressure was observed (5.317 Pa and 6.385 Pa for Re = 500 and 2000, respectively) at the first bifurcation, which is

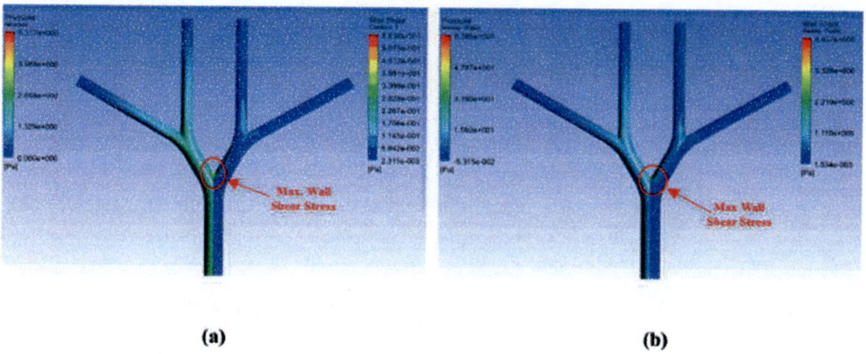

Fig. 5 Distribution of the pressure and wall shear stresses in the airway **a** at Re = 500, **b** at Re = 2000

almost double of that at the second bifurcation. This is because of the impingement of the high momentum flow at the first bifurcation. From the previous study by Longest and Vinchurkar [10], we know that high particle deposition occurs at the high wall shear stress location. So, from Fig. 5, we can identify the high wall shear stress location to be at the first bifurcation. The maximum wall shear stress value is 0.5634 Pa for Re = 500 and 4.437 Pa for Re = 2000.

Figure 6 shows the comparison of the distribution of the skin friction coefficient and wall shear stress on the section A-A′ for FSI case (fluid and solid coupled) as well as for non-FSI case (fluid only) for Re = 2000. We observe the higher values of skin friction coefficient and wall shear stress for FSI case as compare to the non-FSI case. These higher values in the FSI case are due to the higher deformation in section A-A′.

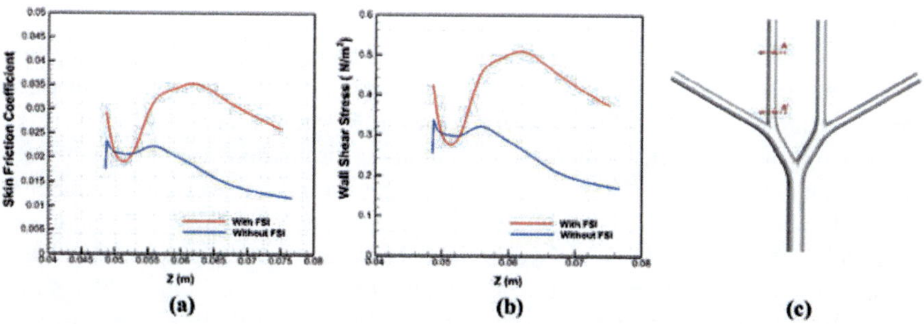

Fig. 6 Distribution on section A-A′. **a** Skin friction coefficient. **b** Wall shear stress. **c** Section A-A

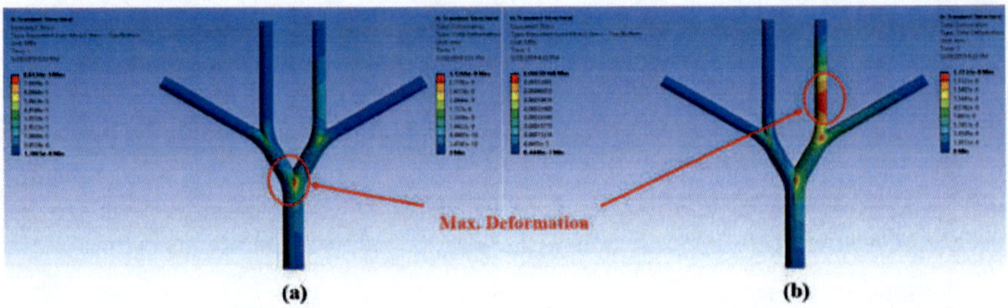

Fig. 7 Distribution of Von-Mises stresses and deformation in the airway **a** at Re = 500, **b** at Re = 2000

3.3.3 Development of Von-Mises Stresses and Deformation

Von-Mises stresses are developed in the airway due to the effect of pressure, and the corresponding deformation is shown in Fig. 7a, b for Re = 500 and Re = 2000. The higher values of Von-Mises stress are observed at the bifurcations.

The maximum value of the Von-Mises stress is found to be 88.534 Pa for Re = 500 and 591.68 Pa for Re = 2000. Further, we can also observe that the maximum deformation for Re = 500 is found to be at the first bifurcation while for higher Reynolds number, Re = 2000, the maximum deformation occurs in the G_{5M} branch.

4 Conclusions

A three-dimensional two-way-coupled FSI study using ANSYS is performed to better understand the airflow patterns inside a human airway for two different Reynold numbers. The related stresses and deformations along with the velocity profiles at different axial locations inside the airway are also reported, and it is observed that the maximum deposition of the particles, corresponding to the maximum wall shear stress, occurs at the first bifurcation for both the Reynold numbers. The walls of the human airway are considered to be elastic in the simulations for obtaining more physically correct results. The maximum deformation for low Reynolds number is observed to be at first bifurcation while for higher Reynolds number, the maximum deformation is observed to be at G_{5M} branch. The present study provides the basis for finding the more efficient ways of medication for the various lung problems like asthma, bronchiole inflammation, chronic bronchitis, and other infectious diseases like influenza and pneumonia.

Acknowledgements The authors would like to acknowledge the High-Performance Computing (HPC) Facility at IIT Kanpur (www.iitk.ac.in/cc) for providing the necessary environment to perform the above study.

References

1. Schroter RC, Sudlow MF (1969) Flow patterns in models of the human bronchial airways. Respir Physiol 7(3):341–355
2. Zhao Y, Lieber BB (1994) Steady inspiratory flow in a model symmetric bifurcation. J Biomech Eng 116(4):488–496
3. Comer JK, Kleinstreuer C, Zhang Z (2001) Flow structures and particle deposition patterns in double-bifurcation airway models. Part 1. Air flow fields. J Fluid Mech 435:25–54
4. Augusto LLX, Lopes GC, Goncalves JAS (2016) A CFD study of deposition of pharmaceutical aerosols under different respiratory conditions. Braz J Chem Eng 33(3):549–558
5. Weibel ER, Cournand AF, Richards DW (1963) Morphometry of the human lung, vol 1. Springer, Berlin
6. Sera T, Satoh S, Horinouchi H, Kobayashi K, Tanishita K (2003) Respiratory flow in a realistic tracheostenosis model. J Biomech Eng 125(4):461–471
7. Croteau JR, Cook CD (1961) Volume-pressure and length-tension measurements in human tracheal and bronchial segments. J Appl Physiol 16(1):170–172
8. Prakash UB, Hyatt RE (1978) Static mechanical properties of bronchi in normal excised human lungs. J Appl Physiol 45(1):45–50
9. Turek S, Hron J (2006) Proposal for numerical benchmarking of fluid-structure interaction between an elastic object and laminar incompressible flow. In: Fluid-structure-interaction. Springer, Berlin, pp 371–385
10. Longest PW, Vinchurkar S (2007) Validating CFD predictions of respiratory aerosol deposition: effects of upstream transition and turbulence. J Biomech 40(2):305–316

Film Cooling Aspects of a Dual Bell Nozzle

Mayank Verma and Ashoke De

Abstract In the field of space science and aerospace technology, altitude adaptive nozzles recently have got more importance. Dual bell nozzle presents a solution for the altitude adaption. Dual bell nozzle has dual operating modes (at low and high altitudes) without any mechanical activation. For future reusable launch vehicles, it is necessary to understand the active control of the working mode of the dual bell nozzle. Film cooling with injection of the secondary fluid provides a solution for the active control of the nozzle. The effectiveness of this secondary injection is investigated here by injecting the secondary fluid at three different locations viz. at the throat, at the base nozzle, and at the inflection point for the low-altitude as well as high-altitude operation mode. A two-dimensional steady RANS simulation has been carried out on an axisymmetric parabolic contour dual bell nozzle. Turbulence is modelled with k-Omega SST model with the updated coefficients. The present study highlights the thermodynamic aspects of the film cooling in a dual bell nozzle.

Keywords Dual bell nozzle · Film cooling · Secondary injection · RANS · CFD

1 Introduction

Designing of an optimal nozzle plays an important phase in the rocket development. Its performance is heavily affected by the nozzle's effectiveness in converting thermal energy to kinetic energy. High expansion rate is highly desirable to increase the vacuum performance of classical convergent-divergent rocket nozzles. A Dual bell nozzle combines two differently designed conventional nozzle into one nozzle. The first nozzle with relatively low expansion ratios is called 'Base nozzle,' while the

---

M. Verma (✉) · A. De
Department of Aerospace Engineering, Indian Institute of Technology Kanpur, Kanpur, Uttar Pradesh 208016, India
e-mail: vmayank@iitk.ac.in

A. De
e-mail: ashoke@iitk.ac.in

L. Venkatakrishnan et al. (eds.), *Proceedings of 16th Asian Congress of Fluid Mechanics*, Lecture Notes in Mechanical Engineering,
https://doi.org/10.1007/978-981-15-5183-3_11

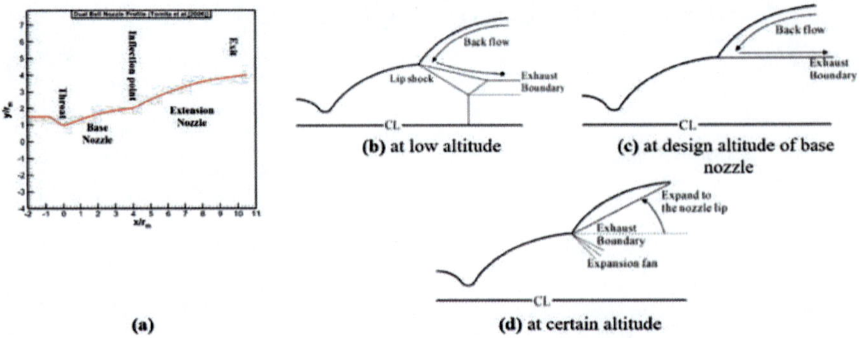

Fig. 1 **a** Dual bell nozzle test article contour. Schematics of the flow development of a dual bell nozzle (**b**, **c** and **d**) [1]

second nozzle with the larger expansion ratio is called 'Extension nozzle' in the literature of nozzles [1–3]. Figure 1a shows the schematic of the nozzle contours.

A dual bell nozzle because of its geometry inherits the dual operating mode. For the low-altitude operation mode, the main flow from the nozzle inlet fills only the base nozzle while the flow separation point stabilizes at the inflection point as depicted in Fig. 1b. For the high-altitude operation mode, the main flow from the nozzle inlet fills both the base nozzle as well as extension nozzle and the separation point shifts at the lip of the extension nozzle. During the ascent of the rocket from sea-level to the space, the working mode transition from the low-altitude operation mode to high-altitude operation mode take places within the dual bell nozzle. This transition from the full flowing base nozzle to the full flowing extension nozzle with the altitude change is purely ruled by the fluid dynamics parameter, i.e., nozzle pressure ratio, NPR, which is the ratio of the combustion chamber pressure and the ambient pressure. A decrease in the nozzle thrust is observed during such transition [2, 3].

For clustered engines, the NPR for each engine at the same altitude may be slightly different resulting in the different transition altitude for each engine. Due to the difference in the transition altitude of the different engines, the resulting imbalance in the thrust of each engine may have severe influence on the vehicle trajectory. Thus, active control of the transition is of vital importance for dual bell nozzles. The injection of the secondary fluid into the nozzle provides an active control on the transition without any mechanical activation. It has been found in the literature that the effectiveness is the inverse function of the injection angle from the main flow. As the injection angle from the main flow increases, the cooling effectiveness decreases due to the increased mixing [4]. Therefore, a majority of studies focus on a parallel secondary injection [5–8]. In the present work, the effectiveness of the film cooling aspect of a dual bell nozzle is studied.

Table 1 Design parameters of dual bell nozzle

Nozzle	Design method	Nozzle length/throat radius	Initial expansion angle (in degree)	Nozzle exit angle (in degree)
Base nozzle	Parabolic	4	25	5
Extension nozzle	Parabolic	6.25	30	5

2 Numerical Set up

2.1 Geometrical Details

To investigate the film cooling aspects of a dual bell nozzle, the nozzle geometry is extracted from the experimental study of Tomita et al. [1]. The base nozzle is a parabolic nozzle with small nozzle length, with large initial expansion angle, and with small nozzle exit angle (B-TO-SLS). The extension nozzle is designed to be a parabolic nozzle with long nozzle length, with large initial expansion angle, and with small nozzle exit angle (E-TO-LLS). Figure 3 shows the contours of the base and extension nozzle. Design parameters for the dual bell nozzle are given in Table 1.

2.2 Computational Details

The computational domain and corresponding boundary conditions are depicted in Fig. 2. The computational domain extends to a distance of $100R_{th}$ in the axial direction and $40R_{th}$ in the radial direction. At the inlet of the nozzle, fixed pressure inlet conditions corresponding to different NPRs are imposed with zero gradient conditions for velocity and temperature. The nozzle walls are made adiabatic with no-slip boundary conditions. At the outlet and the far-field, wave transmissive conditions are imposed so that the no spurious waves could contaminate the solution inside the computational domain.

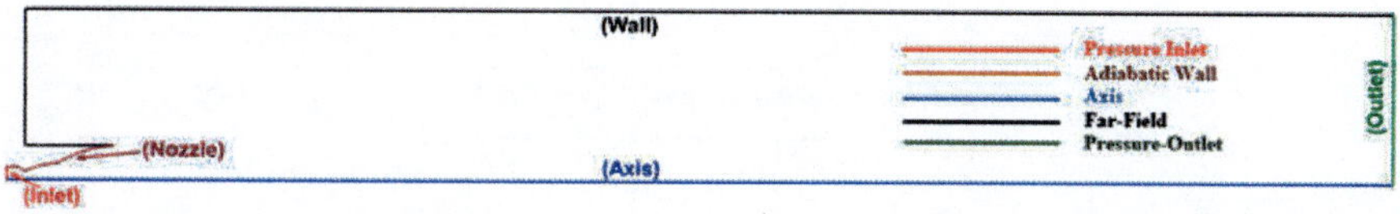

Fig. 2 Computational domain with the corresponding boundary conditions used for the calculations

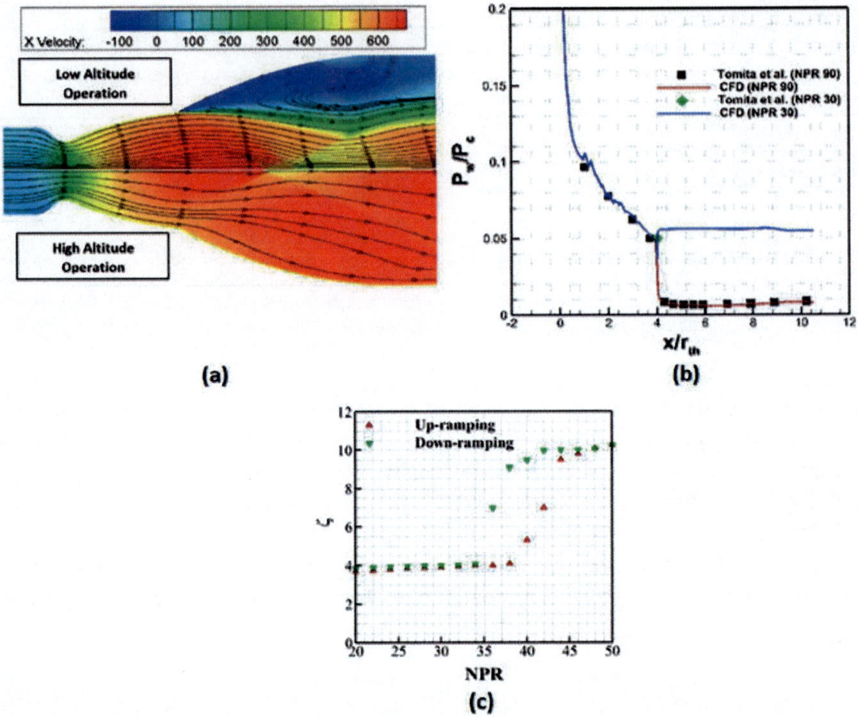

Fig. 3 **a** Schematic of the flow in a dual bell nozzle at low-altitude and high-altitude operation. **b** Normalized wall pressure profile for different NPRs. **c** Hysteresis curve obtained for DBN

2.3 Governing Equations

The Favre averaged, steady equations for mass, momentum, and energy conservation [9, 10], presented below, are solved using a finite volume method along with the perfect gas equation.

$$\frac{\partial \bar{\rho}\tilde{u}_i}{\partial x_i} = 0 \tag{1}$$

$$\frac{\partial\left(\bar{\rho}\tilde{u}_i\tilde{u}_j + \bar{p}\delta_{ij} + \overline{\rho u_i'' u_j''} - \bar{\tau}_{ij}\right)}{\partial x_j} = 0 \tag{2}$$

$$\frac{\partial\left[\bar{\rho}\tilde{u}_j\left(\tilde{h} + \frac{\tilde{u}_i\tilde{u}_j}{2}\right) + \tilde{u}_j\frac{\overline{\rho u_i'' u_j''}}{2}\right]}{\partial x_j} = \frac{\partial\left[-q_j - \overline{\rho u_j'' h''} + \overline{\tau_{ij} u_i''} - \frac{1}{2}\overline{\rho u_j'' u_i'' u_j''}\right]}{\partial x_j}$$

$$+\frac{\partial\left[\tilde{u_i}\left(\overline{\tau}_{ij}-\overline{\rho u_i'' u_j''}\right)\right]}{\partial x_j} \tag{3}$$

$$P=\bar{\rho}R\tilde{T} \tag{4}$$

The symbols (—) and (~) represent the time averaging through Reynolds decomposition and density-weighted time averaging (Favre averaging), respectively. Further details of the different terms in the above equations are provided in [11]. The turbulence is modeled with the help of k-ω SST model with modified co-efficient [12].

3 Results

3.1 *Flow Development in Dual Bell Nozzle*

Flow is developed for two different NPRs corresponding to two different operating modes of the dual bell nozzle and presented in Fig. 3. It can be observed from Fig. 3a that for low-altitude operation mode (at NPR 30), the base nozzle is full flowing and the flow is separated from the inflection point while for the high-altitude operation mode (at NPR 90), both the nozzles are full flowing and flow separation takes place outside the nozzle.

The normalized wall pressure plot corresponding to the two different NPRs is presented in Fig. 3b. The pressure distribution for high-altitude operation is validated against the experimental results by Tomita et al. [1]. It is observed that the pressure in the base nozzle is decreased until the extension nozzle is reached. For low-altitude operation (at NPR 30), the flow separates at the start of the extension nozzle resulting in increased pressure. However, for the high-altitude operation (at NPR 90), the flow separates near the end of the extension nozzle resulting in a further decrease in the pressure at the nozzle wall. Thus, the transition takes place somewhere between NPR 30 and NPR90 (~NPR 38.88 according to the experimental studies by Tomita et al. [1]).

To capture the transition NPR for the DBN, we have performed the steady simulations ranging between NPR 20 to NPR 50. The step size between the simulations is set to 2 bar. Figure 3c reports the flow separation positions (ζ) for different NPRs for up-ramping and down-ramping of the inlet pressure feed. From Fig. 3c, we can observe that the flow separation shifts from base nozzle to extension nozzle somewhere between NPR 35 to NPR 45. Hence, NPR 40 is taken as the transition NPR.

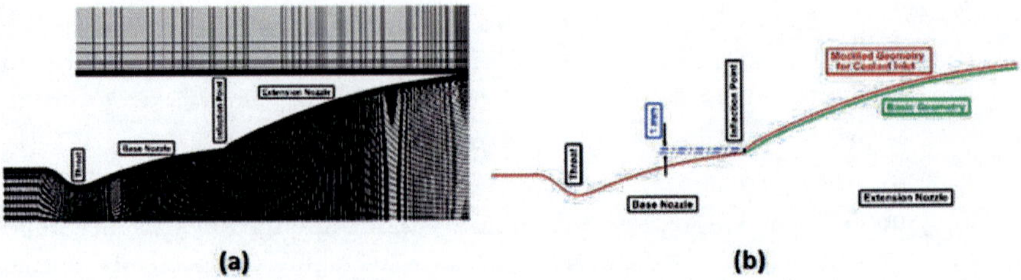

Fig. 4 **a** Meshing view of the dual bell nozzle. **b** Modified geometry schematic for film cooling through secondary injection at inflection point

3.2 *Film Cooling Effects on Dual Bell Nozzle*

For film cooling of the dual bell nozzle, the secondary injection is made at three different axial locations on the nozzle wall. The coolant inlet is created by shifting the downstream curvature in the lateral direction by 1 mm. Thus, the coolant inlet has a height of 1 mm. To get the optimum grid size, three different grid sizes having 0.8, 1.8, and 3 million cells were tested and finally, 1.8 million grid size is selected for the further calculations. The wall is resolved up to y + value of 1 to properly resolve the boundary layer. Figure 4a presents the multiblock mesh created in ICEM CFD [13]. Figure 4b presents the modified geometry for the coolant inlet. Air with the inlet temperature of 500 K is used as a coolant. As the injection angle from the main flow increases, the cooling effectiveness decreases due to the increased mixing [4]. Therefore, the secondary injection is made parallel to the main flow leaving the injection angle as 0°. The effectiveness of the secondary injection of the fluid is studied considering the vital parameters, i.e., the location of the secondary injection for both the operating modes, i.e., at NPR 25 for low-altitude operating mode as well as at NPR 40 for high-altitude operating mode. The Mach number of the injection is 2.4 with the inlet temperature of 500 K. The nozzle main flow is developed with the total temperature of 2842 K while the coolant is injected at 500 K. The mass flow ratio (σ) of the secondary flow with respect to the main flow is 0.11.

3.2.1 Effect of Coolant Injection Location for Low-Altitude Operating Mode

To find the most effective location of the coolant injection, the injection is made at three different nozzle locations viz. at base nozzle throat ($x/r_{\text{th}} = 0$), at base nozzle ($x/r_{\text{th}} = 2$), and at inflection point ($x/r_{\text{th}} = 4$), respectively.

Figure 5 shows the results for the effect of the coolant injection for the low-altitude operating mode (at NPR 25). From Fig. 5a, we see that injection of the coolant at the throat results in the cooling of the base nozzle only while the extension nozzle still experiences high temperature. Also, from Fig. 5b, it can be concluded that injecting the fluid at the throat of the nozzle gives rise to the sudden pressure oscillation near

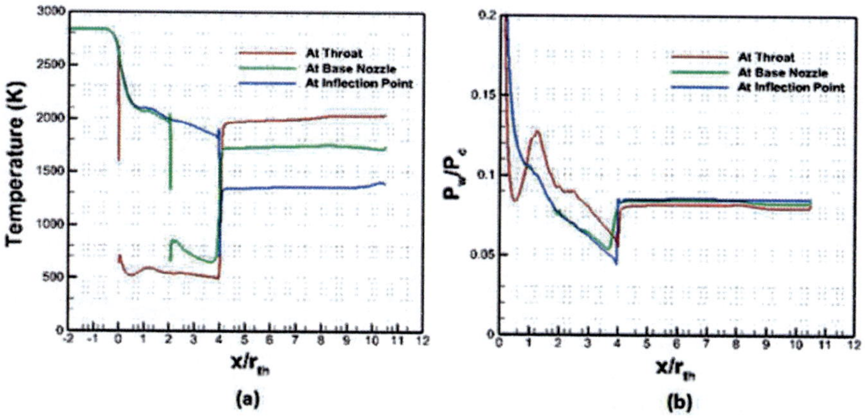

Fig. 5 Effect of the injection location of the coolant for low-altitude operation (at NPR 25). **a** Temperature plot. **b** Normalized wall pressure plot

the throat which results in high lateral loads that are totally undesirable in nozzle flows. Injection of the fluid at base nozzle results in the early separation of flow from the base nozzle wall (at $x/r_{\text{th}} \sim 3.6$), which reduces the axial thrust obtained by the nozzle. Injection of the fluid at the inflection point makes the base nozzle full flowing resulting in the increase of the axial thrust relative to the injection at the base nozzle condition.

3.2.2 Effect of Coolant Injection Location for High-Altitude Operating Mode

The results for coolant injection at the high-altitude operating mode is presented in Fig. 6. NPR 40 is selected for the calculations. Figure 6a presents the temperature contour for the three injection locations. From temperature contours, it can be observed that maximum cooling of the extension nozzle is achieved in the case of the injection at the inflection point. Also, from the pressure plot in Fig. 6b, we can also observe that the injection of the coolant at the inflection point shifts the separation point downstream to $x/r_{\text{th}} \sim 6$ from $x/r_{\text{th}} \sim 5$. It results in a decrease of the axial thrust of the nozzle. Thus, it is clear that with the help of the secondary injection at the inflection region, a full flowing extension nozzle can be achieved at the transition NPR with the effective cooling of the nozzle.

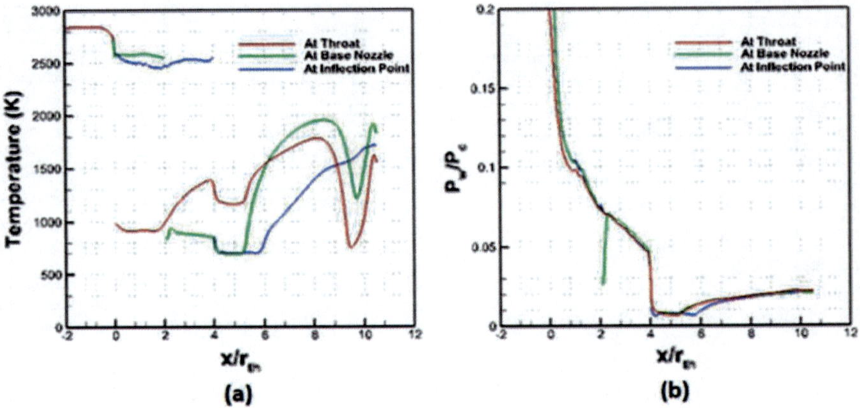

Fig. 6 Effect of the injection location of the coolant for high-altitude operation (at NPR 40). **a** Temperature plot. **b** Normalized wall pressure plot

4 Conclusions

The flow through the dual-bell nozzle for two operational modes is studied using computational methods along with the effect of the secondary injection of the fluid. It is observed that the transition from low-altitude operation mode to the high-altitude operation mode is observed to be approximately at NPR 40. The wall pressure distribution on the nozzle wall from the numerical study for the different NPRs closely matches with the experimental results. From the study, it is observed that the injection of the secondary fluid (with temperature of 500 K) at the inflection point (having main flow temperature of 2842 K) results in the effective cooling of the nozzle wall and the maximum movement of the separation point. So, it can be concluded that the secondary injection not only cools the extension nozzle but also have significant effects on the movement of the separation point. Thus, at transition NPR, the separation point in a Dual Bell Nozzle can be controlled with the help of coolant injection in the inflection region so that an optimum amount of thrust can be obtained.

Acknowledgements Financial support for this research is provided through IITK-Space Technology Cell (STC). The authors would also like to acknowledge the High-Performance Computing (HPC) Facility at IIT Kanpur (www.iitk.ac.in/cc).

References

1. Tomita T, Takahashi M, Sasaki M, Tamura H (2006) Investigation on characteristics of conventional-nozzle-based altitude compensating nozzles by cold-flow tests. In: 42nd AIAA/ASME/SAE/ASEE joint propulsion conference & exhibit, p 4375
2. Tomita T, Takahashi M, Sasaki M (2009) Control of transition between two working modes of a dual-bell nozzle by gas injection. In: 45th AIAA/ASME/SAE/ASEE joint propulsion conference & exhibit, p 4952
3. Tomita T, Takahashi M, Sasaki M (2007) Investigation on characteristics of conventional-nozzle-based altitude compensating nozzles by cold-flow tests (II) side-load characteristics during transition. In: 43rd AIAA/ASME/SAE/ASEE joint propulsion conference & exhibit, p 5472
4. Seban RA et al (1957) Heat transfer to a turbulent boundary layer downstream of an injection slot. American Society of Mechanical Engineers, Paper 57-A-36, 195
5. O'Connor JP, Haji-Sheikh A (1992) Numerical study of film cooling in supersonic flow. AIAA J 2426–2433
6. Juhany KA et al (1994) Influence of injectant mach number and temperature on supersonic film cooling. J Thermophys Heat Transf 59–67
7. Aupoix B (1998) Experimental and numerical study of supersonic film cooling. AIAA J 915–923
8. Arya N, Verma M, De A (2019) Flow evolution in a dual-bell nozzle during different modes of operation with film cooling. In: Proceedings of the 32nd international symposium on shock waves (ISSW32). National University of Singapore, Singapore, July 14–19, 2019, Paper 211
9. Soni RK, De A (2017) Investigation of strut-ramp injector in a scramjet combustor: effect of strut geometry, fuel and jet diameter on mixing characteristics. J Mech Sci Technol 31(3):1169–1179
10. Soni RK et al (2015) Assessment of RANS based models in a supersonic flow. In: AIP conference proceedings. AIP Publishing, New York
11. Wilcox DC (1998) Turbulence modeling for CFD. DCW Industries, La Canada, CA, pp 172–180
12. Yaravintelimath A, Raghunandan BN, Morinigo JA (2016) Numerical prediction of nozzle flow separation: issue of turbulence modeling. Aerosp Sci Technol 50:31–43
13. ANSYS Inc. (2013) CFD user's manual. Ansys Inc., Canonsburg

Wall Distance Computation Based on Higher-Order Variational Reconstruction on Unstructured Grids

Qian-Min Huang and Yu-Xin Ren

Abstract Minimum distance distribution to a solid wall is important for turbulent and peripheral physical models. In this paper, the non-conservational Eikonal equation is solved under the framework of high-order variational reconstruction on unstructured grids to calculate the minimum wall distance distribution. Convectional reconstruction is proposed to realize the upwind treatment for the reconstructed variables. Numerical examples show that the proposed method in this paper can accurately and effectively solve the minimum wall distance distribution for external flow problems with complex geometric configurations on unstructured grids.

Keywords Variational reconstruction · Minimum wall distance · Convectional reconstruction

1 Introduction

Accurately calculating the minimum distance from a space point to a solid wall is of great significance to some fields of computational fluid dynamics, such as turbulence model and mesh generation [1]. The algorithm of minimum wall distance computation has been extensively proposed and studied, which can be classified as the search method [2–4], integral equation method and differential equation method [5–11]. Crude search procedures require $O(n_w, n_g)$ operations, where n_w and n_g are, respectively, the number of grid points on the wall and the number of internal points. Moreover, the results of search method for curve surface are not very accurate because the search method needs to approximate the surface by linear plane. The integral equation method has some problems in the calculation of complex geometric shapes. The differential equation method iteratively calculates the minimum wall distance by solving Eikonal, Poisson, Hamilton/Jacobi equations or the equivalent conservational equations [8–10]. In addition, many studies have transformed the differential equation into the transport equation by introducing pseudotime derivative

Q.-M. Huang (✉) · Y.-X. Ren
Department of Engineering Mechanics, Tsinghua University, Beijing 100084, China
e-mail: kokhuang@163.com

L. Venkatakrishnan et al. (eds.), *Proceedings of 16th Asian Congress of Fluid Mechanics*, Lecture Notes in Mechanical Engineering,
https://doi.org/10.1007/978-981-15-5183-3_12

term, which makes the parallel calculation of wall distance simpler and more efficient [11].

However, numerical instability will appear when the high-order methods are used to solve the transport differential equations, especially for the external flow problems of complex geometric configurations, because most of the boundaries are far-field and their boundary values are not given, those methods have faced challenges in convergence and achieving high-order accuracy. Fares and SchröDer pointed out that for the transport differential equations there exists a sort of nonlinear effect which cannot be analytically analysed, and this effect will have a significant impact on the calculation [1]. Stefan and Ralf proposed a DG method for calculating the minimum wall distance distribution and successfully overcame the instability problem by introducing the artificial viscosity.

Most study has discussed the computation of the minimum wall distance distribution by solving the conservational transport equations, while the study by solving the non-conservational transport equations is very limited. In this paper, the computation of minimum wall distance distribution by solving the non-conservational transport Eikonal equation under the framework of high-order variational reconstruction (VR) on unstructured grids will be conducted and discussed. The results are not only meaningful for the minimum wall distance distribution computation on unstructured grids, but also referable for the computation of non-conservational equations by high-order finite volume method.

2 Methodology

The minimum wall distance distribution can be given by the solution of the Eikonal equation, which takes the form [11]

$$|\nabla\phi| = 1 \tag{1}$$

with homogeneous Dirichlet boundary conditions, $\phi = 0$, imposed on the wall boundary. Similar to Ref. [11], by squaring Eq. (1) and introducing the pseudotime derivative term and the artificial viscosity term, we have the following equation

$$\frac{\partial\phi}{\partial\tau} + \vec{u}\cdot\nabla\phi - \epsilon\phi\Delta\phi = 1, \quad \vec{u} = \nabla\phi, \tag{2}$$

where ∇ and Δ, respectively, denote the gradient and Laplacian operator, and the artificial viscous coefficient $\epsilon = \alpha|1 - |\nabla\phi||$, with $\alpha = 0.1$.

In what follows we will firstly introduce the high-order VR finite volume method (FVM) proposed by Ren, Wang and Pan [12–15], and then propose the concept of convectional reconstruction (CR), which is able to realize the upwind treatment for the reconstructed variables and as a result ensure the feasibility of solving the non-conservational transport Eikonal equation via FVM. Assuming the computational

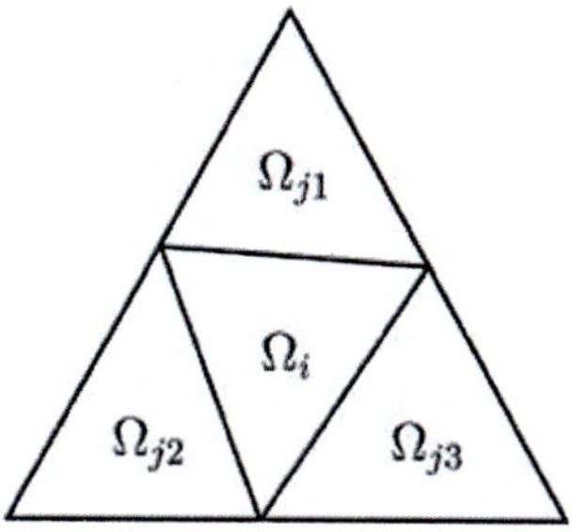

Fig. 1 Reconstruction stencil for control volume Ω_i

domain Ω can be divided into N non-overlapping control volumes, i.e., $\Omega = \bigcup_{i=1}^{N} \Omega_i$. The volume of control volume Ω_i is denoted by $\overline{\Omega}_i$. For arbitrary high-order VR, the reconstruction stencil is compact, i.e. only involving the face-neighbours of the control volume of interest, and denoted by $S_i = \{\Omega_i, \Omega_{j1}, \Omega_{j2}, \Omega_{j3}\}$, as shown in Fig. 1.

Considering the cell centred finite volume method, the cell average of a solution $\phi(\vec{x})$ on the control volume Ω_i can be expressed as

$$\overline{\phi}_i = \frac{1}{\overline{\Omega}_i} \iint_{\Omega_i} \phi(\vec{x}) \mathrm{d}\Omega, \tag{3}$$

The solution $\phi(\vec{x})$ can be approximated by a kth order polynomial

$$\phi_i(\vec{x}) = \overline{\phi}_i + \sum_{l=1}^{N_c(k)} \phi_{i,l} \psi_{i,l}(\vec{x}), \tag{4}$$

where the number of unknown polynomial coefficients $N_c(k) = (k+1)(k+2)/2-1$, $\phi_{i,l}$ is the unknown polynomial coefficients, and $\psi_{i,l}(\vec{x})$ is the zero-mean basis defined by

$$\begin{aligned} \psi_{i,l}(\vec{x}) &= \delta x_i^m \delta x_i^n - \overline{\delta x_i^m \delta x_i^n}, \\ \delta x_i &= \frac{x - x_i}{\Delta x_i}, \delta y_i = \frac{y - y_i}{\Delta y_i}, \overline{\delta x_i^m \delta x_i^n} = \frac{1}{\overline{\Omega}_i} \iint_{\Omega_i} \delta x_i^m \delta x_i^n \mathrm{d}\Omega. \end{aligned} \tag{5}$$

where m and n are the powers of the corresponding Taylor basis, and the length scale Δx and Δy are defined following the proposal of Luo et al. [16]

$$\Delta x_i = \frac{x_{i,\max} - x_{i,\min}}{2}, \Delta y_i = \frac{y_{i,\max} - y_{i.\min}}{2}, \tag{6}$$

where $x_{i,\max}$, $x_{i,\min}$ and $y_{i,\max}$, $y_{i,\min}$ are, respectively, the maximum and minimum coordinate values along the x-direction and y-direction of the control volume.

Theoretical foundation of VR is the functional extremum problem. It is a method to construct the linear equations that are used to determine the unknown coefficients of the reconstructed polynomials by minimizing the cost function of variational principles. Therefore, the specific reconstruction depends on the form of the cost function. The cost function in this paper is defined by an interfacial jump integration (IJI) function, which takes the form

$$I_f = \frac{1}{|S_{ij}|} \int_f \sum_{p=0,q=0}^{p+q\le k} \left[\Delta x_{ij}^p \Delta y_{ij}^q \frac{C_{p+q}^p}{(p+q)!} \left(\frac{\partial^{p+q}\phi_i(\vec{x})}{\partial x^p \partial y^q} - \frac{\partial^{p+q}\phi_j(\vec{x})}{\partial x^p \partial y^q} \right) \right]^2 \mathrm{d}s, \quad (7)$$

where $|S_{ij}|$ is the length of the interface between the left and right control volumes, $\Delta x_{ij} = \frac{1}{2}(\Delta x_i + \Delta x_j)$, $\Delta y_{ij} = \frac{1}{2}(\Delta y_i + \Delta y_j)$. IJI is a measure of the jump of the reconstructed polynomial and its derivatives at the interface. The cost function is taken as the sum of all IJIs in the computational domain, i.e.

$$I = \sum_{f=1}^{N_f} I_f, \quad (8)$$

where N_f denotes the number of interfaces in the computation domain. The constitutive relations of VR can be established by minimizing the cost function, namely

$$\frac{\partial I}{\partial \phi_{i,l}} = 0, \quad l = 1, 2, \ldots, N_c(k), i = 1, 2, \ldots, N. \quad (9)$$

As shown in Fig. 1, the unknown reconstructed polynomial coefficients $\phi_{i,l}$ on the control volume are determined only by the stencil S_i. By substituting Eq. (4) into (7)–(9), the following linear equations can be established

$$\begin{aligned} &\sum_{l=1}^{N_c(k)} \left\{ \left[\sum_{j\in S_i, j\ne i} \frac{1}{|S_{ij}|} \int_{\partial\Omega_i \cap \partial\Omega_j} \sum_{p=0,q=0}^{p+q\le k} \Delta x_{ij}^{2p} \Delta y_{ij}^{2q} \frac{{C_{p+q}^p}^2}{(p+q)!^2} \frac{\partial^{p+q}\psi_{i,l}}{\partial x^p \partial y^q} \frac{\partial^{p+q}\psi_{i,r}}{\partial x^p \partial y^q} \mathrm{d}s \right] \right. \\ &= \sum_{j\in S_i, j\ne i} \left\{ \sum_{l=1}^{N_c(k)} \left[\frac{1}{|S_{ij}|} \int_{\partial\Omega_i \cap \partial\Omega_j} \sum_{p=0,q=0}^{p+q\le k} \Delta x_{ij}^{2p} \Delta y_{ij}^{2q} \frac{{C_{p+q}^p}^2}{(p+q)!^2} \frac{\partial^{p+q}\psi_{i,l}}{\partial x^p \partial y^q} \frac{\partial^{p+q}\psi_{j,r}}{\partial x^p \partial y^q} \mathrm{d}s \right] \phi_{j,l} \right\} \\ &+ \sum_{j\in S_i, j\ne i} \int_{\partial\Omega_i \cap \partial\Omega_j} \frac{\psi_{i,r}}{|S_{ij}|} (\bar{\phi}_j - \bar{\phi}_i) \mathrm{d}s, \quad l = 1, 2, \ldots, N_c(k), r = 1, 2, \ldots, N_c(k), \end{aligned} \quad (10)$$

Equation (10) can be rearranged into the following matrix form

$$\mathbf{A}_i\vec{\phi}_i = \sum_{j\in S_i, j\neq i} \mathbf{A}_j\vec{\phi}_j + \vec{b}_i, \tag{11}$$

where

$$\mathbf{A}_i = \left[\sum_{j\in S_i, j\neq i} \frac{1}{|S_{ij}|} \int\limits_{\partial\Omega_i \bigcap \partial\Omega_j} \sum_{p=0,q=0}^{p+q\leq k} \Delta x_{ij}^{2p} \Delta y_{ij}^{2q} \frac{{C_{p+q}^{p}}^2}{(p+q)!^2} \frac{\partial^{p+q}\psi_{i,l}}{\partial x^p \partial y^q} \frac{\partial^{p+q}\psi_{i,r}}{\partial x^p \partial y^q} \mathrm{d}s \right],$$

$$\mathbf{A}_j = \left[\frac{1}{|S_{ij}|} \int\limits_{\partial\Omega_i \bigcap \partial\Omega_j} \sum_{p=0,q=0}^{p+q\leq k} \Delta x_{ij}^{2p} \Delta y_{ij}^{2q} \frac{{C_{p+q}^{p}}^2}{(p+q)!^2} \frac{\partial^{p+q}\psi_{i,l}}{\partial x^p \partial y^q} \frac{\partial^{p+q}\psi_{j,r}}{\partial x^p \partial y^q} \mathrm{d}s \right],$$

$$\vec{b}_i = \sum_{j\in S_i, j\neq i} \int\limits_{\partial\Omega_i \bigcap \partial\Omega_j} \frac{\psi_{i,r}}{|S_{ij}|} (\bar{\phi}_j - \bar{\phi}_i) \mathrm{d}s. \tag{12}$$

Assembling the constitutive relationships of all elements into the global matrices, we have

$$\mathbf{A}\vec{\phi} = \vec{b}, \tag{13}$$

with

$$\begin{aligned} &\mathbf{A} = \mathbf{D} - \mathbf{L} - \mathbf{U}, \mathbf{D} = \{\mathbf{A}_i\}, \mathbf{L} = \{\mathbf{A}_j, j < i\}, \\ &\mathbf{U} = \{\mathbf{A}_j, j > i\}, \vec{\phi} = \{\vec{\phi}_i\}, \vec{b} = \{\vec{b}_i\}. \end{aligned} \tag{14}$$

It can be proved that the matrix $\mathbf{A}$ established by the VR is symmetrically positive definite and $2\mathbf{D} - \mathbf{A}$ is positive definite. Detailed proof is available in Refs. [13, 14].

In order to realize the upwind treatment for the reconstructed variables, we propose the CR, which is presented as follows:

1. Let the reconstruction expression of the solution $\phi(\vec{x})$ on the control volume Ω_i be

$$\tilde{\phi}_i(\vec{x}) = \bar{\phi}_i + \sum_{l=1}^{N_c(k)} \tilde{\phi}_{i,l}\psi_{i,l}(\vec{x}), \tag{15}$$

2. Define the IJI and cost function

$$\tilde{I}_f = \sum_{j\in S_i, j\neq i} \int\limits_f (\tilde{\phi}_i(\vec{x}_f) - \phi_f(i,j))^2 \mathrm{d}s, \tag{16}$$

where

$$\phi_f(i,j) = \frac{1}{2}[1 + \text{sign}(\vec{u}_f(i,j)\cdot\vec{n}_f)]\phi_i(\vec{x}_f) + \frac{1}{2}[1 - \text{sign}(\vec{u}_f(i,j)\cdot\vec{n}_f)]\phi_j(\vec{x}_f),$$
$$\vec{u}_f(i,j) = \frac{1}{2}\left(\nabla\phi_i(\vec{x}_f) + \nabla\phi_j(\vec{x}_f)\right), \tag{17}$$

3. Solving the functional extremum problem

$$\frac{\partial \tilde{I}}{\partial \tilde{\phi}_{i,l}} = 0, \quad l = 1, 2, \ldots, N_c(k), i = 1, 2, \ldots, N, \tag{18}$$

By substituting Eq. (15) into (16)–(18), we have

$$\sum_{l=1}^{N_c(k)} \left(\sum_{j\in S_i, j\neq i} \int_{\partial\Omega_i \bigcap \partial\Omega_j} \psi_{i,l}\psi_{i,r} ds \right) \tilde{\phi}_{i,l} = \sum_{j\in S_i, j\neq i} \int_{\partial\Omega_i \bigcap \partial\Omega_j} (\phi_{f,j} - \overline{\phi}_i)\psi_{i,r} \mathrm{d}s, \quad r = 1, 2, \ldots, N \tag{19}$$

The boundary treatment is referred to Refs. [12–15]. Since the problem in this paper is a steady one, the so-called reconstruction and time integration coupled iteration method [14] can be executed by using an explicit time integration method, i.e. the third-order TVD RK method is used in this paper, with CFL = 0.6.

3 Results

Three numerical examples are used in this paper, i.e. the circle, NACA0012 and 30p30n air foil. Initial values $\phi_0(\vec{x}) = 0$. The fourth-order VR and CR is used for the spatial discretization. Figure 2 shows the computational grid and results of the circle. In Fig. 2, the numerical results (colour curves) are in good agreement with the analytical results (black curves), indicating the accuracy of the method proposed in this paper for the minimum wall distance distribution computation. Figures 3, 4 and 5, respectively, present the computational grids and results of NACA0012 and 30p30n air foil. These results show that the method proposed in this paper can effectively overcome the numerical instability of the external flow problems of complex geometric configurations on unstructured grids when solving the non-conservational transport Eikonal equation.

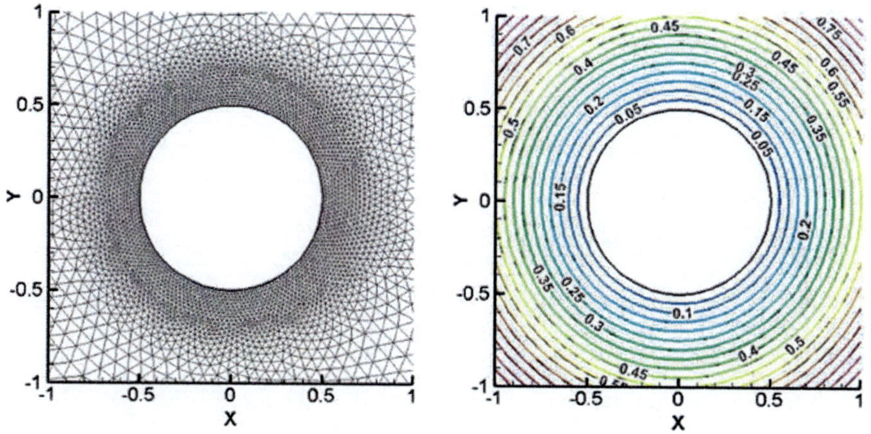

Fig. 2 Computational grid and results of the circle

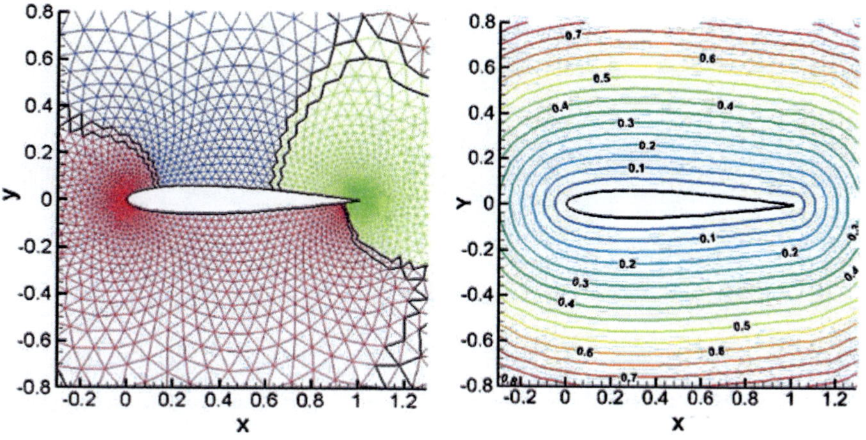

Fig. 3 Computational grid and results of NACA0012 air foil

4 Conclusions

In this paper, the non-conservational transport Eikonal equation has been solved under the framework of high-order FVM on unstructured grids to calculate the minimum wall distance distribution. The numerical examples show that the minimum wall distance of the external flow problems of complex geometric configurations on unstructured grids can be calculated accurately and effectively via the method proposed in this paper. Our study is not only meaningful to the computation of minimum wall distance distribution, but also referable for solving the non-conservational equations by high-order FVM.

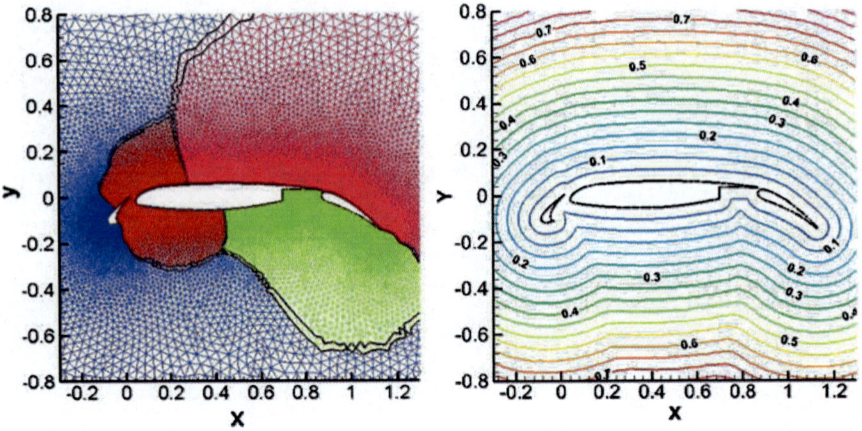

Fig. 4 Computational grid and results of 30p30n air foil

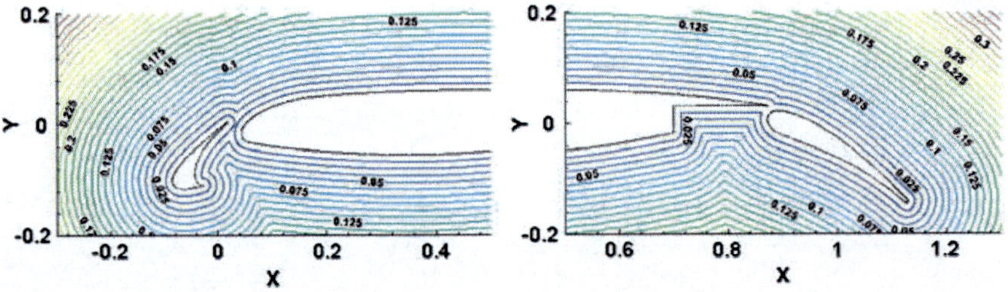

Fig. 5 Local enlargement of results of 30p30n air foil

Acknowledgements This work was supported by the National Natural Science Foundation of China (Nos. 91752114 and 11672160), and the China Postdoctoral Science Foundation under Grant Nos. 2019M660613.

References

1. Fares E, SchröDer W (2002) A differential equation for approximate wall distance. Int J Numer Methods Fluids 39(8):743–762
2. Roget B, Sitaraman J (2013) Wall distance search algorithm using voxelized marching spheres. J Comput Phys 241(5):76–94
3. Zheng Z, Gang W, Mian HH et al (2014) An efficient and fast calculation method of wall distance for hybrid-unstructured grids. In: International Bhurban conference on applied sciences & technology
4. Guo Z, He Z, Xia C et al (2017) KD tree method for efficient wall distance computation of mesh. J Nat Univ Defense Technol 39(4):21–25 (in Chinese)

5. Xia H, Tucker PG (2010) Finite volume distance field and its application to medial axis transforms. Int J Numer Meth Eng 82(1):114–134
6. Paul GTA, Chris LRB, Robert EBD et al (2012) Computations of wall distances based on differential equations. AIAA J 43(3):539–549
7. Xu JL, Yan C, Fan JJ (2011) Computations of wall distances by solving a transport equation. Appl Math Mech 32(2):141–150
8. Tucker PG (2003) Differential equation-based wall distance computation for DES and RANS. J Comput Phys 190(1):229–248
9. Nithiarasu P, Liu CB, Tucker PG (2010) Wall distance calculation using Eikonal/Hamilton-Jacobi equations on unstructured meshes—a finite element approach. Eng Comput 27(5):645–657(13)
10. Xia H, Tucker PG (2011) Fast equal and biased distance fields for medial axis transform with meshing in mind. Appl Math Model 35(12):5804–5819
11. Schoenawa S, Hartmann R (2014) Discontinuous Galerkin discretization of the Reynolds-averaged Navier-Stokes equations with the shear-stress transport model. J Comput Phys 262:194–216
12. Wang Q, Ren YX, Li W (2016) Compact high order finite volume method on unstructured grids I: basic formulations and one-dimensional schemes. J Comput Phys 314:863–882
13. Wang Q, Ren YX, Li W (2016) Compact high order finite volume method on unstructured grids I: basic formulations and one-dimensional schemes. J Comput Phys 314:883–908
14. Wang Q, Ren YX, Pan J et al (2017) Compact high order finite volume method on unstructured grids III: variational reconstruction. J Comput Phys 337:1–26
15. Pan J, Ren YX, Sun Y (2017) High order sub-cell finite volume schemes for solving hyperbolic conservation laws II: extension to two-dimensional systems on unstructured grids. J Comput Phys 338:165–198
16. Luo H, Baum JD, Löhner R (2008) A discontinuous Galerkin method based on a Taylor basis for the compressible flows on arbitrary grids. J Comput Phys 227(20):8875–8893

Numerical Study on Pulsatile Flow of Non-Newtonian Fluid Through Arterial Stenosis

Mohammad Ali, Kazi Shafi Sami, and Amanullah Kabir Tonmoy

Abstract In this research, the pulsatile blood flow through stenosed artery of 60% severity has been investigated. Flow behavior for two non-Newtonian viscosity models is compared. A standard k-ω turbulence model is used for turbulent flow at post-stenotic region. The Wall Shear Stress (WSS), hydrostatic pressure, and streamline contours at different time steps are obtained. A Fluid Structure Interface (FSI) model is used to determine the dynamic structural behavior with fluid flow. The maximum deformation, maximum von-mises stress, and peak principal stress that are obtained for FSI model. Wall shear stress and von-mises stress are found significantly higher at throat region. Peak principal stress is the highest over time at pre-stenotic region. Streamline patterns are observed for different time steps with the change in inlet velocity.

Keywords Non-Newtonian fluid · Fluid structure interface · Wall shear stress

1 Introduction

Atherosclerosis is a disease characterized by the formation of plaque. This plaque causes stenosis which is defined as a partial constriction of blood vessel due to accumulation of LDL and fats resulting in this abnormal growth of tissue. When the plaque is formed, the hemodynamic parameters are significantly changed. The change in hemodynamic parameters play a significant role in enlarging the plaque. Eventually, it leads to cardiovascular diseases. Study of blood flow in stenosed artery is important for interpreting the causes and progression of atherosclerosis. Computational fluid dynamics are used to examine the role of hemodynamics on the localization, development, and progression of arterial stenosis. The results of computational simulations obtained by Stroud et al. [1] should supplement MR and other in vivo diagnostic techniques to provide an accurate picture of the hemodynamics in particular vessels.

M. Ali (✉) · K. S. Sami · A. K. Tonmoy
Department of Mechanical Engineering, Bangladesh University of Engineering and Technology, Dhaka 1000, Bangladesh
e-mail: mali@me.buet.ac.bd

L. Venkatakrishnan et al. (eds.), *Proceedings of 16th Asian Congress of Fluid Mechanics*, Lecture Notes in Mechanical Engineering,
https://doi.org/10.1007/978-981-15-5183-3_13

Casscell [2] observed that increasing degrees of stenosis carries greater risks of atherothrombotic events. Pasterkamp and Smits [3] found arteries often dilated in an attempt to normalize elevated wall shear stresses as atherosclerosis develops, a process known as expansive remodeling. Manica and Bortoli [4] simulated incompressible non-Newtonian flow through channels with expansion using the power law model. Nakamura and Sawada [5] observed that the non-Newtonian property of blood weakens the distortion of flow pattern, pressure, and shear stress at the wall associated with the stenosis. Johnston and Kilpatrick [6] investigated the Newtonian model of blood viscosity as a good approximation in regions of mid-range to high shear, and they advised to use the generalized power law model (which tends to the Newtonian model in those shear ranges in any case) in order to achieve better approximation of wall shear stress at low shear. Gijsen and Allanic [7] found that the non-Newtonian model could not be predicted with a Newtonian model with a constant viscosity. Ballyk and Steinman [8] studied non-Newtonian blood rheology which creates a significant effect on steady flow wall shear stresses, but no significant effect on unsteady flow wall shear stresses. Li and John [9] observed restraining wall motion due to severe stenosis which results in higher blood velocities and higher peak wall shear stress, and localization of hoop stress. Inzitari et al. [10] found that forty-five percent of strokes in patients with asymptomatic stenosis of 60–90% severity are attributable to lacunar stroke. Below 60% severity, the chances of stroke are less. So, an axisymmetric 3D model artery of 60% severity was considered by Sami and Tonmoy [11]. The total length of the model is taken as 102 mm (17D) where diameter $D = 6$ mm. A parabolic velocity profile with velocity pulse is used as inlet boundary condition. At velocity pulse, Reynolds number varies from 200 to 1100 to observe the effect of fluctuating flowrate. For FSI model, one-way coupling method is used. For one-way Fluid Structure Interface (FSI) model, the artery is assumed to be isotopic linear elastic and Reynolds number varies from 240 to 310.

2 Methodology

Numerical simulation is performed by ANSYS Fluent 15.07. Figure 1 shows the shape of the stenosis in the blood vessel which is created by using cosine formula. ANSYS CAD software is used to build the geometry. Discretization of domain is performed in ANSYS mesh module. Pressure-based solver is used. Low Reynold k-ω turbulent model is used for turbulence in post-stenotic region. Simple algorithm

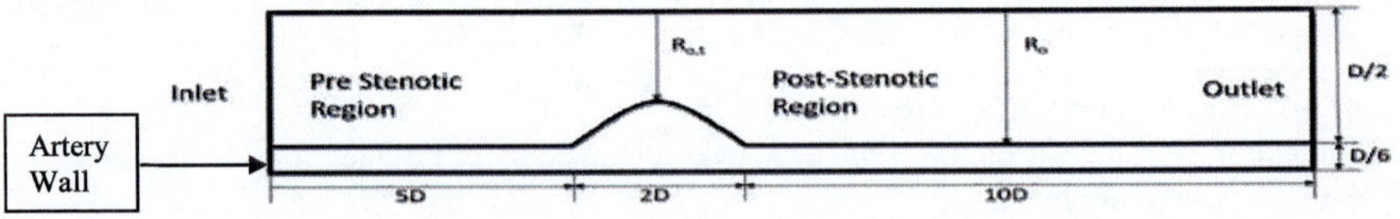

Fig. 1 Schematic of the geometry

is selected for pressure velocity coupling. First-order upwind scheme is employed as a numerical scheme for discretization of the momentum equation. In ANSYS Workbench, the FSI (one-way coupling) analysis can be performed by connecting the coupling participants to a system component called system coupling. A participant either feeds or receives data in a coupled analysis. Here, ANSYS Fluent (participant 1) and ANSYS Transient Structural (participant 2) are acting as coupling participants. The inlet boundary conditions and viscosity equations are given by C-language code using the interface of User Defined Function (UDF).

3 Governing Equations

The following governing equations are used to perform the present numerical study.

Mass conservation or continuity equation:

$$\frac{\partial \rho}{\partial t} + \nabla \cdot (\rho v) = 0 \tag{1}$$

Momentum equation:

$$\frac{\partial \rho \mathbf{v}}{\partial t} + \nabla \cdot (\rho \mathbf{v}\mathbf{v}) = \nabla \cdot \boldsymbol{\tau} + \rho \mathbf{b} \tag{2}$$

Here, b is the body force. Blood is considered to be a single phase, incompressible, homogeneous non-Newtonian fluid. Here, typical value of density is, $\rho = 1050\ \text{kg/m}^3$. Viscosity of the blood is defined as Carreau and cross model.

Cross Model:

$$\mu = \mu_\infty + (\mu_0 - \mu_\infty)\left[1 + \left(\frac{|\dot{\gamma}|}{\gamma_c}\right)^m\right]^{-1} \tag{3}$$

Here, viscosity at zero shear rate, $\mu_0 = 0.0364$ Pa s, viscosity at infinite shear rate, $\mu_\infty = 0.00345$ Pa s, reference shear rate, $\gamma_c = 2.63\ \text{s}^{-1}$ and $m = 1.45$.

Carreau model:

$$\mu = \mu_\infty + (\mu_0 - \mu_\infty)\left[1 + (\gamma_c \dot{\gamma})^2\right]^{\frac{n-1}{2}} \tag{4}$$

Here, $\mu_0 = 0.056$ Pa s, $\mu_\infty = 0.00345$ Pa s, $\gamma_c = 3.313\ \text{s}^{-1}$, and $n = 0.3568$. For FSI model, elastic modulus of the artery wall is given 9.1 MPa. Density of artery wall is $1300\ \text{kg/m}^3$. Poisson ratio of artery wall is 0.49 [12]. As the pumping action of heart is periodic, so an oscillatory blood flow is sufficient to consider. Womersley

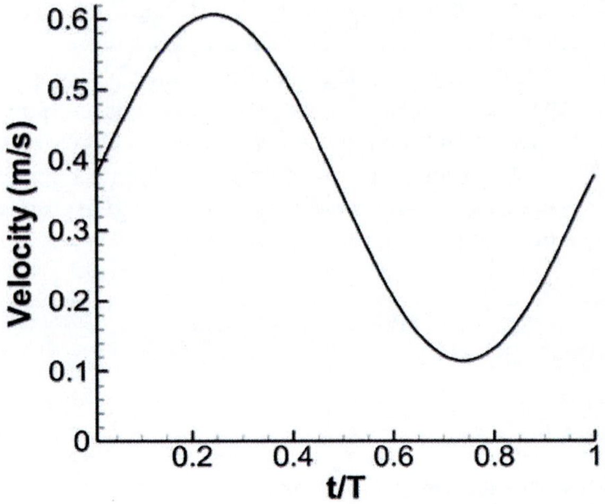

Fig. 2 Velocity distribution with non-dimensional time (t/T) used at inlet

number for this investigation is 4.58. The constant pressure of 75 mm Hg is kept at outlet. Figure 2 shows the velocity pulse at inlet.

4 Results and Discussion

The present research is devoted to compare the two different non-Newtonian models and studies the fluid structure interaction between the blood flow and the plaque in a 3D modeled arterial stenosis. The code validation has been done by comparing computational results with the experimental results obtained by Ahmed and Giddens [13]. From experiment, single stenosis with 75% severity model is considered for numerical study. Steady inlet velocity condition is given. Glycerin-water mixture is used as flowing fluid which is modeled as non-Newtonian fluid using power law for validation. Velocity profiles in the post-stenotic regions are compared for $Z = D$, $2.5D$ where Z is the normalized distance from the center of the stenosis. From the comparison in Fig. 3, it can be shown that axial velocity patterns are very much similar to the experimental pattern for $Z = 2.5D$. But due to turbulence model, the numerical result slightly overpredicts near centerline. A grid independence test has been performed to ensure that the numerical solutions are independent on the choice of the grid arrangements between 627,222 and 986,880 elements.

Wall Shear Stress (WSS) is the force on flowing blood on endothelial surface of vessel. As the shear stress for flow varies from zero at the centerline to a maximum at the wall, the wall shear stress becomes an important parameter to determine to flow characteristics. In this research, high wall shear stress is dominant in high flow velocities at throat region. Low shear stress obtained in the descending zone is due to

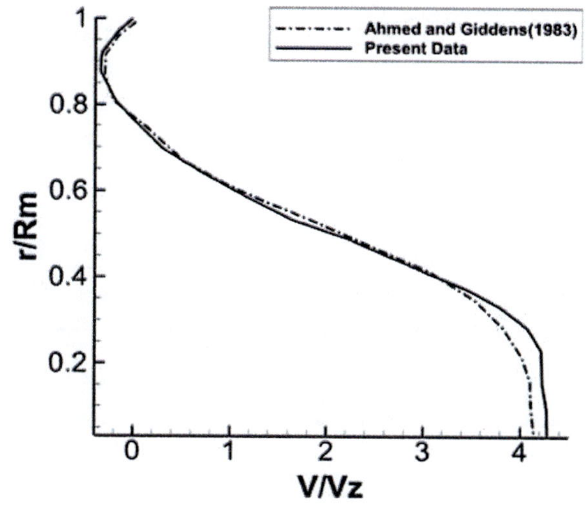

Fig. 3 Validation with the experimental result. When $Z = 2.5D$ for Re $= 500$

increase in area. For both Carreau and cross viscosity model, Figs. 4, 5, and 6 show that the maximum wall shear occurs at throat in the upper boundary. Here, WSS is higher in cross viscosity model. For both models, the WSS is uniformly distributed at pre-stenotic and post-stenotic region for this geometry.

As the wall shear stress is directly proportional to the pressure drop, hydrostatic pressure is an important parameter to show the effect of stenosis. Figures 7, 8, and 9 show the fluctuation of pressure occurred in different time steps due to the effect of pulsatile velocity at inlet. Due to blockage, large pressure drop is observed at throat. The streamline pattern is changed due to the presence of stenosis. As the flow is sinusoidal and time dependent, significant flow variation on the post-stenotic region

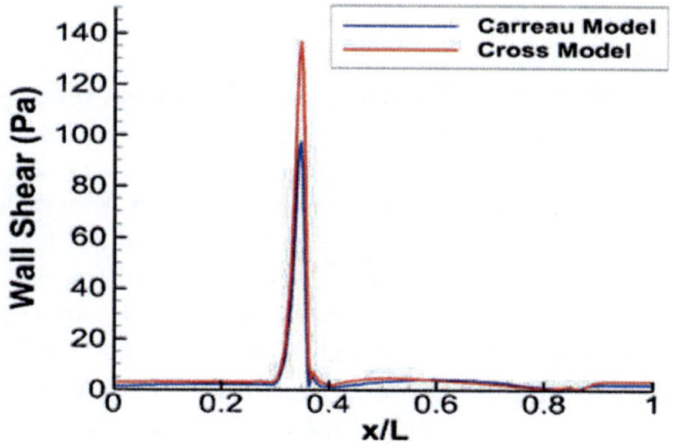

Fig. 4 Wall shear stress distribution at upper wall at $t/T = 0.125$

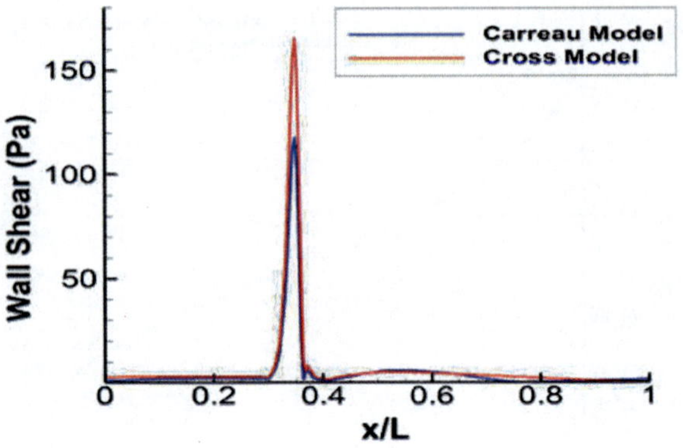

Fig. 5 Wall shear stress distribution at upper wall at $t/T = 0.25$

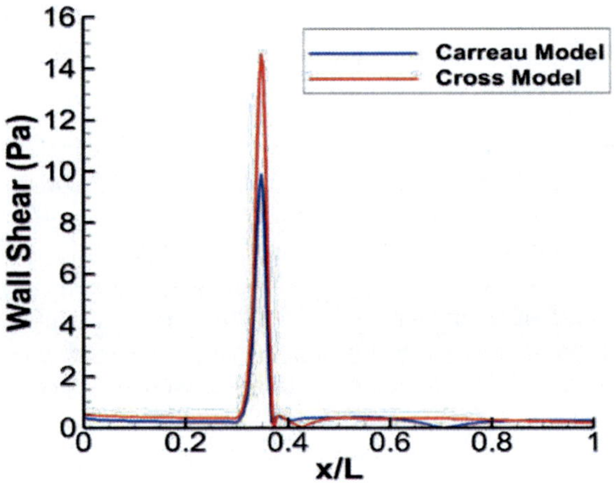

Fig. 6 Wall shear stress distribution at upper wall at $t/T = 0.75$

are observed. Variations of streamlines contours have been investigated to study the flow pattern of blood through the stenotic artery. Figure 10 shows streamline contours for Carreau and cross model vortex at three different time steps. At $t/T = 0.125$, two large vortices length of 0.038 and 0.0465 m from the throat of 60% severities presented in Fig. 10 (a, d). At $t/T = 0.25$, two separated vortex, regions are created near the upper wall in Carreau model for 60% severity presented in Fig. 10b. At Fig. 10e, there is a large vortex region of length 0.0498 and 0.0557 m created at upper and lower wall, respectively, in cross model. At $t/T = 0.75$, the massive changes occur. At Fig. 10 (e, f), there are two vortex rings created at pre-stenotic region near the wall, and combination of large and small vortex rings are observed for Carreau model, and a large vortex ring near the upper wall changes the direction

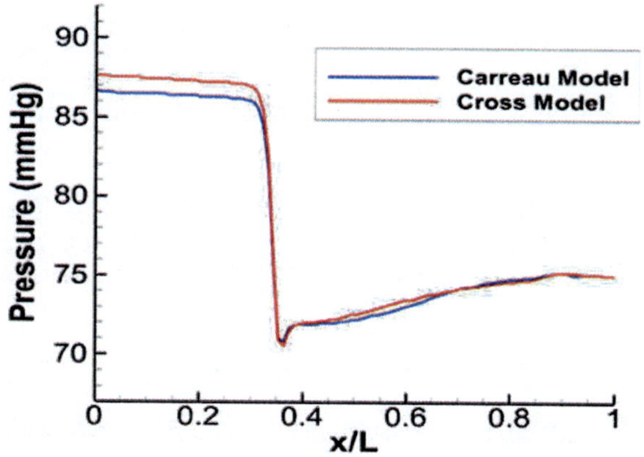

Fig. 7 Centerline pressure distribution at $t/T = 0.125$

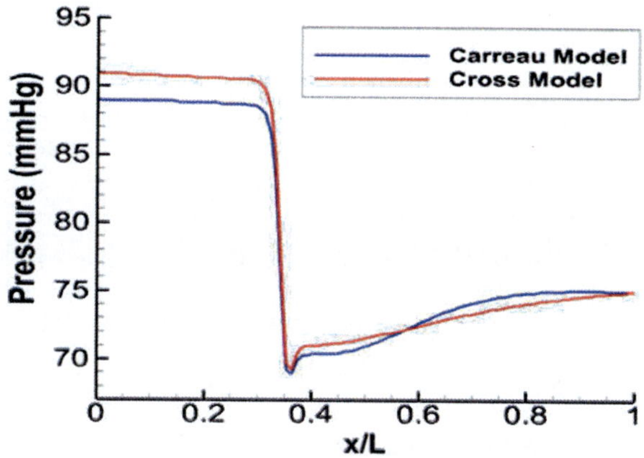

Fig. 8 Centerline pressure distribution at $t/T = 0.25$

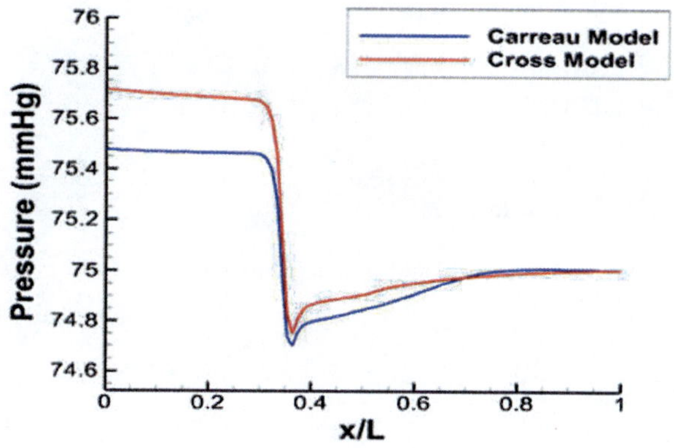

Fig. 9 Centerline pressure distribution at $t/T = 0.75$

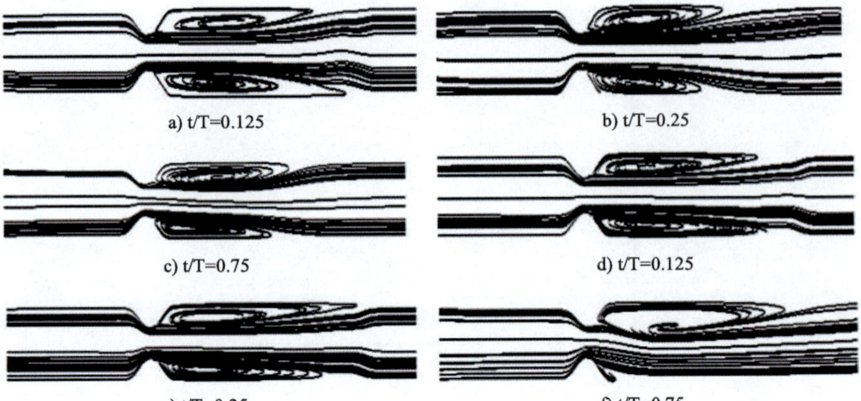

Fig. 10 Streamline contours at different time steps for Carreau model (**a**–**c**) and cross model (**d**–**f**)

of the flow a little for cross model which is unusual. From this observation, Carreau model is more suitable for the fluid structure interface analysis.

Principal stress is one of the most significant parameters in determining the wall rupture. From Fig. 11, the peak principal stress basically occurs when the velocity and pressure gradient are maximum at pre-stenotic region. Figure 13 shows that the total deformation is maximum there at pre-stenotic region because of the high normal stress. Von-mises stress is another vital parameter in determining the risk of rupture. Von-mises stress is the equivalent of uni-axial tensile stress. Figure 12 shows that the von-mises stress is very high at throat region.

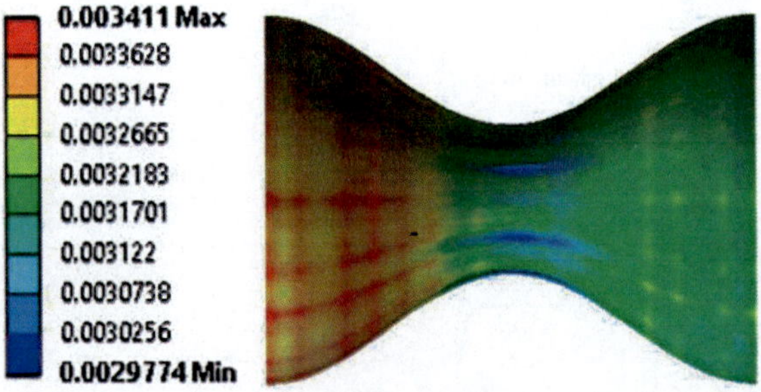

Fig. 11 Maximum principal stress (MPa) over time Figure missing

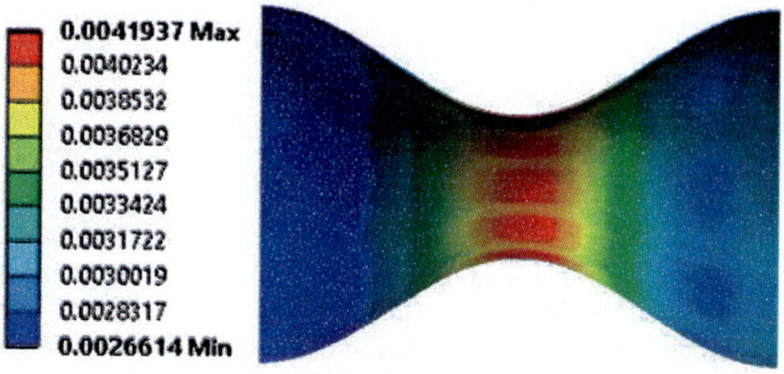

Fig. 12 Maximum von-mises stress (MPa) over time Figure missing

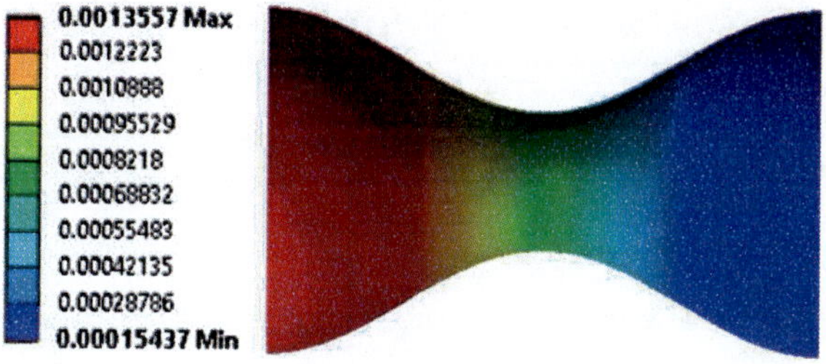

Fig. 13 Total deformation along x-axis (mm) Figure missing

5 Conclusion

A numerical investigation is performed here to analyze the effects of stenosis on flow parameters. For this analysis, the high shear stress at throat will cause the possibility to damage the intimal tissue. At throat, there is higher chance of damage of endothelial wall due to shearing action. Low shear stress at post-stenotic region due to recirculation enhances the possibility of plaque formation in post-stenotic region. Vortex length changes as the Reynolds number changes with the velocity pulse at inlet. Centerline pressure distribution is quite similar for Carreau and cross model. In FSI model, deformation is very small. Both peak principal stress and deformation occur at pre-stenotic region due to high hydrostatic pressure gradient and constriction. Better result will be found considering hyper-elastic artery wall by using Mooney–Rivlin Model.

Acknowledgements This work has been carried out in the department of Mechanical Engineering, Bangladesh University of Engineering and Technology (BUET). The authors gratefully acknowledge the support and facilities provided by BUET.

References

1. Stroud JS, Berger SA, Saloner D (2000) Influence of stenosis morphology on flow through severely stenotic vessels: implications for plaque rupture. J Biomech 33(4):443–455
2. Casscells W, Naghavi M, Willerson JT (2003) Vulnerable atherosclerotic plaque: a multifocal disease. Circulation 107(16):2072–2075
3. Pasterkamp G, Smits PC (2002) Imaging of atherosclerosis. Remodelling of coronary arteries. J Cardiovasc Risk 9(5):229–235
4. Manica R, De BAL (2011) Simulation of incompressible Non-Newtonian flows through channels with sudden expansion using the power-law model. TEMA-Tendencias Matematica Apl Comput 4(3):333–340
5. Nakamura M, Sawada T (1988) Numerical study on the flow of a non-Newtonian fluid through an axisymmetric stenosis. J Biomech 11(2):28–31
6. Johnston PR, Corney S, Kilpatrick D (2004) Non-Newtonian blood flow in human right coronary arteries: steady state simulations. J Biomech 37(5):709–720
7. Gijsen FJH, Allanic E, Van De Vosse FN, Janssen JD (1999) The influence of the non-Newtonian properties of blood on the flow in large arteries: unsteady flow in a 90° curved tube. J Biomech 32(7):705–713
8. Ballyk PD, Steinman DA, Ethier CR (1994) Simulation of non-Newtonian blood flow in an end-to-side anastomosis. Biorheology 31(5):565–586
9. Li MX, Beech-Brandt JJ, John LR, Hoskins PR, Easson WJ (2007) Numerical analysis of pulsatile blood flow and vessel wall mechanics in different degrees of stenoses. J Biomech 40(16):3715–3724
10. Inzitari D, Eliasziw M, Gates P, Sharpe BL, Chan RK, Meldrum HE, Barnett HJ (2000) The causes and risk of stroke in patients with asymptomatic internal-carotid-artery stenosis. North Am Symptomatic Carotid Endarterectomy Trial Collaborators. J Med 342(23):1693–700
11. Sami KS, Tonmoy AK, Characteristics of Non-Newtonian flow through arterial stenosis. B.Sc. Engineering, thesis, Department of Mechanical Engineering, BUET, Bangladesh
12. Jahangiri M, Saghafian M, Sadeghi MR (2015) Numerical study of turbulent pulsatile blood flow through stenosed artery using fluid-solid interaction. Comput Math Methods Med ID 515613, 1–10
13. Ahmed SA, Giddens DP (1983) Flow disturbance measurements through a constricted tube at moderate Reynolds numbers. J Biomech 16(12):955–963

Numerical Simulation of Contaminant Release During Sediment Starting in Dynamic Water Environment

Peng-da Cheng, Xin-guang Zhu, Chun Feng, and Xiao-liang Wang

Abstract In environmental hydrodynamics, the release of contaminant from sediments is one of the main problems. Based on a large number of experimental data provided by water channel experiments, a coupled mechanical model of overlying water, sediment and contaminant is established in this paper. The process of sediment starting and contaminant release is numerically simulated under different velocity conditions of overlying water. The quantitative relationships among velocity, particle volume fraction, contaminant concentration, turbulent kinetic energy and time are obtained by analyzing the relationship between flow field characteristics and contaminant concentration distribution. The results show that contaminant is released rapidly with the suspension of sediment particles and quickly reaches equilibrium concentrations. When the flow field characteristics (Re) change, the contribution of convection and turbulent diffusion to contaminant release process is different. Establishing a quantitative relationship between hydrodynamic conditions and contaminant release can provide support for constructing water pollution model in lake and reservoir areas.

Keywords Sediment contaminant · Velocity · Turbulence kinetic energy · Particle volume fraction · Concentration

P.da Cheng · X. Zhu · C. Feng
Institute of Mechanics, Chinese Academy of Sciences, Beijing 100190, China
e-mail: pdcheng@imech.ac.cn

X. Zhu
School of Engineering Science, University of Chinese Academy of Sciences, Beijing 100049, China

X. Wang (✉)
Beijing Institute of Technology, Beijing 100081, China
e-mail: wangxiaoliang36@bit.edu.cn

L. Venkatakrishnan et al. (eds.), *Proceedings of 16th Asian Congress of Fluid Mechanics*, Lecture Notes in Mechanical Engineering,
https://doi.org/10.1007/978-981-15-5183-3_14

1 Introduction

The water is considered to be a natural complex composed of water, dissolved matter, suspended matter, aquatic organisms and sediment [1]. When contaminants enter the water, they will deposit in the sediment, gradually enrich and make the sediment becoming a reservoir of contaminants. The hydraulic conditions are more complicated in the estuary area. The endogenous release effect of the contaminated sediment is more and more obvious. The movement of the sediment with the water flow is very complicated and has great randomness, and the contaminants combined with sediment migrate along with the movement of sediment in water body. On the one hand, the movement and transport of suspended sediment in the water directly affects the temporal and spatial distribution of contaminants; on the other hand, sediment becomes an important potential "source" or "sink" of contaminants as sediment deposits into the bed, and the mechanism of "source" and "sink" will change under certain hydrodynamic disturbances and environmental conditions [2–4]. Hydrodynamics is an important physical factor affecting the diffusion and migration of sediments at the sediment–water interface. On the one hand, dynamic flow enhances the diffusion and mixing ability near the sediment–water interface. On the other hand, the shear stress and turbulence intensity of the sediment–water interface flow will cause the re-suspension of polluted sediment, which will cause a large release of contaminants to the water body. Previous studies on contaminants release from sediments have focused on relatively static waters such as lakes, and systematic and in-depth analysis have been conducted. However, there are few studies on the release law of sediment contaminants in hydrodynamic water, and most of them are qualitative research. The lack of quantitative analysis can easily lead to the exaggeration or underestimation of the re-suspension release of sediment contaminants. In order to understand the physical process and influencing factors of re-suspension release of contaminant, a coupled mechanical model considering overlying water–sediment–contaminant is established. The relationship between velocity, turbulent kinetic energy and vertical distribution of sediment or contaminant under different hydrodynamic conditions is studied. The relationship between contaminant releasing flux and flow characteristics (Reynolds number) is obtained, and the effects of convection and turbulent diffusion on contaminant release are obtained.

2 Problem Formulation and Solution Methods

2.1 Governing Equations

In this paper, suspended sediment is considered a suspension in the overlying water–sediment–contaminant model. Based on analysis of many experimental data [5, 6], the viscosity of this kind of suspension can be expressed as a function of the particle volume fraction. Therefore, it is considered that suspension (fluid-particle) has a single flow continuum with macroscopic properties (such as density and viscosity). The hypothesis of the model is as follows: (1) The density of each phase is approximately constant. (2) Both phases share the same pressure field. (3) The particle relaxation time is short compared to the time scales of the macroscopic flow.

The mixture of density and viscosity are respectively given by,

$$\rho = (1 - \varphi)\rho_1 + \varphi\rho_2 \tag{1}$$

$$\mu = (1 - \varphi)\mu_1 + \varphi\mu_2 \tag{2}$$

where $\rho_{1,2}$ and $\mu_{1,2}$ are the density and viscosity properties of fluid and particle respectively, φ is the particle volume fraction.

Considering that the suspension can be treated as a continuous medium, the flow is governed by incompressible mass and momentum conservation equations, which require that:

$$\nabla(\rho u) = 0 \tag{3}$$

$$\rho\left(\frac{\partial u}{\partial t} + u \cdot \nabla u\right) = -\nabla p + \nabla \cdot \tau + \rho g \tag{4}$$

Here, u is the velocity field, ρ is the suspension density, t is the time, p is the pressure field, τ is the sum of the viscous and turbulent stresses, and g is a body force per unit mass. The τ is written as for a generalized Newtonian fluid, such that $\tau = \mu\gamma$, where μ is the suspension viscosity and $\gamma = \nabla u + \nabla u^T$. In order to close the governing equations, the k-ε two-equation formulations are adopted, $\sigma_k = 1.0$, $\sigma_\varepsilon = 1.3$, $C_{\varepsilon 1} = 1.44$, and $C_{\varepsilon 2} = 1.92$.

The transport equation for the particles in the flow is ruled by,

$$\frac{\partial \varphi}{\partial t} + u \cdot \nabla \varphi = -\nabla \cdot N_\varphi \tag{5}$$

where N_φ is the total diffusive flux of particles.

The transport equation for the contaminant concentration in the flow is ruled by,

$$\frac{\partial c}{\partial t} + u \cdot \nabla c = -\nabla \cdot N + R \tag{6}$$

where c is the relative concentration and is a dimensionless value, N is the total diffusive flux of concentration, R is a source and sink expression, and u is the velocity field.

The viscosity of a particle suspension is usually written as a function of the local particle volume fraction [6, 7, 9–11]. In this paper, the MPQ model [8, 9] is adopted for suspension.

$$\mu = \mu_1 \left(1 - \frac{\varphi}{\varphi_m}\right) \tag{7}$$

where φ_m is the maximum packing concentration, which for solid particles is approximately 0.62.

2.2 Geometry, Materials, Boundary Conditions and Initial Conditions

Now, we consider the case of the particle suspensions flow through a water channel under pressure-driven. In our previous experiments, the water channel was 0.25 m wide and 3.0 m long, the water depth was set at 0.1 m, the sediment depth was set at 0.08 m. The geometric model used in the numerical simulation is consistent with the experimental size as Fig. 1, and set the left side of water–sediment interface as coordinate origin. In the numerical simulation, the densities of overlying water and sediment are 997 kg/m^3 and 2650 kg/m^3, respectively. The viscosity of the overlying water is set to 1×10^{-3} Pa s. The sediment particle size (D_{50}) is 0.03 mm and the water content is 57.5%.

It is also needed to specify boundary condition; we assume that there is no slip at the water channel wall and zero shear stress at the top surface for flow. For a contaminant (NaCl) concentration calculation, the water channel and the top surface are set to no-flux boundary conditions. We use periodic boundary conditions to the left and right boundaries, i.e., equal velocity, equal concentration, equal volume fraction, equal turbulence kinetic energy and equal turbulence dissipation rate.

For the initial conditions, we set the initial pressure to be related to gravity, the initial velocity is set to 0, the initial particle volume fraction is set to 42.5%, the initial relative concentration of the overlying water is set to 0, and the initial relative concentration of the sediment is set to 1.

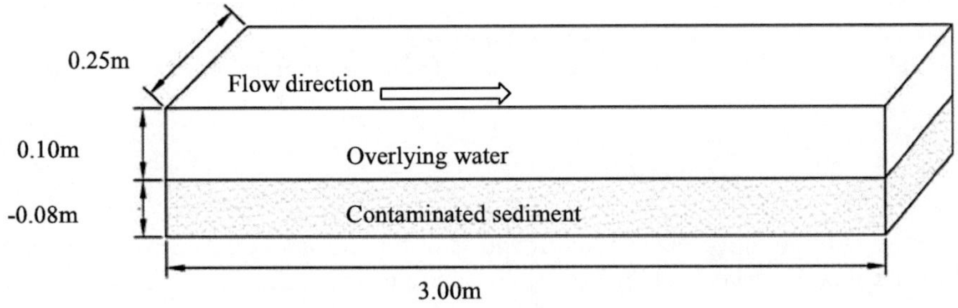

Fig. 1 The geometry of water channel for pressure-driven flow

3 Results

In recent years, the finite-element method (FEM) has developed rapidly and is often used to solve the problem of partial differential equations (PDE) with specific boundary conditions in solid mechanics and fluid mechanics. In this paper, the standard Galerkin finite-element discretization method is used to solve the flow field and contaminant concentration field [12]. In particular, the momentum and turbulence transport equations are dominated by convection and it is well known that a standard Galerkin discretization of such equations leads to oscillations in the solutions. Hence, some form of upwinding is required to suppress these non-physical oscillations. Here, we use a Galerkin least-squares method as described by Hughes et al. [13, 14]. The current study uses a structured grid system of 82,800 cells, which is chosen after a series of grid independence tests.

Under different hydrodynamic conditions, the movement of sediment with water flow is very complicated and has great randomness, and the contaminant will migrate with the movement of sediment in the water. For quantitative analysis, we introduce the concept of average velocity (U), average volume fraction (Φ), relative average concentration (C) and average turbulence kinetic energy (K) as follows:

$$C = \frac{\int_0^\infty ucds}{UH}; \quad \Phi = \frac{\int_0^\infty u\varphi ds}{UH}; \quad U = \frac{\int_0^\infty u^2 ds}{\int_0^\infty udz}; \quad K = \frac{\int_0^\infty ukds}{UH}; \tag{8}$$

where C is the relative average concentration and is a dimensionless value, U is the average speed (unit: m/s), K is the average turbulence kinetic energy (m^2/s^2), Φ is the average volume fraction, u is the flow field velocity (unit: m/s), z is the vertical height from water–sediment interface (unit: m), k is the turbulence kinetic energy (m^2/s^2).

The vertical surface of the center of the overlying water in the geometric model is selected as the research object. The average velocity of the overlying water during the sediment starting is shown in Fig. 2. The average velocities of Case1-6 are 0.03 m/s, 0.08 m/s, 0.13 m/s, 0.23 m/s, 0.35 m/s and 0.50 m/s, respectively. In the initial stage of sediment starting, the average velocity of different cases is very small. With the increase in time, the average velocity of water increases rapidly and reaches a stable speed in a short time.

The vertical distribution of particle volume fraction of sediment at different time is analyzed by using Case 3 and Case 4 data, as shown in Fig. 3. At the initial stage of sediment starting, the distribution of particle volume fraction is similar under different velocity conditions. With the increase in time, the particle volume fraction in overlying water increases rapidly and reaches the same stable value in 2 min, which is consistent with the experimental phenomena. The vertical distribution of sediment particles varies in physical process due to their different properties. For fine sediment ($D_{50} = 0.03$ mm), the spatial structure of sediment is uniform. The balance between upward suspension and downward deposition of sediment leads

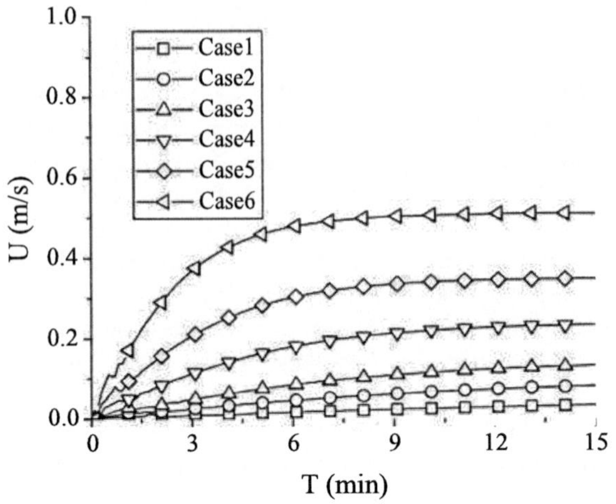

Fig. 2 Velocity changes with time during sediment starting

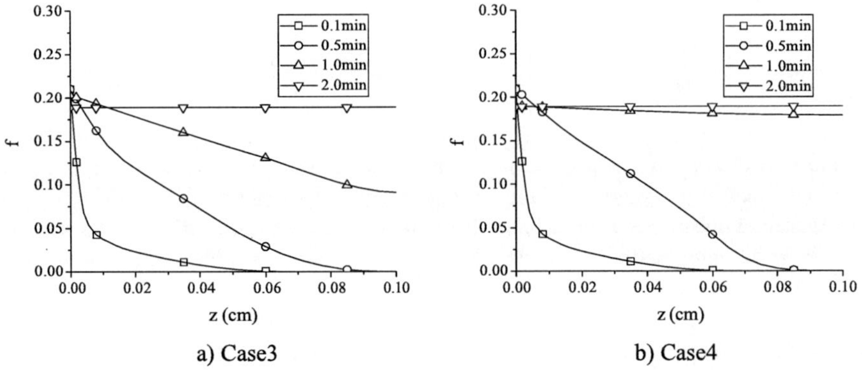

Fig. 3 Vertical distribution of particle volume fraction during sediment starting at different times

to the distribution equilibrium of particle volume fraction. When the hydrodynamic conditions are constant, the equilibrium will not be broken. Considering that the deposition velocity of fine sediment is a function of volume fraction, the deposition velocity is smaller than that of coarse sediment, so the distribution of particle volume fraction along water depth is more uniform.

The sediment particles rapidly entering the overlying water have an impact on the flow characteristics of the overlying water and affect the release of contaminant. Data of Case 1, Case 3, Case 4 and Case 5 are selected to analyze the relationship between average turbulent kinetic energy, relative average concentration of contaminant and time at different velocity. Because of the small deposition velocity of fine sediment, the volume fraction of fine sediment quickly reaches equilibrium and remains stable.

Contaminant in pore water also enters the overlying water rapidly and reachws the same equilibrium concentration soon. It is noteworthy that the average turbulent kinetic energy reaches its peak value after the starting of fine sediment, then decreases rapidly and gradually reaches a stable level over time. As the flow velocity increases, the turbulent kinetic energy increases, and the time for the contaminant to reach equilibrium decreases in the overlying water. During the process of contaminant equilibrium in the overlying water, convection and turbulent diffusion have made their respective contributions (Fig. 4).

When the adsorption and desorption of fine sediments are not considered, convective and turbulent diffusion have different effects on the process of contaminant balance in the overlying water. Analysis of the overlying water contaminant to reach the equilibrium, the relationship between the flux (total flux N_T, convective flux N_C, turbulent diffusion flux N_D) and water flow characteristics (Reynolds number) is shown in Fig. 5. For non-adsorbed media, the total flux of the contaminant increases linearly with the velocity, that is, with the Reynolds number increasing linearly. When the Reynolds number is small (0 < Re < 35,000), the contribution of convection and turbulent diffusion is basically the same during the diffusion of contaminant. When the Reynolds number is large (Re > 35,000), the contribution of turbulent diffusion decreases rapidly, and the contaminant diffusion is mainly dominated by convection.

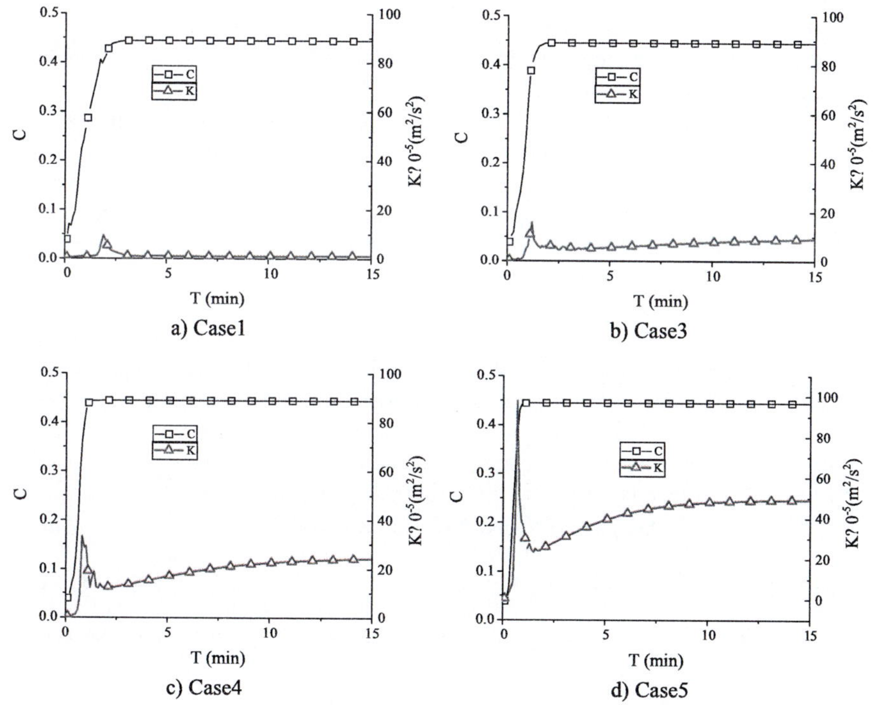

Fig. 4 The curve of average concentration, turbulent kinetic energy and time

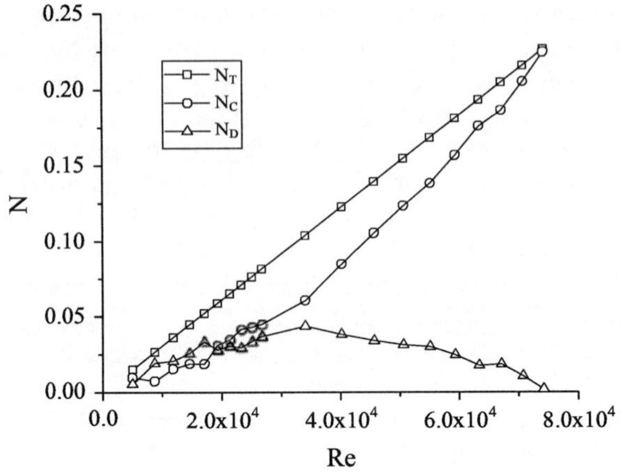

Fig. 5 Curve of contaminant release flux and Reynolds number

4 Conclusions

The re-suspension release process of sediment contaminant is a coupling process consisting of overlying water, sediment and contaminant. The complex flow field characteristics after sediment starting are the main influencing factors for the release of contaminant from sediment re-suspension. The fine sediment is susceptible to the effect of water flow. When re-suspension, the sediment particles volume fraction in overlying water reach equilibrium in short time (<2 min) and remain stable, because of uniform spatial structure and small deposition speed. The sediment particles rapidly entering the overlying water have an impact on the flow characteristics of the overlying water and affect the release of contaminant. The average turbulent kinetic energy reaches its peak value after the starting of fine sediment, then decreases rapidly and gradually reaches a stable level over time. As the flow velocity increases, the turbulent kinetic energy increases, and the time for the contaminant to reach equilibrium decreases in the overlying water. Convection and turbulent diffusion have different effects on the contaminant equilibrium process in the overlying water. When the Reynolds number is small ($0 < Re < 35{,}000$), the contribution of convection and turbulent diffusion is basically the same during the diffusion of contaminant. When the Reynolds number is large ($Re > 35{,}000$), the turbulent diffusion contribution decreases rapidly, and the contaminant diffusion is mainly dominated by convection.

Acknowledgements The authors are grateful to the financial support by National Natural Science Foundation of China (No. 11602278, 11432015, 11872117 and 11802313), National Key R&D Program of China (2018YFC1505504), and “Beijing Institute of Technology Research Fund Program for Young Scholars.”

References

1. Zhang C, Yu ZG, Zeng GM et al (2014) Effects of sediment geochemical properties on heavy metal bioavailability. Environ Int 73(4):270–281
2. Cheng P, Zhu H, Fan J, Fei M, Wang D (2013) Numerical research for contaminant release from un-suspended bottom sediment under different hydrodynamic conditions. J Hydrodynam 25(4):620–627
3. Cheng P, Zhu H, Zhong B, Fei M, Wang D (2014) Sediment rarefaction re-suspension and contaminant release under tidal currents. J Hydrodynam 26(5):827–834
4. Zhu HW, Cheng PD, Li W et al (2017) Empirical model for estimating vertical concentration profiles of re-suspended, sediment-associated contaminants. Acta Mech Sin 33(5):846–854
5. Stickel JJ, Powell RL (2005) Fluid mechanics and rheology of dense suspensions. Annu Rev Fluid Mech 37(1):129–149
6. Hinch EJ (2011) The measurement of suspension rheology. J Fluid Mech 686:1–4
7. Krieger IM, Dougherty TJ (1959) A mechanism for non-Newtonian flow in suspensions of rigid spheres. Trans Soc Rheol 3:137–152
8. Maron SH, Pierce PE (1956) Application of ree-eyring generalized flow theory to suspensions of spherical particles. J Colloid Sci 11(1):80–95
9. Quemada D (1977) Rheology of concentrated disperse systems and minimum energy dissipation principle. Rheol Acta 16(1):82–94
10. Mendoza CI, Santamari A-Holek I (2009) The rheology of hard sphere suspensions at arbitrary volume fractions: an improved differential viscosity model. J Chem Phys 130(4):044904
11. Shewan HM, Stokes JR (2015) Analytically predicting the viscosity of hard sphere suspensions from the particle size distribution. J Nonnewton Fluid Mech 222:72–81
12. Ignat L, Pelletier D, Ilinca F (2000) A universal formulation of two-equation models for adaptive computation of turbulent flows. Comput Methods Appl Mech Eng 189(4):1119–1139
13. Franca LP, Frey SL (1992) Stabilized finite element methods: II. The incompressible Navier-Stokes equations. Comput Methods Appl Mech Eng 99(2–3):209–233
14. Hughes TJR, Franca LP, Hulbert GM (1989) A new finite element formulation for computational fluid dynamics: VIII. The galerkin/least-squares method for advective-diffusive equations. Comput Methods Appl Mech Eng 73(2):173–189

Thermal Behavior of a Completely Voided Coolant Channel for Indian PHWR Under Slumped Fuel Pin Condition: Experimental and Numerical Approach

Ketan Ajay, Ravi Kumar, Akhilesh Gupta, and Deb Mukhopadhyay

Abstract Loss of coolant accident (LOCA) along with the unavailability of emergency core coolant system (ECCS) may cause severe damage in the channel like slumping of the fuel pins inside the pressure tube (PT). The thermal behavior of the coolant channel under this condition is a major concern for the safety of the nuclear reactor. To understand this phenomenon, an experimental and numerical study of an Indian pressurized heavy water reactor (PHWR) housing disassembled fuel pins has been performed. The fuel bundle of 37 pins configuration was used. The present investigation is aimed to capture the steady-state temperature profile over a clad tube, pressure tube and calandria tube under completely voided condition. ANSYS Fluent 19.0 was used to carry out the numerical analysis. Close agreement was obtained between experimental and numerical results. Results showed that a significant temperature gradient was developed in the PT and the fuel pins of the outer ring that were in contact with the PT. It was also found that the moderator acts as a good heat sink as it removes 89% decay heat.

Keywords PHWR · LOCA · Slumped fuel pins PT · CT · Moderator

1 Introduction

The Indian PHWR has a horizontal fuel channel and each channel has an assembly of short fuel bundles enveloped by pressure tube (PT). Each fuel pin consists of pellets of natural uranium dioxide, which are placed inside the clad tube made of zircaloy-4. The PT is surrounded by a coaxial calandria tube (CT). The PT and CT are made from annealed zircaloy-4 and zirconium 2.5% niobium, respectively. The pressurized heavy water coolant flows inside the PT. The coolant system removes

K. Ajay (✉) · R. Kumar · A. Gupta
Department of Mechanical and Industrial Engineering, Indian Institute of Technology Roorkee, Roorkee, India
e-mail: kajay@me.iitr.ac.in

D. Mukhopadhyay
Reactor Safety Division, Bhabha Atomic Research Centre, Trombay, Mumbai, India

L. Venkatakrishnan et al. (eds.), *Proceedings of 16th Asian Congress of Fluid Mechanics*, Lecture Notes in Mechanical Engineering,
https://doi.org/10.1007/978-981-15-5183-3_15

the nuclear fission heat and transfers it to the secondary side of the steam generator for the generation of useful electrical power. The annular gap between the PT-CT is filled with inert CO_2 gas. The CT is submerged in the pool of heavy water called moderator maintained at a temperature of 65 °C. In the case of LOCA, along with the unavailability of ECCS, the channel experiences a significant heat up as coolant flow through the channel reduces [1]. The temperature of the channel continues to rise due to the generated decay heat from the fuel pins. The channel is subjected to a variety of deformations, as the accidental conditions (temperature, pressure, liquid stratification and moderator environment) may be different [2]. The PT can undergo deformation like ballooning sagging or both, which results in physical contact with CT. Due to which there is a sharp increase in the heat fluxes at the outer surface of CT [3, 4]. Under more extreme conditions, the mechanical integrity of the fuel bundle is breached, which results in the slumping of fuel pins over the bottom surface of PT [5]. There is an increase in the heat transfer from fuel pins to PT due to the presence of high contact conductance. Understanding of the heat transfer characteristics of the channel under this condition is essential from a reactor safety point of view. To facilitate this, an experimental and numerical study was performed on horizontal fuel channel housing disassembled fuel pins under inert environment condition. The fuel bundle of 37 pins configuration was used for the analysis.

2 Experimental Setup and Procedure

Experimental setup to study the thermal behavior of a horizontal channel housing slumped fuel pins is shown in Fig. 1. The test section consists of PT, CT and slumped fuel bundle of 37 pins configuration. The cross-sectional view of the test section is

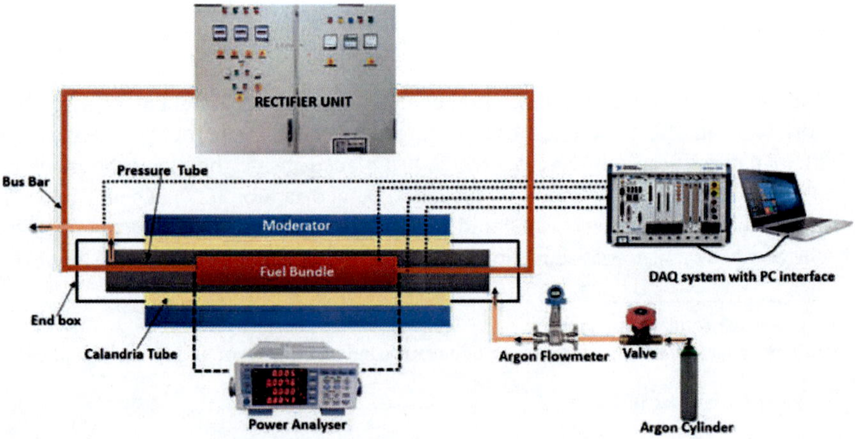

Fig. 1 Schematic of an experimental setup

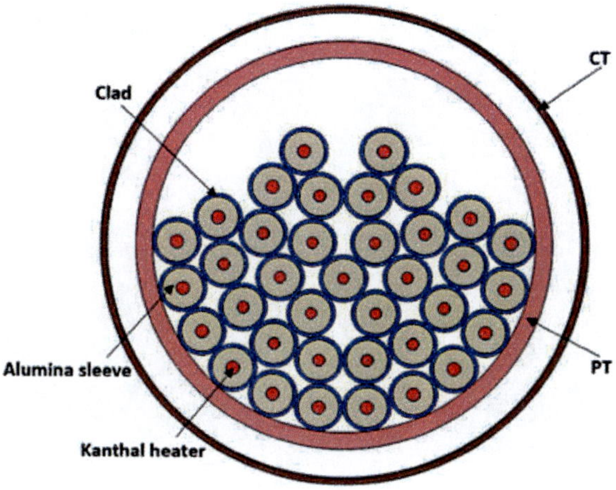

Fig. 2 Cross-sectional view of the test section

Table 1 Specifications of the test section

Component	Material	Dimension
Fuel pin	Zircaloy-4	ID = 11.20 mm; OD = 13.04 mm; $L = 1000$ mm
Pressure tube	Zirconium 2.5% niobium	ID = 103.80 mm; OD = 112.40 mm; $L = 1600$ mm
Calandria tube	Zircaloy-2	ID = 128.30 mm; OD = 132.10 mm; $L = 1450$ mm
Tank	Stainless steel	$L = 1400$ mm; $W = 500$ mm; $H = 750$ mm

shown in Fig. 2. The specifications of the test section are mentioned in Table 1. The CT of length 1450 mm was fixed horizontally to the tank wall with the help of flanges and silicon rubber packing. The tank was filled with water so that CT was submerged in it, which simulated the moderator condition. The fuel bundle in a disassembled state having a 1000 mm length was placed inside the PT. The PT of 1600 mm was further placed inside the CT in such a way that it was extended by 75 mm in length from both ends of the CT. The annular space between the PT-CT was filled with argon gas using end boxes attached at both ends of the tank. This gas-filled gap was adjusted through the lead screws provided at each of the end boxes. The PT was also filled with argon gas. The structure of the 37-pin fuel bundle has four rings of fuel pins under regular operation: one fuel pin in the first ring (1), six fuel pins in the second ring (2–7), twelve fuel pins in the third ring (8–19) and eighteen fuel pins in the fourth ring (20–37). The arrangement of disassembled fuel bundle simulator used in this study is shown in Fig. 3.

Each fuel pin consisted of Kanthal heater placed inside the alumina sleeve, which was covered with the clad tube. The power ratio of 0.715: 0.748: 0.843: 1 was maintained among fuel pins of ring-1, 2, 3 and 4 by adjusting the different diameters of Kanthal heater. The Kanthal heaters of the fuel bundle simulator were fixed in

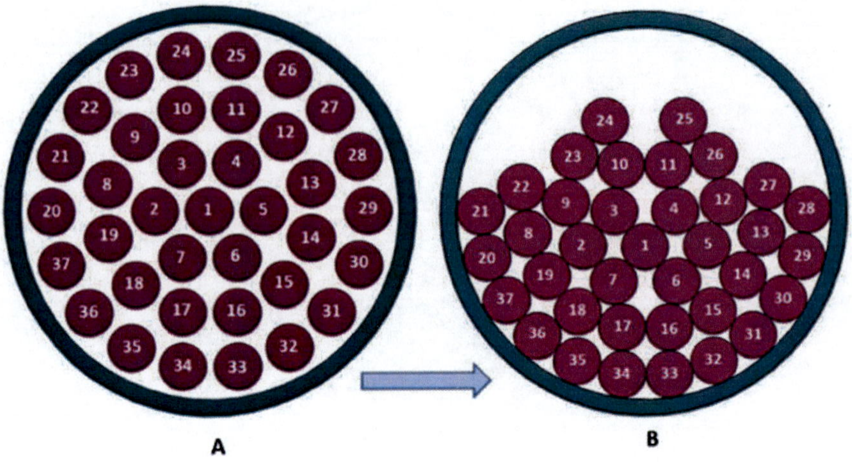

Fig. 3 Arrangement of 37 fuel pins: **a** assembled fuel pins, **b** disassembled fuel pins

a parallel arrangement with the help of current distributor headers. The decay heat generated by the fuel was simulated by passing direct current (DC) from thyristor controlled rectifier (7000 A/70 VDC) to the fuel bundle simulator.

In the experiment, initially, the tank was filled with water in such a way that CT was submerged in it. The water bath was heated to a temperature of 65 °C by using six immersion heater rods each of 1.5 kW. The argon gas was then filled inside the PT to create an inert environment. The annular gap between PT-CT was also filled with argon gas. The Joule heating of the fuel pins was then carried out using DC power. The experiment was carried until the center fuel pin attain a steady-state temperature of 550 °C.

The axial and circumferential position of the thermocouple used in the test section is shown in Fig. 4. The K-type thermocouple of 0.5 mm diameter was used to measure the temperature of fuel pins, PT and copper bus bar. While J-type thermocouple having a diameter of 1.0 mm diameter was used to measure the temperature of CT and moderator. All thermocouples were ungrounded, minerally insulated and covered with Inconel sheath. The temperature measurement uncertainty for K-type thermocouple was ±1.4 °C at 850 °C and the for J-type thermocouple was ±1.5 °C at 500 °C. The power analyzer was used to measure the power given to the test section.

3 Numerical Methodology

The 3D numerical analysis of channel housing disassembled fuel pins under steady-state condition was performed using ANSYS Fluent 19.0. The geometrical model of the computational domain is shown in Fig. 2. The computational domain consisted of PT, CT and fuel pins assembly having alumina sleeve and Kanthal heater as

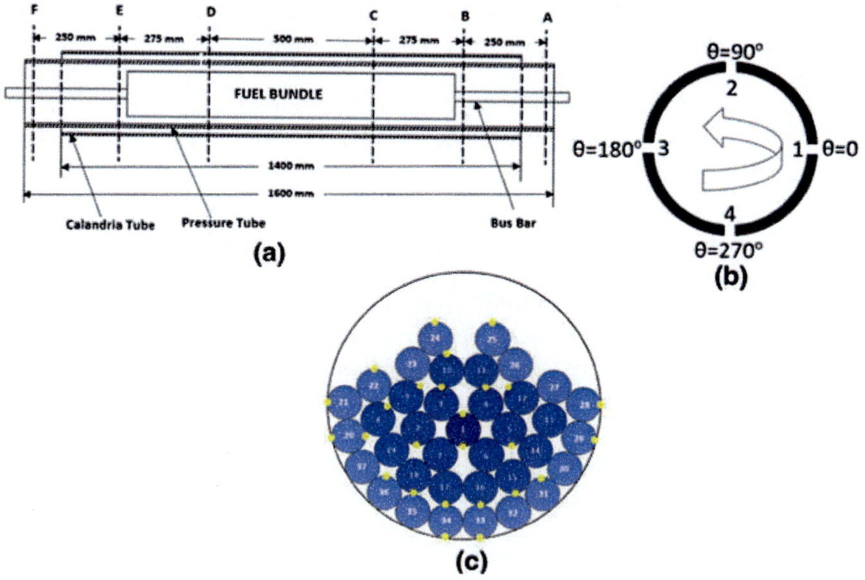

Fig. 4 Thermocouples position in the test section: **a** axial position, **b** circumferential position in PT and CT, **c** circumferential position in fuel pins

used in the experiment. The argon gas inside the annulus space between PT-CT and inside the PT was also considered in the computational analysis. The length of 1 m was used for each component of the test section. Regarding mesh generation strategy, the inflation layer was used to capture the temperature gradient in the fluid-solid interfaces accurately. In the solid domain, a hexahedral dominant mesh was used. The simple scheme was adopted to perform steady-state analysis. The mass momentum, energy and radiation equations were discretized using the second-order upwind scheme.

Continuity equation

$$\nabla \cdot (\rho V) = 0 \tag{1}$$

Momentum equation

$$\nabla \cdot (\rho V V) = -\nabla p + \nabla \cdot \tau + \rho b \tag{2}$$

Energy equation

$$\nabla \cdot (\rho e V) = Q - \nabla \cdot q + \nabla \cdot (V \cdot \tau) \tag{3}$$

The radiative transfer equation (RTE) for an absorbing, emitting and scattering medium at position **r** in the direction **s** is

$$\frac{dI(r,s)}{ds} + (\alpha + \sigma_s)I(r,s) = an^2\frac{\sigma T^4}{\pi} + \frac{\sigma_s}{4\pi}\int_0^{4\pi} I(r,s')\phi(s,s')d\Omega' \qquad (4)$$

The contact conductance of 2.0 kW/m^2-°C between fuel pin to fuel pin and between fuel pin to PT was considered. Since at high-temperature radiation mode of heat transfer is predominant; therefore, radiation heat transfer was modeled using the discrete ordinates (DO) model. The effect of gravity was considered for natural convection. The temperature-dependent thermal properties were used for materials of PT, CT, fuel pin and for the argon gas. These properties have been adopted from the MATPRO. The Dirichlet boundary condition (constant surface temperature) was used on the outer surface of the CT. The decay power of the fuel bundle simulator was modeled by giving an energy source in the heater rods. The total power of 2.9 kW is distributed in the fuel pins of first, second, third and fourth rings of fuel bundle simulator in a particular ratio as described above.

4 Results and Discussion

The temperature of the fuel pins under the slumped condition obtained through the experiment is shown in Fig. 5. Since there is a minimal axial variation in the temperature of the fuel pin between sections C and D, so the graph has been plotted by taking the average temperature at these sections as shown in Fig. 5.

The temperature of the fuel pin is found to decrease from center fuel pin to outer fuel pins. The center fuel pin attained maximum temperature among the other fuel pins. It is because of the resistance offered by the surrounding fuel pins to transfer its decay heat. There is also an insignificant circumferential temperature gradient in the center fuel pin as the percentage temperature between the top and bottom nodes is 0.2% only. The presence of uniform contact conductance results in a uniform temperature in each of the fuel pins of ring-2 and ring-3.

The temperature of each fuel pin of the outer ring is lowest among the other fuel pins due to enhanced heat transfer through conduction to the surface of the PT. The significant circumferential temperature gradient is developed in the fuel pins of the outer ring, as seen in the fuel pin 20. The percentage temperature difference of 7.1% is obtained between the nodes adjacent to PT and its diametrically opposite node facing the fuel pin of ring-3. Similar temperature results for the slumped fuel pins were predicted from the numerical analysis, as shown in Fig. 6. It can be seen that here also center fuel pin attained the highest temperature while the lowest temperature is obtained by the fuel pins of the outer ring. The

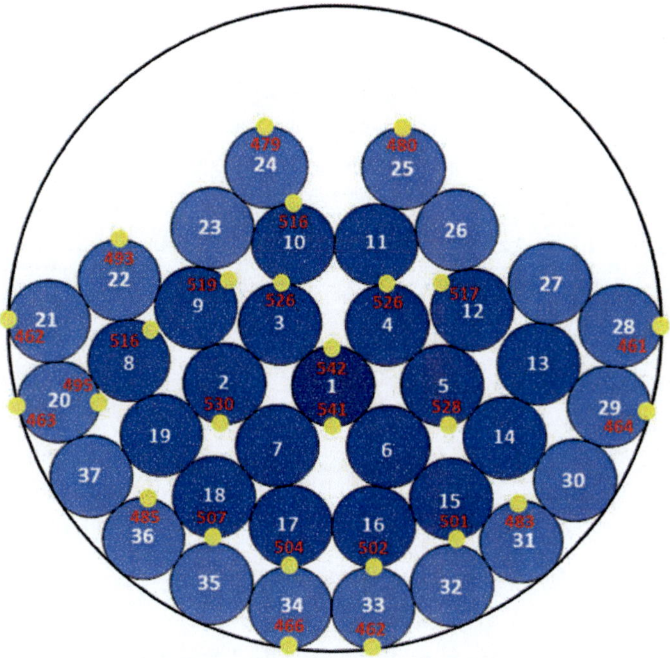

Fig. 5 Temperature distribution in the fuel pins

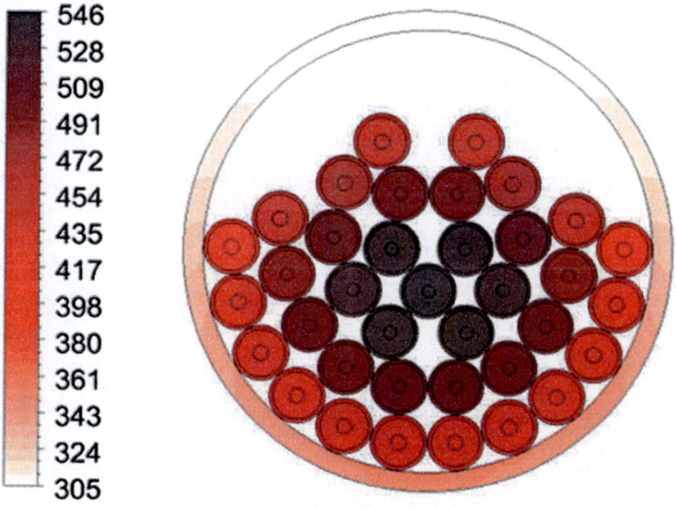

Fig. 6 Temperature contours in fuel pins and PT

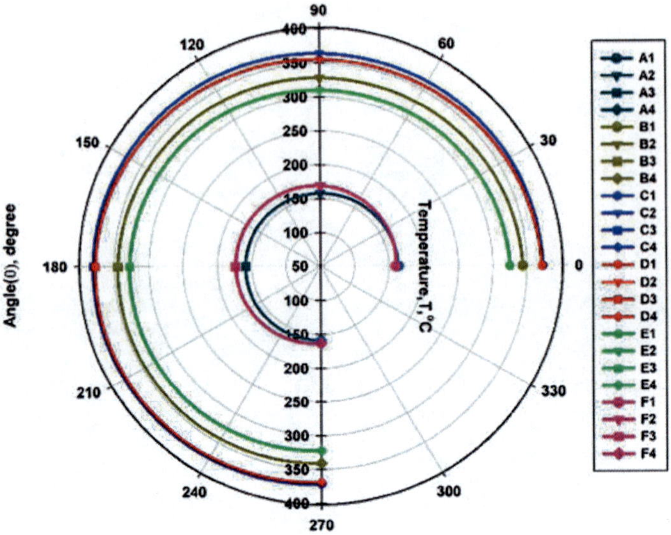

Fig. 7 Circumferential temperature distribution in the PT

maximum percentage temperature difference of 4.1% is found between the experimental and simulated results. The circumferential temperature distribution in the PT for all axial sections (A, B, C, D, E and F) from experimental data is shown in Fig. 7. It is seen that temperature is maximum in sections C and D.

It is also observed that for all axial positions, the temperature of the PT is found to decrease from the upper half to the lower half. This is because of the enhanced heat transfer through conduction from the fuel pins to the lower half of the PT. The maximum percentage temperature difference of 4.2% is found been top and bottom nodes. The numerical results also exhibit a similar temperature trend in the PT, as shown in the figure. Figure 8 shows the circumferential variation in the temperature of the CT. The insignificant temperature variation is found along its circumference and length, which highlights the effectivities of the moderator as a heat sink under critical break LOCA condition. The heat balance is performed to estimate the amount the heat flow in the different parts of the test section. The details of the heat balance are shown in Table 2.

5 Conclusions

From both experimental and numerical results, it is found that center fuel pin attained the highest temperature and developed an insignificant temperature gradient. The significant temperature gradient is found in the fuel pins of the outer ring that are in contact with the PT. The temperature difference of around 30–40 °C is seen between

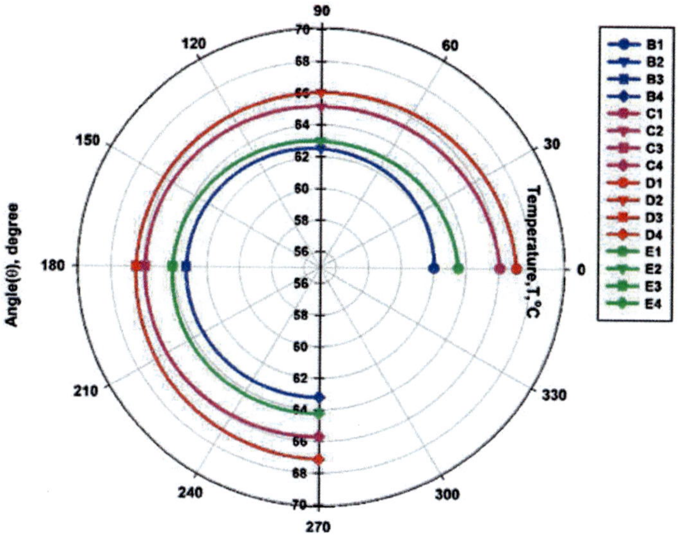

Fig. 8 Circumferential temperature distribution in the CT

Table 2 Heat balance sheet

Parameter	Heat (W)
Heat supplied to fuel bundle	2867.00
Heat loss through conduction from copper bus bar	183.41
Heat transferred to PT	2683.59
Axial heat loss from PT	51.55
Heat transferred to CT	2632.04
Axial heat loss from CT	0.51
Heat transferred to moderator	2557.32
Unaccounted heat	74.21
Percentage unaccounted heat (%)	2.59

the node facing the PT and its diametrically opposite node facing the fuel pin of the third ring. The temperature distribution in the PT is greatly influenced by the slumped fuel pins. The temperature difference of 30–50 °C is found between the top and bottom nodes of the PT. The moderator acts as an effective alternate heat sink as 89% decay heat of the fuel gets transferred to the moderator during a critical break accidental scenario.

References

1. Gupta SK, Dutta BK, Venkat Raj V, Kokodkar A (1997) A study of the Indian PHWR reactor channel under prolonged deteriorated flow conditions. International Atomic Energy Agency, Vienna (Austria), pp 331–359, 475
2. Shewfelt RSW, Layall LW, Godin DP (1984) High temperature creep model for Zr2.5 wt% Nb pressure tubes. J Nucl Mater 125:228–235
3. So CB, Gillespie GE, Moyer RG, Litke DG (1987) The experimental determination of circumferential temperature distributions developed in pressure tube during slow coolant boil down. In: Proceedings of the CNS 8th annual conference, Saint John, pp 241–248
4. Shoukri M, Chan AMC (1987) On the thermal analysis of pressure tube/calandria tube contact in CANDU reactors. Nucl Eng Des 104:197–206
5. Ajay K, Kumar R, Mukhopadhyay D, Gupta A, Das AK (2019) Experimental investigation of radiation heat transfer in coolant channel under impaired cooling scenario for Indian PHWR. Nucl Eng Des 347:45–52

A Micro Multi-blade Vertical Axis Wind Turbine for Built-Up Areas

Bavin Loganathan, Harun Chowdhury, Hamed Allehibi, Firoz Alam, and Akshoy Ranjan Paul

Abstract Rising greenhouse gas emission and global warming compel the world community to look for renewable energy sources to generate power. The renewable wind energy provides an alternative to fossil fuel. A significant research attention has been placed on the use of vertical axis wind turbine due to its low-wind operational capability. However, the effectiveness of traditional vertical axis wind turbines (VAWT) in built-up areas is below expectation. Therefore, the primary objective of this study is to develop a single stage drag-based multi-blade micro vertical axis wind turbine which can generate a reasonable amount of torque and power in complex wind conditions in built-up areas.

Keywords Vertical axis wind turbine · Wind tunnel · Power · Torque · Built-up area

1 Introduction

In recent time, significant emphasis is put on power generation from clean and renewable energy sources. Harnessing wind energy to generate power can be materialized using horizontal axis and vertical axis wind turbines. The vertical axis wind turbine has drawn notable attention to researchers and engineers due to its cost, simplicity and low wind operation capability. However its effective use in built-up areas is constrained due to its less than expected in-situ performance [1, 2].

In urban built-up areas, the atmospheric wind becomes highly turbulent and exhibits significant fluctuations in speeds and high variability of wind direction caused by the urban structures and buildings [2, 3]. Under such wind conditions, horizontal axis wind turbines are not effective power generators [4, 5]. Despite having

B. Loganathan · H. Chowdhury · H. Allehibi · F. Alam (✉)
School of Engineering, RMIT University, Melbourne, VIC 3001, Australia
e-mail: firoz.alam@rmit.edu.au

A. R. Paul
Department of Applied Mechanics, Motilal Nehru National Institute of Technology Allahabad, Prayagraj, India

L. Venkatakrishnan et al. (eds.), *Proceedings of 16th Asian Congress of Fluid Mechanics*, Lecture Notes in Mechanical Engineering,
https://doi.org/10.1007/978-981-15-5183-3_16

some advantages (fewer moving parts, lower tip speed ratio, quieter, lower cost, and insensitive to wind direction) over horizontal axis wind turbines, most vertical axis wind turbines (Darrius or Savonius type) do not produce appreciable power in built-up areas [6, 7].

In order to increase the torque and power of a drag-based turbine, the effect of gap between two cylindrical halves has been investigated and reported in the literature [8, 9]. The gap allows the air to pass through this gap from the concave section facing the wind to the concave section of the blade opposing the wind as shown in Fig. 1. In this study, a newly designed multi-blade turbine is shown in Fig. 2. A top view of such turbine's schematic is shown in Fig. 3.

Despite some studies have been undertaken to enhance the power output of Savonius type wind turbines using straight blades, no work has been reported in the public domain on a single stage multi-blade drag-based micro vertical axis wind turbine (mVAWT) with semi-circular shaped blades. Moreover, no information is available about the optimal clearance between two subsequent blades of a multi-blade drag-based vertical axis wind turbines. Therefore, the primary objective of this study is to develop an empirical relation between the number of blades, turbine effective radius, blade radius and clearance between two subsequent blades to determine the optimal blade configuration of a single stage drag based multi-blades vertical axis wind turbine.

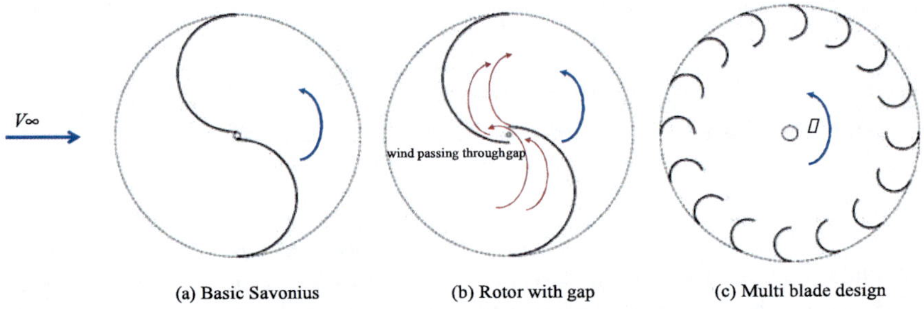

Fig. 1 Various Savonius type drag based vertical axis wind turbines adapted from [10]

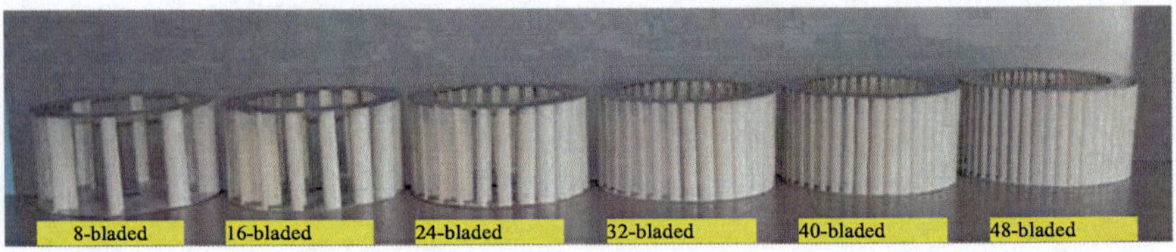

Fig. 2 Side view of 8, 16, 24, 32, 40 and 48-bladed prototype turbines adapted from [11]

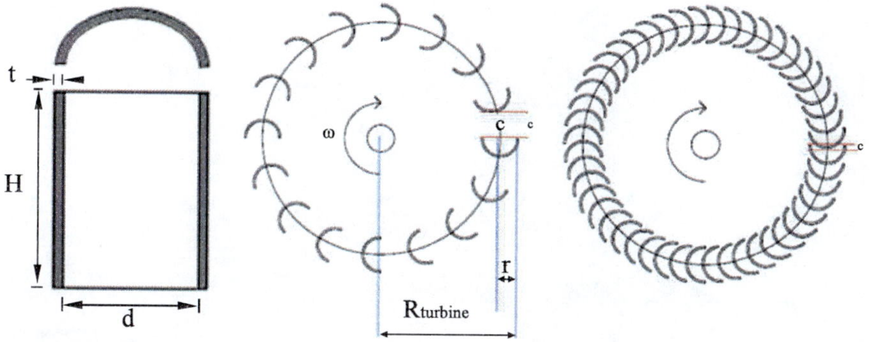

Fig. 3 Schematics of 16 and 48-bladed wind turbine adapted from [10]

2 Methodology

Six prototype turbines' blade numbers are: 8, 16, 24, 32, 40 and 48. Micro vertical axis wind turbine of 300 mm diameter with various blade configurations has been designed and manufactured. The blade shape of each turbine is semi-circular with zero twisting. It is single stage turbine. The blades are installed around the circumference of the turbine by providing equal space with a clearance 'c' between two subsequent blades. The blades are secured by two circular discs, made of Suntuf™ clear acrylic rigid plastics (top & bottom) as shown in Fig. 3. The height of the turbine is 160 mm with 40 mm diameter blades. The blades are made of PVC to provide structural rigidity.

A vertical support and data acquisition system was developed as shown in Fig. 4 to position the turbine in the test section of the wind tunnel. The setup was positioned 200 mm above the tunnel floor to avoid the tunnel boundary layer effect. A torque sensor (T20WN made by HBM) and a mechanical brake system have been integrated with the support system which is shown in Fig. 4. The maximum capacity of the torque transducer is 5 Nm with 0.01% accuracy. Data logging software supplied by the torque transducer manufacturer was used to acquire the data (rotational speed and torque). Torque transducer was calibrated before and after each set of test conducted.

The experimental investigation was carried out in RMIT Wind Tunnel. The tunnel is a closed return circuit wind tunnel with a turntable to yaw a suitable model to the wind direction. It has a rectangular test section with 3 m wide, 2 m high and 9 m long. The cross-sectional area of the rectangular test section is 6 m^2. The maximum speed of the tunnel in the test section is approximately 145 km/h. Its turbulence intensity is around 2%. The tunnel was calibrated prior conducting the experimental investigation and the mean air speeds at the entry and at the test position in the tunnel were measured with a modified National Physical Laboratory (NPL) ellipsoidal head pitot-static tube connected through flexible tubing with a Baratron® pressure sensor made by MKS Instruments, USA.

The wind tunnel tests were conducted at a range of wind speeds (4.5–8.5 m/s with an increment of 1 m/s) for all prototype models. At each speed, the turbine

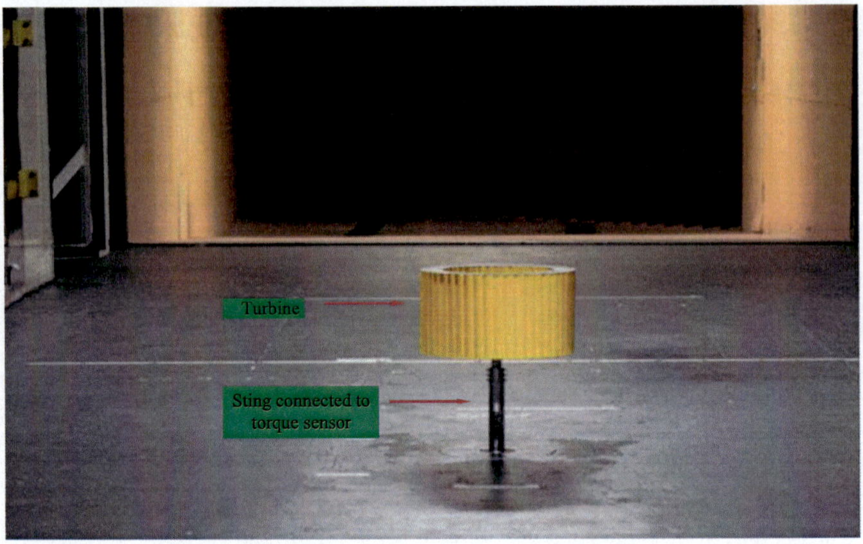

Fig. 4 Experimental setup inside the wind tunnel

rotational speed (rpm) and torque (T) were acquired for all speeds. Measurements were taken three times for each configuration at all wind speeds. The acquired data was averaged to minimise the error. The minimum wind speed (4.5 m/s or 16.2 km/h) was constrained by the ability of the turbine to overcome bearing frictional and inertial resistance. The upper limit of wind speed (8.5 m/s or 30.6 km/h) was limited by the structural rigidity and vibration of the turbine. It may be mentioned that the average wind speed in built-up area at 10 m or below generally does not exceed to 8.5 m/s [3].

3 Results and Discussion

For each prototype turbine, the torque (T) and rotational speeds (ω) were measured for all wind speeds as mentioned in Sect. 2. The output power (P) generated by each turbine was determined using equation

$$P = T\omega \tag{1}$$

where:

$P =$ Mechanical power (W)

$T =$ Torque (Nm).

$\omega =$ Rotational speed (rad/s).

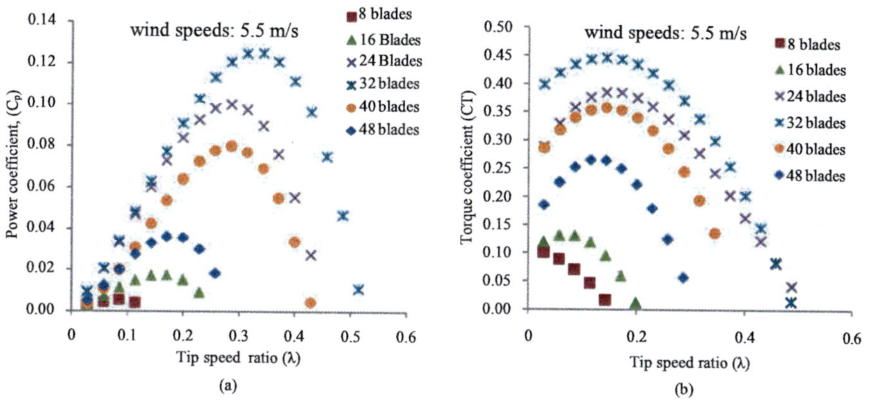

Fig. 5 Power as a function of rotational speed at all configurations

The mechanical power (W) as a function of rotational speed (ω) for all wind speeds have been plotted. The plot for 5.5 m/s wind speed for all prototype turbines is shown in Fig. 5. This is the average wind speed generally encountered in the built-up areas below 10 m height. The maximum power outputs for 8, 16, 24, 32, 40 and 48 blades turbine are 0.0261, 0.0864, 0.490, 0.6120, 0.3900 and 0.1776 respectively. The 32 blades turbine produces maximum power (over 23 times more) compared to 8 blades turbine. The maximum power for 8 blades turbine occurs at 30 rpm while 32 blades turbine at 120 rpm. The rotational speed has increased four times. At other speed range, the 32 blades turbine produces higher power consistently.

One of the important performance indicators of a turbine is the power coefficient. The power coefficient (C_P) is the ratio of mechanical power turbine produced and the wind power as shown in Eq. (2).

$$C_P = \frac{\text{Mechanical Power}}{\text{Wind Power}} - \frac{T_\omega}{\frac{1}{2}C_D \rho v^3 A} \tag{2}$$

where T is the torque, ω is the rotational speed, ρ is air density, A is the projected frontal area of turbine ($A = 2R_{\text{rotor}} \times H$), the turbine effective diameter is d ($d = 2R_{\text{rotor}}$) and the blade height is H. The tip speed ratio (λ) is defined as the ratio of the blade tip speed (ωR_{rotor}) and free stream wind speed, (v_∞) as shown in Eq. 3. Where ω is rotational speed of the turbine and R_{rotor} is the effective turbine radius.

$$\lambda = \frac{\omega R_{\text{rotor}}}{v_\infty} \tag{3}$$

The power coefficients and tip speed ratios for all turbine configurations have determined at all wind speeds. The power coefficient versus tip speed ratio for all prototype turbines at 5.5 m/s is shown in Fig. 6a. The power coefficient increases

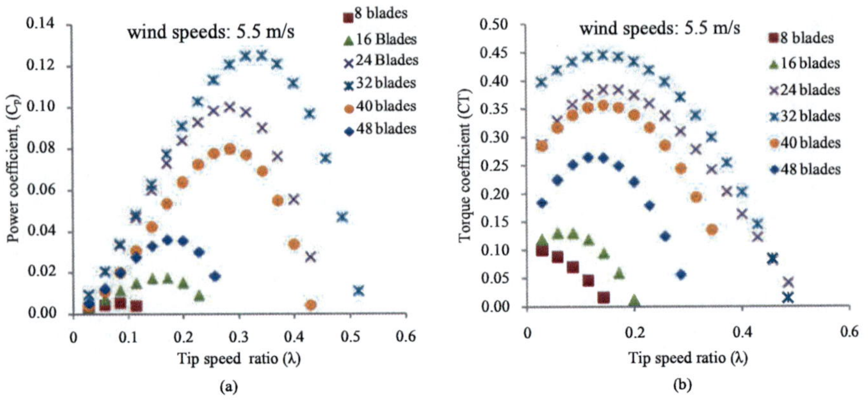

Fig. 6 **a** Power coefficient as a function of Tip speed ratio of number of blades; **b** torque coefficient versus tip speed ratio of different number of turbine

with the increase of tip speed ratio (λ) until the peak level is obtained and then it decreases; and finally becomes zero as the turbine attains its constant maximum rotational speed. The maximum power coefficient of 0.13 is obtained by the 32 blades turbine at $\lambda = 0.34$ as shown in Fig. 6a. Hau [12] reported that a drag based single stage vertical axis wind turbine can have the maximum theoretical power coefficient of around 0.20.

Torque coefficient (C_T) is used to assess the torque generated by a wind turbine. Wind turbines with higher value of C_T generally have higher torque values at low wind speeds. This higher torque allows spinning the turbine at low wind speeds which is very much desirable [13]. The Torque coefficient is defined as the ratio of mechanical torque and the wind torque as shown in Eq. 4.

$$C_T = \frac{T}{\frac{\rho}{2} V^2 A R} \tag{4}$$

where, T is the torque, ρ is air density, A is projected frontal area of turbine ($A = 2R_{\text{rotor}} \times H$), turbine effective diameter ($d = 2R_{\text{rotor}}$), turbine radius (R_{rotor}) and V is wind velocity.

The torque coefficient variation with tip speed ratio for all configurations is shown in Fig. 6b. The torque coefficients achieved at 5.5 m/s wind speed are 0.100 at $\lambda = 0.029$, 0.132 at $\lambda = 0.086$, 0.385 at $\lambda = 0.143$, 0.447 at $\lambda = 0.143$, 0.58 at $\lambda = 0.143$, and 0.265 at $\lambda = 0.114$ for 8 blades, 16 blades, 24 blades, 32 blades, 40 blades and 48 blades configurations respectively. It is interesting to note that variable torque coefficients were obtained at approximately the same tip ratio for 24, 32 and 40 blades wind turbines. The highest torque coefficient is obtained for the 32 blades turbine. The lowest torque coefficient is obtained from the 8 blades configuration. In short, based on torque coefficient values, the ranking is as follows: highest 32 blades turbine followed by the 24, 40, 48, 16 and 8 blades turbine. The 32 blades turbine is expected to perform well at lower wind speeds (<4.5 m/s).

A schematic flow pattern around a 16 and 32-bladed turbine in rotation is shown in Figs. 7 and 8. It is noted that the flow in the midpoint section of the 16-bladed turbine is not stationary compared to the 32-bladed turbine. It is also observed that the flow at rear of the 16-bladed turbine is interacting with the rotating boundary layer requiring huge power for the turbine to continue its rotation speed when compared to the 32-bladed turbine. Hence, the power extracted from the incoming airflow in the front blades are being spent by the rear blades resulting in lower power efficiencies. It is clearly shown that the optimal gap between the two subsequent blades allows for efficient interaction between the two flows (existing wind and rotating boundary layer) allowing for efficient power generation (Table 1).

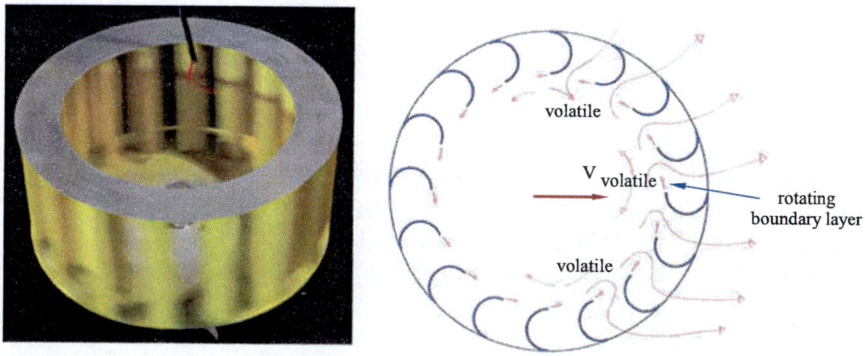

Fig. 7 A schematic flow pattern around a 16-bladed turbine in rotation

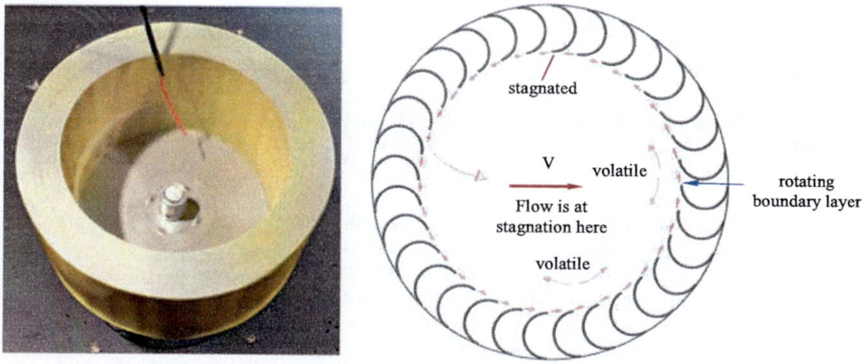

Fig. 8 A schematic flow pattern around a 32-bladed turbine in rotation

Table 1 Production cost for one300 mm diameter turbine

Materials	Specifications	Quantity	Unit price (A$)	Total cost (A$)
PVC pipes	40 mm diameter with 1 m length	6	2.50	15.00
Acrylic sheets	2400 mm × 1200 mm	2	55.00	110.00
General purpose adhesive	30 ml/tube	3	3.5	10.5
Aluminium flange	1 m diameter with 6 m length	1	10	10
Steel Shaft		1	20	20
Bearings		1	15	15
Labour	For manufacturing (in hours)	1	100	100
Total				280.50

4 Conclusions

The study reveals that the 300 mm turbine with 32 blades configuration produces the highest power at all wind speeds (5.5–8.5 m/s). The turbine produces a maximum power of 0.61 W. The power coefficient (C_P) variation with tip speed ratio (λ) for 300 mm diameter with 32 blades at 5.5 m/s wind speed is 0.125 at $\lambda = 0.343$. This indicates that the 300 mm diameter turbine with 32 blades can rotate faster at low wind speeds which are typically encountered in built-up areas.

References

1. Pagnini LC, Burlando M, Repetto MP (2015) Experimental power curve of small-size wind turbines in turbulent urban environment. Appl. Energy 154:112–121
2. Wekesa DW et al (2016) Experimental and numerical study of turbulence effect on aerodynamic performance of a small-scale vertical axis wind turbine. J. Wind Eng. Ind. Aerodyn. 157:1–14
3. A. Webb, *The viability of domestic wind turbines for urban Melbourne*. Sustainability Victoria, 2007
4. Lubitz WD (2014) Impact of ambient turbulence on performance of a small wind turbine. Renew. Energy 61:69–73
5. Yoon S-H, Lim H-C, Kim D-K (2013) Study of several design parameters on multi-blade vertical axis wind turbine. Int. J. Precis. Eng. Manuf. 14(5):831–837
6. Ahmadi-Baloutaki M, Carriveau R, Ting DS-K (2015) Performance of a vertical axis wind turbine in grid generated turbulence. Sustain. Energy Technol. Assessm. 11:178–185
7. Loganathan B et al (2015) An experimental study of a cyclonic vertical axis wind turbine for domestic scale power generation. Procedia Eng. 105:686–691
8. Kamoji M, Kedare SB, Prabhu S (2009) Experimental investigations on single stage modified Savonius rotor. Appl. Energy 86(7–8):1064–1073
9. Saha U, Thotla S, Maity D (2008) Optimum design configuration of Savonius rotor through wind tunnel experiments. J. Wind Eng. Ind. Aerodyn. 96(8–9):1359–1375
10. Loganathan B et al (2019) Design of a micro wind turbine and its economic feasibility study for residential power generation in built-up areas. Energy Procedia 160:812–819

11. B. Loganathan, *Aerodynamic study of single stage multi-blade drag-based vertical axis wind turbines*. 2018, RMIT University, Australia, pp. 1–153
12. Erich H (2000) Wind Turbines: Fundamentals, Technologies, Application, Economics. Springer, New York
13. Ikegaya N et al (2016) Effect of turbulent flow on wall pressure coefficients of block arrays within urban boundary layer. Build Environ. 100:28–39

Motion of a Solid Sphere Launched Vertically Upward in Water and Deformation of Water Surface by the Sphere

Kotaro Takamure and Tomomi Uchiyama

Abstract A solid sphere is launched vertically upward from water toward the air–water interface, and the modulation of the sphere's motion and the behavior of the air–water interface induced by the sphere are investigated. Spheres with three different densities ρ_p are used: the almina sphere ($\rho_p = 3900$ kg/m^3), teflon sphere ($\rho_p = 2180$ kg/m^3), and duracon sphere ($\rho_p = 1360$ kg/m^3). Through a visualization experiment on the motion of a sphere in water, it was demonstrated that the launch velocity of the sphere can be set with high accuracy. When the sphere passes through the air–water interface, it forms a complex interface while entraining water. Additionally, spheres with small ρ_p were demonstrated to not always vertically collide with the air–water interface; this causes different behaviors of the spheres in the air.

Keywords Gas–liquid–solid three-phase flow · Deformation of interface · Visualization

1 Introduction

Elucidating the motion of an object and the collision between the object and an air–water interface can provide useful knowledge in the industrial field. Information on the above phenomena can be used in a wide range of applications, such as reducing wave resistance caused by interference between an object and an interface, and predicting flow in a fluidized bed.

In existing researches, a solid sphere has often been used for examining the effect of an object on air–water interface interference. Most researches using spheres have been conducted via free fall from the air to the water surface. These studies have clarified the details of the modulation of the sphere's motion and the behavior of the air–water interface [1–5]. However, research examining the case in which a sphere rises from the water to the air– water interface is limited to the study on a buoyant

K. Takamure (✉) · T. Uchiyama
Institute of Materials and Systems for Sustainability, Nagoya University Furo-Cho, Chikusa-ku, Nagoya 464-8601, Japan
e-mail: kotaro.takamure@imass.nagoya-u.ac.jp

L. Venkatakrishnan et al. (eds.), *Proceedings of 16th Asian Congress of Fluid Mechanics*, Lecture Notes in Mechanical Engineering,
https://doi.org/10.1007/978-981-15-5183-3_17

rising sphere by Truscott et al. [6]. Therefore, the sphere's motion and the air–water interface behavior caused by the rising sphere are not as well understood as free fall behavior.

In this study, three types of spheres (alumina, teflon, and duracon) with different densities are launched vertically upward from water toward the air–water interface. The modulation of the sphere's motion and the behavior of the air–water interface are investigated.

2 Experimental Set up

A schematic of the experimental equipment is presented in Fig. 1. Water is stored in a transparent acrylic tank that has a test section of 0.4 m × 0.4 m × 0.4 m. A cylinder with a length of 67 mm and an internal diameter of 19.3 mm is installed at the center of the bottom of the tank, and a solid sphere with diameter d is placed at the top of the cylinder. An iron rod with a cylindrical shape with a length of 170 mm and diameter of 19.1 mm is placed inside the cylinder. The lower end of the iron rod and a coil spring are connected in series. The distance between the bottom of the sphere and the tip of the iron rod is 25 mm. The iron rod is pressed down vertically to compress the coil spring, and it is fixed with a trigger. When pulling the trigger to release the coil spring at $t = 0$, the iron rod moves upward in the cylinder, and the sphere is launched into the water because the upper end of the iron rod strikes the sphere. The launch speed of the sphere can be controlled by the compression displacement of the coil spring. In this study, three types of spheres with different densities ρ_p are used: an almina sphere ($\rho_p = 3900$ kg/m^3), teflon sphere ($\rho_p = 2180$ kg/m^3), and duracon sphere ($\rho_p = 1360$ kg/m^3). The diameter d of these spheres is 25.4 mm.

Figure 2 illustrates the acrylic tank (test section) and sphere launch cylinder. The coordinate origin is at the center of the sphere at $t = 0$. The coordinate system has an

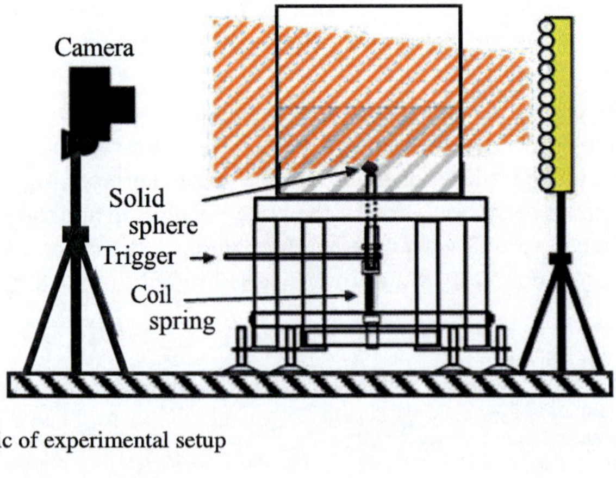

Fig. 1 Schematic of experimental setup

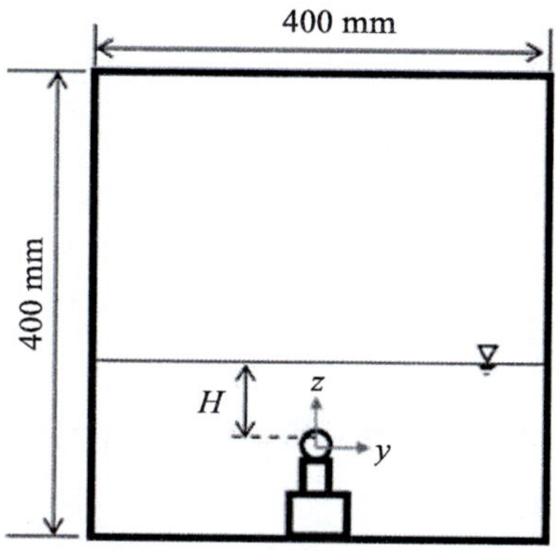

Fig. 2 Schematic of acrylic tank and sphere launcher

x-y axis in the horizontal direction and a z-axis in the vertical direction. The distance H from the coordinate origin to the water surface is 850 mm ($H/d = 3.4$). The inside of the acrylic tank is illuminated by a light fixture (Godox) with 1024 LEDs arranged in a plane. To equalize the light intensity inside the acrylic tank, a diffuser sheet is attached to the wall of the tank on the lighting side. The motion of the launched sphere and the behavior of the air–water interface are captured by a high-resolution camera (SONY, RX100v). The spatial resolution is set to 1920 × 1080 dpi, and the frame rate is set to 920 fps.

3 Results

Firstly, the time variations of a sphere's displacement were measured. It should be noted that 20 repeated measurements were performed for each sphere. To compare the results of repeated measurements, a virtual origin z_0 was defined for z_p. Figure 3 presents an example of the time change of z_p immediately after launch. The maximum value of time changes of z_p (i.e., the vertical position at which the sphere velocity is maximum) is z_0, while time is t_0 and the sphere velocity is w_0.

Figure 4 illustrates the relationship between t_0 and w_0 for three types of spheres. The results of 20 repeated measurements are distributed as indicated by the open symbols. Figure 4 also presents averaged values and standard deviations for repeated measurements. The variation of each sphere is within 4%, confirming the reproducibility of this experiment.

Figure 5 presents the visualization results when using the teflon sphere with $\rho_p = 2180$ kg/m^3 as an example. The time at which the sphere passes the air–water interface ($z/d = 3.3$) is denoted t_1. The time is normalized by t^* ($=t_1 - t_0$). The Reynolds number based on w_0 and d is 6.6×10^4. The launched sphere travels vertically in the water ($(t - t_0)/t^* < 0.7$) and interferes with the air–water interface

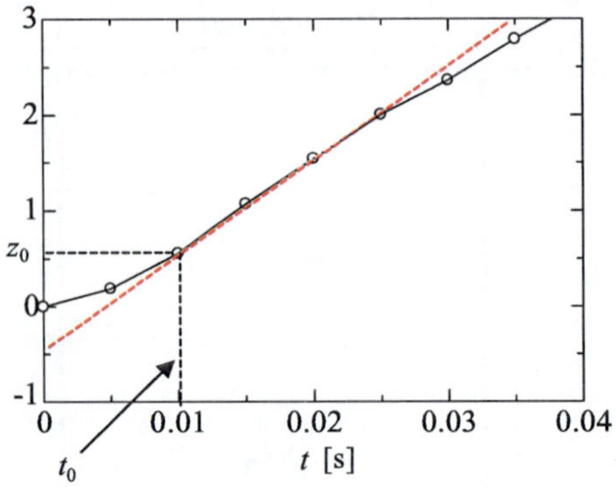

Fig. 3 Example of time variation of z_p immediately after launch

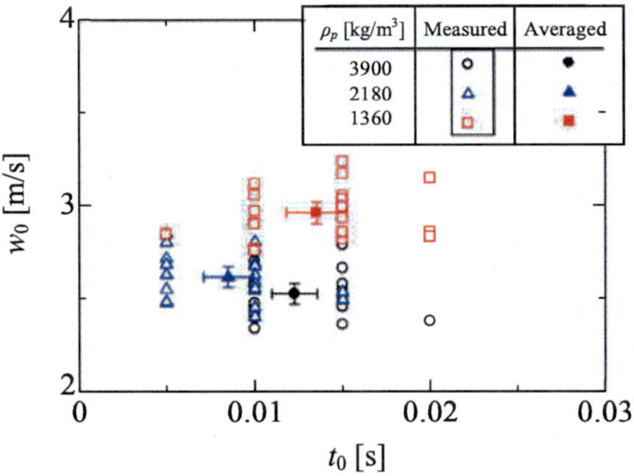

Fig. 4 Relation between w_0 and t_0

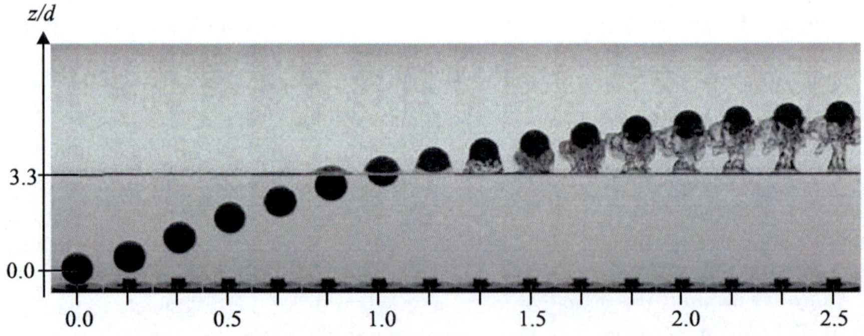

Fig. 5 Visualization images of launched sphere with $\rho_p = 2180$ and air–water interface

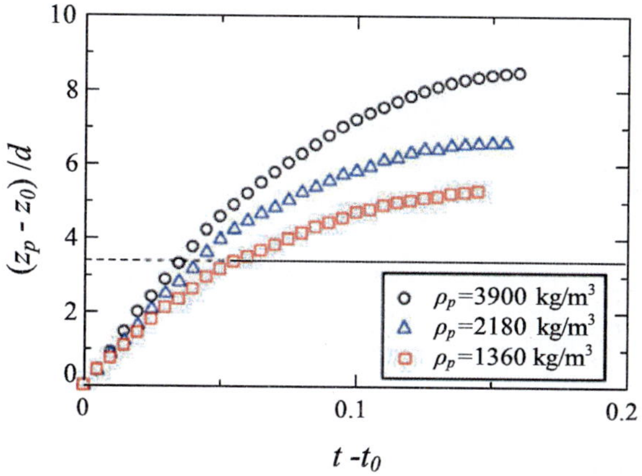

Fig. 6 Time variation of the vertical displacement of sphere

$(0.7 < (t - t_0)/t^* < 1.2)$. When this interference occurs $(1.2 < (t - t_0)/t^* < 1.5)$, the air–water interface behind the sphere deforms in a complex manner. Even when the sphere is completely separated from the air–water interface $((t - t_0)/t^* > 1.5)$, the water is entrained by the sphere, and the lower half of the sphere is surrounded by water. In addition, a water column can be observed between the sphere and the interface. The diameter of the water area around the sphere is approximately $2d$, and the width of the water column is approximately $0.5d$.

Figure 6 illustrates the time variation of displacement z_p for three types of spheres with different density ρ_p. It should be noted that one measurement result is provided for each sphere. The initial velocity of the three spheres is almost equal. However, in water ($z/d < 3.3$), a lower density ρ_p leads to a lower velocity. This is because the lower the density ρ_p is, the smaller the inertial force applied to the sphere is, and the greater the flow resistance of water is. Due to this speed reduction, a lower ρ_p leads to a lower maximum height of the sphere.

The coefficient of the variation of z_p that occurs in 20 repeated experiments is presented in Fig. 7. The time is normalized by t^* $(= t_1 - t_0)$. The coefficient of variation begins to increase after the sphere is launched, and reaches a maximum value at the air–water interface $((t - t_0)/t^* = 1)$. This is due to the fact that the shape of the sphere becomes unclear, as the air–water mixed flow covers the sphere at the air–water interface. It should be noted that the coefficient of variation increase as ρ_p decreases in the region in which the coefficient of variation takes the maximum value. The coefficient of variation at $(t - t_0)/t^* > 2$ also increases as the ρ_p of the sphere decreases. It is clear that a lower ρ_p leads to a lower reproducibility of the behavior of the sphere when passing through the air–water interface.

The trajectories of the three spheres in the y-z plane are presented in Fig. 8. Each figure displays the results of four repeated experiments. Alumina spheres with ρ_p = 3900 kg/m³ always rise in an almost vertical line. Teflon spheres with ρ_p = 2180 kg/m³ rise almost vertically; however, after passing through the air–water

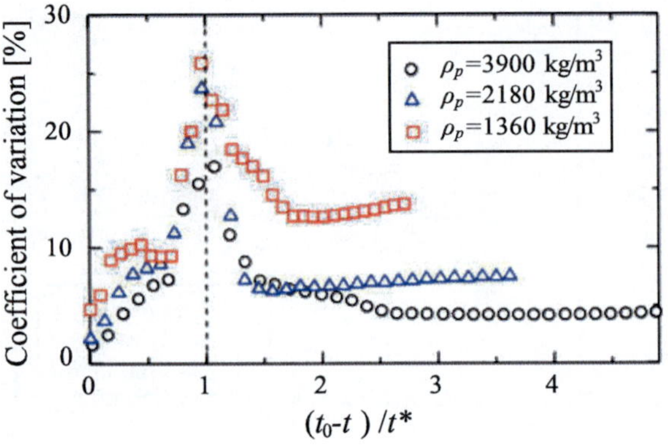

Fig. 7 Time variation of coefficient of variation of vertical displacement of sphere

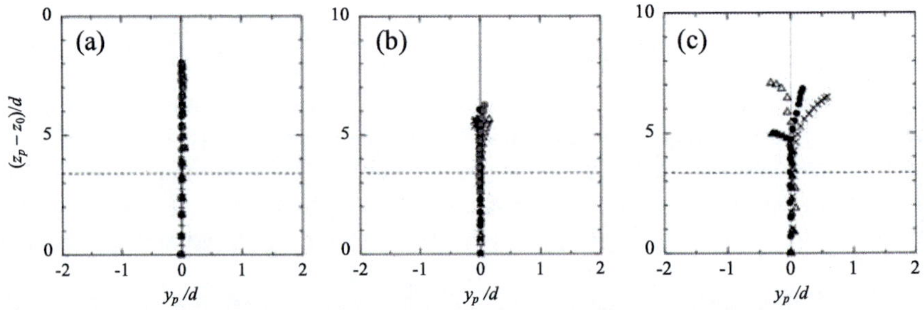

Fig. 8 Trajectory of spheres. **a** $\rho_p = 3900$ kg/m^3, **b** $\rho_p = 2180$ kg/m^3, and **c** $\rho_p = 1360$ kg/m^3

interface, their trajectory changes its direction slightly to the left and right. A similar phenomenon is also observed when a buoyancy sphere is used [6]. Duracon spheres with $\rho_p = 1360$ kg/m^3 are already scattered to the left and right in the water, and after passing through the air–water interface, their trajectory greatly deviates from the vertical line. As illustrated in Fig. 8, the behavior of the alumina sphere with $\rho_p = 3900$ kg/m^3 and the duracon sphere with $\rho_p = 1360$ kg/m^3 greatly differs after passing through the air–water interface. Therefore, the visualized images are compared before and after the spheres pass through the air–water interface. Figure 9 presents the results for a sphere with $\rho_p = 3900$ kg/m^3. When the sphere reaches the air–water interface ($(t - t_0)/t^* = 1.0, 1.2$), the interface swells symmetrically on the left and right of the center of the sphere. Thereafter, the sphere pulls the interface ($(t - t_0)/t^* = 1.5$), and the water droplets scatter around the sphere ($(t - t_0)/t^* = 2.3, 3.5, 4.5$). Here, the scattering of the water droplets is almost symmetrical with respect to the sphere. This signifies that the sphere is vertically incident on the air–water interface.

Figure 10 presents an example of the imaging results when a sphere with $\rho_p =$ 1360 kg/m^3 passes through the interface. Figure 10a is a visualized image of the time

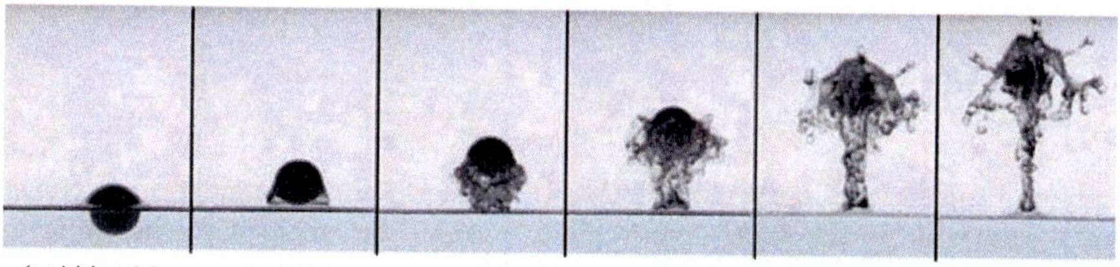

$(t\text{-}t_0)/t^* = 1.0$ 1.2 1.5 2.3 3.5 4.5

Fig. 9 Motion of sphere with $\rho_p = 3900$ kg/m^3 and behavior of air–water interface

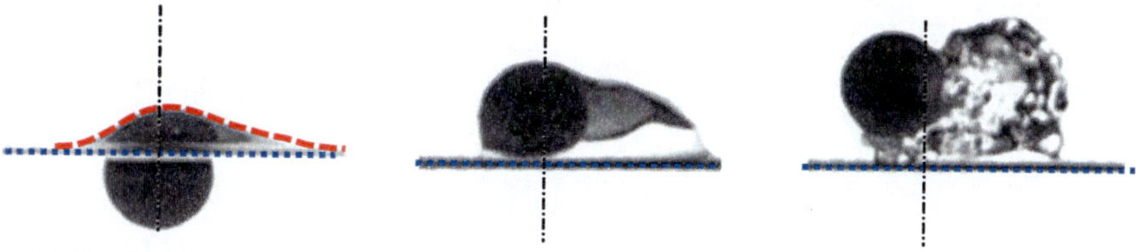

(a) $(t\text{-}t_0)/t^* = 0.9$ (b) $(t\text{-}t_0)/t^* = 1.1$ (c) $(t\text{-}t_0)/t^* = 1.3$

Fig. 10 Motion of sphere with $\rho_p = 1360$ kg/m^3 and behavior of air–water interface (1)

($(t - t_0)/t^* = 0.9$) when the upper surface of the sphere collides with the air–water interface. As represented by the dashed line, the air–water interface on the top of the sphere is asymmetric with respect to the vertical line. As illustrated in Fig. 10b, c, the shape of the water interface carried by the sphere is also largely asymmetric after the sphere passes through the air–water interface. Figure 11 is a visualized image captured under the same conditions as Fig. 10. It should be noted that the results presented in Figs. 10 and 11 are obtained in independent experiments. As in Fig. 10, the asymmetry of the air–water interface is also seen in Fig. 11. However, water is entrained in the direction opposite to that in Fig. 10; thus, the reproducibility of the experiment is not seen.

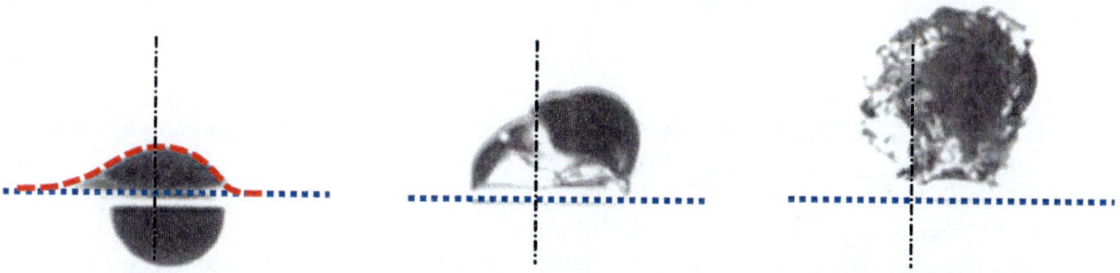

(a) $(t\text{-}t_0)/t^* = 0.9$ (b) $(t\text{-}t_0)/t^* = 1.1$ (c) $(t\text{-}t_0)/t^* = 1.3$

Fig. 11 Motion of sphere with $\rho_p = 1360$ kg/m^3 and behavior of air–water interface (2)

As illustrated in Figs. 10 and 11, when the duracon sphere ($\rho_p = 1360$ kg/m^3) passes through the water surface, the air–water interface on the top of the sphere is asymmetrical. This is due to the fact that the sphere does not collide vertically with the air–water interface. The Reynolds number of the sphere immediately before collision with the air–water interface is Re = 2.3 × 10^4. At this Reynolds number, a turbulent vortex with large regular fluctuations is emitted behind the sphere placed in a uniform flow [7]. Therefore, it is expected that the same structure as the vortex shedding described in [7] can be observed behind the duracon sphere in this experiment. Because the duracon sphere has the lowest density in this experiment, it is strongly affected by the horizontal force of the vortex shedding. Thus, duracon spheres do not collide perpendicularly with the air–water interface.

4 Conclusions

Spheres with a diameter d of 25.4 mm are launched vertically upward from water toward the air–water interface, and the modulation of the spheres' motion and the behavior of the air–water interface are investigated. Three types of spheres with different densities ρ_p are used: the almina sphere ($\rho_p = 3900$ kg/m^3), teflon sphere ($\rho_p = 2180$ kg/m^3), and duracon sphere ($\rho_p = 1360$ kg/m^3). A visualization experiment on the motion of a sphere in water demonstrates that the launch (initial) velocity of a sphere can be set with high accuracy. When the sphere passes through the air–water interface, it forms a complex interface while entraining water. In addition, it is demonstrated that spheres with small densities ρ_p do not always collide vertically with the air–'water interface, leading to different behaviors of the spheres in the air.

References

1. May A, Woodhull JC (1948) Drag coefficients of steel spheres entering water vertically. J Appl Phys 19:1109–1121
2. Dyke MV (1971) Functional dependence of drag coefficient of a sphere on Reynolds number. Phys Fluids 14:1038–1039
3. Bivin YK, Glukhov YM, Permyakov YV (1986) Vertical entry of solids into water. Fluid Dyn 20:835–841
4. Shi H-H, Kume M (2001) An experimental research on the flow field of water entry by pressure measurements. Phys Fluids 13:347–349
5. Wang J, Faltinsen OM, Lugni C (2019) Unsteady hydrodynamic forces of solid objects vertically entering the water surface. Phys Fluids 31:027101
6. Truscott TT, Epps BP, Munns RH (2016) Water exit dynamics of buoyant spheres. Phys Rev Fluids 1:074501
7. Sakamoto H, Haniu H (1990) A study on vortex shedding from spheres in a uniform flow. Trans ASME 112:386–392

Study on Washing Ability of Water Flows

Takuto Sasaki, Yu Nishio, and Takanobu Ogawa

Abstract This study investigates the flow on an inclined wall, especially focusing on the relationship between the wall shear stress and the flow behavior. Flows of a wall jet impinging on an inclined plate and rivulet flows is experimentally observed. An aqueous luminous paint is applied to the wall, and the other experiment is also carried out for a visualization of the washing ability of the water flow. The experimental result shows that the paint located at the side of the straight rivulets disappears as soon as the water passes through the region. A numerical simulation is also performed under the same condition as the first experiment. It is found that high wall shear stress regions lie in the side of the straight rivulets and the outer region of the bend in the meandering ones. This result agrees with the experimental observations. A velocity profile in the meandering rivulet indicates that the high shear region comes from the high speed region of the bend and the liquid being thinner at the edge of the rivulet.

Keywords Experiments · Numerical simulation · Rivulet flow · Wall shear stress magnitude

1 Introduction

A beverage can is used as a container of soft or alcoholic drink. The can is washed by much water to remove tiny particles or plastic pieces right before it is filled with a drink in the packaging process. The method of washing the can is simple: A water jet is ejected into the can, flows along the can surface, and finally sweeps away the small undesirable objects. The process, however, requires much water, and the reduction of the water consumption is demanded in beverage industries. The industries have tried to improve the washing process using a trial and error method, but it would seem that there is room for further improvement of the washing process using a method based on a physical investigation of washing the inside of the can.

T. Sasaki (✉) · Y. Nishio · T. Ogawa
Department of System Design Engineering, Seikei University, 3-3-1, kichijoji-kitamachi, Musashino-shi, Tokyo 1808633, Japan
e-mail: dm196308@cc.seikei.ac.jp

L. Venkatakrishnan et al. (eds.), *Proceedings of 16th Asian Congress of Fluid Mechanics*, Lecture Notes in Mechanical Engineering,
https://doi.org/10.1007/978-981-15-5183-3_18

The phenomenon related to the washing process, that is a particle resuspension, has been investigated by many research groups [1]. Some researches focus on the hydrodynamic forces such as drag and lift which provide roll or sliding motion with the particles [2, 3].

Phares et al. [4] experimentally showed the importance of the shear stress on the particle resuspension in an air flow under a simple flow configuration such as a two-dimensional channel flow. Their result was followed by a numerical simulation performed by Iimura et al. [5]. There, however, is less study investigating the effects of thin or rivulet water flows which often appear in the washing process on the particle resuspension.

This study aims to investigate the flow phenomena relevant to washing process to obtain the effective washing method with a small amount of water, focusing on the relationship between the wall shear stress and the flow behavior.

2 Visualization of a Water Flow on a Flat Plate

First, we study a water flow on a flat plate instead of an actual beverage can to understand the fundamental aspects of the flow and its wall shear stress. Figure 1 shows an experimental setup for flow visualization. A water is issued at the flow rate of 3.0m^3/min. from a nozzle located 150 mm above an acrylic plate, which is inclined at an angle $\theta = 15°$. The water from the nozzle falls on the plate and spreads over it and then flow down. This motion of the water is recorded at 60 FPS with a camera (CASIO, EOS Kiss X4). The experiment is performed in a dark room, and multiple lights are set around the plate for better visibility of the water surface. The recorded images are post-processed to enhance the water surface shape.

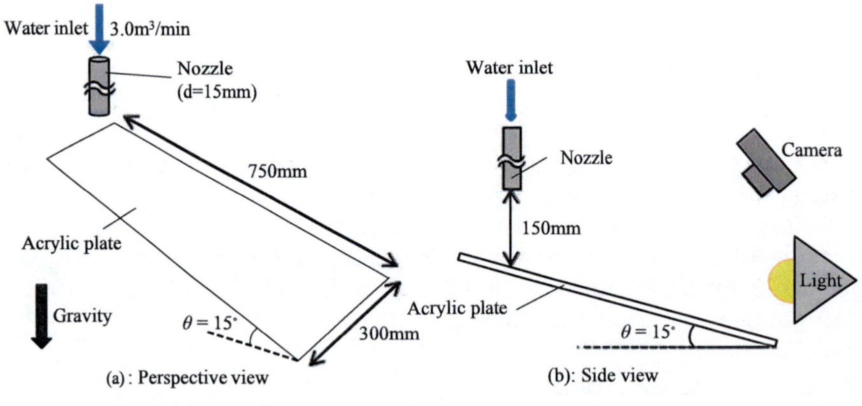

Fig. 1 Visualization experimental equipment of the water flow

3 Numerical Simulation

The same flow field as described in the previous section is numerically simulated by solving the 3D unsteady Navier–Stokes eqs. We use a finite volume method with the second-order upwind scheme and the second-order time integration, and the time step is 2.0×10^{-4} s. The VoF method is used for handling the free surface.

Figure 2a shows the computational domain, which consists of the region for the water from the nozzle and the region over the plate. The computational mesh distribution in the cross section is shown in Fig. 2b. A hexahedral mesh is used and is refined to about 0.1 mm along the paths of the water, and the mesh size elsewhere is about 20 mm. The total number of the computational mesh is about 8 million.

Table 1 shows the physical properties used in the simulation. The contact angle between the acrylic plate and the water shows hysteresis and is taken into account in the simulation. We measure the advancing and receding contact angles by another experiment, and they are shown in Table 2.

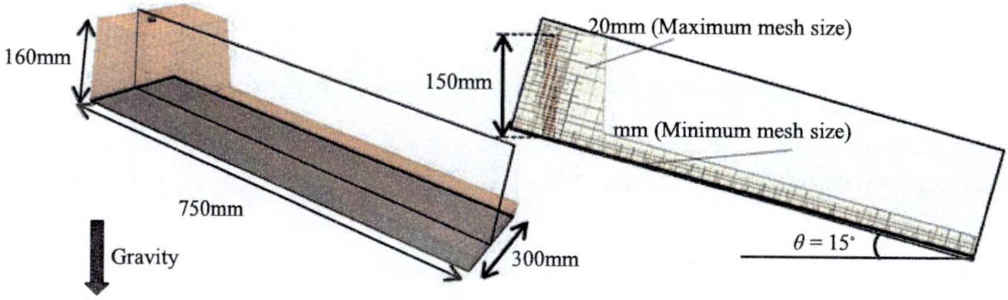

(a): Perspective view (b): Cross-sectional view

Fig. 2 Computational domain

Table 1 Physical property values

	Water	Air
Density (kg/m^3)	997.6	1.18
Viscosity coefficient (Pa s)	8.89×10^{-4}	0.186×10^{-4}
Surface tension (N/s)	7.4×10^{-2}	

Table 2 Dynamic contact angle

	Advancing	Reciding
Contact angle (°)	93.0	46.0

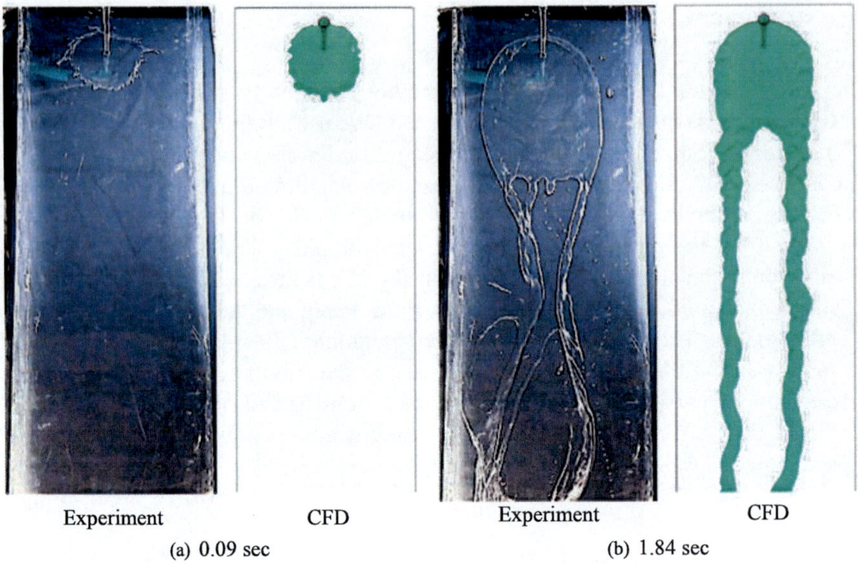

Fig. 3 Experimental and numerical result of the flow over the flat plate

4 Experimental and Numerical Results

Figure 3 shows the shape of the water surface flowing on the flat plate. At $t = 0.25$ s after the water from the nozzle impinges on the flat plate, it spreads forming a thin disk-like shape, as shown in Fig. 3a. At the tip of the disk, the fingering instability appears. The numerical result is qualitatively consistent with the experimental one, but the fingering is more intense in the experiment.

Then, the water flows down the plate splitting into two rivulets, as shown in Fig. 3b. In the experiment, these rivulets wind unsteadily. In the numerical result, however, this winding of the rivulet cannot be observed, and the rivulet is wider than the experimental result. The excess numerical diffusion might smear the water surface, and thus, the fingering and the winding of the rivulets become less pronounced.

5 Washing Ability of the Water Flow

We study the washing ability of the water flow on a flat plate again both experimentally and numerically.

In the experiment, we directly evaluate the washing ability by visualizing stains applied on the flat plate being washed out by the water flow over the plate. Figure 4 shows the experimental setup. We use aqueous fluorescent ink as stains and apply it on the acrylic plate. The water issued from the nozzle, as described in Sec. 2, flows

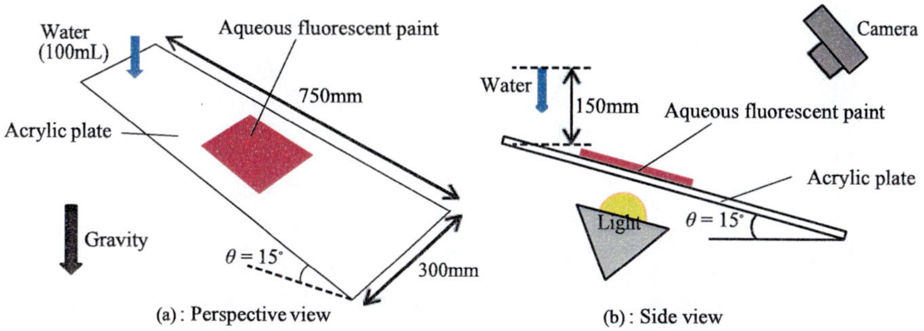

Fig. 4 Experimental equipment conducted to investigate wall shear stress

down the plate and washes away the stains. The stains are illuminated by the light underneath the plate, and the motion of the water and the stains being washed away are recorded with a camera placed above. The regions where the stain is removed can be seen as white regions. The other conditions of the flow field are the same as those in Sect. 2.

Figure 5 shows the time sequence of the stain being washed by the water flow. At $t = 0.5$ s (Fig. 5b), the water over the plate forms straight rivulets, and the stain is wiped at the edges of the rivulets. Then, as shown in Fig. 5c, some of the rivulet begins to wind. In the winding rivulets, the stain is more washed away at the outer edge of the winding rivulet than at the inner edge. In this flow field, the wall shear stress plays a dominant role in washing the stain, and this result shows that the water flow on the plate has higher wall shear stress at its edges.

Now, we plot the wall shear stress of the numerical result in Fig. 6. The enlarged view of the rivulet (Fig. 6b) shows that the wall shear stress becomes large at the edge of the rivulet, and this agrees with the experimental result. The rivulet sometimes winds, and the shear stress of a wound rivulet is shown in Fig. 6c. The highest shear stress is observed at the outer side of the wound rivulet.

Now, we observe velocity profiles for a detailed discussion. Figure 7a shows the wall shear stress distribution of the winding rivulet and the velocity magnitude in the cross section of the rivulet. The region shown in Fig. 7a corresponds to that of Fig. 6c.

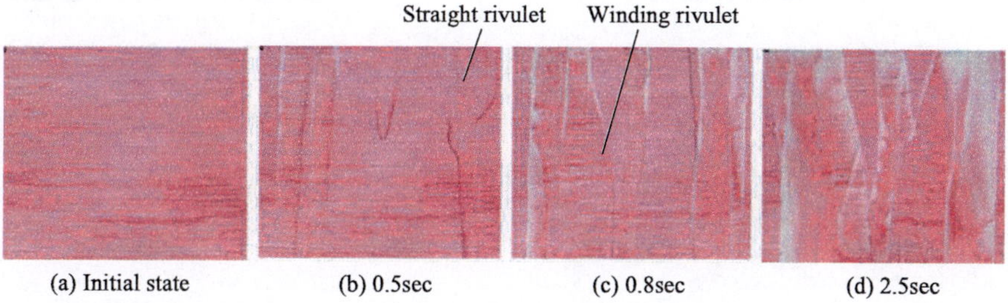

Fig. 5 Results of experiments conducted to investigate wall shear stress

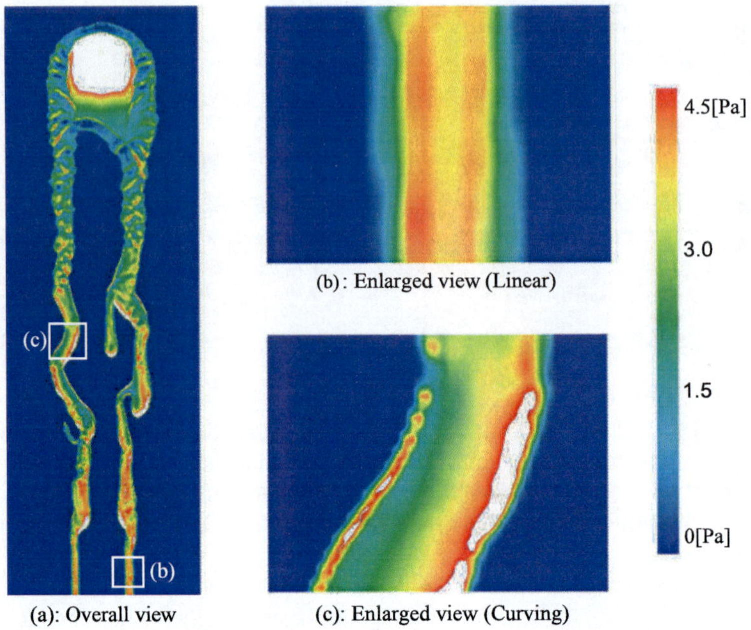

Fig. 6 Wall shear stress of the numerical result at $t = 1.3882$ s

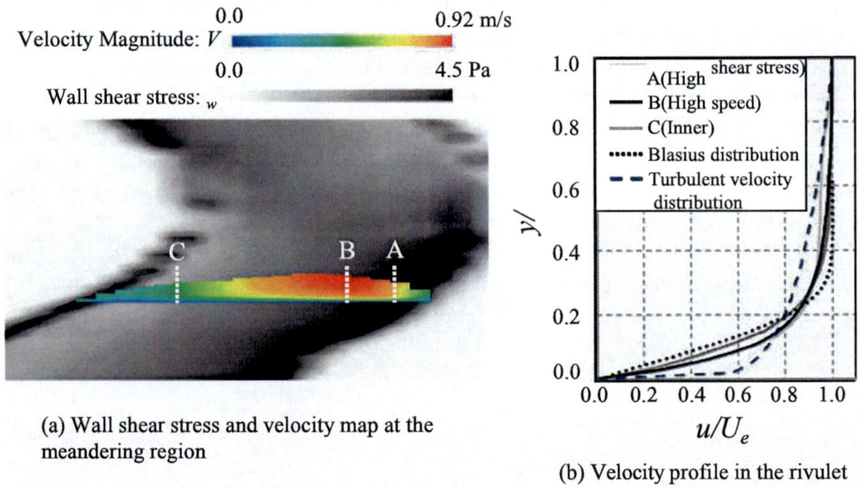

Fig. 7 Velocity profiles and wall shear stress of the meandering rivulet.

The velocity magnitude distribution indicates that the high speed region lies in the outer rim of the bend, which is thicker than the inner region due to the centrifugal force.

Figure 7b shows the velocity profiles at the high wall shear region, the highest speed region, the inner region of the winding rivulets, and these locations are denoted by A, B, and C in Fig. 7a, respectively. The vertical axis represents the height from the wall normalized by those of the rivulet, and the horizontal axis is the velocity normalized by those at the surface of the rivulet. The velocity profiles of the Blasius solution and the turbulent boundary layer are also plotted in Fig. 7b. The experimental profiles are close to those of the Blasius solution, and this water flow seems to be laminar.

The velocity profile C at the inner region is obviously different from the others. A comparison of two profiles A and B in Fig. 7b shows that the two curves agree with each other near the wall. This agreement in the velocity profile near the wall means that the non-dimensional wall shear stresses are almost the same. On the other hand, the dimensional quantity, that is the wall shear stress, depends on the edge velocity and the thickness of the rivulets. Thus, it is found that the edge velocity and the thickness are some of the most important parameters on the wall shear stress in the rivulet flows and the high shear region in this flow comes from the high speed region of the bend and the liquid being thinner at the edge of the rivulet.

6 Conclusions

Flows of a wall jet impinging on an inclined plate and rivulet flows were experimentally and numerically investigated. The flow shape in the numerical result agrees with the experimental one. An aqueous luminous paint was applied to the wall, and a similar experiment to the first one was carried out to validate the numerical observation. The experimental result shows that the paint located at the side of the straight rivulets disappears as soon as the water passes through. The numerical results show that the region with a high wall shear stress was located in the both sides of the straight rivulets and the outer region of the bend in the meandering ones. It is found that the edge velocity and the thickness are some of the most important parameters on the wall shear stress in the rivulet flows.

References

1. Sehmel GA (1980) Particle resuspension: a review. Environ Int 4:107–127
2. Nicholson KW (1988) A review of particle resuspension. Atmos Environ 22:2639–2651
3. Vainshtein P, Ziskind G, Fichman M, Gutfinger C (1997) Kinetic model of particle resuspension by drag force. Phys Rev Lett 78:551–554

4. Phares DJ, Smedley GT, Flagan RC (2000) Effect of particle size and material properties on aerodynamic resuspension from surfaces. J Aerosol Sci 31:1335–1353
5. Iimura K, Watanabe S, Suzuki M, Hirota M, Higashitani K (2009) Simulation of entrainment of agglomerates from plate surfaces by shear flows. Chem Eng 64:1455–1461

On the Local Axisymmetry of a Vortex

Václav Kolář and Jakub Šístek

Abstract The local 3D cylindrical axisymmetry of a vortex is examined. Simple scalar quantities with a clear geometrical interpretation are introduced as suitable analytical tools for visualization of the local axisymmetry and skewness of the flow in the domain characterized by complex eigenvalues of the velocity-gradient tensor. Both complex and real eigenvectors are employed, including the real-valued dual-eigenvectors representing the orthogonal elongated directions in the swirl plane. The proposed quantities are applied to an impulsively started incompressible flow around an inclined flat plate.

Keywords Vortex · Axisymmetry · Velocity-gradient tensor · Complex eigenvectors · Dual-eigenvectors

1 Introduction

The physical reasoning for vortex identification includes a number of "required properties" of both Eulerian and Lagrangian nature, see, among others, the recent review paper by Epps [1], and thesis of Holmén [2]. Up to now, in view of numerous vortex-identification applications, the central position still belongs to Eulerian local region-type methods (i.e., aiming at vortex region) based on the velocity-gradient tensor $\nabla\mathbf{u}$, though these criteria are not providing the same results in different rotating frames (cf. the requirement of frame indifference according to Haller [3]). The widely used 3D vortex-identification criteria include Q (Hunt et al. [4]), λ_2 (Jeong and Hussain [5]), Δ (Dallmann [6], Vollmers et al. [7], Chong et al. [8]), and λ_{ci} (Zhou et al. [9]). According to Chakraborty et al. [10] and Gao and Liu [11], these pointwise criteria derived from $\nabla\mathbf{u}$ can be mostly regarded as eigenvalue-based criteria, though the

V. Kolář (✉)
Institute of Hydrodynamics of the Czech Academy of Sciences, 16612 Prague, Czech Republic
e-mail: kolar@ih.cas.cz

J. Šístek
Institute of Mathematics of the Czech Academy of Sciences, 11567 Prague, Czech Republic

L. Venkatakrishnan et al. (eds.), *Proceedings of 16th Asian Congress of Fluid Mechanics*, Lecture Notes in Mechanical Engineering,
https://doi.org/10.1007/978-981-15-5183-3_19

quantity λ_2 can be expressed in terms of the eigenvalues of $\nabla\mathbf{u}$ in the special case only when the eigen-basis vectors are orthonormal [10]. Further, the eigenvector-based definition of the recently introduced vortex vector has been proposed in [11]. It is not without interest that the line-type vortex-identification schemes (i.e., aiming at vortex skeleton) may also successfully employ both the eigenvalues and eigenvectors of $\nabla\mathbf{u}$ as shown in the algorithm of Sujudi and Haimes [12].

The present study concentrates on the property of axisymmetry of $\nabla\mathbf{u}$, or more precisely, on the measures of departure from the local flow axisymmetry of a vortex expressed via the eigenvectors of $\nabla\mathbf{u}$ in the region in which the eigenvalues of $\nabla\mathbf{u}$ are complex (a pair of complex conjugate eigenvalues occurs). To study the symmetry property of a vortex is not new. Nakayama [13] examined in detail the properties of symmetry (axisymmetry or skewness) of the vortical flow in the swirl plane, that is, in the plane spanned by the complex eigenvectors of $\nabla\mathbf{u}$. The vortex properties of "swirlity" and "sourcity" in a planar cross-section have been introduced by decomposing the flow-field into azimuthal and radial parts [13]. For the application of a symmetry (or skewness) eigenvector-based measure in the swirl plane, see Nakayama et al. [14]. However, the main difference in the present contribution lies in simultaneous approach to all the three eigen-basis vectors including the real eigenvector. Consequently, its corresponding skewness relative to the swirl plane, or relative to the complex eigenvectors (contained in the swirl plane), is taken into account and simple analytical tools are proposed. That is, here, we deal with the full local 3D *cylindrical axisymmetry* of vortical flows while previous studies [13, 14] dealt with the symmetry aspects in the swirl plane only.

2 Departure from the Local Cylindrical Axisymmetry

The ideal local cylindrical axisymmetry of the flow described in terms of $\nabla\mathbf{u}$ and conveniently expressed in the frame of strain-rate principal axes reads (without loss of generality and applicability to compressible flows, the strain rate is assumed deviatoric; vorticity and strain-rate components are denoted through ω and σ, respectively)

$$\nabla\mathbf{u} \equiv \begin{pmatrix} -\sigma & -\omega & 0 \\ \omega & -\sigma & 0 \\ 0 & 0 & 2\sigma \end{pmatrix}. \tag{1}$$

Such a local axisymmetry expressed by $\nabla\mathbf{u}$ is fulfilled provided that (i) the strain-rate tensor (the symmetric part of $\nabla\mathbf{u}$) is axisymmetric, and (ii) the vorticity vector coincides with the strain-rate axis of axisymmetry (recall that the vorticity tensor is given by the antisymmetric part of $\nabla\mathbf{u}$). The relative motion near a point according to the axisymmetric strain-rate tensor in (1) represents a uniaxial isochoric stretching (or contraction) coupled with a uniform radial contraction (or stretching).

Let us show that the necessary and sufficient condition for an arbitrary input $\nabla \mathbf{u}$ with non-zero vorticity to be axisymmetric is the orthogonality of eigen-basis vectors. This orthogonality condition is equivalent to the condition that $\nabla \mathbf{u}$ represents a normal matrix. That is, redenoting $\nabla \mathbf{u}$ simply by $\mathbf{G}$,

$$\mathbf{G}\mathbf{G}^{\mathrm{T}} = \mathbf{G}^{\mathrm{T}}\mathbf{G} \text{ or } (\mathbf{G}\mathbf{G}^{\mathrm{T}} - \mathbf{G}^{\mathrm{T}}\mathbf{G}) = \mathbf{0}. \tag{2a, b}$$

Let us examine the 3×3 matrix $(\mathbf{G}\mathbf{G}^{\mathrm{T}} - \mathbf{G}^{\mathrm{T}}\mathbf{G})$ in the reference frame of strain-rate principal axes for not necessarily divergence-free input $\mathbf{G}$ with non-zero vorticity (i.e., representing an arbitrary asymmetric tensor)

$$\mathbf{G} \equiv \begin{pmatrix} \sigma_1 & -\omega_3 & \omega_2 \\ \omega_3 & \sigma_2 & -\omega_1 \\ -\omega_2 & \omega_1 & \sigma_3 \end{pmatrix}, \tag{3}$$

$$\mathbf{G}\mathbf{G}^{\mathrm{T}} - \mathbf{G}^{\mathrm{T}}\mathbf{G} = \begin{pmatrix} 0 & 2\omega_3(\sigma_1 - \sigma_2) & 2\omega_2(\sigma_3 - \sigma_1) \\ 2\omega_3(\sigma_1 - \sigma_2) & 0 & 2\omega_1(\sigma_2 - \sigma_3) \\ 2\omega_2(\sigma_3 - \sigma_1) & 2\omega_1(\sigma_2 - \sigma_3) & 0 \end{pmatrix}. \tag{4}$$

According to the structure of the symmetric matrix (4), the condition (2a, b) is satisfied only by the ideal local cylindrical axisymmetry of the velocity-gradient tensor expressed by (1). Note that the principal strain-rate differences in (4) describe just a deviatoric motion assumed already for simplicity in (1).

The structure (1) has a pair of complex conjugate eigenvalues. For an arbitrary asymmetric input $\mathbf{G}$, the existence of complex conjugate eigenvalues ensures the largest volumetric region of a vortex core among the widely used local region-type criteria Q, λ_2, Δ and λ_{ci} for all practical purposes [10] while a priori excluding non-zero vorticity non-swirling regions like shear layers. Consequently, in a local reference frame moving with the examined point, the streamline pattern is spiralling or closed on a plane spanned by the complex eigenvectors (the so-called swirl plane). Within the critical-point theory, these points are elliptic ones (focus or center). The corresponding eigen decomposition of the velocity-gradient tensor in the domain characterized by complex conjugate eigenvalues (shortly, the complex domain) reads where $(\lambda_{cr} \pm i\lambda_{ci}, \lambda_r)$ denote the eigenvalues and $(\mathbf{v}_{cr} \pm i\mathbf{v}_{ci}, \mathbf{v}_r)$ stand for the associated eigenvectors. The condition that the discriminant of the characteristic equation for $\mathbf{G}$ is positive, $\Delta > 0$, implies that $\mathbf{G}$ has complex eigenvalues ($\lambda_{ci} > 0$). For incompressible flows, the discriminant Δ reads $\Delta = (Q/3)^3 + (R/2)^2$, where Q and R are the second and third invariants of $\mathbf{G}$ [8]. For compressible flows, the same condition holds for deviatoric quantities [15].

$$\mathbf{G} = \begin{pmatrix} | & | & | \\ \mathbf{v}_{cr} + i\mathbf{v}_{ci} & \mathbf{v}_{cr} - i\mathbf{v}_{ci} & \mathbf{v}_r \\ | & | & | \end{pmatrix} \begin{pmatrix} \lambda_{cr} + i\lambda_{ci} & 0 & 0 \\ 0 & \lambda_{cr} - i\lambda_{ci} & 0 \\ 0 & 0 & \lambda_r \end{pmatrix}$$

$$\begin{pmatrix} | & | & | \\ \mathbf{v}_{cr} + i\mathbf{v}_{ci} & \mathbf{v}_{cr} - i\mathbf{v}_{ci} & \mathbf{v}_r \\ | & | & | \end{pmatrix}^{-1} \tag{5}$$

The departure from the local 3D cylindrical axisymmetry (1) leading to the inequality $(\mathbf{G}\mathbf{G}^{\mathrm{T}} - \mathbf{G}^{\mathrm{T}}\mathbf{G}) \neq \mathbf{0}$ can be examined through the non-orthogonality of eigen-basis vectors within the region of complex conjugate eigenvalues $(\lambda_{cr} \pm i\lambda_{ci})$. However, the pair of associated complex conjugate eigenvectors $(\mathbf{v}_{cr} \pm i\mathbf{v}_{ci})$ is not unique as multiplying them with any complex number yields another valid pair. Following the related *asymmetric tensor* analysis in [16, 17], the swirl plane spanned by the complex eigenvectors can be conveniently expressed in terms of the real-valued *dual-eigenvectors* representing the orthogonal elongated directions of the local linearization near a point in the complex domain. For the determination of dual-eigenvectors, see Appendix.

The first-choice approach toward expressing the departure from the 3D cylindrical axisymmetry by a single scalar function is proposed in terms of cosines of angles between three pairs of eigenvectors (using a generalized dot product for vectors with complex entries). A pair of complex conjugate eigenvectors is given in terms of the dual-eigenvectors in the form $(\boldsymbol{a}^* \pm i\boldsymbol{b}^*)$, see Appendix. Denoting the three mutual angles between eigenvectors (where the three eigenvectors are simply labeled 1, 2, 3) ε_{12}, ε_{23}, and ε_{31}, expressing the cosine measure of non-orthogonality, and conveniently shifting the negative of it for an easier isosurface visualization to a measure of orthogonality, denoted as $\delta_{\cos}$, we obtain

$$\delta_{\cos} = 1 - \sqrt{\left[(\cos\varepsilon_{12})^2 + (\cos\varepsilon_{23})^2 + (\cos\varepsilon_{31})^2\right]/3}. \tag{6}$$

Due to its different nature, we may need to distinguish between (i) the skewness in the swirl plane and (ii) the real eigenvector skewness relative to the swirl plane. For this purpose, an alternative axisymmetry measure is proposed below. The main reason is as follows: even the ideal cylindrical vortex exhibits (outside the vortex axis) local shearing due to a relative rotational motion of concentric layers and the resulting complex-eigenvector skewness in the swirl plane may be therefore considered as an inherent vortex property. On the other hand, the non-orthogonality of the real eigenvector relative to the swirl plane stems from the 3D shearing effects of different geometrical configuration than the "inherent local shearing" in the swirl plane (reflected by the planar elliptic critical points). For more general 3D approach to combining (superimposing) shearing components in terms of the *purely* asymmetric tensors interpreted through material shearing elements of the relative motion near a point as planes, lines, or points, see Kolář [18].

According to the Appendix, the swirl plane can be characterized by the mutually orthogonal major and minor dual-eigenvectors. The ratio $|\boldsymbol{b}^*|/|\boldsymbol{a}^*|$ of minor to major dual-eigenvector magnitudes has been employed in [13, 14] to describe the relative axisymmetry or skewness in the swirl plane in the range (0, 1). This ratio can be related to $\delta_{\cos}$: if denoting the angle between complex eigenvectors $(\boldsymbol{a}^* \pm i\boldsymbol{b}^*)$ simply

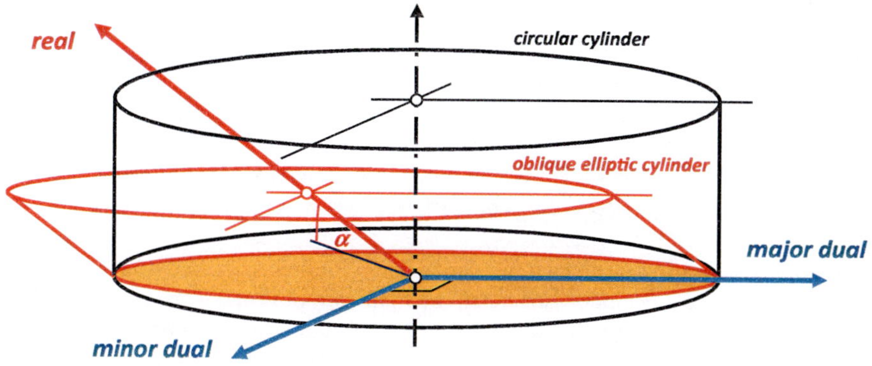

Fig. 1 Scheme of eigenvectors for the derivation of the volume measure δ_{vol} (eigenvector length not in scale)

as ε, $|b^*|/|a^*| = \tan(\varepsilon/2)$. Further, we can equivalently interpret this ratio of dual-eigenvector magnitudes as the area ratio of the skewed elliptical area to the full circular area (of the radius of major dual-eigenvector magnitude). Note that the ideal circular geometry near a point is associated with the local rigid body rotation in the swirl plane while the skewed elliptical shape indicates an inherent shearing component in the swirl plane. The skewness is proportional to the shearing component relative to the swirling component.

The planar axisymmetry or skewness expressed through the area ratio can be easily extended to a 3D cylindrical axisymmetry or skewness by means of the volume ratio. In the present context, this ratio is conveniently expressed in the range (0, 1) by the relative volume of an oblique elliptic cylinder to the volume of an ideal circular cylinder, as shown in Fig. 1. The relative volume of an oblique elliptic cylinder is given by the relative elliptical base area $|b^*|/|a^*|$ times the relative oblique-cylinder height. For a unit height of the circular cylinder, the relative height of the oblique elliptic cylinder is given by $\sin\alpha$ where α denotes just the angle of the real eigenvector (of arbitrary direction!) relative to the swirl plane, as indicated in Fig. 1. The proposed volume measure δ_{vol} of cylindrical axisymmetry then reads

$$\delta_{vol} = \left(|b^*|/|a^*|\right) \cdot \sin\alpha. \tag{7}$$

3 Application

Both measures, δ_{cos} and δ_{vol}, have been applied to an impulsively started incompressible flow around a flat plate (aspect ratio 2) at an angle of attack of 30 deg solved numerically for Reynolds number Re = 300. Details about the numerical simulation of this problem can be found in Šístek and Cirak [19]. Figure 2 shows the isosurface outcome of $\delta_{\cos}$ (virtually embedded in the complex domain) alongside with the isosurface of the complex domain at $\lambda_{ci} = 0.1$ (a low threshold of λ_{ci} is necessary to avoid a noisy region close to zero isovalue). Unlike the quantity λ_{ci} representing the local swirling strength (strictly said, the time period for completing one revolution of the streamline on the plane spanned by the complex eigenvectors is given by $2\pi/\lambda_{ci}$ [10]), the quantity $\delta_{\cos}$ reflects the geometry of $\mathbf{G}$ and its isosurface is always a subset of the complex domain, $\lambda_{ci} > 0$.

Figure 3 shows the results for δ_{vol} alongside with the isosurface of the complex domain at $\lambda_{ci} = 0.1$. Figure 4 separates the main components of the multiplication (7) to distinguish both the skewness in the swirl plane by $|\boldsymbol{b}^*|/|\boldsymbol{a}^*|$ and the real eigenvector skewness relative to the swirl plane by $\sin\alpha$. The observed values of α

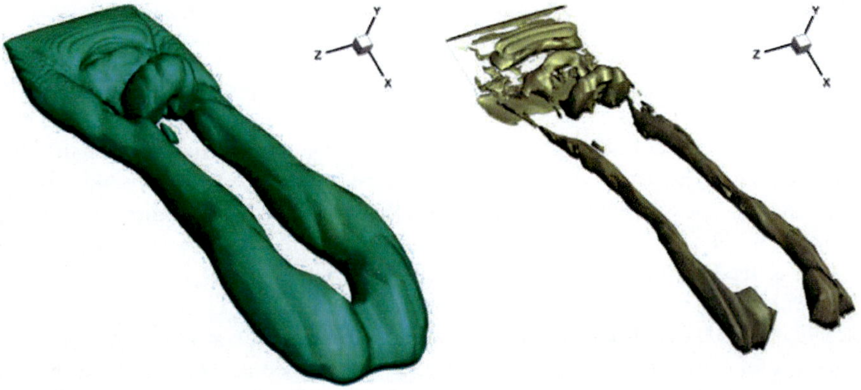

Fig. 2 Flow around a flat plate: Complex domain (isovalue $\lambda_{ci} = 0.1$, left) and $\delta_{\cos}$ (isovalue 0.55, right)

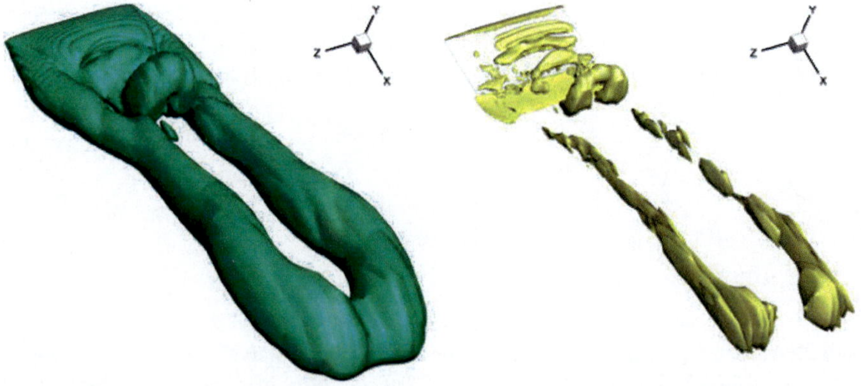

Fig. 3 Flow around a flat plate: Complex domain (isovalue $\lambda_{ci} = 0.1$, left) and δ_{vol} (isovalue 0.45, right)

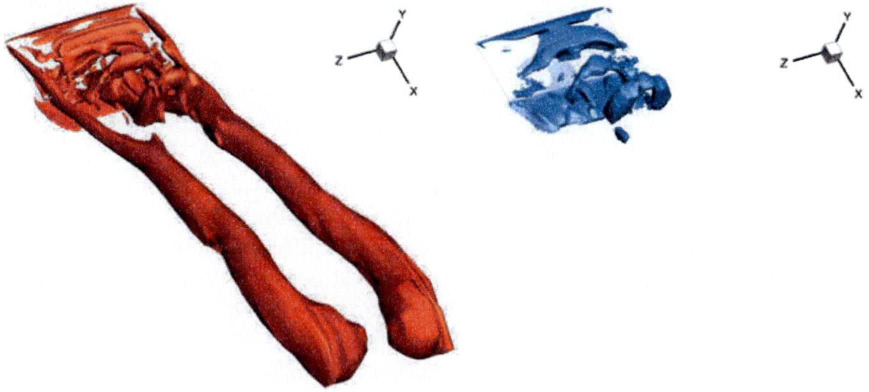

Fig. 4 Flow around a flat plate: The ratio $|\boldsymbol{b}^*|/|\boldsymbol{a}^*|$ (isovalue 0.5, left) and $\sin\alpha$ (isovalue 0.9, right)

are relatively significant, typically ranging between $\pi/4$ and $\pi/2$. The threshold 0.9 employed for $\sin\alpha$ in Fig. 4 corresponds to $\alpha = 64$ deg. Interestingly, the near wake structures, with the exception of longitudinal "vortex legs" persisting far downstream, are characterized by relatively higher values of α. However, the longitudinal vortices tend to exhibit relatively significant skewness of the real eigenvector to the swirl plane. A plausible explanation could be a longitudinal shearing motion oriented downstream in the main flow direction associated with a strong axial component of the resulting helical motion similarly as for wingtip vortices.

Figure 5 compares the outcome for both measures dealing with cylindrical axisymmetry, δ_{vol} and $\delta_{\cos}$, indicating a fairly good similarity.

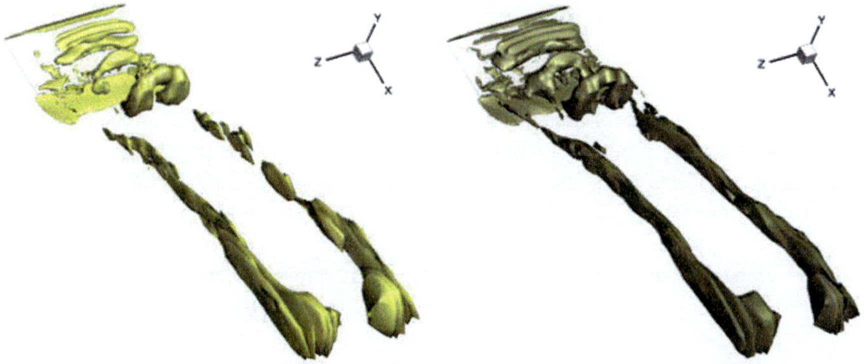

Fig. 5 Flow around a flat plate: Comparison of δ_{vol} (isovalue 0.45, left) and $\delta_{\cos}$ (isovalue 0.55, right)

4 Conclusions

The property of the local 3D cylindrical axisymmetry of a vortex is examined as an extension of previous studies dealing with the symmetry aspects in the swirl plane only. Simple scalar quantities, cosine measure $\delta_{\cos}$ defined by (6) and volume measure δ_{vol} defined by (7), are proposed as suitable analytical tools for visualization in the complex domain of the velocity-gradient tensor. Both measures have a clear geometrical interpretation and provide reasonably similar results as shown in Fig. 5. The components of the volume measure δ_{vol} are also depicted separately in Fig. 4 to distinguish between complex-eigenvector skewness (in the swirl plane) given by $|\boldsymbol{b}^*|/|\boldsymbol{a}^*|$ and the real eigenvector skewness (to the swirl plane) given by $\sin\alpha$. It is shown that, unlike near wake structures, the longitudinal vortices persisting far downstream exhibit relatively significant skewness of the real eigenvector to the swirl plane due to longitudinal shearing motion. This motion, oriented in the main flow direction, is associated with a strong axial component of the resulting helical motion similarly as it can be found for wingtip vortices. Finally, the role of the quantity $(\mathbf{G}\mathbf{G}^{\mathrm{T}} - \mathbf{G}^{\mathrm{T}}\mathbf{G})$ used at the beginning of Sect. 2 needs further investigation in the complex domain (where the discriminant Δ is positive). The calculation does not need the eigen decomposition, however, it lacks a clear interpretation for $(\mathbf{G}\mathbf{G}^{\mathrm{T}} - \mathbf{G}^{\mathrm{T}}\mathbf{G}) \neq \mathbf{0}$.

Acknowledgements This work was supported by the Czech Science Foundation through Grant 18-09628S, and by the Czech Academy of Sciences through RVO:67985874 and RVO:67985840.

Appendix

The determination of dual-eigenvectors is briefly summarized according to [16]. Assume a real matrix with a pair of complex conjugate eigenvalues as in (5). The complex eigenvectors are not unique since multiplying either of them with a complex number yields another set of eigenvectors. However, if $(\boldsymbol{a} + i\boldsymbol{b})$ is an eigenvector corresponding to the eigenvalue $(\lambda_{cr} + i\lambda_{ci})$ then $(\boldsymbol{a} - i\boldsymbol{b})$ is an eigenvector corresponding to $(\lambda_{cr} - i\lambda_{ci})$. The multiplication by a complex number reads.

$$(\cos\phi + i\sin\phi)(\boldsymbol{a} + i\boldsymbol{b}) = (\boldsymbol{a}\cos\phi - \boldsymbol{b}\sin\phi) + i(\boldsymbol{a}\sin\phi + \boldsymbol{b}\cos\phi) = \left(\boldsymbol{a}^* + i\boldsymbol{b}^*\right). \tag{8}$$

The need to define a unique pair of complex eigenvectors leads to the natural choice of a pair of real-valued vectors $\left(\boldsymbol{a}^*, \boldsymbol{b}^*\right)$, the so-called dual-eigenvectors, that are mutually perpendicular. To satisfy the orthogonality condition $\boldsymbol{a}^* \cdot \boldsymbol{b}^* = 0$, it requires leading to a quadratic equation for $\tan\phi$ of the form

$$(\boldsymbol{a}\cos\phi - \boldsymbol{b}\sin\phi) \cdot (\boldsymbol{a}\sin\phi + \boldsymbol{b}\cos\phi) = 0, \tag{9}$$

$$\boldsymbol{a} \cdot \boldsymbol{b}(\tan\phi)^2 + \left(|\boldsymbol{b}|^2 - |\boldsymbol{a}|^2\right)\tan\phi - \boldsymbol{a} \cdot \boldsymbol{b} = 0. \tag{10}$$

This quadratic equation has usually two distinct solutions which are mutually perpendicular. To cope with the problem of two different ways to align them we choose the solution $(\boldsymbol{a}^*, \boldsymbol{b}^*)$ for which $|\boldsymbol{a}^*| \geq |\boldsymbol{b}^*|$. Then $\boldsymbol{a}^*$ denotes the major dual-eigenvector and $\boldsymbol{b}^*$ denotes the minor dual-eigenvector.

References

1. Epps BP (2017) Review of vortex identification methods. AIAA Paper 2017–0989
2. Holmén V (2012) Methods for vortex identification. Master's Theses in Mathematical Sciences LUTFMA-3239–2012, Lund University Libraries
3. Haller G (2005) An objective definition of a vortex. J Fluid Mech 525:1–26
4. Hunt JCR, Wray AA, Moin P (1988) Eddies, streams, and convergence zones in turbulent flows. Report CTR-S88, Center for Turbulence Research, Stanford, pp 193–208
5. Jeong J, Hussain F (1995) On the identification of a vortex. J Fluid Mech 285:69–94
6. Dallmann U (1983) Topological structures of three-dimensional flow separation. TR-221-82-A07, DLR, German Aerospace Center, Goettingen
7. Vollmers H, Kreplin H-P, Meier HU (1983) Separation and vortical-type flow around a prolate spheroid—evaluation of relevant parameters. In: Proc. of the AGARD Symp. on Aerodynamics of Vortical Type Flows in Three Dimensions AGARD-CP-342, Rotterdam, Netherlands, April 1983
8. Chong MS, Perry AE, Cantwell BJ (1990) A general classification of three-dimensional flow fields. Phys Fluids A 2(5):765–777
9. Zhou J, Adrian RJ, Balachandar S, Kendall TM (1999) Mechanisms for generating coherent packets of hairpin vortices in channel flow. J Fluid Mech 387:353–396
10. Chakraborty P, Balachandar S, Adrian RJ (2005) On the relationships between local vortex identification schemes. J Fluid Mech 535:189–214
11. Gao Y, Liu C (2018) Rortex and comparison with eigenvalue-based vortex identification criteria. Phys Fluids 30(8):085107
12. Sujudi D, Haimes R (1995) Identification of swirling flow in 3-D vector fields. AIAA Paper 1995-1715
13. Nakayama K (2014) Physical properties corresponding to vortical flow geometry. Fluid Dyn Res 46(5):055502
14. Nakayama K, Ohira Y, Yamada S (2014) A new parameter in vortex identification and visualization: symmetry of vortical flow. ASME Proc. - Paper IMECE2014-39859
15. Kolář V (2009) Compressibility effect in vortex identification. AIAA J 47(2):473–475
16. Zheng X, Pang A (2005) 2D asymmetric tensor analysis. In: Proc IEEE Visualization 2005, Oct 2005
17. Zhang E, Yeh H, Lin Z, Laramee RS (2009) Asymmetric tensor analysis for flow visualization. IEEE Trans Vis Comput Graph 15(1):106–122
18. Kolář V (2007) Vortex identification: new requirements and limitations. Int J Heat Fluid Flow 28(4):638–652
19. Šístek J, Cirak F (2015) Parallel iterative solution of the incompressible Navier-Stokes equations with application to rotating wings. Comput Fluids 122:165–183

A Computational Analysis of Fluid Mixing Characteristics Inside a Microchannel by Pulsating Streams

Dhiman Roy, Taufiq Hasan, A. B. M. Toufique Hasan, and A. K. M. Sadrul Islam

Abstract In a microfluidic channel, mixing becomes troublesome as it is obviously constrained by mass diffusion at low Reynolds number. This work computationally analyzes the use of pulsating streams for possible enhancement of fluid mixing in a microchannel at Reynolds number of 0.2. A simple two-dimensional T-shaped microchannel is considered for this study. Two pulsatile streams are kept out of phase by 90° while the pulsation frequency is kept constant at 7.5 Hz. However, the ratio of pulse amplitude to base velocity has been varied in the range of 5.0–12.5. Moreover, the ratio of pulse amplitudes of two streams has been changed from 1.0 to 2.0. It is observed that an increment of pulse amplitude to base velocity ratio results an increase in the degree of mixing inside the microchannel. In addition, fluctuation characteristics in mixing degree decrease with an increase in pulse amplitude. In case of varied pulse amplitudes of two streams, results showed that the quality of mixing is enhanced by increasing the ratio until 1.7. Finally, the required minimum length of a microchannel for a specific degree of mixing is identified from the results of the present work.

Keywords Pulsating streams · Microfluidics · Micromixing · Computations

D. Roy · T. Hasan · A. B. M. Toufique Hasan (✉)
Department of Mechanical Engineering, Bangladesh University of Engineering and Technology (BUET), Dhaka 1000, Bangladesh
e-mail: toufiquehasan@me.buet.ac.bd

A. K. M. Sadrul Islam
Department of Mechanical Engineering (Retired), Bangladesh University of Engineering and Technology (BUET), Dhaka 1000, Bangladesh

L. Venkatakrishnan et al. (eds.), *Proceedings of 16th Asian Congress of Fluid Mechanics*, Lecture Notes in Mechanical Engineering,
https://doi.org/10.1007/978-981-15-5183-3_20

1 Introduction

Microfluidics refers to such devices which control and manipulate fluid flows with length scales less than a millimeter. Microfluidics is getting special attention and experiencing massive progress for its impressive potentials in many advanced technical applications. Such applications in DNA microarrays, drug invention, and lab-on-a-chip devices require mixing of various reagents. In these microscale devices, fluid flow is laminar and mixing of reagents mostly relies on diffusion which makes it very challenging. Moreover, it is required to have a good quality of mixing within a standard time limit, rational channel length, and at a cost-effective power consumption [1].

The principle of minimizing the diffusion distance is applied in many micromixers to ensure rapid mixing. It can be done using miniature agitators. Other techniques involving external energy source or by smart geometrical manipulations which induce secondary flow and promote chaotic advection. The quality of mixing can be enhanced significantly by combining these both methods. Micromixers can be classified as active and passive. Active micromixers use an external source of energy such as ultrasonic vibration, electro-osmosis, magnetic-force-based techniques, and so on. This type of micromixers need an additional control system to regulate the streams to be mixed. On the other hand, in passive micromixers, mixing relies on molecular diffusion assisted by geometrical manipulations. Due to complexity in manipulating microscale geometries, passive micromixers are sometimes very difficult to manufacture [2]. Glasgow and Aubry [3] proposed that the microfluidic mixing can be greatly enhanced by using pulsating steams. They studied fluid mixing under pulsating flow condition in a T-channel. It is the combined effect of geometrical manipulation and external excitation. Fluid is pulsed from two inlets to periodically distort the interface between the reagents to enhance mixing. Glasgow et al. [4], and Goullet et al. [5] also studied mixing behavior using time pulsing in different channel geometries focusing on mixing in a T-channel configuration by varying some dimensionless parameters and studied their effect on the mixing process. The results from the studies [3–5] do not account for some important factors like flow reaching the quasi-stationary state, appropriate time step size, sufficiently small mesh size, proper channel length, energy consumption, etc. With a view to refining these findings, Cheaib et al. [6] investigated the effects of pulse frequency, phase difference, sidewalls, and Reynolds number on the degree of mixing.

In this present work, computational fluid dynamics (CFD) simulations are employed to investigate the effect of a wide variety of pulse amplitudes on the quality of mixing and the mixing length. This also focuses on the effect of different ratios of pulse amplitudes of the two pulsating streams imposed at the inlets of the microchannel.

2 Numerical Method and Validation

In this study, mixing between two fluids in a T-channel by pulsating streams is studied. The inlet velocities of the two streams are defined as

$$V_1(t) = \bar{V} + \delta V_1 \sin(2\pi f t) \tag{1}$$

$$V_2(t) = \bar{V} + \delta V_2 \sin(2\pi f t + \phi) \tag{2}$$

where $\bar{V}$ is the base velocity, f is the pulsating frequency, and ϕ is the phase shift. δV_1 and δV_2 are the amplitude of pulsations of stream 1 and stream 2, respectively. Quality of the mixing process is quantified by the average degree of mixing. It is expressed as

$$\bar{\mu} = 1 - \sqrt{\frac{\int_A (\xi - \xi_{\max})^2 \vec{V} \cdot \hat{n}\, dA}{\int_A \xi_{\max}^2 \vec{V} \cdot \hat{n}\, dA}} \tag{3}$$

where A is the channel cross section, $\hat{n}$ is the unit vector normal to differential area dA, $\vec{V}$ is the velocity vector, ξ is the mass fraction, and $\xi_{\max}$ is mass fraction of perfect mixing ($\xi_{\max} = 0.5$). Moreover, $\bar{\mu} = 1.0$ is the complete mixing whereas $\bar{\mu} = 0$ means no mixing.

The two-dimensional geometry used for the study is shown in Fig. 1a. The inlets are set as velocity inlet while the exit is set as pressure outlet. The waveform of the pulsating velocities applied at the inlets is shown in Fig. 1b and given by Eqs. (1) and (2). Here, t represents time and T stands for pulsation period. A mixture of two fluids with identical properties like water is considered as working fluid. The density and kinematic viscosity of water are set at 1000 kg/m^3 and 0.001 kg/ms, respectively. The diffusion co-efficient is set at 10^{-10} m^2/s. The diffusion co-efficient of biomolecules

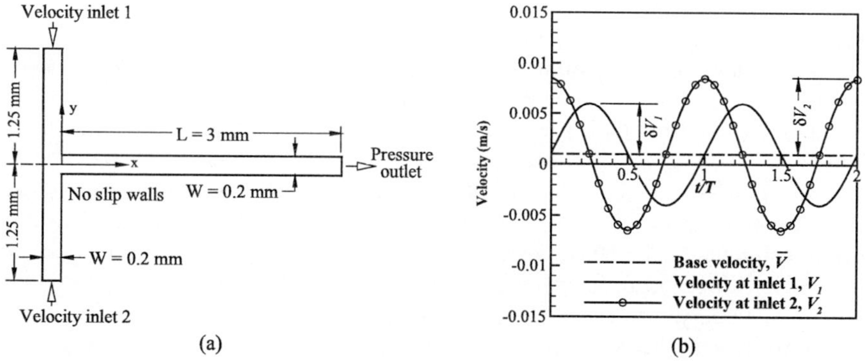

Fig. 1 **a** Geometry and boundary conditions. **b** Pulsating velocities at the inlets

in water is of the order of 10^{-10} [7]. A structured grid is employed throughout the computational domain. The total number of cells is about 44,000.

The following dimensionless parameters are introduced to non-dimensionalize the governing equations for the present problem

$$v^* = \frac{v}{U} \quad x^* = \frac{x}{L} \quad y^* = \frac{y}{W} \quad t^* = tf \quad t^* = tf$$
$$p^* = \frac{p}{\eta UL/W^2} \quad c^* = \frac{c}{\bar{c}}$$

where (*x*, *y*) are coordinates along the mixing channel and along the channel width, respectively. The star superscript is used for dimensionless variables. $U = 2\bar{V}$ is the characteristic speed, L is the channel length, W is the channel width, t represents time, f is the pulse frequency, p represents pressure, c is the concentration of a species, and $\bar{c}$ is the reference concentration.

The governing equations are presented in dimensionless form for parallel flow in the *x*-direction, i.e., $v = (u, 0, 0)$.

Continuity equation $$\frac{\partial u^*}{\partial x^*} = 0 \tag{4}$$

Momentum equation $$\mathrm{St}\,\mathrm{Re}\frac{\partial u^*}{\partial t^*} = -\frac{\partial p^*}{\partial x^*} + \frac{\partial^2 u^*}{\partial y^{*2}} \tag{5}$$

Mass transfer equation $$\mathrm{St}\frac{\partial c^*}{\partial t^*} + \theta u^* \frac{\partial c^*}{\partial x^*} = \frac{1}{\mathrm{Pe}}\frac{\partial^2 c^*}{\partial y^{*2}} \tag{6}$$

where u is x velocity, and $\theta = W/L$ is the width to length ratio. Peclet number, $\mathrm{Pe} = UW/D$, Reynolds number, $\mathrm{Re} = UW/\upsilon$, and Strouhal number, $\mathrm{St} = fW/U$; υ is the kinematic viscosity.

Numerical solutions of the governing equations are carried out using ANSYS Fluent [8]. Simulations were run for sufficiently long time to ensure that the flow had reached the quasi-stationary state, when the solution does not vary between two consecutive cycles.

The validation of the current study has been done by comparing the computational results with the experimental results obtained by Glasgow and Aubry [3]. A photograph taken from their experimental setup is shown in Fig. 2a. Aqueous solutions from the inlets flow at constant velocities of 0.001 m/s. To observe the mixing behavior, flow from inlet B is dyed. Figure 2b represents the present simulation result. There is an excellent agreement between the two. The performance of the present computational method is also verified by comparing with simulation results obtained by Cheaib et al. [6]. The comparison between the two is shown in Fig. 3a. At quasi-stationary state, the degree of mixing does not experience variation between two consecutive cycles but it varies over a pulsation cycle as shown in Fig. 3b.

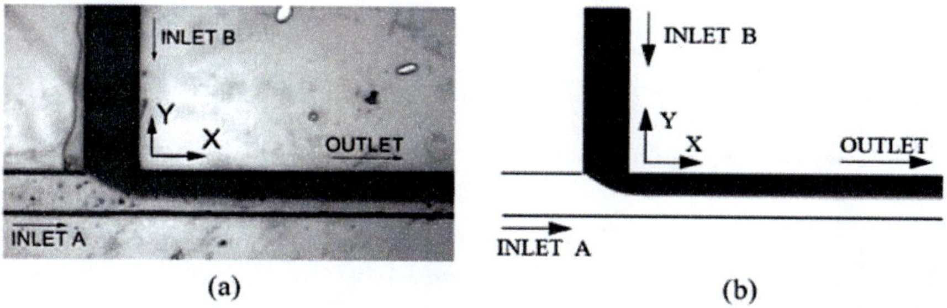

Fig. 2 Mass fraction contour from **a** the experiment of Glasgow and Aubry [3] and **b** present simulation

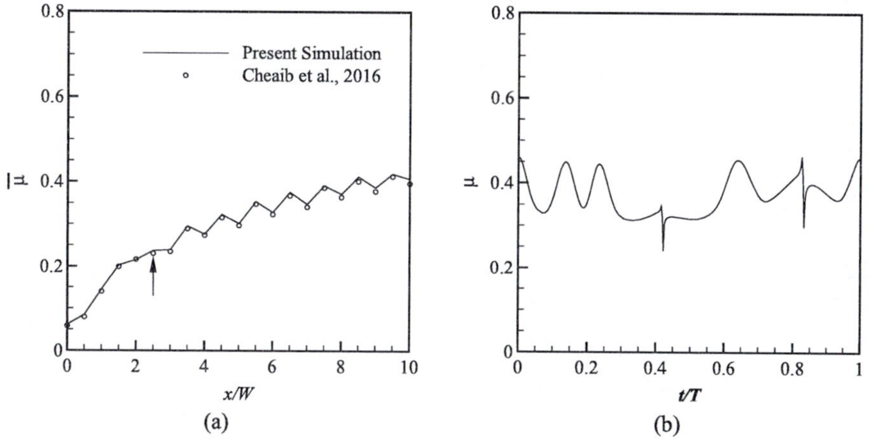

Fig. 3 **a** Average degree of mixing along the channel obtained from present study and Cheaib et al. 2016 [6]. **b** Time history of degree of mixing over a cycle at quasi-stationary state at location $x/W = 2.5$

3 Results and Discussion

The effects of both the same amplitudes and different amplitudes at the inlets on the mixing quality are presented in this paper. Other important parameters like Reynolds number, i.e., base velocity, Strouhal number, i.e., pulse frequency, and phase difference are kept constant at $\mathrm{Re} = 0.2$, $\mathrm{St} = 1.5$, and $\phi = 90°$.

In Fig. 4a and b, comparison of mass fraction contour is made between two pulse amplitudes. At the mixing channel, the interface between the two fluids (red and blue) is stretched by pulsating streams which improves the diffusion process significantly. Moreover, stretching creates chaotic advection which enhances the mixing process.

The higher the pulse amplitude, the higher the stretching. A concave-shaped blue region can be observed at the end of the mixing channel in Fig. 4b. This is known

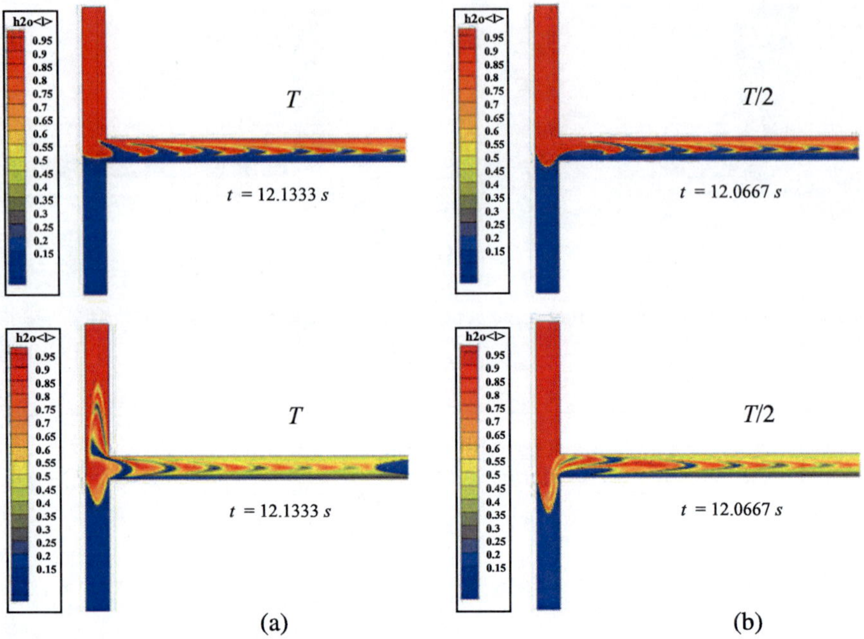

Fig. 4 Mass fraction contour of water at $T/2$ intervals of a pulsation cycle for **a** $\delta V/\vec{V} = 5$ and **b** $\delta V/\vec{V} = 12.5$ where Re = 0.2, St = 1.5

as the end effect due to back flow and the applied boundary condition at the exit of the mixing channel ($x/W = 15$). It creates a drop in average degree of mixing at the end of the channel. To avoid the end effect on average mixing degree, the mixing length is selected considering the distance covered by the flow in reverse direction during a pulsation cycle.

At the quasi-stationary state, the flow is fully developed and the velocity profile is like the Poiseuille profile along the entire mixing channel. It varies with time due to pulsating streams at the inlets which is represented in Fig. 5a and b. The shape of the velocity profile depends on the pulse amplitude.

For same pulse amplitudes at the inlets, the average degree of mixing increases down the channel which is shown in Fig. 6a. Diffusion of mass through the interface between the streams gets more time to take place while the streams move down the channel. This improves the mixing quality down the channel. In addition, increased pulse amplitude results in better quality of mixing, i.e., higher degree of mixing. At low $\delta V/\vec{V}$, poor mixing is obtained. A rapid improvement is found between $\delta V/\vec{V} = 5$ and $\delta V/\vec{V} = 7.5$. Further increment in pulse amplitude improves the mixing process but the rate of improvement is lowered.

Effect of different pulse amplitudes at the inlets on the mixing quality is also checked. It is done by calculating the average degree of mixing while varying the pulse amplitude at the second inlet and at the first inlet, it is kept constant at

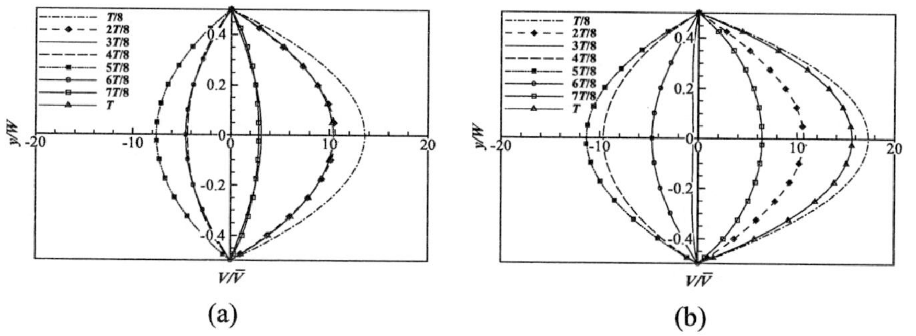

Fig. 5 Velocity profile for the case **a** $\delta V/\vec{V} = 5$ and **b** $\delta V_2/\delta V_1 = 1.7$ where $\delta V_1/\vec{V} = 5$

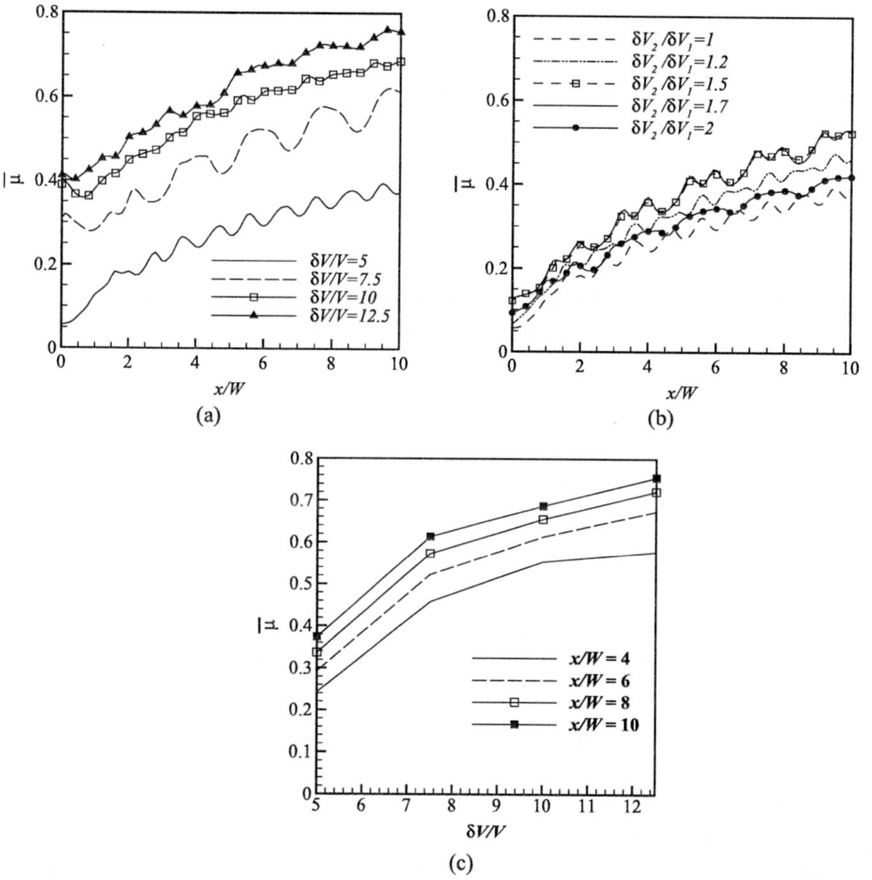

Fig. 6 Effect of **a** same pulse amplitude **b** different pulse amplitudes at the inlets where $\delta V_1/\vec{V} = 5$

$\delta V/\vec{V} = 5$. From Fig. 6b, it is evident that the quality of mixing can also be enhanced by increasing the ratio of the pulse amplitudes. However, there is an optimal value of the ratio of the pulse amplitudes for which best mixing scenario is achieved.

The average degree of mixing for different pulse amplitudes at four locations along the channel length is presented in Fig. 6c. It is observed that the slope of the graph decreases gradually which indicated less improvement in the degree of mixing with increase in pulse amplitude. The minimum required length of the mixing channel can also be predicted from this graph. For example, for required average degree of mixing 0.5 and $\delta V/\vec{V} = 10$, the channel length of $4W$ is more than enough.

4 Conclusions

Pulsating streams create a stretched interface region between the fluids to be mixed which enhances mass diffusion and improves the mixing quality. It is observed that the degree of mixing experiences both spatial and temporal fluctuations at the quasi-stationary state due to pulsating flow condition. Therefore, the temporal average of the degree of mixing is considered to express the quality of mixing. For this study, Strouhal number, i.e., pulsation frequency is kept constant. The major findings of the present study are:

- At low Reynolds numbers, when the pulsation frequency is small (Re < 1 and Re.St < 1) the velocity profile approaches the Poiseuille profile. The degree of mixing is not affected by Re n this regime.
- It is found that the average degree of mixing, i.e., mixing quality is very sensitive to pulse amplitude. An increase in pulse amplitude to base velocity ratio improves the quality of mixing.
- The quality of mixing can also be enhanced by increasing the ratio of pulse amplitudes of the two streams up to an optimal value. Below and above of this ratio result poor mixing in the microchannel.
- With increased pulse amplitude-to-base velocity ratio required average degree of mixing can be achieved within a shorter microchannel.

Future study of the effects of Strouhal number might extend the explanations of mixing characteristics in this regime.

References

1. Karniadakis G, Beskok A, Aluru N (2006) Microflows and nanoflows: fundamentals and simulation. Springer Science & Business Media
2. Capretto L, Cheng W, Hill M, Zhang X (2011) Micromixing within microfluidic devices. Springer, Berlin, Heidelberg 304:27–68
3. Glasgow I, Aubry N (2003) Enhancement of microfluidic mixing using time pulsing. Lab Chip 3(2):114–120

4. Glasgow I, Lieber S, Aubry N (2004) Parameters influencing pulsed flow mixing in microchannels. Anal Chem 76(16):4825–4832
5. Goullet A, Glasgow I, Aubry N (2005) Dynamics of microfluidic mixing using time pulsing. Discrete Cont Dyn Syst, Suppl, pp 327–336
6. Cheaib F, Kekejian G, Antoun S, Cheikh M, Lakkis I (2016) Microfluidic mixing using pulsating flows. Microfluid Nanofluid 20(5):70
7. Atkins P, De Paula J (2011) Physical chemistry for the life sciences. Oxford University Press, USA
8. ANSYS, Inc., Canonsburg, PA, USA (www.ansys.com)

Stochastic Modeling of Particle Dispersion in Wall-Bounded Turbulent Flows

Arjav Thakkar and Amitabh Bhattacharya

Abstract Particle laden turbulent flows are ubiquitous in both nature and technological applications. In this work, we devise methods for evolving dispersion of particles in wall-bounded turbulent flows using Reynolds Averaged Navier Stokes (RANS) or Large Eddy Simulation (LES), where the sub-grid fluid fluctuations at particle locations require modeling. We derive a new Langevin equation for evolving fluid fluctuations, which takes into account non-local wall reflection terms. We also formulate a method for evolving the particle position, in which the Langevin equations for fluid fluctuations are solved on a fixed grid, and are then interpolated onto the particle locations. The particle deposition rates for a large range of Stokes' number are predicted by our method and it compares well with experimental results of Liu-Agrawal.

Keywords Langevin equations · Particle-laden turbulent flows

1 Introduction

Liquid or solid particles, carried by a turbulent air stream, tend to deposit on the bounding walls. The study of such particle laden wall bounded flows is of particular interest in numerous industrial applications, and also in atmospheric science. Applications range from removal of airborne particulate from atmosphere (to ground) to improving fuel injection spray patterns in internal combustion engines. The governing equations for interaction of particles with fluid, in particle laden flows, are long known. However, it is still too expensive to simulate many practically pertinent regimes from first principles. Hence, particle laden flows are generally treated with reduced order modeling for simulating practical scenarios.

There are two established techniques for simulating particle dispersion in turbulent flows. The first one is the 'two-fluid', or the Eulerian-Eulerian approach, wherein both the phases (carrier and dispersed) are treated as interpenetrating and interacting continua. Here, averaged [1] equations of momentum and energy are solved for

A. Thakkar (✉) · A. Bhattacharya
Indian Institute of Technology, Bombay, India
e-mail: arjavthakkar79@gmail.com

L. Venkatakrishnan et al. (eds.), *Proceedings of 16th Asian Congress of Fluid Mechanics*, Lecture Notes in Mechanical Engineering,
https://doi.org/10.1007/978-981-15-5183-3_21

both the phases separately. The second approach is the Lagrangian Particle Tracking (LPT) or the Eulerian-Lagrangian Approach wherein the carrier phase is evolved via any single phase turbulent flow simulation method like Reynolds Averaged Navier Stokes (RANS) or Large Eddy Simulation (LES), and the dispersed phase is treated as an ensemble of particles. The two-fluid approach relies on extensive modeling assumptions, which, therefore, motivates us to pursue the LPT approach instead.

There are two common approaches in LPT used for approximating the fluid velocity 'seen' by a solid particle. The first approach specifies the velocity 'seen' as the sum of the local mean fluid velocity and a random fluctuating velocity. The fluctuating component is generally selected from a Gaussian distribution having a zero mean and a variance related to the turbulent velocity scale. Such techniques are denoted as 'Random Walk Models' [2–4]. In the second approach, either the total fluid velocity is directly determined by solving a stochastic differential equation (SDE), or the fluctuating part is determined from an SDE and added to local mean velocity. These models are known as Langevin models [5]. In this work, a Langevin model is used, and the fluctuating component of velocity is obtained by solving an SDE derived by authors. The inspiration for deriving this equation comes from Durbin's [6] and Pope's work [7]. The particle dispersion problem is then solved for particle deposition in vertical turbulent pipe flow. The results are validated with the experimental results of Liu-Agarwal [8].

2 Methodology and Results

Dispersion of solid particles in a background flow requires the knowledge of the position ($\boldsymbol{x}_p$) and the velocity ($\boldsymbol{V}$) of the particle at each time instant. The governing equations for motion of sphere in non-uniform flow, considering all possible external forces, were given by Maxey and Riley [9]. The external force terms in evolution of velocity include pressure-gradient, added-mass, drag, Basset force and buoyancy terms. However, for particles that are much denser than the fluid ($\rho_p \gg \rho_f$), only drag and buoyancy terms remain significant. The corresponding equations of motion will be:

$$\frac{\mathrm{d}\boldsymbol{x}_p}{\mathrm{d}t} = \boldsymbol{V} \text{ and } \frac{\mathrm{d}\boldsymbol{V}}{\mathrm{d}t} = \frac{\boldsymbol{U}^* - \boldsymbol{V}}{\tau_p} + g \tag{1}$$

Here, $\boldsymbol{U}^*$ represents the local fluid velocity at the instantaneous particle location. The drag term is written using the particle relaxation time, τ_p, given as [10],

$$\tau_p = \frac{4\rho_p}{3\rho_f} \frac{d_p}{C_D|\boldsymbol{U}^* - \boldsymbol{V}|} \text{ where } C_D = \begin{cases} \frac{24}{\mathrm{Re}_p}[1 + 0.15\mathrm{Re}_p^{0.687}], & \mathrm{Re}_p < 1000 \\ 0.44, & \mathrm{Re}_p \geq 1000 \end{cases} \tag{2}$$

Direct Numerical Simulation (DNS) is computationally expensive, and number of grid points required for DNS simulation scales as $\sim \mathrm{Re}_L^{2.2}$ [11]. Hence, RANS, LES or some hybrid RANS-LES simulations can be used to represent background fluid flow. For such approaches, we only have ensemble averaged velocity field with RANS or spatially filtered velocity field with LES. This averaged/filtered fluid velocities U_i are interpolated to particle locations and then an extra random noise, which takes care of the effect of the sub-grid scale is added to get $\boldsymbol{U}^*$. The sub-grid scale velocity or velocity fluctuation (u_i^*) is obtained by solving the Eq. 3, derived by authors.

$$\begin{aligned} \mathrm{d}u_i^* = & \left[-\frac{C_1}{2(k/\epsilon)} + \frac{0.125\sqrt{k}}{2x_2}\right]u_i^*\mathrm{d}t + \left[C_2 - 1 - \frac{0.015k^{3/2}}{\epsilon x_2}\right]u_k^*\partial_k U_i \mathrm{d}t \\ & + \left[C_3 + \frac{0.015k^{3/2}}{\epsilon x_2}\right]u_k^*\partial_i U_k \mathrm{d}t \\ & + \frac{\partial \tau_{ij}}{\partial x_j}\mathrm{d}t + \sqrt{C_0\epsilon}\mathrm{d}W_i(t) + \sqrt{C_s\epsilon}M_{ik}\mathrm{d}W_k'(t) \end{aligned} \tag{3}$$

The above equation is an SDE where first four terms represent the drift part and last two terms represent the diffusion part of the fluctuating velocity field. The above equation for the fluctuating velocity is a type of Langevin equation. The anisotropy of Reynolds stresses due to wall effect (pressure reflection) is captured using the terms involving wall normal distance (x_2) in the above equation, which is not found in any SDEs present in the literature. Here, k is turbulent kinetic energy and ϵ is dissipation rate of turbulent kinetic energy. $W_i(t)$is the Wiener stochastic process and $\mathrm{d}W_i(t)$ are its increments with the following properties,

$$\overline{\mathrm{d}W_i} = 0; \overline{\mathrm{d}W_i d W_j} = \mathrm{d}t\delta_{ij}; \overline{u_j \mathrm{d}W_i} = 0 \tag{4}$$

Thus, $\mathrm{d}W(t) = \sqrt{\mathrm{d}t}\zeta$ with ζ as standardized Gaussian random variable $\left(\overline{\zeta} = 0; \overline{\zeta^2} = 1\right)$. C_1, C_2, C_3, C_s are constants from Launder Reece Rodi-Quasi Isotropic (LRR-QI) turbulence model [12]. The value of C_0 is derived to be,

$$C_0 = \frac{2}{3}\left[C_1 - \frac{0.125k^{3/2}}{\epsilon x_2} - 1 + (C_2 + C_3)\frac{P}{\epsilon}\right] - \frac{C_s}{3}M_{kk}^2 \tag{5}$$

Here, P is half the trace of production stress tensor (P_{ij}), and it is assumed equal to ϵ (which is the case for most of the regions in channel flow). The only term now remaining to evaluate is M_{ij}. While deriving Eq. 3, a constraint comes up as following,

$$M_{ij}^2 - \frac{1}{3}M_{kk}^2\delta_{ij} = -\frac{k}{\epsilon}S_{ij} \text{ where } S_{ij} \text{ is the strain rate tensor} \tag{6}$$

Using the linear algebra, M_{ij} is given by,

$$\boldsymbol{M} = \boldsymbol{Q}.\text{diag}\big[0, (\lambda_1 - \lambda_2)^{1/2}, (\lambda_1 - \lambda_3)^{1/2}\big].\boldsymbol{Q}^T \tag{7}$$

where $\boldsymbol{Q}$ is a unitary matrix of eigenvectors diagonalizing the symmetric tensor $\frac{k}{\epsilon}S_{ij}$

$$\frac{k}{\epsilon}S_{ij} = \boldsymbol{Q}.\text{diag}[\lambda_1, \lambda_2, \lambda_3].\boldsymbol{Q}^T \tag{8}$$

It can be shown that $M_{kk}^2 = 3\lambda_1$ for $\lambda_1 > \lambda_2 > \lambda_3$, which is then used in Eq. 5. Since the stochastic processes are not differentiable in time, therefore, Eq. 3 is converted to a difference equation and implemented in MATLAB for turbulent channel flow. To validate the above approach, we have taken the mean flow data from DNS dataset of Javier Jimenez [13] for channel flow at $\text{Re}_\tau = 950$. The fluctuations are obtained by solving Eq. 3 and it is made sure that they are statistically stationary. Figure 1a represents the normalized (in '+' units) RMS velocity fluctuation in flow direction obtained from the Langevin equation and is compared with DNS data. The match with DNS data indicates the accuracy of the Langevin model. The obtained fluctuations are correlated to get normalized Reynolds stress (uv^+) and the comparison with DNS data is shown in Fig. 1b. The Reynolds stress from Langevin equation is generated with and without considering wall effect. It is depicted from Fig. 1b that the wall effect is indeed important to capture the Reynolds stress. This result shows the importance of derived Langevin equation.

We next numerically evolve the particle dispersion in a fully developed turbulent pipe flow. The evolution equations for particles and fluctuating fluid velocity in vertical pipe flow are solved in Cartesian domain. The pictorial computational domain for the deposition problem is shown in Fig. 2. Initially, the solid particles are randomly distributed at the inlet, within the pipe ($r < R$), as shown in Fig. 2. Each particle is advected and diffused in the domain as per Eq. 1. When the particles reach a radial distance equal to R, they are taken out of the computational domain. This absorption

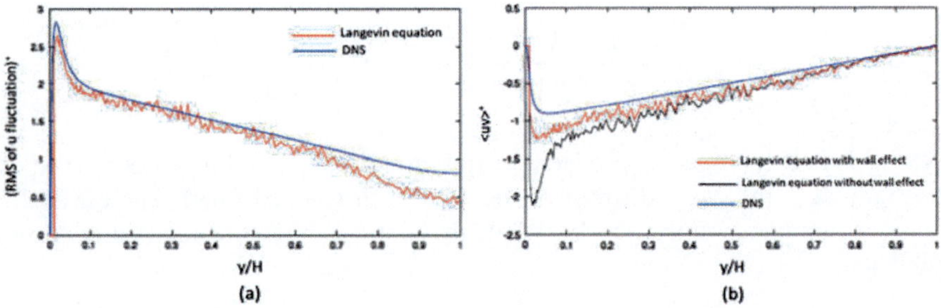

Fig. 1 Comparison of **a** normalized RMS velocity fluctuations in flow direction **b** uv^+ with DNS data at $\text{Re}_\tau = 950$

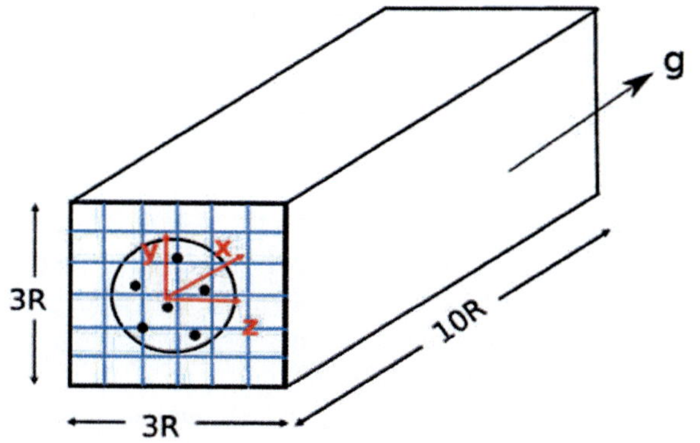

Fig. 2 Pictorial representation of computational flow domain for deposition problem

boundary condition mimics the experiments by Liu-Agarwal [8], who used olive oil particles in their experiments, which tended to stick to the wall.

The domain in Y- and Z-directions is uniformly divided with 300×300 grid points to get accurate values for interpolation. The values of U, k and ϵ, as a function of radius, are read from DNS data input files and are interpolated to the grid points, considering origin as the center of imaginary pipe. The interpolation method used here is Piecewise Cubic Hermite Interpolating Polynomial (PCHIP). The values of average flow variables on grid points lying outside radius R are assigned to their values at the wall. As shown in Fig. 2, gravity is applied along the streamwise direction, to mimic vertical pipe flow. The force from the particles on the fluid is not important since the mass loading of the particles in the experiment is low. Hence, the DNS data from single phase turbulent flow simulation is directly used here.

The bulk Reynolds number of the fluid flow in the simulation corresponds to $Re_\tau = 300$, which is consistent with the bulk Reynolds number of 10,000 used in Liu-Agarwal's experiments [8]. The density of the particles considered is 920 kg/m^3, which is the same as in experiments. The density of the fluid is taken as 1.3 kg/m^3 and its kinematic viscosity is taken as 1.6e−5 m^2/s. The diameter of the particles is decided on the basis of required time scale τ_p^+, which is the particle relaxation time non-dimensionalized using ϑ and u_τ. τ_p^+ can also be interpreted as Stokes' number where flow time scale is determined by ϑ and u_τ, which happens to be the case for turbulent flows. The time step is taken to be 0.3 in plus units, which is 10 times smaller than the lowest τ_p^+ considered. The deposition velocity (V_d) is defined as the ratio of the rate of particle deposition per unit area to the bulk mean concentration (C) of particles. Thus, for deposition in a pipe of diameter D,

$$V_d = \frac{1}{C(\pi DL)} \frac{\mathrm{d}N(t)}{\mathrm{d}t} \tag{9}$$

The number of particles (N) in the flow is bulk mean concentration times the volume of the pipe. Substituting for N in Eq. 9,

$$V_d = \frac{D}{4}\frac{\mathrm{d}(lnC)}{\mathrm{d}t} \tag{10}$$

$$\therefore V_d^+ = \frac{D}{4u_\tau}\frac{\mathrm{d}(lnN)}{\mathrm{d}t} \tag{11}$$

where ln$C$ in Eq. 10 is replaced by ln$N$ in Eq. 11, since the volume of pipe remains constant. To obtain deposition velocities, firstly a graph of natural logarithm of number of particles remaining in the flow versus time is made. The slope of this curve will give us the value of $d(\ln N)/\mathrm{d}t$. This slope, when multiplied by $D/4u_\tau$, will give us the deposition velocity in plus (+) units.

Initially, 1000 particles are introduced at the inlet, and they are held at their position until a statistically stationary state is reached by evolving the Langevin equation (Equation 3) at the particle locations. The velocity of the particles is initialized to the mean fluid velocity U at the inlet. The forcing from particles onto the fluid is not considered due to low particle loading. Once the correct RMS values of velocity fluctuation and Reynolds stress are obtained at the particle positions, they are allowed to advect using Eq. 1. The velocity $\boldsymbol{U}^*$ in Eq. 1 is obtained using the sum of interpolated mean velocity U at the particle position and the fluctuations obtained at the particle position using Eq. 3. The inputs to Eq. 3 like k, ϵ, wall normal distance, mean velocity gradients are interpolated from the grid points to particle positions. These interpolations are linear as much finer grid is used in the YZ-plane. The simulations are evolved for particles of different sizes, having relaxation time scales τ_p^+ of 3.0, 10.0. 20.6, 29.6, and 46.3. This covers a large range of Stokes' number. The graph of ln$N$ versus time for $\tau_p^+ = 10$ is shown in Fig. 3, which also depicts that the slope

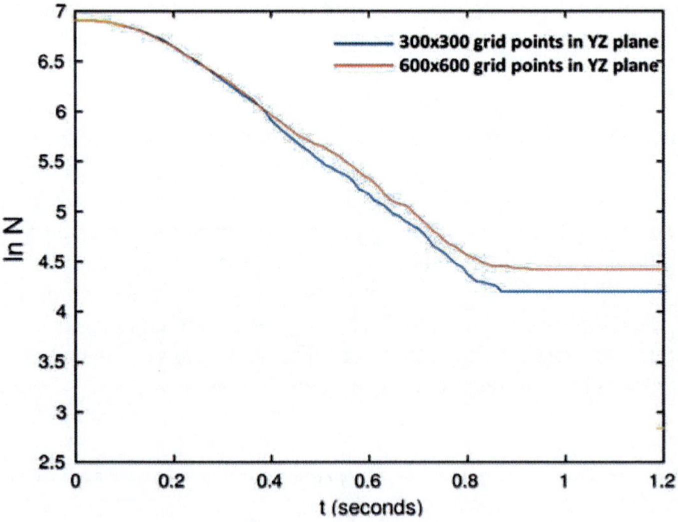

Fig. 3 ln N versus time for $\tau_\mathrm{p}^+ = 10$

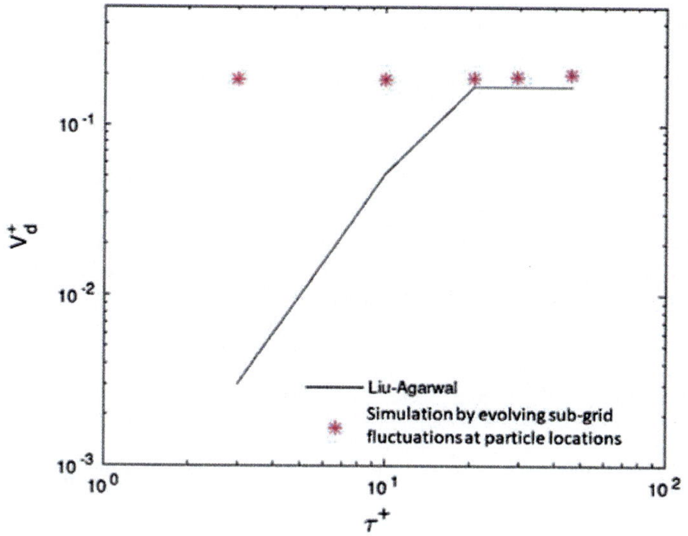

Fig. 4 Comparison of deposition velocity with literature

is not changing significantly by increasing the number of grid points in *YZ*-plane, proving grid independence. Using Eq.11, we get the value of deposition velocity, and Fig. 4 shows the V_d^+ versus τ_p^+.

Figure 4 shows good agreement for heavier particles, but the deposition velocity for lighter particles is highly over predicted for the following reason. Initially, the fluid fluctuations are allowed to reach statistically stationary state and then the particles are advected. However, after completing one time step, the new fluid fluctuations obtained at new particle locations do not reach a statistically stationary state. As a result, these fluid fluctuations do not yield correct Reynolds stresses, unlike what was obtained for stationary particles. This error will go on increasing with increasing number of time steps. Hence, we do not obtain correct deposition velocities. The deposition velocity predicted by our simulation matches the experimental results better for larger particle relaxation time, since high Stokes' number particles are relatively insensitive to fluctuations in fluid velocity. The low Stokes' number particles are significantly affected by the incorrect fluid fluctuations, and hence their deposition velocities show significant mismatch.

To improve the value of fluid fluctuations at the particle locations, we independently evolve fluctuating velocity field at fictitious fluid points, located on a regular grid, using Eq. 3. These fluid points are distinct from the advecting particles, and will always yield correct statistically stationary velocity fluctuations. The velocity fluctuations at the position of solid particles are then interpolated from this stationary points. To perform this task, a periodic 2D box (assuming axi-symmetry) of size Rx4R is considered and a grid of 318x11 points is used. The pictorial representation is shown in Fig. 5. The average correlation length at $\mathrm{Re}_\tau = 300$ is calculated to

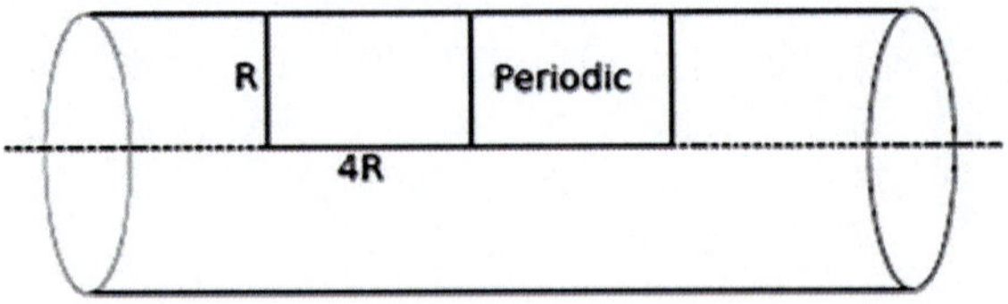

Fig. 5 Pictorial representation for interpolating from Langevin equations

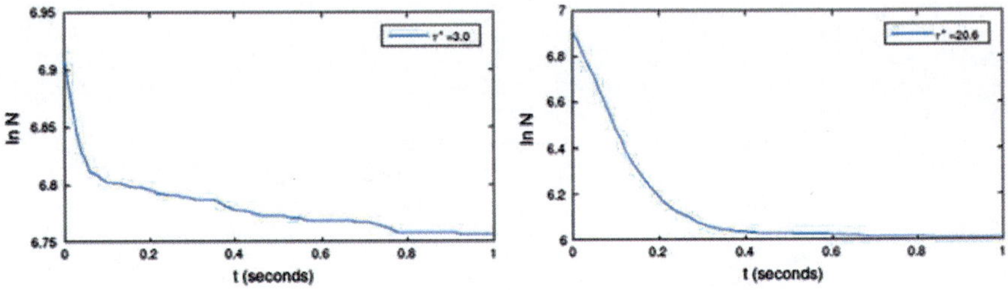

Fig. 6 ln$N$ versus time for $\tau_p^+ = 3$ (left) and $\tau_p^+ = 20.6$ (right), using interpolation from stationary points approach

be 0.317R. Therefore, the grid points in periodic box in axial direction are kept at a distance of 0.4R, as the SDE would give uncorrelated fluctuations. Since the points at which the fluctuations are being calculated are stationary, the divergence of Reynolds stress term (fourth term in Eq. 3) should not be present (due to local homogeneity). If this term is present, it brings in a mean spurious drift. The simulations are repeated with the method discussed above. The results are presented in Figs. 6 and 7. It is seen from Fig. 7 that a very good match is obtained between the simulations and Liu-Agarwal experimental data.

3 Conclusions

In this work, we first derived Langevin equations for fluid fluctuations which can be used for inhomogeneous wall-bounded turbulent flows. Here, we have incorporated non-local wall effects, which give a very good match with DNS data for predicting Reynolds stresses in turbulent channel flow. We then proceeded to conduct simulations of particle laden turbulent flows with a large range of Stokes' number, in which the position of the particles is explicitly evolved and the fluid fluctuation is used to obtain the slip velocity between the particle and the ambient. Evolving the Langevin equations for fluid fluctuations along the particle path grossly over predicts the deposition velocity for lighter particles, since the fluid fluctuations do not reach a statistically stationary state here. We, therefore, devised a new approach, in which the Langevin equations for fluid fluctuations are evolved separately at stationary points.

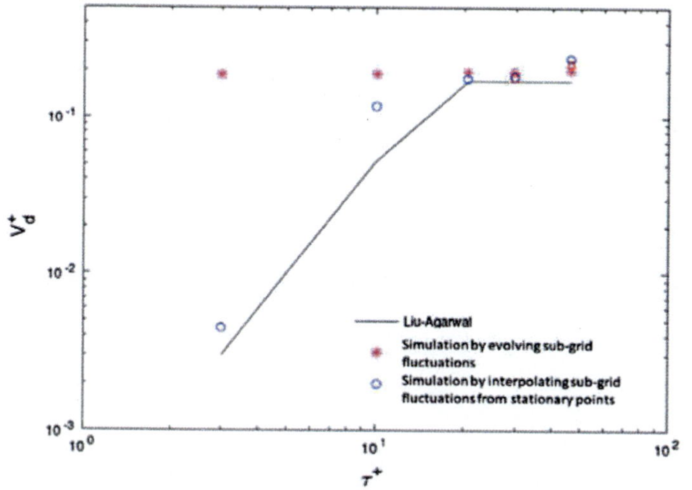

Fig. 7 Comparison of deposition velocity with literature for interpolation from stationary points approach

These fluid fluctuations are then linearly interpolated onto solid particles. The new method yields very good agreement for the prediction of particle deposition velocity with respect to experimental measurements. The work presented here is, therefore, relevant toward accurately simulating particle dispersion in wall bounded flows. In the future, we will test this method in more complex particle laden turbulent flows.

Acknowledgements The authors would like to thanks Professor Partha Sarathi Goswami, Chemical Engineering Department, IIT Bombay for his valuable suggestions during the course of this work.

References

1. Drew DA, Segel LA (1971) Averaged equations for two-phase flows. Stud Appl Math 50(3):205–231
2. Yuu S, Yasukouchi N, Hirosawa Y, Jotaki T (1978) Particle turbulent diffusion in a dust laden round jet. AIChE J 24(3):509–519
3. Gosman AD, Loannides E (1983) Aspects of computer simulation of liquid-fueled combustors. J Energ 7(6):482–490
4. Ormancey A, Martinon J (1984) Prediction of particle dispersion in turbulent flows. Phys Chem Hydrodynam 5:229–244
5. Pope SB (1983) A Lagrangian two-time probability density function equation for inhomogeneous turbulent flows. Phys Fluids 26(12):3448–3450
6. Durbin PA, Speziale CG (1994) Realizability of second-moment closure via stochastic analysis. J Fluid Mech 280:395–407
7. Haworth DC, Pope SB (1986) A generalized Langevin model for turbulent flows. Phys Fluids 29(2):387–405

8. Liu BY, Agarwal JK (1974) Experimental observation of aerosol deposition in turbulent flow. J Aerosol Sci 5(2):145–155
9. Maxey MR, Riley JJ (1983) Equation of motion for a small rigid sphere in a nonuniform flow. Phys Fluids 26(4):883–889
10. Rose PN (1961) Drag forces in a hydraulic model of a fluidized bed- Part II. Trans Inst Chem Eng 39:175–180
11. Chapman DR (1979) Computational aerodynamics development and outlook. AIAA J 17(12):1293–1313
12. Launder BE, Reece GJ, Rodi W (1975) Progress in the development of a Reynolds-stress turbulence closure. J Fluid Mech 68(3):537–566
13. Jimenez J (2008) Turbulent channel flow DNS data. http://torroja.dmt.upm.es/channels/data/statistics/

Thermal Mixing of Shear-Thinning and Newtonian Fluids in a T-Channel Using Impinging Streams

Anamika Maurya, Naveen Tiwari, and R. P. Chhabra

Abstract In this work, thermal mixing of shear-thinning (i.e. CMC solutions) and Newtonian fluids has been numerically investigated in a rectangular T-channel for a specified heat flux boundary condition at the mixing zone walls. The influence of the Reynolds number ($10 \leq \text{Re} \leq 50$), power-law index ($0.6161 \leq n \leq 1$), Nusselt number (a dimensionless form of convective heat transfer coefficient for external air flow, $10^3 \leq \text{Nu}_o \leq 10^4$), and ambient temperature ($-2.7 \leq \theta_a \leq 1.3$) is studied on the mixing behaviour. The flow is assumed to be steady, laminar, and incompressible. The new results are presented in terms of isotherm contours, mixing index, and required channel length to achieve complete mixing. The mixing index decreases along channel length for both shear-thinning and Newtonian fluids. Better mixing is seen at high power-law index, ambient temperature, and Nusselt number and low Reynolds numbers. The length required to achieve the 95% of the ambient temperature or the exact ambient temperature is shorter for mildly shear-thinning fluids (high power-law index) and Nusselt number, and low Reynolds numbers. Also, for the case of heating (high ambient temperature), the complete mixing demands a shorter channel length than that in the case of cooling.

Keywords Mixing characteristics · Ambient temperature · Nusselt number · Channel length · Degree of mixing

1 Introduction

Thermal mixing through impinging streams is of practical importance when it is required to mix highly viscous fluids, for instance, polymer solutions, polymers melts, liquid foodstuffs, etc. Additional applications are also found in diverse fields

A. Maurya (✉) · N. Tiwari · R. P. Chhabra
Indian Institute of Technology Kanpur, Kanpur 20816, India
e-mail: mauryaanamika1112@gmail.com

R. P. Chhabra
Indian Institute of Technology Ropar, Ropar, Punjab 140001, India

L. Venkatakrishnan et al. (eds.), *Proceedings of 16th Asian Congress of Fluid Mechanics*, Lecture Notes in Mechanical Engineering,
https://doi.org/10.1007/978-981-15-5183-3_22

like drying, cooling, extraction, absorption, dust collection, micro mixers, etc. A conventional process of mixing of viscous fluids has the possibility of a loss of quality, product through contamination, and it requires high energy consumptions which can be overcome by mixing process through impinging streams. For the easy handling and processing of the homogeneous mixtures of liquids, it is an essential unit operation in many industries. Thermal mixing through the laminar flow is desirable whenever there is a requirement to avoid high impingement pressures. Also, it is an easy and simple way of heating and cooling in a restricted region due to a high heat transfer rate in the impingement zone. Other advantages of mixing through laminar jet impingement include the low pumping cost, controllability, uniformity, and no mechanical devices like impellers. Much research effort has been expended in exploring the characteristics of the two-dimensional opposite jet mixer [1–3], and it is instructive to briefly review the salient features of the previous studies.

Devahastin and Mujumdar [4] investigated the mixing of fluids in the steady laminar regime numerically using the multiple inlet streams in a novel 2D in-line mixer for adiabatic channel wall boundary conditions. They reported that the mixing increases by offsetting the top and bottom inlet jets because of the intense mixing regions formed in between the inlet streams jets. Later, Niamnuy and Devahastin [5] experimentally investigated the mixing characteristics in an in-line mixer for different operating conditions and geometry for adiabatic channel walls. They reported the better mixing with increasing Reynolds number using one set of three inlet streams in the impingement zone and its vicinity. Further, Roy et al. [6] experimentally investigated the laminar mixing behaviour in a T-junction using the Laser Doppler Velocimeter (LDV) for a 2D steady flow of Newtonian fluid (air). They used adiabatic as well as constant heat flux boundary conditions on the channel walls and $H/W = 2$. They have reported the velocity and temperature profiles across the channel height in the mixing zone. Poh et al. [7] investigated the effect of Reynolds number numerically, jet exit velocity and nozzle-to-impingement target spacing for a single axi-symmetric semi-confined laminar jet flow of a carboxymethyl cellulose (CMC) solution in water which behaves as a power-law fluid, impinging normally on a plane wall maintained at a constant temperature. They found an enhancement in the Nusselt number values for the power-law fluids than that for Newtonian fluid due to augmentation of the jet velocity. Recently, Srisamran and Devahastin [8] studied numerically the mixing behaviour of confined impinging streams of shear-thinning fluids for a 2D, laminar flow using temperatures as a passive tracer for an adiabatic wall boundary condition. They reported the mixing characteristics for a wide range of Reynolds number (10–200) and flow behaviour index (0.6161–1). They reported shorter channel length for perfect mixing with increasing value of the flow behaviour index and decreasing Reynolds number. More recently, Mashaei et al. [9] have reported a numerical investigation of mixing characteristics of 2D, laminar, confined impinging flow to see the effect of impingement angle, momentum ratio, and flow behaviour index for an adiabatic wall boundary condition. They mentioned that the mixing behaviour improved with the increasing Reynolds number, decreasing momentum ratio while the impingement angle exerts an insignificant effect. Thus, it is fair to conclude that much of the literature relates to the adiabatic wall boundary condition prescribed on the channel walls of the mixer. Thus, the present work focuses on the specified heat flux boundary condition on the mixing channel walls, which allows the control of convection between the system and surrounding. Also, the surrounding temperature is varied in this work, which affects the internal temperature field in the mixing channel and allows the heating or cooling of the fluid according to the ambient temperature. Thus, the objective of the present work is to investigate the mixing behaviour for a 2D, steady, confined laminar flow of the shear-thinning, and Newtonian fluids as a function of Reynolds number ($10 \leq \mathrm{Re} \leq 50$), power-law index ($0.6161 \leq n \leq 1$), Nusselt number ($10^3 \leq \mathrm{Nu}_o \leq 10^4$) and ambient temperature ($-2.7 \leq \theta_a \leq 1.3$) for a specified heat flux boundary condition at the mixing zone walls. Also, reported here are the channel lengths required to achieve the complete mixing over the range of conditions spanned herein.

2 Problem Formulation and Governing Equations

The present work has been investigated for both Newtonian (i.e. water) and shear-thinning fluids (i.e. CMC, carboxymethyl cellulose, solutions). The physical properties of water (at an average temperature of two inlet streams) and CMC solutions are taken from the work of Srisamran and Devahastin [8]. The values of the power-law constants such as consistency coefficient (m) and flow behaviour index (n) are taken from the Poh et al. [7]. The present flow configuration consists of two different inlet streams entering into the system, as shown in Fig. 1 in the opposite directions but along the same axis or at an angle. The incoming two streams result in a zone called as impingement zone (a relatively narrow zone) which promotes intense mixing; thereby, enhancing the convective transport processes in this region. Subsequently, the streams exit the system symmetrically on either side of the impingement zone. Since the flow is symmetric about $x = 0$ plane, only half domain was considered

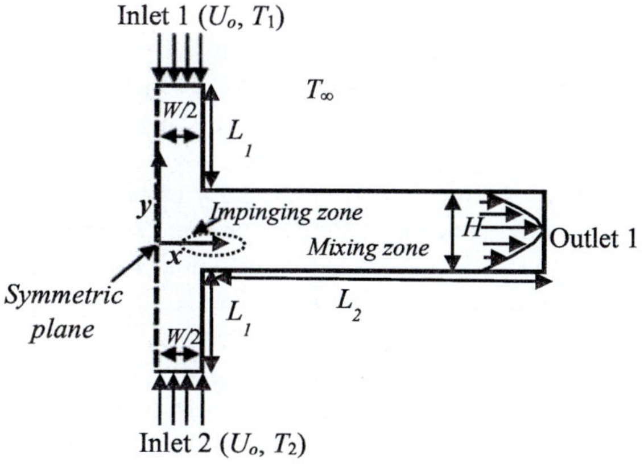

Fig. 1 Schematic diagram of the computational domain of the impinging streams

herein to reduce the computational efforts. The ratio of channel height and width is chosen as H/W is 2.0 where H is channel height (used as the length scale), and W is the width of the channel. Two fluids of identical physical properties but at different temperatures enter the system along the same axis but in opposite directions with velocity U_o, which is used as a velocity scale in this study. The inlet 1 and inlet 2 fluid temperatures are T_1 and T_2 ($>T_1$), respectively. The thermo-physical properties of the fluids, namely density (ρ), thermal conductivity (k_{fluid}), heat capacity (C_p), and power-law model parameters (m, n) are considered to be independent of temperature because the maximum temperature difference in the present work is 10 K ($\Delta T = T_2 - T_1$). Also, the Brinkman number is only of the order of (~10^{-5}) for the range of conditions, and it is thus safe to neglect the viscous dissipation effects in the energy equations. The thermal resistance of the channel wall is assumed to be negligible for heat transfer across it. The exiting streams loss or gain energy depending upon the ambient air temperature. Within the framework of these assumptions, the governing equations in their non-dimensional forms are written as:

$$\textbf{Continuity equation}: \quad \nabla \cdot \boldsymbol{U} = 0 \tag{1}$$

$$\textbf{Momentum equation}: \quad (U \cdot \nabla)U = -\nabla P + \frac{1}{\text{Re}} \nabla \cdot \tau \tag{2}$$

$$\textbf{Thermal energy equation}: \quad (U \cdot \nabla)\theta = \frac{1}{\text{Re} \times \text{Pr}} \nabla^2 \theta \tag{3}$$

The non-dimensional groups present in above Eqs. (1–3), i.e. Reynolds number, $\text{Re} = [(\rho U_o^{2-n} W^n)/(8^{n-1} m)\ (4n/(3n + 1)^n]$ and Prandtl number, $\text{Pr} = [(mC_p/k_{\text{fluid}})\ (U_o/W)^{n-1}]$ are based on the channel height (H). The mean velocity has been calculated by using the following expression: $U_o = [(((8^{n-1} m\ \text{Re})/(\rho W^n))^{1/(2-n)})((3n + 1)/(4n))^n]$. The non-dimensionalization scale for pressure and viscous stress components is ρU_o^2 and $m(U_o/H)^n$, respectively. The non-dimensional temperature θ is defined as $(T - T_1)/(T_2 - T_1)$. The value of channel height (H) and width (W) is 1 m and 0.5 m, respectively, in the present work. However, the exact values of the physical parameters are of no particular significance as the final results are reported in a dimensionless form; thereby, enhancing their utility. For an incompressible fluid: $\tau = 2\eta\dot{\gamma}$ where $\dot{\gamma} = \frac{\nabla U + (\nabla U)^T}{2}$ which is the rate of the strain tensor. For a power-law fluid, the scalar viscosity function, η is written as follows: $\eta = m\ (I_2/2)^{(n-1)/2}$ where m is the flow consistency index and n is the fluid behaviour index. Clearly, $n = 1$ indicates the Newtonian fluid behaviour, whereas $n < 1$ represents shear-thinning fluids and $n > 1$ denotes shear-thickening fluid behaviour. The second invariant of the rate of the strain tensor, I_2 is given as:

$$I_2 = \sum_i \sum_j \dot{\gamma}_{ij} \dot{\gamma}_{ji}; \qquad (i, j) = (x, y) \tag{4}$$

Finally, the following boundary conditions have been used in this work:

$$\textbf{Inlet } 1(y = L_1) : U_x = 0;\; U_y = -1;\; \theta = 0; \textbf{ Inlet } 2(y = -L_1) : U_x = 0;\; U_y = 1; \theta = 1 \tag{5}$$

$$\textbf{Side walls}(x = W/2; \pm H/2 \le y \le \pm(H/2 + L_1)) : U_x = 0;\; U_y = 0;\; \frac{\partial \theta}{\partial Y} = 0 \tag{6}$$

$$\textbf{Mixingzones wall}(W/2 < x < (L_2 + W/2), y = \pm H/2) : U_x = 0;\; U_y = 0;\; -\frac{\partial \theta}{\partial Y} = Bi\,(\theta_w - \theta_a); \tag{7}$$

$$\textbf{Along the plane of symmetry}(x = 0) : U_x = 0;\; \frac{\partial \phi}{\partial X} = 0; \textbf{ Outlet } (x = L_2) : \frac{\partial \phi}{\partial X} = 0 \tag{8}$$

where ϕ represents the velocity and temperature field, θ_w [$= (T_w - T_1)/(T_2 - T_1)$] non-dimensional wall temperature, θ_a [$= (T_\infty - T_1)/(T_2 - T_1)$] non-dimensional ambient temperature, and Bi is the Biot number defined as $Bi = (h_{\text{air}}H/k_{\text{fluid}})$. Furthermore, the degree of mixing is defined in terms of the mixing index [8] as follows: Mixing index, $\text{Mi} = (S_T/\Delta T)$ where S_T is the standard deviation of the fluid temperature across the channel height at a specific axial location and ΔT is the temperature difference of the two inlet streams. As the combined flow moves towards the exit of the channel, an increasing degree of uniformity in temperature is expected. Thus, complete mixing is indicated by $\text{Mi} = 0$.

3 Numerical Solution Methodology

For both solving the equation (Eqs. 1–3) and meshing of the computational domain (Fig. 1a), the finite element-based solver COMSOL Multiphysics, with the direct linear solver PARDISO (Parallel Direct Solver) has been used herein which is based on the LU matrix deposition method. In this work, a 2D, stationary and "laminar single-phase fluid flow" module in conjunction with the "heat transfer in fluids" is used. The built-in meshing function was used to create the fine non-uniform quadrilateral grids to capture the steep gradients in the flow domain. The grid expansion ratios were specified to resolve the thin boundary layers with adequate accuracy. The relative convergence criterion of 10^{-5} is used herein. To ensure the accuracy and reliability of the present numerical scheme, the choice of parameters (channel length, grid structure, etc.) has been made very carefully. In this work, the inlet and mixing zone channel length are fixed as 5 and 1000 m which is found to be sufficiently large for the flow to be hydrodynamically fully developed for the range of conditions spanned here. Several simulations have been carried out to obtain the optimized grid structure (total number of grid elements is 360,770), which was kept very fine in the impingement zone and at the mixing zone walls.

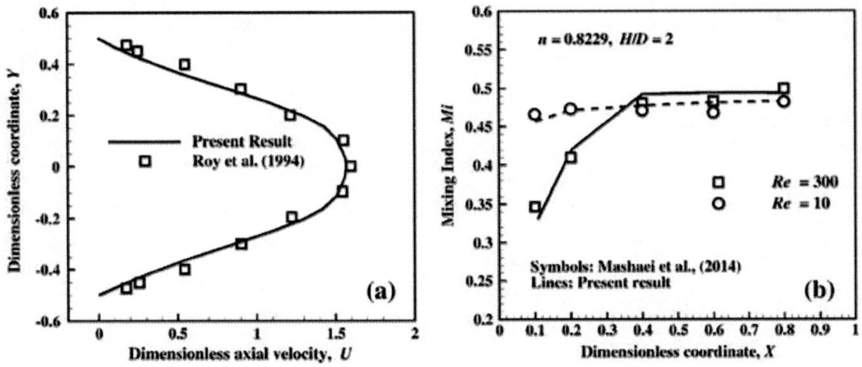

Fig. 2 **a** Validation of dimensionless axial velocity profiles at $X = 5$ and Re = 500 with that of Roy et al. [7]. **b** Validation of the impinging zone mixing index with that of Mashaei et al. [3] at $n = 0.8229$ and Re = 10, 300

4 Results and Discussions

4.1 Validation of Results

Before presenting the new results obtained in this work, it is desirable to validate the numerical solution procedure used herein with the available literature for the stream impingement. Figure 2a shows a comparison of the present dimensionless axial velocity profile with the experimental results of Roy et al. [7] for a 2D, steady, laminar, mixing behaviour of Newtonian fluid as discussed earlier. Figure 2a shows a satisfactory validation of the present simulated results with the experimental results. Similarly, the impingement zone mixing index has been validated with that of the Mashaei et al. [10] for a CMC fluid, which behaves as a power-law fluid, Fig. 2b. The results are again seen to be in good agreement ($\pm 1.5\%$ deviations), which is believed to be due to the different numerical schemes used in the two studies.

4.2 Isotherm Contours and Mixing Index

It is customary to plot isotherm contours to delineate the regions of hot/cold temperature and it is significant for the handling and processing of temperature sensitive materials and their transportation at different temperatures. Figure 3a shows representative isotherm contours (dimensionless temperature, θ) in the vicinity of the impingement zone for extreme values of the Reynolds number (10 and 50) and power-law index (0.6161 and 1) at $\mathrm{Nu}_o = 4 \times 10^3$ and $\theta_a = -2.7$ (cooling case). Figure 3a clearly shows that the two incoming streams impinge on each other and relatively narrow zone is formed in which the temperature gradient becomes very

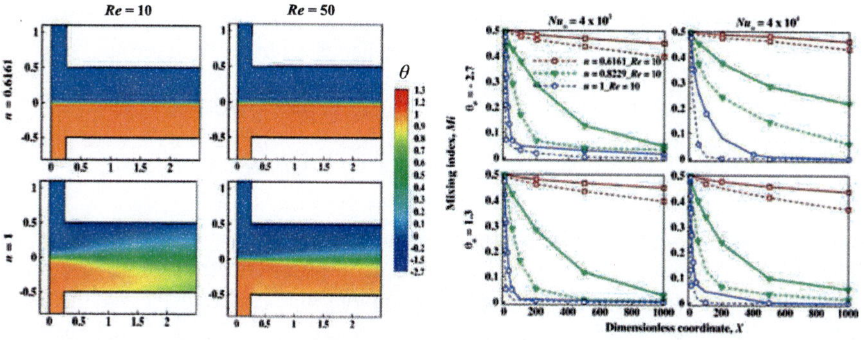

Fig. 3 **a** Representative isotherm (dimensionless temperature, θ) contours in the vicinity of the impingement zone at $\mathrm{Nu}_o = 4 \times 10^3$ and $\theta_a = -2.7$ for extreme values of Reynolds number. **b** Mixing index behaviour for extreme values of Reynolds number (Re = 10 and 50), ambient temperature ($\theta_a = -2.7$, 1.3), convective heat transfer coefficient ($\mathrm{Nu}_o = 4 \times 10^3$ and 4×10^4), and for various power-law index

sharp and hence the heat transfer is very high in this region. Thus, a better mixing is expected in a relatively shorter channel length. The temperature gradient (between the fluid streams, and fluid and ambient fluid, i.e. air) decreases along the channel which results in an expansion of the narrow zone in a cone-like shape. Figure 3a represents the isotherm contours at $\theta_a = -2.7$ which is smaller than the temperature of the inside fluid (for instance, water in case of Newtonian); thus, loss of energy occurs from the system to the surrounding. At high power-law index (closer to the Newtonian fluid behaviour, i.e. $n = 1$), the temperature distribution across the channel height increased due to the increased residence time of the fluid, and hence the cone-like shape is more pronounced at high n. It is also seen that at the channel end, the fluid temperature approaches the ambient temperature indicating complete mixing of the fluid at low Re and high n. For $n = 0.6161$, the process fluid temperature never reaches the surrounding fluid indicating the requirement of a very long channel length for complete mixing. In case of $\theta_a = 1.3$ (heating case), the fluid temperature is lower than that of the ambient air; thus, it gains energy from the ambient fluid (figure not shown here for the sake of conciseness). It has been observed that the increasing convective heat transfer coefficient and ambient temperature leads to shorter channel length to reach the ambient temperature (i.e. flat temperature profile across the channel or complete mixing).

Furthermore, Fig. 3b shows the mixing behaviour of the impinging streams for extreme values of Reynolds number, Re = 10 and 50, ambient temperature, $\theta_a = -2.7$, 1.3, convective heat transfer coefficient, $\mathrm{Nu}_o = 4 \times 10^3$ and 4×10^4, and for various values of the power-law index. In general, the mixing index is found to decrease along the channel length. The better mixing (i.e., Mi = 0) has been observed at high power-law index, ambient temperature and Nusselt number, and at low Reynolds numbers. It is clear from Fig. 3b that the channel length of $L = 1000$ m is not sufficient to achieve complete mixing for all concentrations of CMC fluids except Newtonian fluid. Thus, it is essential to investigate the role of channel length needed for the complete fluid mixing, as discussed in detail in the next section.

4.3 Required Channel Length to Achieve Complete Mixing

In this section, the mixing zone channel length has been calculated (normalized by H) in two parts such as the length needed to achieve a temperature that is 95% of the ambient temperature ($L_{0.95\theta a}$) and the actual length required to achieve the exact ambient temperature ($L_{\theta a}$). The effect of Re, θ, n and Nu_o on the mixing behaviour is also studied, (Table 1). To calculate this length, the mixing zone length has been fixed arbitrarily as 30,000 m. Then, the length of channel has been noted at which the internal fluid average temperature across the channel height becomes 95% of the ambient temperature and exactly the same as ambient temperature. The ratio of the lengths have also been calculated to see by what factor the longer channel length is required at high Reynolds number than at low Reynolds numbers. The key results can be summarized as follows. In general, a longer channel length is needed to achieve complete mixing (i.e. flat temperature profile across the channel height or to gain ambient temperature) at high Reynolds numbers than at low Reynolds numbers, by as much as a factor of ~1.7 for a given power-law index, ambient temperature, and Nusselt number. This may be attributed to the smaller residence time at high Reynolds numbers. This ratio increases with power-law index for a give Reynolds number, ambient temperature, and Nusselt number while it is almost equal for $L_{0.95\theta a}$ and $L_{\theta a}$. Similarly, as the power-law index decreases (i.e. increasing the concentration of the CMC solution), the channel length needed for complete mixing increases. This is also due to the decrease in residence time of the fluid. It can be noted from Table 1 that for $n = 0.6161$, the channel length is not sufficient to achieve the complete mixing (Table 1). Furthermore, ambient temperature and Nusselt number show the similar effect as with an increase in these parameters, the length necessary for the flat temperature profile across the channel height decreases, irrespective of values of the Reynolds number and power-law index. Thus, for the case of heating (i.e. $\theta_a = 1.3$), a shorter channel length is needed for complete mixing than that for the case of cooling (i.e. $\theta_a = -2.7$). These results are in-line with the isotherm contours shown earlier.

5 Conclusions

The present work deals with the thermal mixing of shear-thinning fluid (CMC solutions) and Newtonian fluids in a 2D, steady, confined impacting rectangular T-channel for a specified heat flux boundary condition at the mixing zone walls. The effect of Reynolds number ($10 \leq \text{Re} \leq 50$), power-law index ($0.6161 \leq n \leq 1$), Nusselt number (a dimensionless form of convective heat transfer coefficient for external

Table 1 Mixing zone length required to achieve the ambient temperature (perfect mixing) at extreme values of Reynolds number, Re = 10 and 50, ambient temperature, $\theta_a = -2.7$ and 1.3 and convective heat transfer coefficient, $Nu_o = 4 \times 10^3$ and 4×10^4 for values of the power-law index

n	Re = 10		Re = 50		$\frac{(L_{0.95\theta_a})_{Re=50}}{(L_{0.95\theta_a})_{Re=10}}$	$\frac{(L_{\theta_a})_{Re=50}}{(L_{\theta_a})_{Re=10}}$
	$L_{0.95\theta_a}$	L_{θ_a}	$L_{0.95\theta_a}$	L_{θ_a}		
$Nu_o = 4 \times 10^3$ and $\theta_a = -2.7$						
0.6161	–	–	–	–	–	–
0.7051	–	–	–	–	–	–
0.8229	13,799	23,471	–	–	–	–
0.9512	3401	5599	15,821	25,000	4.65	4.46
1	592	1009	2957	5035	5.00	4.99
$Nu_o = 4 \times 10^4$ and $\theta_a = -2.7$						
0.6161	–	–	–	–	–	–
0.7051	6675	11,798	23,120	–	3.46	–
0.8229	1645	2905	6456	11,410	3.92	3.92
0.9512	425	757	1878	3322	4.39	4.42
1	73	129	362	645	4.99	4.98
$Nu_o = 4 \times 10^3$ and $\theta_a = 1.3$						
0.6161	–	–	–	–	–	–
0.7051	–	–	–	–	–	–
0.8229	10,951	16,120	–	–	–	–
0.9512	2681	4271	12,439	19,794	4.64	4.63
1	468	745	2338	3743	5.00	5.02
$Nu_o = 4 \times 10^4$ and $\theta_a = 1.3$						
0.6161	–	–	–	–	–	–
0.7051	5145	8564	17,830	25,000	3.47	3.47
0.8229	1268	2112	4976	8283	3.92	3.92
0.9512	326	546	1447	2709	4.44	4.41
1	56	93	278	467	4.98	4.99

air flow, $10^3 \leq Nu_o \leq 10^4$), and ambient temperature ($-2.7 \leq \theta_a \leq 1.3$) has been analysed in terms of isotherm contours, mixing index, and channel length required to achieve complete mixing. The results show that the complete mixing can be achieved only for Newtonian fluid for the simulated channel length while a longer channel length is needed as the power-law index, ambient temperature, and Nusselt number decrease. The length required to achieve the 95% of the ambient temperature or the exact ambient temperature is shorter for the low Reynolds number, and high power-law index and Nusselt number. A shorter channel length is needed for the case of heating than that of the case of cooling.

References

1. Chang YR, Chen KS (1995) Measurement of opposing heated line jets discharged at an angle to a confined cross flow. Int J Heat Mass Transfer 37:2935–2946
2. Monclova LA, Forney LJ (1995) Numerical simulation of a pipeline tee mixer. Ind Eng Chem Res 34:1488–1493
3. Maurya A, Tiwari N, Chhabra RP (2019) Thermal mixing of impinging laminar streams of shear-thinning fluids. Heat Transfer Eng. 41: 1576-1595
4. Devahastin S, Mujumdar AS (2001) A numerical study of mixing in a novel impinging stream in-line mixer. Chem Eng Process 40:449–470
5. Niamnuy C, Devahastin S (2005) Effect of geometric and operating conditions on the mixing behaviour of an in-line impinging stream mixer. Chem Eng Sci 60:1701–1708
6. Roy JC, Bertrand C, Le Palec G (1994) Numerical and experimental study of mixed and forced convection in a junction. Int J Heat Mass Transfer 37:1985–2006
7. Poh HJ, Kumar K, Chiang HS, Mujumdar AS (2004) Heat transfer from a laminar impinging jet of a power law fluid. Int Comm Heat Mass Transfer 31:4884–4892
8. Srisamran C, Devahastin S (2006) Numerical simulation of flow and mixing behaviour of impinging streams of shear-thinning fluids. Chem Eng Sci 61:4884–4892
9. Mashaei PR, Hosseinalipour SM, Esmailpour K (2014) Numerical investigation of thermal mixing of shear-thinning fluids in one-way opposing jets. JCARME 3:95–103

Third-Order WENO Schemes on Unstructured Meshes

Dasika Sunder, Dipak Vaghani, and Ratnesh Shukla

Abstract We develop a third-order accurate finite volume weighted essentially non-oscillatory (WENO) scheme for numerical solution of hyperbolic conservation laws on unstructured quadrilateral meshes. The high-order non-oscillatory discretization methodology combines third- and second-order accurate reconstructions through linear weights that are positive and adds up to unity. The reconstruction procedure is based on a constrained least-squares approach that leads to a significant reduction in discretization errors. Results from a range of test cases are provided to assess the convergence attributes and test the shock capturing capability of the proposed scheme.

Keywords WENO · Compressible flows · Unstructured grids

1 Introduction

In this paper, a third-order weighted essentially non-oscillatory (WENO) scheme is developed on quadrilateral unstructured meshes for hyperbolic conservation laws. WENO discretization methods have become a popular choice for approximating general system of hyperbolic conservation laws. The inherent high-order discretization associated with the WENO reconstruction procedure minimizes numerical dissipation and enables resolution of the intricate flow features such as propagating vortices and unsteady slip lines on relatively coarse grids. The non-oscillatory feature of the WENO schemes ensures that discontinuous features such as shocks are captured with high-fidelity. For smooth solutions, standard second-order accurate schemes based on TVD discretization typically require many more mesh cells to

D. Sunder (✉) · D. Vaghani · R. Shukla
Department of Mechanical Engineering, Indian Institute of Science, Bangalore 560012, India
e-mail: dasikasunder@iisc.ac.in

R. Shukla
e-mail: ratnesh@iisc.ac.in

L. Venkatakrishnan et al. (eds.), *Proceedings of 16th Asian Congress of Fluid Mechanics*, Lecture Notes in Mechanical Engineering,
https://doi.org/10.1007/978-981-15-5183-3_23

achieve similar resolution and hence WENO schemes based on high-order reconstruction tend to be more efficient than low-order schemes in terms of CPU time and memory requirements. Examples of flows where WENO schemes have been shown to be remarkably successful include shock-vortex interaction, high-speed turbulent flows, reacting flows, and compressible multi-component flows.

In the current work, we develop a new third-order accurate finite volume WENO scheme. The reconstruction in the scheme is based on a constrained least-squares method as opposed to the least-squares method used in previous works [1–8]. The constrained least-squares approach leads to a significant reduction in errors with a marginal increase in the cost of computation. The adaptive nature of the reconstruction allows the scheme to gradually degrade to lower-order accuracy in the presence of discontinuities. Also, the linear weights can be arbitrarily chosen provided they add up to unity and are non-negative.

2 Methodology

We consider the inviscid compressible flow of an ideal gas. The governing two-dimensional Euler's equations for such flows in conservative form is given by:

$$\frac{\partial}{\partial t}\int_{V} U \mathrm{d}V + \int_{V} F \cdot \mathrm{d}S \tag{1}$$

where V is the control volume and S is the surface bounding the volume. $U = (\rho, \rho u, \rho v, E)$ denotes the vector of the conserved variables and $F = \left(\rho u, \rho u^2 + p, \rho u v, u(E + p)\right)i + \left(\rho u, \rho u v, \rho v^2 + p, v(E + p)\right)j$ is the flux vector. Here, u and v are the velocity components along x- and y-directions, ρ denotes the density with p as the pressure and E as the total energy. On using an appropriate N-point Gaussian quadrature for the surface integral, the following semi-discrete form of the governing equations is obtained:

$$\frac{\mathrm{d}\overline{U}_i}{\mathrm{d}t} = L\left(\overline{U}_i\right) = -\frac{1}{V_i}\sum_{f=1}^{N_f} |S_f| \sum_{k=1}^{N_{gp}} w_k \mathrm{F} \cdot \mathrm{n}_k \tag{2}$$

In the above equation, V_i is the volume of the control volume, N_f is the total number of faces on the control volume, $|S_f|$ is the surface area of a face, N_{gp} is the number of Gauss quadrature points on the face, and w is the weight assigned to each quadrature point. The flux F at the quadrature point should be a proper upwind flux. In the current work, rotated HLLC Riemann solver [9] is used for computing the flux. The above semi-discrete equation can be integrated in time using the third-order accurate SSP Runge–Kutta scheme [10]:

$$
\begin{aligned}
\overline{U}_i^1 &= \overline{U}_i^n + \Delta t L\left(\overline{U}_i^n\right) \\
\overline{U}_i^2 &= \frac{3}{4}\overline{U}_i^n + \frac{1}{4}\overline{U}_i^1 + \frac{1}{4}\Delta t L\left(\overline{U}_i^1\right) \\
\overline{U}_i^{n+1} &= \frac{1}{3}\overline{U}_i^n + \frac{2}{3}\overline{U}_i^2 + \frac{2}{3}\Delta t L\left(\overline{U}_i^2\right)
\end{aligned} \tag{3}
$$

In order to find the values of the conserved variable at the quadrature points with third-order accuracy, we assume that $u = u(x, y)$ is represented using a third-order (second degree) two-dimensional polynomial. The reconstruction polynomial in a cell j_0 can be obtained by following procedure:

1. Select a central spatial stencil containing enough distinct cells to obtain a third-order polynomial. In two-dimensions, the third-order polynomial consists of six degrees of freedom. Therefore, apart from the cell j_0, there should be atleast five cells in the stencil. A reference stencil is shown in Fig. 1. In order to obtain the third order reconstruction polynomial $u^{r3}(x, y)$ on the stencil we maintain the cell average of the conserved quantity exactly on the cells j_0 (for making the scheme conservative), and its face neighbor cells j_1, j_2, j_3 and j_4 (for increasing the accuracy of the scheme). On the rest of the cells, the cell average id maintained in a least-squares sense.

2. From the bigger stencil shown in Fig. 1, we choose a set of four smaller stencils for obtaining the second-order reconstruction polynomials $u^{r2:k}(x, y), k = 1, 2, 3, 4$. Note that, in two dimensions, the second-order polynomial has three degrees of freedom. Therefore, apart from cell j_0, there must be atleast two cells in each stencil. For a mesh consisting of quadrilaterals, each cell has four face

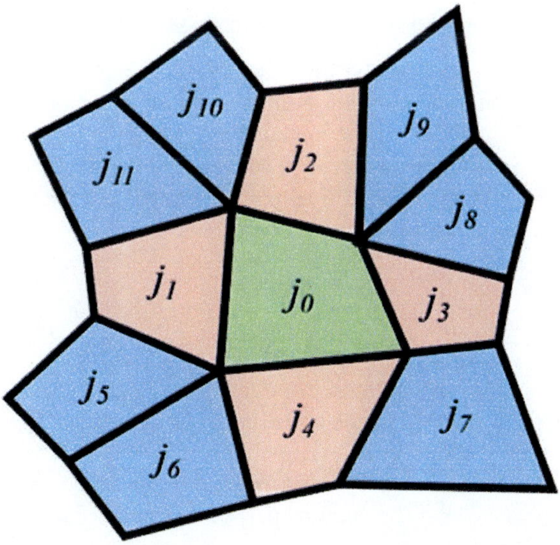

Fig. 1 Stencil for third-order reconstruction

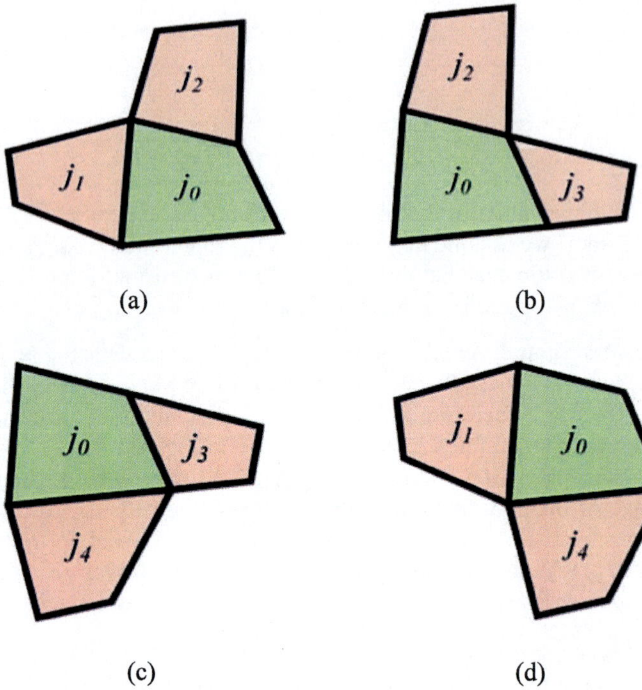

Fig. 2 Stencils for second-order reconstruction

neighbors. Therefore, we will take four stencils, each consisting of two face neighboring cells that share a common vertex as shown in Fig. 2. Since the stencils cover all the directions, robust shock capturing capabilities are obtained.

3. Let γ denote the linear weight assigned to each stencil, IS denote the corresponding smoothness indicator and w denote the nonlinear weight. We have followed the same method for finding the smoothness indicator as in [5] so, we will not repeat here. The superscripts denote the order of the stencils. To preserve accuracy at the critical points, we will use the WENO-Z [11] recipe for finding the nonlinear weights.

$$\tau = \frac{\sum_{k=1}^{4} \left| IS^{r3} - IS^{r2:k} \right|}{4}, w^r = \gamma^r \left(1 + \frac{\tau^2}{IS^r + \varepsilon}\right), w^r = \frac{w^r}{w^{r3} + \sum w^{r2:k}} \tag{4}$$

4. The last part of the above equation is to normalize the nonlinear weights, so that they add to unity. The parameter $\in$ in the above equation is a small number to avoid division by zero. In all the numerical tests, this number has been taken as 10^{-12}. The final step is to hybridize the third-order and the second-order

reconstruction polynomials. To do this, we first assign linear weights to each stencil. Each stencil is given a linear weight: $\gamma^{r3} = \alpha$, $\gamma^{r2:k} = \frac{1}{4}(1-\alpha)$. The parameter α is taken as 0.85 throughout the paper. Smoothness indicators and nonlinear weights for each stencil can be calculated for each stencil as specified in [5]. If w represents the nonlinear weight associated with a stencil, the final reconstruction polynomial can be written as:

$$u(x, y) = \frac{w^{r3}}{\gamma^{r3}}\left(u^{r3}(x, y) - \sum_{k=1}^{4} \gamma^{r2:k} u^{r2:k}(x, y)\right) + \sum_{k=1}^{4} w^{r2:k} u^{r2:k}(x, y) \quad (5)$$

3 Results

In this section, we will illustrate some results using the above scheme for two-dimensional Euler equations.

3.1 Order Property: Isentropic Vortex Test

In this test case considered originally by Balsara and Shu [8], an isentropic vortex initially centered at $(x_0, y_0) = (0, 0)$ is advected at an angle of 45° through a periodic square domain. Periodic boundary conditions are used on all the four sides of the domain. Initially, a uniform coarse mesh is considered with $h = 0.3125$. The mesh is then distorted by randomly perturbing all the vertices of the grid as shown in Fig. 3. The distorted mesh is then refined successively to check the convergence rate. The final time is set to $t = 10$ which corresponds to one advection of the vortex through the domain.

As can be seen from Table 1, the scheme achieves the desired optimal convergence rate of three.

3.2 Two-Dimensional Riemann Problems of Lax and Liu

Next, we consider the thirteenth configuration of the two-dimensional Riemann problems of Lax and Liu [12]. This configuration comprises of two shocks and two contact waves. Figure 4 depicts the computed density field at a final time $t = 0.3$. The vortices formed due to the operation of Kelvin–Helmholtz instability at the contact discontinuity can be clearly seen. The low dissipation of the reconstruction method enables resolution of these dissipation sensitive intricate flow features.

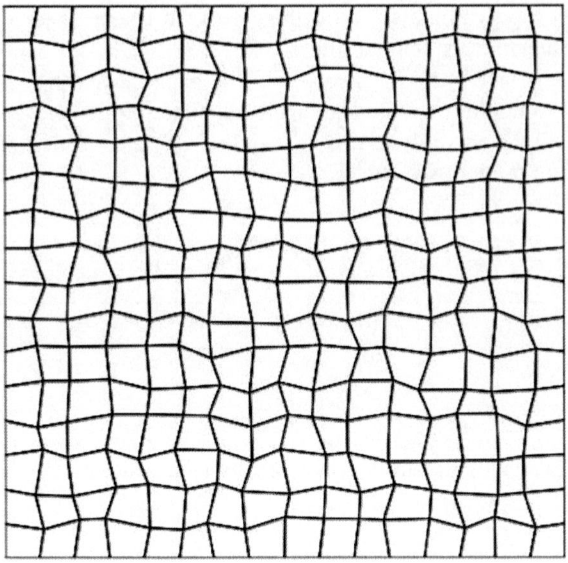

Fig. 3 Sample mesh for isentropic vortex test

Table 1 Convergence table for third-order scheme (errors are based on density). h denotes the distance between two consecutive mesh points on the boundary

h	L_2 Error	L_2 Order	L_∞ Error	L_∞ Order
0.3125	9.30692×10^{-3}		7.42625×10^{-2}	
0.1562	1.81551×10^{-3}	2.36	1.26745×10^{-2}	2.55
0.07812	2.88546×10^{-4}	2.78	1.77233×10^{-3}	2.84
0.03906	3.40563×10^{-5}	2.96	2.29116×10^{-4}	2.95

3.3 Double Mach Reflection of a Strong Shock

Next, we consider the double Mach reflection (DMR) configuration originally proposed by Woodward and Collela [12]. The problem consists of a strong Mach 10 shock that is initially incident on a wedge that makes an angle of 60° with the plane of the shock. We compute the solution to this problem on an unstructured mesh with 600,000 cells to a final time of 0.2. The density contours computed using the current WENO scheme is depicted in figure. One can observe that the physically unstable features of the flow, namely the slip lines and the wall jet, exhibit an instability growth leading to rollup (Fig. 5).

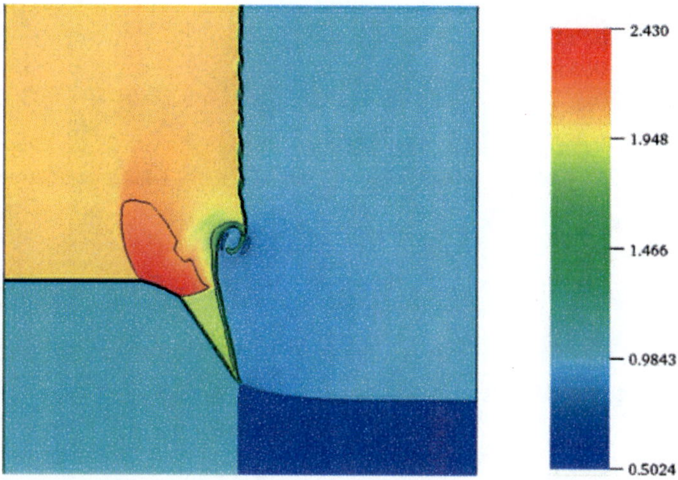

Fig. 4 Two-dimensional Riemann configuration of Lax and Liu

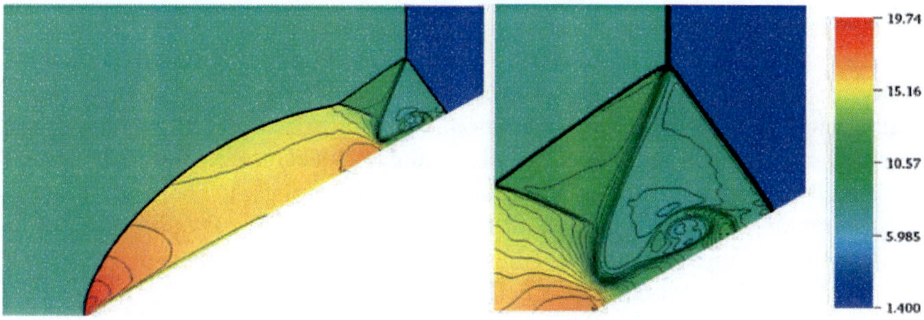

Fig. 5 Density field for the double Mach reflection of a strong shock

3.4 Shock Diffraction from a Circular Cylinder

In this test case, a Mach 10 shock hits a stationary circular cylinder of unit radius. Initially, a bow shock is reflected from the cylinder when the incident shock hits the cylinder. As the incident and the reflected shocks meet, two Mach stems along with contact discontinuities emanating from the triple points form. These shocks and contact discontinuities undergo multiple interactions and reflections lead to formation of complex vortical structures in the cylinder wake. Clearly, the proposed WENO scheme captures these unsteady features quite well (Fig. 6).

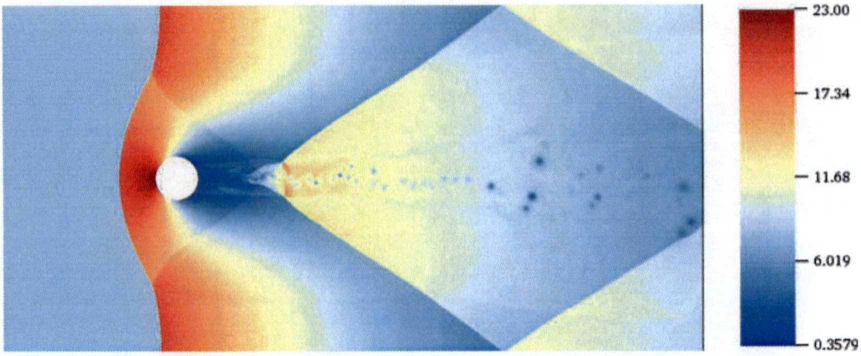

Fig. 6 Interaction of a Mach 10 shock with a stationary cylinder (Density field) at $t = 5$. Nearly three million cells are used for the computation

4 Conclusions

A third-order accurate finite volume adaptive WENO scheme for unstructured quadrilateral meshes was developed. The principal advantage of these WENO schemes is that the computation of the optimal positive linear weights is not required and the reconstruction polynomial retains higher order accuracy throughout the cell, hence maintaining a very good subcell resolution. In the proposed scheme, we have used convex combination of five quadratic polynomials and a cubic polynomial to achieve the desired convergence rate and handle strong shocks robustly. The reconstruction of the polynomials is performed using a constrained least-squares approach. Unlike the least-squares approach in which the cell averages of all the cells in the stencil are only maintained approximately, the cell averages of four face neighbors of a cell are maintained exactly in the present constrained-least squares approach. This approach leads to a more accurate solution with only a slight increase in computational cost. Numerical tests indicate that the WENO scheme proposed in this work achieves the expected convergence rates for problems that involve smooth solutions. The reconstruction procedure was shown to robustly handle strong shock interactions such as those that arise in the double Mach reflection configuration.

Acknowledgements The implementation of the current WENO scheme on unstructured grids was performed in the open-source framework deal.II [13].

References

1. Zhao F, Pan L, Wang S (2018) Weighted essentially non-oscillatory scheme on unstructured quadrilateral and triangular meshes for hyperbolic conservation laws. J Comput Phys 374:605–624
2. Tsoutsanis P, Titarev VA, Drikakis D (2011) Weno schemes on arbitrary mixed-element unstructured meshes in three space dimensions. J Comput Phys 230(4):1585–1601
3. Dumbser M, Käser M (2007) Arbitrary high order non-oscillatory finite volume schemes on unstructured meshes for linear hyperbolic systems. J Comput Phys 221(2):693–723
4. Dumbser M, Boscheri W, Semplice M, Russo G (2017) Central weighted eno schemes for hyperbolic conservation laws on fixed and moving unstructured meshes. SIAM J Sci Comput 39(6):A2564–A2591
5. Dumbser M, Balsara DS, Toro EF, Munz C-D (2008) A unified framework for the construction of one-step finite volume and discontinuous galerkin schemes on unstructured meshes. J Comput Phys 227(18):8209–8253
6. Liu Y, Zhang Y-T (2013) A robust reconstruction for unstructured weno schemes. J Sci Comput 54(2):603–621
7. Hu C, Shu C-W (1999) Weighted essentially non-oscillatory schemes on triangular meshes. J Comput Phys 150(1):97–127
8. Balsara DS, Shu C-W (2000) Monotonicity preserving weighted essentially non-oscillatory schemes with increasingly high order of accuracy. J Comput Phys 160(2):405–452
9. Huang K, Wu H, Yu H, Yan D. Cures for numerical shock instability in hllc solver. Int J Numerical Methods Fluids 65(9):1026–1038
10. Gottlieb S, Ketcheson D, Shu C-W (2011) Strong stability preserving Runge-Kutta and multistep time discretizations. World Scientific
11. Castro M, Costa B, Don WS (2011) High order weighted essentially non-oscillatory WENO-Z schemes for hyperbolic conservation laws. J Comput Phys 230(5):1766–1792
12. Lax PD, Liu X-D (1998) Solution of two-dimensional riemann problems of gas dynamics by positive schemes. SIAM J Sci Comput 19(2):319–340
13. Bangerth W, Hartmann R, Kanschat G (2007) Deal.ii; a general-purpose object-oriented finite element library. ACM Trans Math Softw 33(4)

Computational Study on the Flow Characteristics in a Film Cooled Dual-Bell Nozzle

Martin Raju, V. V. Ijas Muhammed, Abhilash Suryan, and Heuy Dong Kim

Abstract A dual-bell nozzle comes under the classification of an altitude compensating nozzle. The absence of any moving parts makes the dual-bell nozzle a better substitute for the existing bell-shaped nozzles. In this study, a numerical simulation is carried out on a 2D axisymmetric dual-bell nozzle model to find the nozzle wall pressure distribution at different nozzle pressure ratio (NPR) which is validated with the experimental results from previous literature. The main flow is modeled as LO2/LH2 combustion mixture and secondary flow is modeled as GH2. Film cooling is introduced at the inflection point of the dual-bell nozzle to study the influence of the GH2 cooling film to the flow characteristics inside the nozzle. The influence of secondary coolant injection to the movement of separation location and film cooling effectiveness is studied by varying the coolant flow rate.

Keywords Dual-bell nozzle · Nozzle pressure ratio · Film cooling

1 Introduction

The new era of cost-effective space exploration encourages the need of single stage to orbit (SSTO) mission launch vehicles. The area ratio of conventional bell nozzle is limited due to asymmetrical flow separation due to high over-expansion at low altitude for high area ratio nozzle. Therefore, multistaging is required in cases where conventional bell nozzles are used. A dual-bell nozzle (DBN) is a blend of two bell nozzles with different area ratios. It consists of a base nozzle and an extension nozzle separated by an inflection region. Here, the advantages of a low area ratio and high area ratio nozzle can be exploited by allowing the flow to get attached to the base

M. Raju (✉) · V. V. I. Muhammed · A. Suryan
Department of Mechanical Engineering, College of Engineering, Trivandrum 695016, India
e-mail: martinr@cet.ac.in

H. D. Kim
Department of Mechanical Engineering, Andong National University, Andong 760749, Korea

L. Venkatakrishnan et al. (eds.), *Proceedings of 16th Asian Congress of Fluid Mechanics*, Lecture Notes in Mechanical Engineering,
https://doi.org/10.1007/978-981-15-5183-3_24

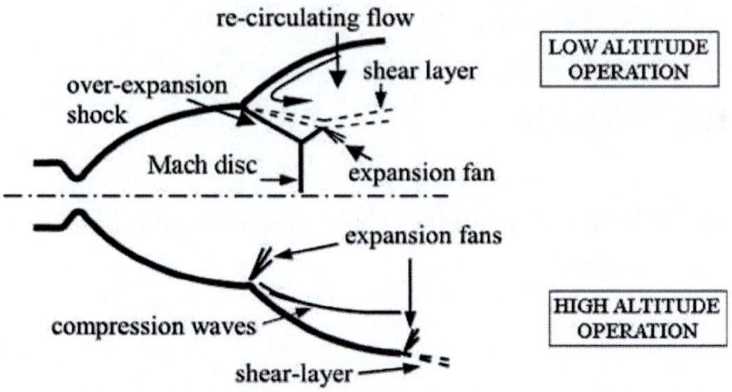

Fig. 1 Modes of operation in a dual-bell nozzle

nozzle at low altitude operation, and at high altitude, the flow gets attached to the extension nozzle part of the nozzle, thereby allowing complete expansion [1, 2].

Figure 1 shows the shear layer caused by the recirculating flow and main flow which contributes to aspiration drag in a dual-bell nozzle for low altitude operation mode. No shocks are present inside the nozzle at high altitude operation unlike in low altitude operation. The transition from low altitude to high altitude is an important performance phenomenon because early transition can negatively affect the efficiency of the dual-bell nozzle.

In modern high-performance liquid rocket engines, the combustion chamber and nozzle walls are subjected to high pressure and temperature combustion gas. Therefore, inadequate cooling of the walls may lead to catastrophic engine failure. The present computational study evaluates the flow characteristics inside dual-bell nozzle of a liquid oxygen–liquid hydrogen (LOx-LH2) SSME engine with film cooling at inflection of the dual-bell nozzle. Here, fuel is used as the coolant and its influence on separation location movement and cooling effectiveness is analyzed.

2 Theory and Methods

The flow characteristics in a dual-bell nozzle can be influenced by parameters like types of nozzle contour [3], area ratio of the base nozzle and extension nozzle, nozzle length [4], inflection geometry of the nozzle [5], inflection angle of the extension part of nozzle which is a function of Prandtl–Meyer angle [6], gas density on shock strength [7] and Reynolds number of the flow [8].

The numerical simulation was performed on a two-dimensional axisymmetric dual-bell nozzle using ANSYS FLUENT software. The geometry parameters of the dual-bell nozzle which is being validated are given in Table 1. The experimental data showed the variation of the wall pressure of the nozzle through the axial direction at different NPR values [9, 10].

Table 1 Dual-bell nozzle geometry parameter

Name	Symbol	Value
Radius of throat	R_{th}	10 mm
Base nozzle area ratio	$\varepsilon b = A_b/A_{th}$	11.3
Extension nozzle area ratio	$\varepsilon e = A_e/A_{th}$	25.6
Length of Base nozzle	L_b/R_{th}	6.2
Length of extension nozzle	L_e/R_{th}	8.3
Total nozzle length	L_t/R_{th}	14.5
Inflection angle	α_i	7.2

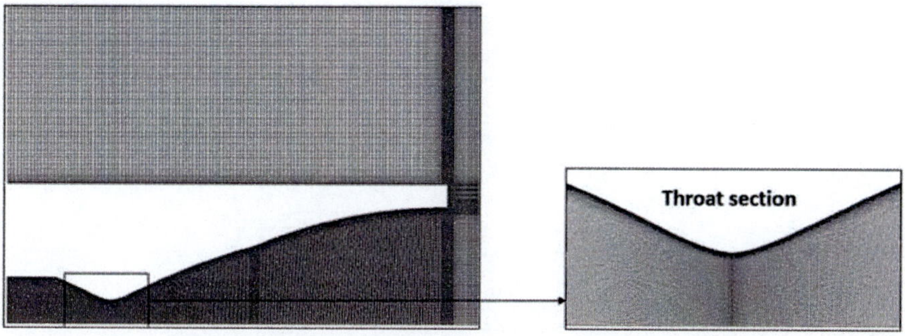

Fig. 2 Generated mesh with 3.5×10^5 elements

The present study uses a truncated ideal contour (TIC) base nozzle and a positive pressure (PP) extension. A structured mesh is generated, and refined grids were given near nozzle wall and regions where gradients are believed to be high. Grids with total number of nodes of 350,000 and y + of 5 were generated as shown in Fig. 3 (Fig. 2).

Different turbulence model was used in the present investigation for validation. A pressure inlet and exit are considered, and total temperature of the inlet and far-field boundaries was set to 300 K. An extended domain was considered to avoid any influence of the free stream boundary condition to the flow inside the nozzle. The numerical model solves Reynolds-averaged Navier–Stokes (RANS) equations with different turbulence model for validation.

The film cooling effectiveness (Θ) is calculated using the relation given as,

$$\Theta = \frac{T_w(x) - T_{w,f}(x)}{T_w(x) - T_f} \tag{1}$$

In the above equation, (x) is the wall temperature at point 'x' without film cooling, $T_{w,f}(x)$ is wall temperature with film cooling and Tf is the coolant temperature which is taken as 120 K for GH2 in an SSME rocket [12]. This study was conducted at nozzle pressure ratio where full flow was attained inside the dual-bell nozzle so

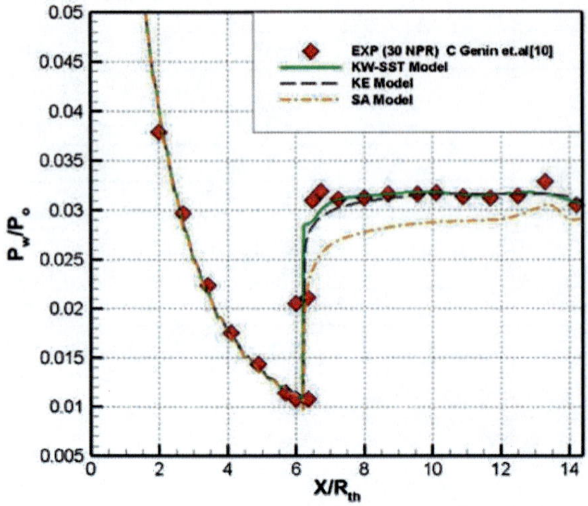

Fig. 3 Wall pressure variation in axial direction of nozzle for NPR = 30

Table 2 Different values of mass ratio

Coolant flow rate (g/s)	Main flow rate (g/s)	MR
61.24	918.6	15
45.93	918.6	20
36.74	918.6	25

that the effect of film coolant on the extension part of the dual-bell nozzle could be studied. The study was done for three different mass ratio (MR) which is the ratio of mass flow rate of main flow to the secondary coolant flow as shown in Table 2. The MR was varied by changing only the mass flow rate of the coolant so that the influence of coolant on reduction of wall temperature could be studied.

3 Results

3.1 Validation

The validation of the dual-bell nozzle geometry was done by comparing the numerical results of wall pressure variation in the axial direction of the nozzle with the experimental data [9, 10]. The experimental data from reference paper quoted above shows that transition occurs at NPR = 30 where separation location is found at location X/Rth = 6.2 which is the point of inflection. Therefore, the experimental data of wall pressure distribution in axial direction of the dual-bell nozzle at transition nozzle pressure ratio (NPRtr) is taken for numerical validation.

The numerical results for pressure variation through the dual-bell axis matched with the experimental data as shown in Fig. 3. Different turbulence models were used and it was found that k-ω SST gave better matching results to the experimental data. The validation of film cooling in the same geometry is equally important; therefore, it was clearly shown by Siba et al. [11] that the numerical model which solves RANS with *k-ω* SST turbulence model has successfully predicted the interaction of secondary and main supersonic flow by comparing data obtained through the numerical model to the experimental data.

3.2 Flow Phenomenon with and Without Secondary Coolant Injection

For the sake of understanding the flow phenomenon with and without cooling for NPR = 5 and NPR = 20, a side by side comparison of results for with coolant injection and without coolant injection are shown in Fig. 4. In Fig. 4, the Mach contour for without coolant injection is shown above the axis and Mach contour for with coolant injection is shown below the axis.

At NPR = 5, there is no significant change in shock pattern while comparing with and without coolant injection Mach contour. Secondary coolant injection recirculation region is captured successfully as shown in Fig. 4 for case with film cooling.

There is significant change in shock pattern when NPR = 20. The Mach contour obtained at 20 NPR with and without cooling depicts the early transition of separation location from base part of the nozzle to the inflection region due to secondary film coolant injection at the inflection.

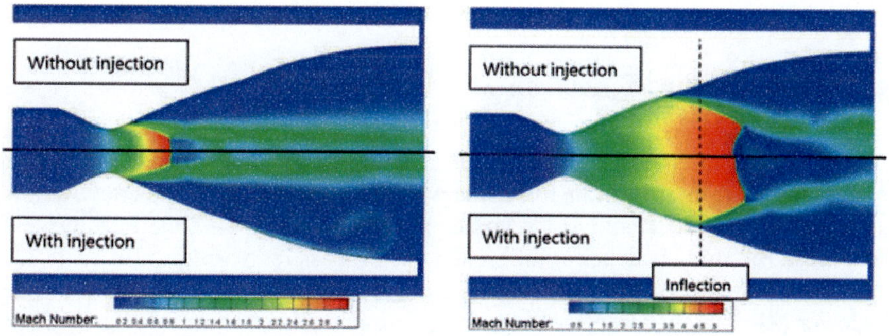

Fig. 4 Mach contour at NPR = 5 (left) and NPR = 20 (right)

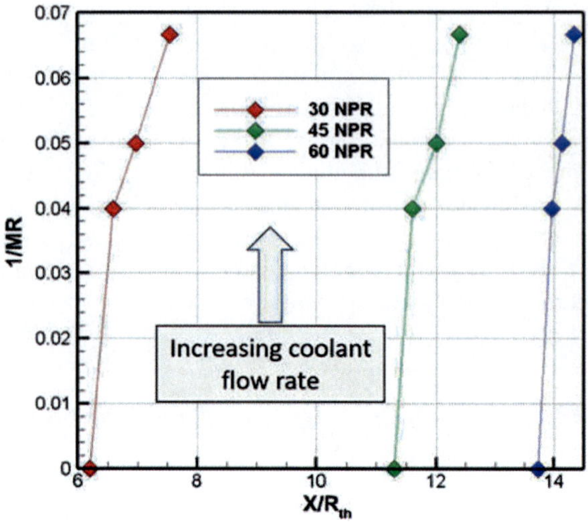

Fig. 5 Separation location movement

3.3 *Movement of Separation Location*

A study is conducted to study the influence of coolant flow rate on the movement of separation location. This influence is studied for three different cases of NPR, i.e., 30, 45, and 60.

The inverse of MR in y-axis depicts the increase in coolant flow rate. The separation location moves downstream of the dual-bell nozzle as the coolant flow rate increases for three different cases of NPR as shown in Fig. 5. The movement is found to be highest for 30 NPR case, then decreases for 45, and further decreases for 60 NPR. Therefore, the study shows that at lower NPR, influence of coolant flow rate in the movement of separation location is higher, and at higher NPR, the influence of coolant flow rate is low.

3.4 *Film Cooling Effectiveness*

The film cooling effectiveness is calculated using Eq. (1). The effectiveness of coolant to remove heat from the nozzle extension wall is shown in Fig. 6. The effectiveness is highest when $\Theta = 1$, and as we go downstream of the nozzle, the effectiveness decreases. The study is done for different cases of mass ratio as shown in Table 1.

The cooling effectiveness of GH2 on the extension wall of the dual-bell nozzle for lower MR is high. The effectiveness of the film coolant reduces downstream of the nozzle due to inadequate mixing length of the film coolant stream. There is a limitation in increasing the coolant flow rate as it can affect the specific impulse of the rocket.

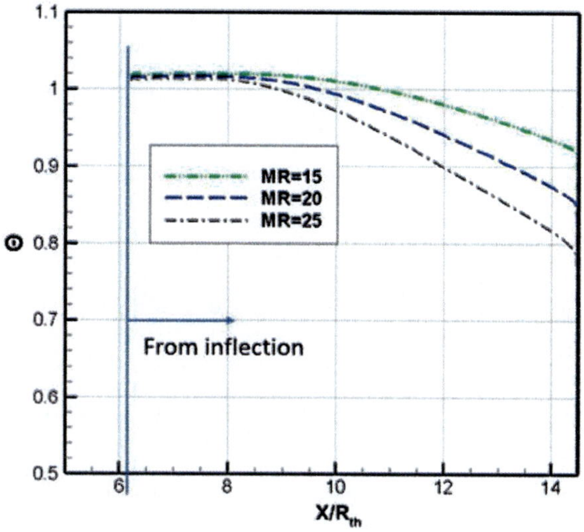

Fig. 6 Film cooling effectiveness from inflection

Further studies are being carried out to understand the influence of coolant flow rate to the specific impulse loss of the rocket.

4 Conclusion

The results for the dual-bell nozzle geometry have been carried out using the commercial ANSYS FLUENT software. The numerical simulation is conducted for different nozzle pressure ratios and is compared with the available experimental data for validation of the geometry and numerical model. The flow pattern with and without secondary film cooling was studied to visualize the secondary flow recirculation region and the early transition of separation location to inflection region. The influence of secondary coolant flow rate on the movement of separation location showed that at lower NPR, the influence is highest. Film cooling effectiveness of the film coolant on the nozzle extension wall shows that higher concentration of fuel helps in effective cooling of the nozzle wall.

References

1. Foster C, Cowles F (1949) Experimental study of gas-flow separation in over expanded exhaust nozzles for rocket motors. Jet Propulsion Lab Progress Report, 4-103
2. Horn M, Fisher S (1994) Dual-bell altitude compensating nozzles. NASA CR-194719
3. Nasuti F, Onofri M, Martelli E (2005) Role of wall shape on the transition in dual-bell nozzles. J Propulsion Power 21(2):243–250

4. Martelli E, Nasuti F, Onofri M (2007) Numerical parametric analysis of dual-bell nozzle flows. AIAA J 45(3):640–650
5. Génin C, Stark R (2011) Experimental investigation of the inflection geometry on dual bell nozzle flow behavior. In: 47th AIAA/ASME/SAE/ASEE joint propulsion conference & exhibit, 31 July–03 August
6. Otsu H, Miyazawa M, Nagata Y (2005) Design criterion of the dual-bell nozzle contour. In: 56th International Astronautical Congress of the International Astro-nautical Federation, the International Academy of Astronautics and the Inter-National Institute of Space Law, AIAA, October
7. Verma SB, Stark R, Haidn O (2012) Gas density effects on dual-bell transition behavior. J Propulsion Power 28(6) November–December
8. Verma SB, Stark R, Haidn O (2013) Reynolds number influence on dual-bell transition phenomena. J Propulsion Power 29(3):602–609
9. Nürnberger-Génin C, Stark R (2010) Experimental study on flow transition in dual bell nozzles. J Propulsion Power 26(3) May–June
10. Schneider D, Génin C (2015) Numerical Investigation of flow transition behavior in cold flow dual bell rocket nozzles. In: Propulsion and energy forum, 27–29 July, Orlando, FL
11. Choudhury SP, Suryan A, Pisharady JC, Jayashree A, Rashid K (2018) Parametric study of supersonic film cooling in dual bell nozzle for an experimental air–kerosene engine. Aerosp Sci Technol 78:364–376
12. Huzel DK, Huang DH (1992) Modern engineering for design of liquid-propellant rocket engines. In: Progress in astronautics and aeronautics, vol 147, AIAA, Washington, DC, ISBN 1-56347-013-6

Application of Adaptive Optics for Flow Visualization in Open-Air

Yuki Yoshitomi, Kyojiro Enami, Shin Oya, Yutaka Hayano, Takeo Minezaki, Noriaki Miura, and Toshiharu Mizukaki

Abstract Objective of this research is high spatial resolution acquisition of measurement image in telescopic visualization measurement. Telescopic visualization measurements cannot be performed under strong atmospheric disturbances because of the spatial resolution degraded. So that, an adaptive optics system combining a deformable mirror and a wavefront sensor is used to reduce the influence of atmospheric disturbance. The visualization measurement uses a background-oriented schlieren method that has attracted attention for use in open-air. In this paper, the phase plate was used to conduct the experiment indoors. The phase plate was designed and manufactured from the result of measuring the atmospheric disturbance. In addition, the phase plate was combined into the adaptive optics system and experiments were performed.

Keywords Background-oriented schlieren · Deformable mirror · Wavefront sensor

1 Introduction

Recently, by the increase of global air demand, optimization of arrival and departure intervals at airports is required. At Tokyo International Airport (Haneda Airport) the aim is to increase the number of arrivals and departures. The cause of the time separation between aircrafts is the wake turbulence of the aircraft. For example, according to International Civil Aviation Organization (ICAO) regulations, a small

Y. Yoshitomi (✉) · K. Enami · T. Mizukaki
Tokai University, 4-1-1 Kitakaname, Hiratuka 259-1292, Kanagawa, Japan
e-mail: yoshitomi-y@mzkklab.com

S. Oya · Y. Hayano
National Astronomical Observatory of Japan, 2-21-1 Osawa, Mitaka 181-8588, Tokyo, Japan

T. Minezaki
The University of Tokyo, 2-21-1 Osawa, Mitaka 181-0015, Tokyo, Japan

N. Miura
Kitami Institute of Technology, 165 Koen-cho kitami, Hokkaido 090-8507, Japan

L. Venkatakrishnan et al. (eds.), *Proceedings of 16th Asian Congress of Fluid Mechanics*, Lecture Notes in Mechanical Engineering,
https://doi.org/10.1007/978-981-15-5183-3_25

aircraft following a large aircraft requires a time separation of 159 s. Therefore, the time interval can be optimized if the developmental aspect of the wake turbulence of the aircraft can be visualization measured with high spatial resolution, and the optimization of the departure and arrival intervals can be expected. Recently, for research purpose, the LIDAR has been used to detect the turbulence caused by a large aircraft at take-off and landing at a few airports. Spatial resolution of the LIDAR is about 30 m at most. Thus, authors initiated the research for precise detection of the wake turbulence by optical flow visualization method, background-oriented schlieren (BOS) [3]. On the other hand, under the strong atmospheric disturbances such at an airport, spatial resolution decreases due to atmospheric disturbances, especially for telescopic imaging. In this research, a deformable mirror and a wavefront sensor were installed in optics of BOS to improve the images taken under strong atmospheric disturbances. In this report, a phase plate, which is a thick resin disk with a density gradient, was designed to simulate atmospheric turbulence for an experiment in laboratory. The phase plate was evaluated by measuring laser beam scintillation whether atmospheric turbulence was reproduced.

2 Methodology

2.1 Adaptive Optics

The concept of adaptive optics [6, 7] was proposed in 1953 for astronomer Babcock to achieve the diffraction-limited angular resolution of terrestrial telescopes. Figure 1 shows that schematic of adaptive optics . Adaptive optics is a technique to correct the light wavefront that is disturbed by disturbance. A wavefront sensor (WFS) and a deformable mirror are used to correct the light wavefront shape affected by atmospheric disturbances. Wavefront correction is performed by repeatedly performing control to feedback information obtained by the wavefront sensor to the deformable mirror (DM). The wavefront sensor is a device for measuring the phase plane of light and reconstructs the wavefront from the information obtained from the wavefront sensor using Zernike polynomial. Here, it uses the Shack–Hartmann wavefront sensor. The Shack–Hartmann wavefront sensor has a microlens array in front of the sensor. The wavefront is divided by the microlens array, and the inclination of the wavefront is obtained from the movement amount of the focal point in each lens. The deformable mirror can deform the shape of the surface and perform wavefront correction by deforming the mirror so that the wavefront error of the distorted wavefront becomes smaller when reflected by the deformable mirror. Here, it uses a MEMS deformable mirror that deforms a mirror by electrostatic force in a method using microelectromechanical systems (MEMS) technology.

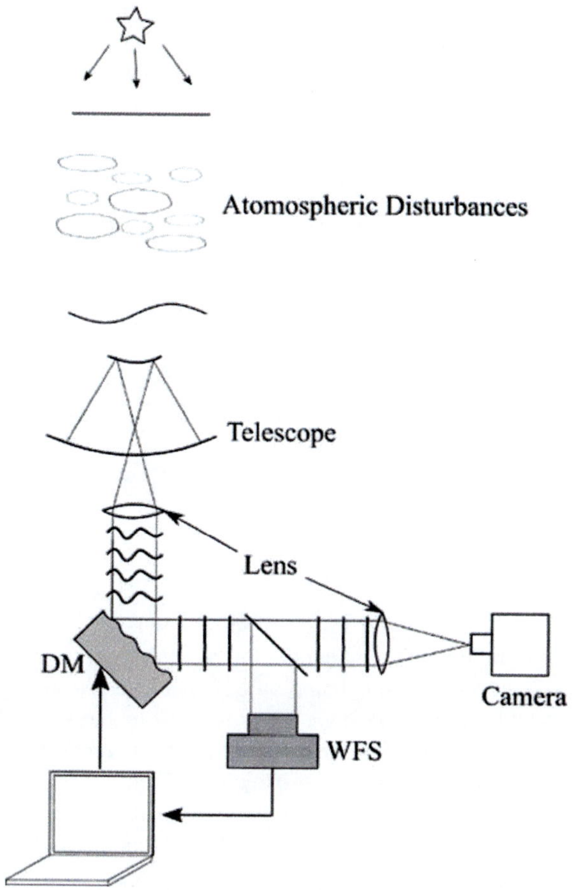

Fig. 1 Schematic of adaptive optics: wavefront sensor, deformable mirror

2.2 Background-Oriented Schlieren Method

Background-oriented schlieren (BOS) method [3, 3] was presented at a conference in 1998 and was registered as a patent in 1999 by Meier, a member of the Deutsches Zentrum für Luft- und Raumfahrt e.V. Figure 2 shows that schematic of BOS. The background-oriented schlieren method is a visualization method that records the refractive index distribution in the flow field as the distribution of the amount of movement of the background placed behind the phenomenon. For this reason, the background pattern is often used to analyze the movement of the background, and in many cases, random dot patterns are used. For recording the phenomenon, take one set of images for each measurement. One is an image in anticipation of the measurement part in the reference state where no phenomenon occurs. The other is an image in anticipation of the phenomenon. Each of these is called a reference

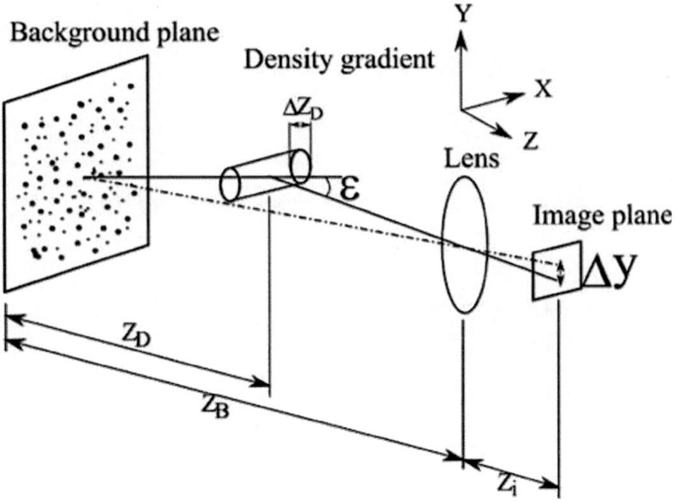

Fig. 2 Schematic of principle of background-oriented schlieren

image and a measurement image. An angle of refraction is given to the light beam by the refractive index gradient in the phenomenon passing from the background to the image recording element, and a small amount of movement is observed in part of the background image. This refraction angle is estimated from the amount of movement of the background image, and the density distribution is reconstructed. To obtain the refraction angle ε from the amount of background movement, assuming that the paraxial optical system and the declination are small, the amount of background movement Δy on the image recording element can be expressed as Eq. (1) by the phenomenon.

$$\Delta y = Z_D \cdot M \cdot \varepsilon \tag{1}$$

$$M = Z_i / Z_D \tag{2}$$

Here, the background-phenomenon distance Z_D, the background-lens distance Z_B, the focal length Z_i of the imaging lens, and the magnification of the optical system M. In addition, ε can be expressed as Eq. (3) if the refractive index n_0 of the initial state of the medium, the refractive index change n due to the phenomenon, and the depth ΔZ_D in the direction of the line of sight of the phenomenon.

$$\varepsilon = \frac{1}{n_0} \int_{Z_D - \Delta Z_D}^{Z_D + \Delta Z_D} \frac{\delta n}{\delta y} \mathrm{d}z \tag{3}$$

Furthermore, given the density ρ of the medium, the refractive index n, and the Gladstoe–Dale constant K, it is possible to relate Δy to the density gradient generated by the phenomenon from (4) of Gladstoe–Dale equation.

$$n = K \cdot \rho + 1 \tag{4}$$

3 Experimental Setup

Figure 3 shows that schematic of the optical system used in this experiment. Beam diameter of 4.4 mm was formed when light was incident on a beam expander consisting of lenses **L1** (ϕ = 25.4 mm, f = 150 mm) and **L2** (ϕ = 25.4 mm, f = 25.4 mm). The phase of the wavefront was disturbed by the light incident on the phase plate. This makes it possible to simulate the effects of atmospheric disturbances indoors as well. The light reflected by the deformable mirror (THORLABS, MEMS type, Actuator Array 14 × 14, Clear Aperture 4.4 × 4.4 mm) was split by a beam splitter into a wavefront sensor (THORLABS, 1440 × 1080 pixel, Microlens Arrays 47 × 35) and a camera (Allied Vision, 1280 × 1024 pixel, 168 fps) for measurement. The phase plate had a phase distribution as shown in Fig. 4. White was a positive phase and black was a negative phase. The phase of incident light can be disturbed by rotating the phase plate. In this experiment, laser (THORLABS, CPS635R, 635 nm) was targeted for wavefront correction. It corrects the wavefront that has passed through the phase plate.

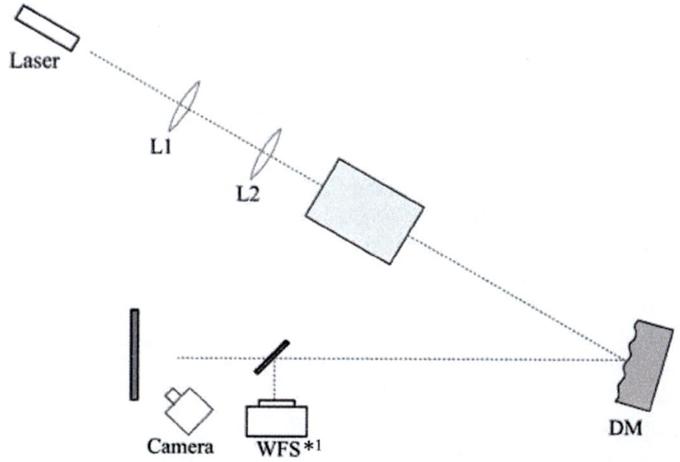

Fig. 3 Schematic of experimental optical system: *[1]Wavefront sensor; *[2]Deformable mirror

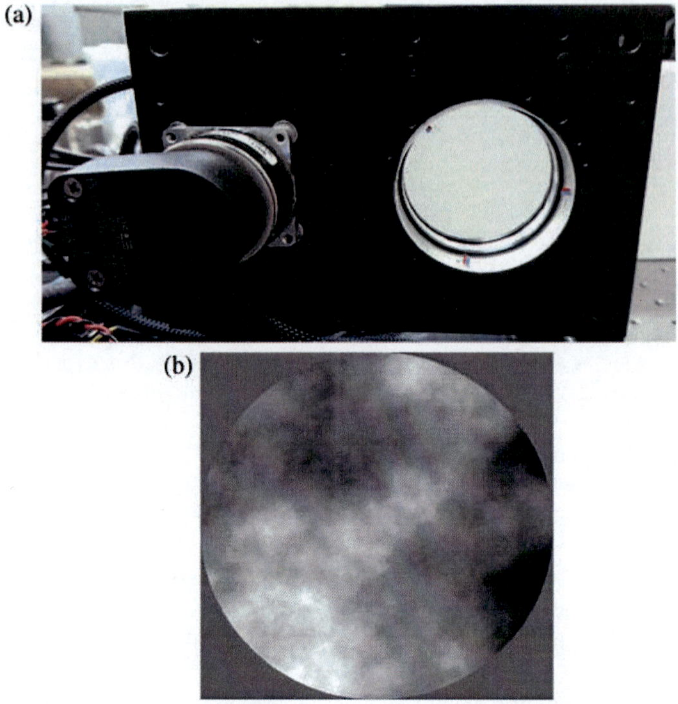

Fig. 4 **a** Phase plate device; **b** Phase pattern

4 Results

The atmospheric disturbance was measured when manufactured the phase plate. The freed length was measured as 5 mm. So that the phase plate manufactured to freed length 5 mm. The freed length can make constant by changing the number of rotations according to the size and position of the beam. The phase plate was combined into the optical system to perform a wavefront correction using a laser. As a result, when a deformable mirror and a wavefront sensor were combined, the light intensity of the laser increased . Therefore, a wavefront correction was confirmed by the deformable mirror and wavefront sensor.

5 Conclusions

The purpose of this research is to make a visualization measurement under strong atmospheric disturbances near the ground. To confirm that it is possible to remove the influence of atmospheric disturbances using existing adaptive optics. In this paper, the phase plate was designed and manufactured from the result of measuring the atmospheric disturbance. An experiment was carried out by combining the phase plate into the adaptive optics system. The result is:

- We manufactured the phase plate with a fried length (r_0) 5 mm from the results measurement of atmospheric disturbance.
- The phase plate was combined into an adaptive optics system and the experimental conditions were $r_0 = 5$ mm.
- A telescope with an effective diameter 203 mm attaches to the optical system, and telescopic measurements are planned open-air.

Acknowledgements This work was supported by JSPS KAKENHI Grant Number 17H03483.

References

1. Hayano Y, Astronomy ITS (2004) Adaptive optics. Information Process Soc Jpn 45:1239–1244
2. Hayano Y (2014) Compensation technique of light wave front disturbed by atmospheric fluctuation. Optical Electro-Optical Eng Contact 52:12–24
3. Mizukaki T (2012) Quantitative visualization by using background-oriented schlieren technique. Explosion 22:6–12
4. Arimoto Y (2007) Adaptive optics for free space optical communication systems. Optical Electro-Optical Eng Contact 45:236–243
5. Miura N (2015) Adaptive optics for solar observation using image correlation. Japanese J Optics Publ Optical Soc Jpn 44:379–383
6. Hattori M, Hayano Y (2015) Adaptive optics in subaru telescope and the application to microscopy. Japanese J Optics Publ Optical Soc Jpn 44:370–378
7. Tyson RK (2016) Principles of adaptive optics, 4th edn. CRC Press
8. Odagiri T, Mizukaki T (2016) A feasibility study of the quantitative visualization of shock waves under 1/100th atmospheric pressure environment. Proc School Eng Tokai Univ 56:103–109

DNS of an Oscillating Shear Layer Between Two Parallel Couette Flows

Kalluri Manohar Teja and Vagesh D. Narasimhamurthy

Abstract A non-planar mixing layer observed between parallel Couette flows by Narasimhamurthy et al. (Phys Rev E 85:036,302, 2012) is considered. Direct numerical simulation is chosen, and simulations are run in order to determine the critical Reynolds number at which the interface between the co-flowing laminar and non-laminar flow becomes unstable exhibiting a meandering motion. The necessary conditions required to trigger the shear-layer instability were also discussed. Different combinations of Reynolds numbers are chosen keeping the Reynolds number ratio between the laminar and non-laminar flows as constant. Preliminary results indicate that the onset of instability occurs, and a meandering motion is observed at the interface when Reynolds number for the non-laminar flow corresponds to 650.

Keywords Direct numerical simulation · Turbulent non-turbulent interface · Instability · Meandering motion

1 Introduction

In our day-to-day life, we often come across many phenomena where fluids (either same or different) with different velocities interact with each other. These are called the mixing layer flows and are often included in the class of free shear flows, other class being the wall-bounded flows, where the flow is restricted in one or many directions by a solid surface. Typical free shear flow class includes jets, wakes, and mixing layers, while the wall-bounded flows include boundary layer flows and Couette flows.

In turbulent free shear flows, the flow is free from solid walls and is often surrounded by quiescent fluid or fluid with negligible co-, counter, or cross-flow.

K. Manohar Teja · V. D. Narasimhamurthy (✉)
Department of Applied Mechanics, Indian Institute of Technology Madras, Chennai 600036, India
e-mail: vagesh@iitm.ac.in

K. Manohar Teja
e-mail: manohartejakalluri25@gmail.com

L. Venkatakrishnan et al. (eds.), *Proceedings of 16th Asian Congress of Fluid Mechanics*, Lecture Notes in Mechanical Engineering,
https://doi.org/10.1007/978-981-15-5183-3_26

The surrounding fluid is free to entrain from all directions. The flow grows as it moves along the stream-wise direction, engulfing the fluid surrounding it. This process occurs through two phenomena, namely nibbling and engulfment [3]. Engulfment refers to large-scale ingestion of surrounding fluid into the turbulent flow. The viscosity has minimal or no effect in this process; it is the inertial motion that encourages this process. On the other hand, nibbling is a viscous process, where the small scale eddies at the interface grow outward, thus pushing the boundaries of the interface at the micro-level. Together, the nibbling and engulfment process encroach the surrounding fluid, and the region of turbulent flow augments.

In the present study, a co-flowing Couette flow arrangement is chosen. The bottom plate at $z = 0$ is kept stationary. The top plate at $z = 2h$ is split into two halves, and the plates are moved at different velocities. The flow configuration is given in Fig. 2. This parallel configuration leads to a co-flow of fluid with different Reynolds numbers adjacent to each other. This creates an instability at the interface resulting in a meandering motion [2]. The hydrodynamic instability observed here is similar to the Kelvin-Helmholtz instability. The uniqueness of the current problem lies in the wall-bounded feature of the flow. Due to the presence of walls at the top and bottom, the entrainment in the wall-normal direction is arrested, and the only direction in which entrainment is allowed is the span-wise direction. Another feature of the flow is the variation of mixing layer width along the wall-normal direction and its non-planar nature. The current problem was inspired from the ship-to-ship cargo transfer operation, where the daughter vessel is aligned and moored toward the mother vessel. During this process, the daughter vessel is either decelerated or accelerated to match the speed of the mother vessel. During this process, it is very important to study the hydrodynamic forces between the two ships. This real-world problem mimics a co-flowing Couette flow, where the keel of a flat bottom ship is assumed to be the moving plate of Couette flow.

Narasimhamurthy et al. [2] used this flow configuration with Reynolds number of 1300 on one side and 260 on the other side. An observable instability is triggered at the interface resulting in a meandering motion. Figure 1b shows their simulation result. In Narasimhamurthy et al. [1], a numerical simulation with similar flow configuration but with laminar flow Reynolds numbers of 260 on one half and 130 on the other half is considered. Besides, an analytic solution to the problem was also derived. Here, there was no flow instability, and the meandering motion at the interface was absent. The results of the simulations can be observed from Fig. 1a. Upon observing both the

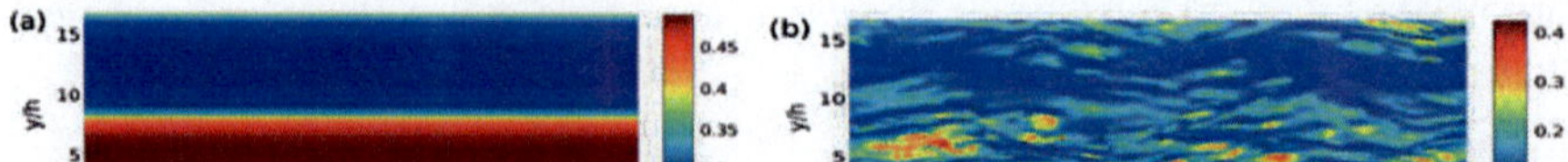

Fig. 1 Instantaneous velocity contours of two co-flowing Couette flows at the mid-plane $z = h$: **a** $\mathrm{Re}_1 = 260$ (laminar), $\mathrm{Re}_2 = 130$ (laminar) [1]; **b** $\mathrm{Re}_1 = 1300$ (turbulent), $\mathrm{Re}_2 = 260$ (laminar) [2]

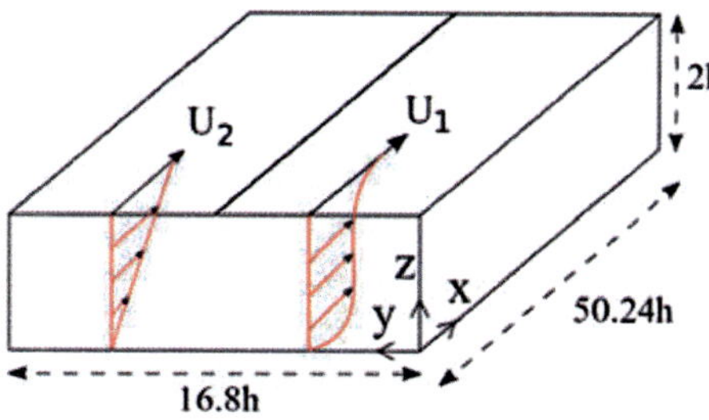

Fig. 2 Flow configuration and computation domain

cases, one can hypothesize that instability does not trigger when adjacent flows are in the laminar regime, but triggers when a turbulent flow co-exists with a laminar flow. This, however, demands a further concrete investigation, and the current objective is toward this end. We are particularly curious to know the earliest Reynolds number and the conditions under which the interface becomes unstable, and the onset of shear-layer instability occurs.

2 Methodology

In the present DNS, the full Navier Stokes equation was solved using a parallel finite volume code, MGLET [4]. The code uses a staggered Cartesian grid arrangement. The spatial discretization of convective and diffusive fluxes is carried out using a second-order central differencing scheme. The momentum equations are advanced in time by a fractional time stepping using a second-order explicit Adams Bashforth scheme. The pressure Poisson equation is solved by a full multi-grid method based on point-wise velocity–pressure iterations. The computation domain size is taken as $50.24h$, $16.80h$, $2h$ along the stream-wise (X), span-wise (Y), and wall-normal (Z) directions, respectively, where h is half channel height. A grid consisting of 256, 256, and 64 points is taken along the X, Y, and Z directions, respectively. Both the domain size and grid are taken in accordance with Narasimhamurthy et al. [1]. The grid along stream-wise and span-wise directions is uniform while the grid along the wall normal is made non-uniform. At the walls, a fine mesh is taken and mesh becomes coarser as we move away from the walls. The boundary conditions on the top and bottom of the domain that corresponds to moving and stationary plates, respectively, are taken as no-slip, and a periodic boundary condition is given on the other sides. The top plate is split into two halves, and both the halves are given different velocities. Thus, the flow imitates a scenario where there are alternative plates moving at high and low speeds. The flow configuration is shown in Fig. 2.

In the present work, a constant Reynolds number ratio (Re_1/Re_2) is taken, and simulations are run for different Reynolds number combinations to find the critical Reynolds number combination at which instability occurs. Note that the Reynolds number ratio is kept constant at five in all cases. It is well known that for a Couette flow phenomenon, a flow Reynolds number from 300 to 370 corresponds to a lower critical transition state, and flows with Reynolds number greater than 500–600 are fully turbulent [5, 6].

In the present work, the characteristic dimension and velocity for defining the Reynolds number of the flow are taken as h and half the velocity of the corresponding moving plate, respectively. Hence, the Reynolds number of the flow for the first half is $\mathrm{Re}_1 = U_1h/2\nu$, and for the second half is $\mathrm{Re}_2 = U_2h/2\nu$. The simulations are first run with a higher Reynolds number (Re_1) on both halves. After the flow has reached a fully developed state, the Reynolds number on the lower side (Re_2) is reduced as per the desired Re combination by reducing the plate velocity.

3 Results

After a preliminary study with various combinations of Re_1 and Re_2 (by keeping their ratio 5 in all the cases), it was observed that $\mathrm{Re}_1 = 600$ exhibits transitional features. With this as the base state, Re_2 was set to 120. The corresponding results are shown in Fig. 3a. As can be seen, the absence of meandering motion is apparent (is similar to Fig. 1a).

A plane Couette flow is then simulated at Reynolds number 650 on both sides, and the flow exhibits turbulent features. Taking the turbulent plane Couette flow of Re = 650 as the base state, Reynolds number on one of the halves is reduced to 130 and simulated. The instantaneous velocity contours are shown in Fig. 3b. The presence of meandering motion is striking.

As discussed earlier, with $\mathrm{Re}_1 = 600$ and $\mathrm{Re}_2 = 120$, the flow becomes laminar on both sides when the base state of the flow system is either in the transitional or laminar regime. The system does not develop any instability at the interface even though there is a significant difference in Reynolds number between the flows (Fig. 3a).

Upon a modest increase in Reynolds number, i.e., $\mathrm{Re}_1 = 650$ and $\mathrm{Re}_2 = 130$, the flow system develops shear-layer instability at the interface. The base state turbulent flow at Re 650 develops instability at the interface and thus an observable meandering motion. The corresponding instantaneous enstrophy contours are shown in Fig. 4b. In addition, a comparison is made between the current study results of $\mathrm{Re}_1 = 650$ and $\mathrm{Re}_2 = 130$ with the results of Narasimhamurthy et al. [2]. From Fig. 4, it is observed

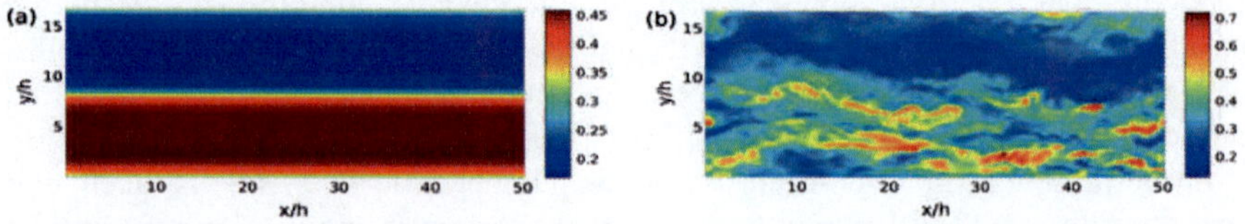

Fig. 3 Instantaneous velocity contours for two co-flowing Couette flow at mid-plane $z = 1h$: **a** $\mathrm{Re}_1 = 600$ and $\mathrm{Re}_2 = 120$ (initial base state taken as 600 on both sides) **b** $\mathrm{Re}_1 = 650$ and $\mathrm{Re}_2 = 130$ (initial base state as 650 on both sides)

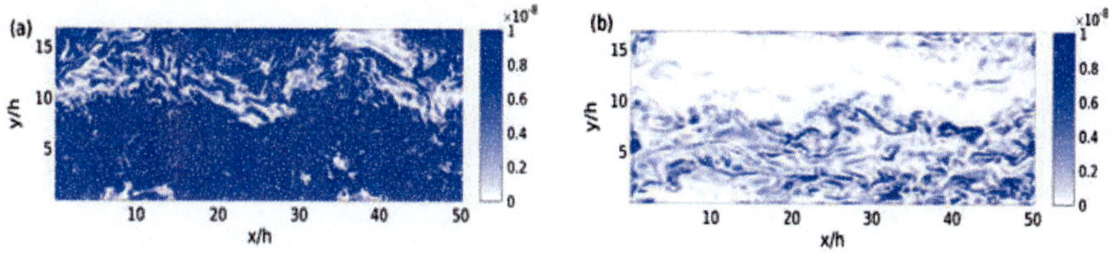

Fig. 4 Instantaneous enstrophy contours for two co-flowing Couette flow at the mid-plane ($z = 1h$): **a** $Re_1 = 1300$ and $Re_2 = 260$ [2]; **b** $Re_1 = 650$ and $Re_2 = 130$(present case)

that enstrophy is much intense in $Re_1 = 1300$ and $Re_2 = 260$ case compared to the $Re_1 = 650$ and $Re_2 = 130$ case. Also, there is a stronger influence of turbulent flow ($Re_1 = 1300$) on its co-flowing counterpart ($Re_2 = 260$) leading to higher amplitude oscillation when compared to the present case where the wave amplitude is modest. In addition, the enstrophy structures appear more elongated on the 650 Reynolds number side with negligible presence on the 130 Reynolds number side.

In Fig. 5, the fluctuating span-wise velocities are compared, and it is observed that the magnitude is more pronounced with alternating beating pattern in the $Re_1 = 1300$ and $Re_2 = 260$ case when compared to the present $Re_1 = 650$ and $Re_2 = 130$ case.

The time history plot of stream-wise velocity is shown in Fig. 6. The figure indicates the temporal variation of stream-wise velocity all along the span of the domain. From the figure, the retardation of fluid flow velocity in the second half of the flow (y/h from 8.4 to 16.8), i.e., the $Re_2 = 130$ side, can be observed while the first half (y/h from 0 to 8.4), i.e., the $Re_1 = 650$ side remains more or less unaltered.

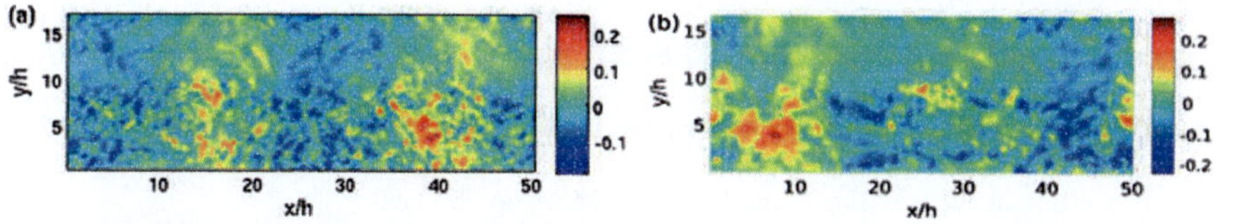

Fig. 5 Fluctuating span-wise velocity contours: **a** $Re_1 = 1300$ and $Re_2 = 260$ [2]; **b** $Re_1 = 650$ and $Re_2 = 130$ (present case)

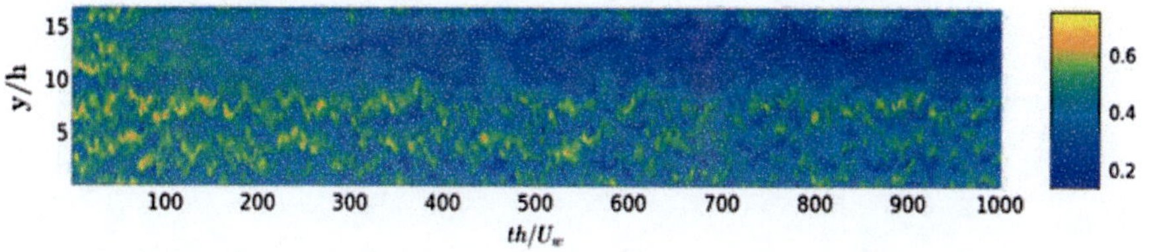

Fig. 6 Spatio-temporal plot of stream-wise velocity. The above span-wise data is recorded at $x = 25.6h$ and $z = 1h$

This warrants spectral analysis of temporal data to reveal quantitative nature of the flow system.

4 Conclusions

The following conclusions can be drawn from the preliminary work done:

- The base state should be turbulent to trigger instability in the flow arrangement.
- The minimum Reynolds number at which instability occurs is around 650.
- Under no circumstances, a laminar or transitional base state is able to trigger or sustain the flow instability.

A further analysis is to be done to explore the quantitative data. The question to be pondered over is whether a turbulent base state would encourage the instability even at a smaller Reynolds number. The effect of Reynolds number ratio is another parameter of concern. We aim to explore the problem further toward this end.

Acknowledgements The first author would like to thank Shashi Kumar J and Karthikeyan J (CFTC lab, IIT Madras) for taking their valuable time in helping and teaching necessary concepts and techniques during all stages of learning process for the work. We thank P.G. Senapathy Center for Computing Resource, IIT Madras, for the grant of computing time.

References

1. Narasimhamurthy VD, Ellingsen SÅ, Andersson HI (2012) Bilateral shear layer between two parallel Couette flows. Phys Rev E 85:036302
2. Narasimhamurthy VD, Andersson HI, Pettersen B (2014) Novel features of a fully developed mixing-layer between co-flowing laminar and turbulent Couette flows. Phys Fluids 26:031703
3. da Silva CB, Hunt JCR, Eames I, Westerweel J (2014) Interfacial layers between regions of different turbulence intensity. Annu Rev Fluid Mech 46:567–590
4. Manhart M (2004) A zonal grid algorithm for DNS of turbulent boundary layers. Comput Fluids 33:435–461
5. Tuckerman LS, Barkley D (2011) Patterns and dynamics in transitional plane Couette flow. Phys Fluids 23:041301
6. Schneider TM, Lillo FD, Buehrle J, Eckhardt B, Dörnemann T, Dörnemann K, Freisleben B (2010) Transient turbulence in plane Couette flow. Phys Rev E 81:015301

Investigation of Flow Separation Lines Over a Finite Wing

G. Vasanth Kumar and Rinku Mukherjee

Abstract Flow over airfoil is extensively studied to identify point of flow separation, transition, re-attachment (formation of laminar separation bubble), and to understand relation between aerodynamic forces generated by airfoil with change in angles of attack. Extension of flow separation concept to finite wing is not straight forward as flow has an additional degree of freedom to move when it detaches from surface of wing. A preliminary investigation of location of flow separation lines over finite wings is carried out using numerical simulations in ANSYS (FLUENT).

Keywords Boundary layer flows · Finite wing · Stall · Separation line · CFD

1 Introduction

Fundamental research in area of aerodynamics begins with understanding nature of flow over a 2D cylinder, axisymmetric objects, at plate, and combination of these fundamental geometries in different arrangement with respect to flow. One of the important parameters in study of flow around afore-mentioned geometries is varying Reynolds number to quantify transition of from laminar to turbulent flow. Next to Reynolds number, one is interested to know the effect of change in angle of attack (AOA) on flow field in order to study formation of laminar separation bubble, total flow separation to name a few. Complexities in fluid dynamics arise from the interplay of parameters that govern the flow—for instance, flow at low Reynolds number at high AOA shows markedly different behaviour from flow at high Reynolds number at high AOA [1].

G. Vasanth Kumar (✉) · R. Mukherjee
Applied Mechanics, IIT Madras, Chennai, India
e-mail: am18d007@smail.iitm.ac.in

R. Mukherjee
e-mail: rinku@iitm.ac.in

L. Venkatakrishnan et al. (eds.), *Proceedings of 16th Asian Congress of Fluid Mechanics*, Lecture Notes in Mechanical Engineering,
https://doi.org/10.1007/978-981-15-5183-3_27

Extending study to three-dimensional flow poses new sets of challenges as flow has an additional degree of freedom to manoeuvre over the surface of object. Additionally, three-dimensional flow separation is markedly different from two-dimensional case—in 2D flows, flow separation is identified by shear stress near wall going to zero and then sign reversal due to change in flow direction. One cannot expect such a direct indication in case of 3D flow [2]. In fact, three-dimensional flow separation occurs along lines and not at some points like in 2D case [3].

Three-dimensional flow separation is studied by studying topology of wing surface and using concepts borrowed from nonlinear dynamics (nodes, saddle points, etc.) as discussed in ref.[3]. Chapman [4] carried out exhaustive work of classification of flow separation in 3D case. Present work is an effort to study the separation lines post-stall over a three-dimensional wing having a cross-section of NACA 0012 using numerical simulations.

2 Methodology

Linear relation between lift coefficient and angle of attack is not valid in post-stall region. Onset of turbulence at Reynolds number of order 10^5 (external flow) translates to fact that one is facing a completely three-dimensional, nonlinear flow field where there is no simple relation between force generated and increase in angle of attack. As a preliminary investigation of flow behaviour at high angles of attack, numerical simulations are carried out in commercially available ANSYS package (v 18.2) for a Reynolds number beyond the transition regime for external flow (Reynolds number calculated based on wing chord length) in this case close to million.

This section very briefly discusses the dimensions of three-dimensional wing, computational domain chosen to study flow variables, mesh generation to study turbulence near the boundary layer. Attention is given to ensure the wall $y+$ value remains small near the wing surface to capture the small scales of turbulence in boundary layer. Turbulence model used for flow simulation is also brie y discussed along with the use of solver for governing Navier-Stokes equation.

A rectangular wing geometry shown in Fig. 1a with NACA 0012 aerofoil as cross-section is modelled to study the three-dimensional flow. Chord length of wing is taken to be 0.3 m and half-span of the wing is 1 m. Taking into consideration the symmetry of aircraft wings, only one half of the wing is simulated to reduce the computational cost. Flow domain is shown in Fig. 1b—the domain extends 3 m in front of leading and 3 m behind the trailing edge of the wing.

The region next to wing surface is modelled with very next elements to account for wall $y+$ to capture small scales of turbulence. Around 7 million elements and 3 million nodes were generated after creation of mesh. Numerical simulation was carried out in commercial package FLUENT using coupled solver and SST k-ω turbulence model. Inlet velocity used for simulation is 20 m/s (incompressible flow regime) with angles of attack varying from 0° to 16°. Convergence criteria for equations solved numerically were set to 10^{-5}.

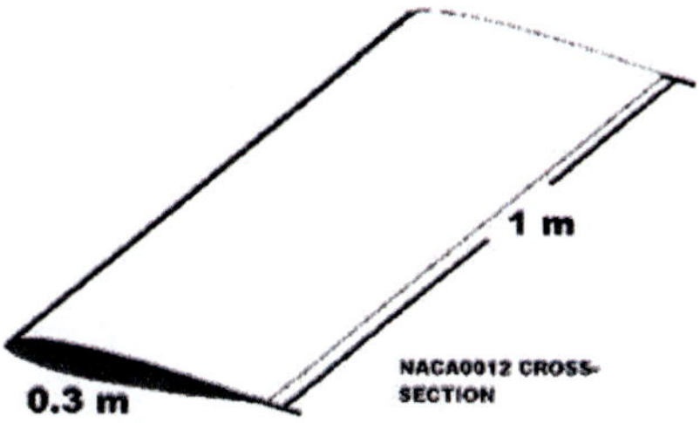

(a) Geometry of wing

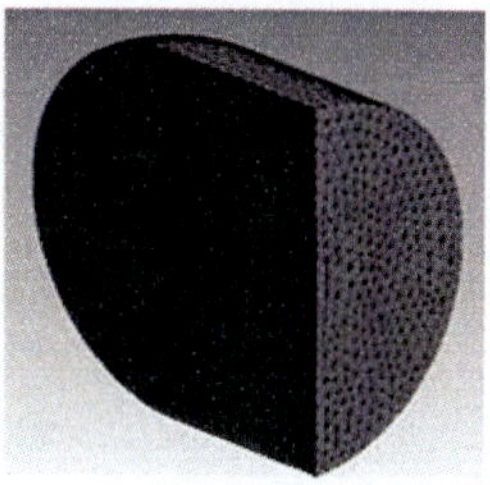

(b) Mesh Generated over ow domain

Fig. 1 Geometry and meshing

3 Results

3.1 Grid Independence and Wall y+

To validate numerical simulation, grid independence study was carried out by changing number of mesh elements and thereby refining wall *y*+ values. Figure 2 shows lift coefficient at 7° AOA and the value is within bounds of 0.51–0.54 for very coarse mesh and very fine mesh, respectively. Relatively fine mesh with 7 million mesh element was used in obtaining results for different AOA. Wall *y*+ value for the generated mesh is shown in Fig. 3.

Fig. 2 Grid independence study

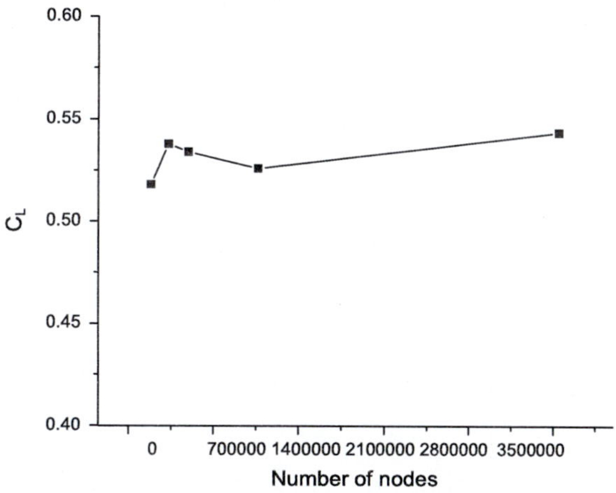

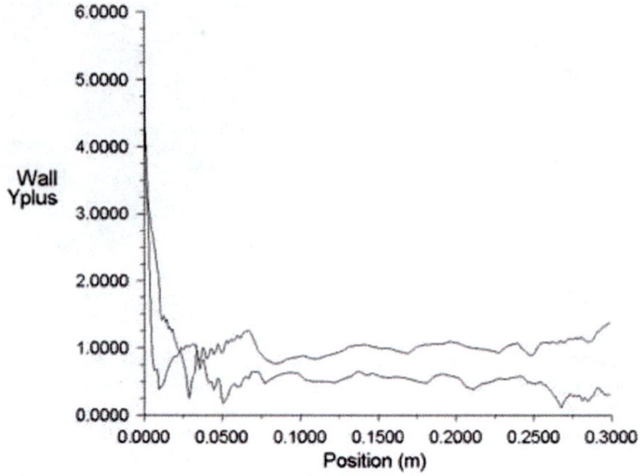

Fig. 3 Wall $y+$ along a section of wing

3.2 *Identifying Flow Separation Patterns*

Ascertaining validity of numerical results using grid independence, simulations were run to identify stall angle for given wing geometry. During run, it was found that flow separation accompanied by decrease in lift coefficient occurs at 15° AOA as seen in Fig. 4. Velocity vector plot at mid-section of wing at 15° is plotted in Fig. 5 to illustrate flow separation.

The objective of present study is to identify separation lines over wing surface and mechanism of complete stall over a wing. NACA 0012 airfoil has trailing-edge stall [5] meaning flow separates near trailing edge first and then proceeds towards leading edge.

After identifying stall angle, wall shear stress and pathlines were plotted over surface of wing close to stall and post-stall. Figure 6 shows wall shear stress at 9° AOA (6a), 13° AOA (6b), 14° AOA (6c), and 15° AOA (6d). Maximum shear stress

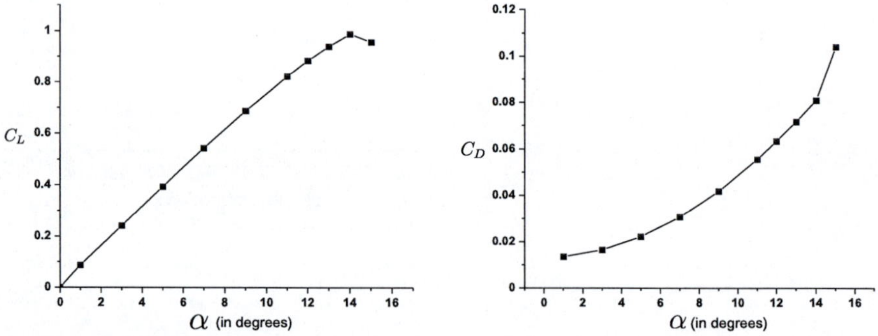

Fig. 4 Lift and drag variation with change in angles of attack

Fig. 5 Velocity vectors over a section of wing at 15° AOA

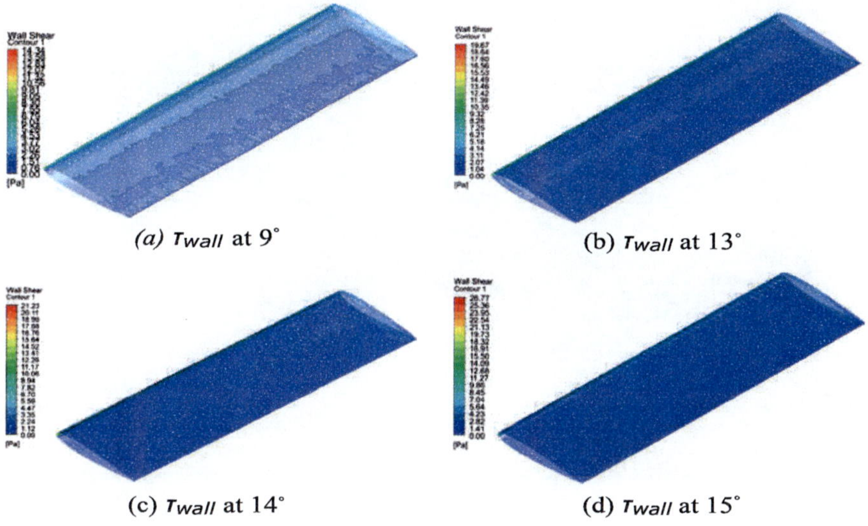

Fig. 6 Wall shear stress at different AOA

is recorded near leading edge and there is a gradual increase in value of maximum shear with increase in AOA. Series of images of wall shear stress also indicates a pattern of decreasing activity in flow near trailing edge of wing, slowly varying from 9° to 15° AOA. This can be explained by trailing-edge stall characteristic of NACA 0012 wing geometry as given in literature.

To identify lines of stall, effort is made to study flow pattern over surface of wing close to trailing edge. During computation, it was observed that the time for convergence of solution for 14° AOA to be significantly higher than the cases for lower AOA. Considering fact that same mesh elements and boundary conditions were used for solving the problem, it was necessary to investigate surface streamlines for 14° AOA. Figure 9 shows regions of flow curling to form elliptic contours near trailing edge close to root of wing. Even though there is no decrease in lift at 140° AOA, flow structure is clearly disturbed. This disturbance becomes even more evident at 15° AOA when wing has stalled. Figure 10 shows surface streamlines for 15° AOA where one can see the contours from 14° AOA becoming more prominent and moving towards leading edge (Figs. 9, 10 and 11).

If one observes Figs. 7 and 8, inference can be made about flow separation starting near root region of trailing edge of wing at 13° AOA which was absent in 9° AOA. Slowly separation region develops till a complete stall happens as shown in Fig. 5. It must be noted that no slip boundary condition is not violated here as the surface streamlines are plotted with a very small offset from surface of wing. Pathlines corresponding to surface streamlines for 15° AOA are plotted in Fig. 12.

Fig. 7 Surface streamline at 9°

Fig. 8 Surface streamline at 13°

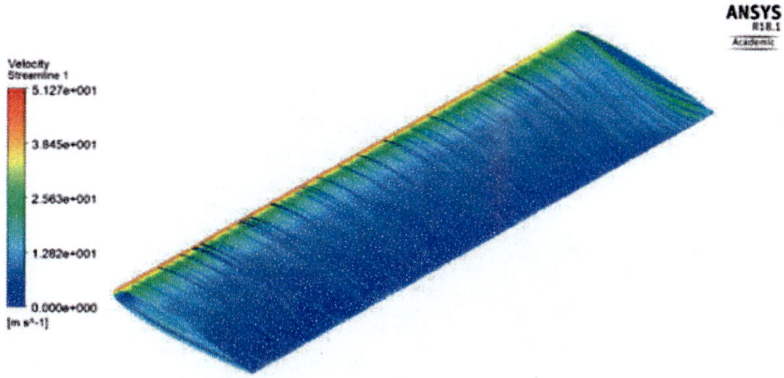

Fig. 9 Surface streamline at 14°

Fig. 10 Surface streamline at 15°

Fig. 11 Surface streamline at 16°

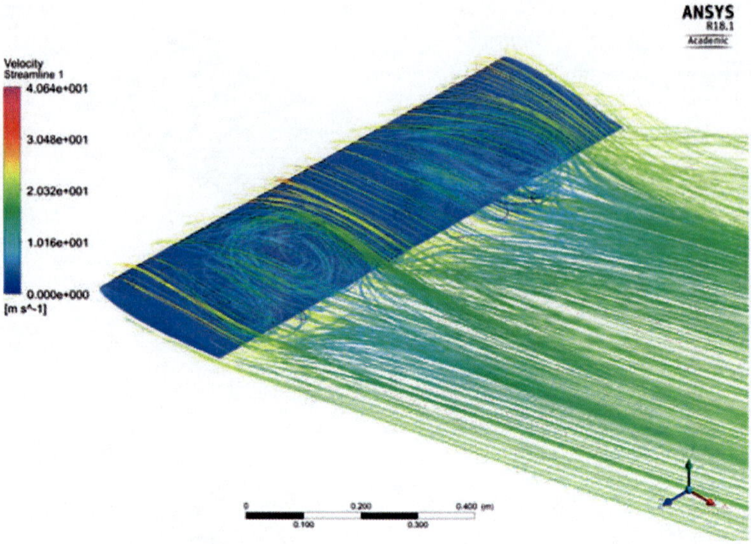

Fig. 12 Pathlines for 15° AOA

4 Conclusion

In this present paper, numerical simulations were run with an objective to visualize process of stall by studying wall shear stress and surface streamlines. NACA 0012 aerofoil was chosen as standard case since nature of stall for this aerofoil is known and it is clear that the theory can be extended to three-dimensional geometry with NACA 0012 as cross-section. Inlet velocity of 20 m/s was chosen so that the results presented in this paper can be replicated even in low-speed wind tunnels using any of the surface flow visualization methods. Effort was made to ensure that numerical simulations are valid by performing grid independence study and getting wall $y+$ value close to 1 as mentioned in literature involving turbulent flows. Results presented in this paper are for steady case at nearly 0.4 million Reynold number. Further investigation with transient case and for different Reynolds number and different wing cross-sections needs to be carried out to identify a signs of flow separation even before a complete stall occurs.

References

1. David Fisher F, Brent Cobleigh R, Daniel Banks W, Robert Hall M, Richard Wahls W (1988) Reynolds number effects at high angles of attack
2. Patel VC (1993) Three-dimensional flow separation. Sadhana 18(3):553–574. https://doi.org/10.1007/BF02744368

3. Surana A, Grunberg O, Haller G (2006) Exact theory of three-dimensional flow separation. Part 1. Steady separation. J Fluid Mech 564:57–103. https://doi.org/10.1017/S0022112006001200
4. Chapman GT, Yates LA (1991) Topology of flow separation on three-dimensional bodies. Appl Mech Rev 44(7):329–345. https://doi.org/10.1115/1.3119507
5. Petrilli JL, Paul RC, Gopalarathnam A, Frink NT (2013) A CFD database for airfoils and wings at post-stall angles of attack. https://arc.aiaa.org/doi/abs/10.2514/6.2013-2916

Scale-Aware Overlap in Turbulent Wall Jets

Abhishek Gupta, Harish Choudhary, A. K. Singh, Thara Prabhakaran, and S. A. Dixit

Abstract Data from experiments on two-dimensional turbulent wall jets suggest existence of two distinct layers, the wall (inner) layer, and the jet (outer) layer, each having its own universal scaling independent of the local Reynolds number Re_τ. This view is distinct from most earlier approaches that are either not clear about what the outer flow is, or consider the wall jet to be comprising of two regions—the region below the velocity maximum is a turbulent boundary layer having its own two-layer structure with a logarithmic overlap, and the region above the velocity maximum is a half free jet. These regions are smoothly patched at the velocity maximum. The present view considers the outer flow to be a universal full jet (rather than a half jet), centered at the velocity maximum, which overlaps with the universal inner wall flow. The hypothesis of a scale-aware overlap of these universal scaling regions leads to the prediction of an Re_τ-dependent power-law velocity profile in the overlap layer. Further, an intermediate variable approach is shown to effectively absorb this Re_τ dependence leading to a universal power-law profile for mean velocity in the overlap layer. Experiments show strong support for this description.

Keywords Wall jets · Turbulence · Mean velocity · Overlap · Scaling

1 Introduction

Two-dimensional, fully turbulent wall jets developing on flat surfaces (henceforth wall jets) have close resemblance to many flows of engineering importance and have therefore been a subject of research interest for quite some time [1–14]. These flows are characterized by a non-monotone velocity profile with a velocity maximum occurring close to the surface on either side of which the velocity decreases in the

A. Gupta · H. Choudhary · T. Prabhakaran · S. A. Dixit (✉)
Indian Institute of Tropical Meteorology, Pashan, Pune 411008, India
e-mail: sadixit@tropmet.res.in

A. Gupta · A. K. Singh
Department of Physics, Institute of Science, Banaras Hindu University, Varanasi 221005, India

L. Venkatakrishnan et al. (eds.), *Proceedings of 16th Asian Congress of Fluid Mechanics*, Lecture Notes in Mechanical Engineering,
https://doi.org/10.1007/978-981-15-5183-3_28

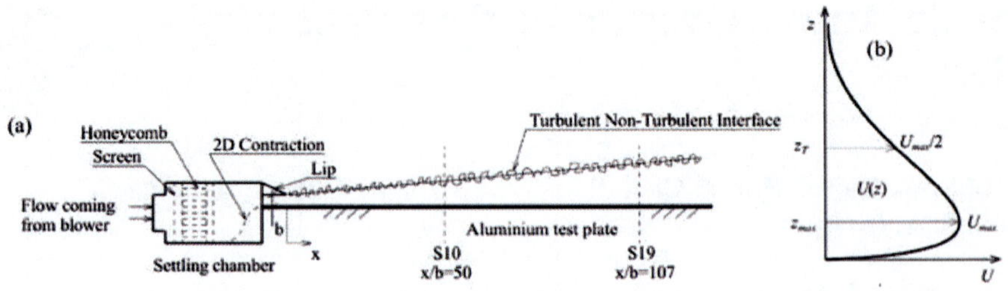

Fig. 1 **a** Side view of the wall jet setup at the Fluid Dynamics Laboratory (FDL), IITM, Pune. Region of interest (ROI) is from stations S10 to S19 where fully developed two-dimensional mean flow is obtained. **b** Typical mean velocity profile in a turbulent wall jet with essential definitions

wall-normal direction (see Fig. 1). Scaling of the mean velocity profile in wall jets [1, 3, 11, 12, 14] is challenging, and typically, two types of approaches exist. The first [12, 14] considers the wall jet comprising of turbulent boundary layer below the velocity maximum and a half free jet above the velocity maximum. This approach rests (implicitly [12] or explicitly [14]) on the hypothesis that the asymptotic state of the wall jet flow far downstream is a half free jet, and therefore, wall jet flow can be considered to be a perturbation of the asymptotic half free jet flow. Overlap layer in this approach corresponds to the overlap of inner and outer scaling regions in the *boundary layer* portion of the flow [12, 14]. The second [11] approach considers wall jet flow made of Re_τ-dependent inner and outer scaling regions whose overlap yields a Re_τ-dependent overlap power law profile. This approach assumes *constant shear stress layer* to occur in both the inner as well as the outer equations in the limit of infinite Reynolds number. However, experimental evidence for the occurrence of or the tendency toward such a *constant stress layer* in wall jets is very slender [9, 15].

In the present work, it is shown that the experimental data support a two-layer structure of a wall jet flow with universal (independent of Re_τ) inner and outer scaling layers (Sects. 3.1 and 3.2); the outer layer, contrary to previous studies, is a *turbulent full free jet* solution of the outer layer equations. Next, it is hypothesized that a *scale-aware* overlap occurs between these universal layers which predict a non-universal (Re_τ-dependent) power-law profile in the overlap layer (Sect. 3.3). The Re_τ dependence of the overlap profile may be effectively absorbed into an intermediate variable that arises in the analysis. This yields a universal overlap power-law velocity profile (Sect. 3.3). Experimental data strongly support this structure of wall jets. In this paper, $\mathrm{Re}_\tau = U_\tau z_T/\nu$ where z_T is the height from the wall (in the outer region beyond the velocity maximum) where mean velocity equals half of the maximum value ($U = U_{\max}/2$), ν is fluid kinematic viscosity, and $U_\tau = \sqrt{\tau_w/\rho}$ is surface friction velocity wherein τ_w is the wall shear stress, and ρ is the density of fluid.

2 Experimental Set Up

Figure 1a shows the wall jet setup constructed at the Fluid Dynamics Laboratory (FDL), Indian Institute of Tropical Meteorology (IITM), Pune, India. A settling chamber admits air from a blower and discharges it through a two-dimensional nozzle (width $L = 300$ mm, height $b = 10$ mm) tangentially onto the test surface. The test surface is a flat, straight, and polished aluminum plate (width 600 mm, length 1500 mm, and thickness 6 mm). Velocity profiles at nozzle exit are close to the top-hat profile (not shown) and uniform across the entire width except for small portions near the ends. The size of the room is large enough so that the secondary flow effects are negligible. Mean velocity profiles at different stream-wise and span-wise stations are measured using a Pitot tube (1.2 mm outer diameter) and a single-component hotwire probe (Pt-Rh Wollaston wire; core diameter 5 μm and active length 0.8 mm) for two nozzle exit Reynolds numbers $Re_j = U_j b/\nu = 10{,}244$ and 21,228 where U_j is the nozzle exit velocity in the potential core (see Table 1). Hotwire sensor is operated at an overheat ratio of 0.6 using the Dantec StreamLine Pro constant temperature anemometer system, and the anemometer output is acquired at 10 kHz using Dantec's StreamWare Pro software. Hotwire sensor is calibrated in situ against Pitot tube by recording mean anemometer voltage over a range of blower speeds at a fixed height from the wall (near the velocity maximum). Data with significant calibration drifts are discarded. Pitot tube and hotwire measurements agree well with each other, and mean flow is found to be two-dimensional over the extent $50 \leq x/b \leq 107$ and $-6 < y/b < 6$; x, y, and z are the strea-mwise, span-wise, and wall-normal coordinates, respectively. Wall shear stress τ_w is measured using Oil Film Interferometry (OFI) [16].

3 Scaling of Mean Velocity Profiles

3.1 Data

Table 1 lists various important parameters for the mean velocity datasets used for testing the scaling hypothesis presented in Sect. 3.3. Two-nozzle Reynolds numbers (four measurement locations in each case) from the present measurements are included. Two other datasets are included—EKP from [9] and SC from [2]. EKP data are available at the ERCOFTAC classic collection database, and SC data have been digitized from the plots given in [2]. For EKP experiments working fluid is water, mean velocity is measured using high-resolution Laser Doppler Velocimetry (LDV), and τ_w is computed from velocity gradient in the viscous sublayer. For SC experiments, working fluid is air, mean velocity is measured using hotwire anemometry, and τ_w is not measured in these experiments. Figure 1b shows a schematic velocity profile in the wall jet flow along with important length and velocity scales.

Table 1 Experimental data used in this study. Present 1 and Present 2 are the present experiments. EKP and SC are datasets from [2, 9], respectively,

Experiments	b(mm)	$\nu \times 10^5$ ($m^2 s^{-1}$)	Re_j	U_j ($m s^{-1}$)	x/b	U_τ (ms^{-1})	U_{max} (ms^{-1})	Z_T (mm)
Present 1	10	1.5	10,244	15.37	50	0.4860	7.60	40.56
					65	0.4222	6.74	51.60
					85	0.3712	5.93	67.14
					107	0.3224	5.24	80.64
Present 2	10	1.5	21,228	31.84	50	0.9373	16.29	40.92
					65	0.8158	14.37	51.86
					85	0.7079	12.54	67.71
					107	0.6081	11.20	80.76
EKP	9.6	0.104	9600	1.04	40	0.0340	0.57	34.85
					70	0.0253	0.42	57.20
					100	0.0202	0.34	83.69
					150	0.0158	0.27	115.60
SC	25.4	1.5	42,839	25.30	30	–	17.35	59.55
					36	–	15.94	72.79
					42	–	14.55	80.12

Note that z_T and $U_{\max}$ are the outer (jet part) length and velocity scales, respectively; the inner (wall part) scales are ν/U_τ and U_τ.

3.2 Outer and Inner Scaling Regions

We initially investigate the scaling of mean velocity profiles in classical outer (jet, $U/U_{\max}$ vs. η) and inner (wall, U_+ vs. z_+) coordinates; $\eta = z/z_T$, $U_+ = U/U_\tau$, and $z_+ = zU_\tau/\nu$. Figure 2a shows mean velocity profile data plotted in outer (jet) coordinates. All the data exhibit excellent scaling (collapse) in the outer region including the velocity maximum (at and above the dashed line in Fig. 2a), over the entire range of nozzle Reynolds numbers listed in Table 1. Figure 2b shows the same data scaling in inner (wall) coordinates in the near-wall region up to $z_+ \approx 12$ (below the dashed line in Fig. 2b); see also [10]. Thus, the mean velocity data support, at least to the lowest order (in a relevant small parameter such as skin friction coefficient or inverse of local Reynolds number Re_τ), the classical inner and outer scaling laws, respectively, of the form

$$U_+ = f(z_+), \tag{1}$$

$$U/U_{\max} = g(\eta). \tag{2}$$

Note that Eqs. (1) and (2) are conceptually different than those proposed by [11, 12, 14] since Eq. (2) represents the *full jet profile* (self-similar solution to the outer

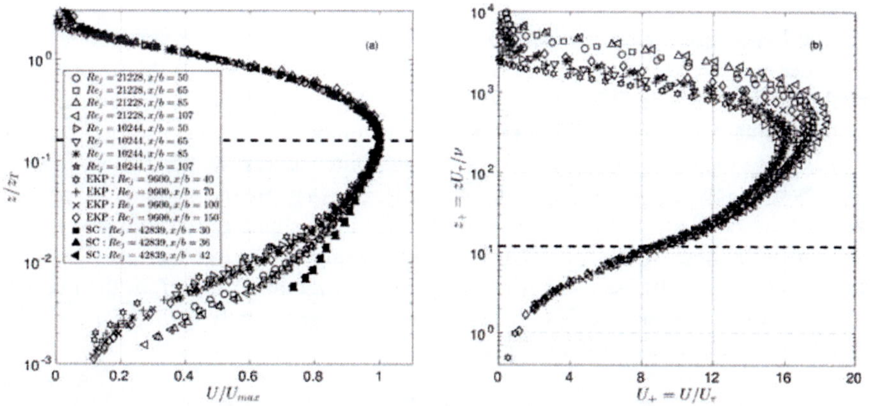

Fig. 2 **a** Outer and **b** inner scaling of mean velocity profiles in wall jets in classical outer and inner coordinates. Data are shown for a range of nozzle Reynolds numbers $9600 \le \mathrm{Re}_j \le 42{,}839$ and stream-wise locations $30 \le x/b \le 150$. Dashed lines in plots **a**, **b**, respectively, mark the locations of $U/U_{\max} \approx 1$ and $z_+ \approx 12$

equations), and there is no Re_τ dependence involved. Notice, therefore, that Eq. (2) will no longer support the assumption of *constant stress layer* for the outer equations as the wall is approached in the outer coordinates $\eta \to 0$ (see Sect. 1). With Eqs. (1) and (2), we now proceed to formulate and apply our hypothesis of *scale-aware overlap* as follows.

3.3 Scale-Aware Overlap: Hypothesis, Analysis, and Testing

The overlap layer in wall jets is expected to occur below the velocity maximum between the dashed lines of Fig. 2a, b. Matching U and $\partial U/\partial z$ from the inner (1) and outer (2) scaling laws leads to

$$\frac{\mathrm{d}\ln f}{\mathrm{d}\ln z_+} = \frac{\mathrm{d}\ln g}{\mathrm{d}\ln \eta}, \tag{3}$$

in the overlap layer. It is known that the solution to (3) is invariant under the transformation $z \to z + \alpha$ where α is an arbitrary shift in the origin for z [11]. Alternatively, (3) also permits the transformation $U \to U + \beta$, where β is an arbitrary shift in the origin for U. Physically either of these shifts accounts for a finite slip velocity at the wall since the overlap layer profile cannot satisfy the inner (no slip) boundary condition due to matching of the inner layer with the outer layer. Therefore, in what follows, we replace U by $\tilde{U} = U + \beta$, so that $\tilde{f} = \tilde{U}/U_\tau = f + \beta_+$ and $\tilde{g} = \tilde{U}/U_{\max} = g + (\beta/U_{\max})$. Since the left and right sides of (3) are purely functions of z_+ and η, respectively, each side must be equal to a universal constant, say A. Integrating $d\ln\tilde{f}/d\ln z_+ = A$ leads to $\ln\tilde{f} = A\ln z_+ + B'$ in the overlap layer.

The hypothesis of *scale-aware overlap* is concerned with the non-universality (Re_τ dependence) of the constant of integration B'. According to this hypothesis, while B' must be constant with respect to z_+ in the overlap layer, nothing precludes it from being a function of the local Reynolds number Re_τ. A justification for this comes from the observation that the inner and outer layers (length scales) in wall jets develop downstream at different rates [13], i.e., they are *out-of-sync*. When two such layers overlap, scaling properties of each layer could get corrupt due to the other resulting in the Re_τ-dependence of both scalings in the overlap layer. In other words, although f in the inner region is purely a function of z_+, f in the overlap layer inherits Re_τ dependence due to *scale-aware overlap* with the outer layer.

Rewriting $B' = A\ln B + \ln D$ yields,

$$\tilde{f} = Dz_+^A B^A, \tag{4}$$

where $B(\mathrm{Re}_\tau)$ and $D(\mathrm{Re}_\tau)$ are unknown functions, still to be determined. Similarly, $d\ln\tilde{g}/d\ln\eta = A$ leads to

$$\tilde{g} = E\eta^A C^A, \tag{5}$$

where $E(\mathrm{Re}_\tau)$ and $C(\mathrm{Re}_\tau)$ are functions unknown as yet. Clearly, (4) and (5) do not scale the mean velocity profiles at different Re_τ in the overlap layer in classical inner and outer coordinates as is clear from Fig. 2a, b. However, if one rewrites (4) and (5) as

$$\left(\tilde{f}/D\right) = (z_+ B)^A, \tag{6}$$

$$(\tilde{g}/E) = (\eta C)^A, \tag{7}$$

, then $\tilde{f}/D$ and $\tilde{g}/E$ could become *universal* functions of z_+B and ηC, respectively.

In order to proceed further, it is necessary to determine functional forms of B, C, D, and E. Toward this, note that z_+B and ηC are both akin to the so-called intermediate variable that is commonly used in the asymptotic analysis of multi-scale problems [11]. Physically, the role of function $B(C)$ is to "rescale" $z_+(\eta)$ appropriately so that the overlap region remains in focus, i.e., $z_+B(\eta C)$ remains of $\mathcal{O}(1)$ although $z_+ \to \infty(\eta \to 0)$ as $\mathrm{Re}_\tau \to \infty$. Therefore, by the definition of an intermediate variable, $z_+B \sim \eta C$ and assuming $B \sim (\mathrm{Re}_\tau)^m$ and $C \sim (\mathrm{Re}_\tau)^n$, one obtains a constraint $1 + m = n$ on the constants m and n. It is known that the overlap layer in canonical wall-bounded turbulent flows occurs beyond $z_+ \sim \sqrt{\mathrm{Re}_\tau}$ (the location of maximum Reynolds shear stress) where mean advection and Reynolds stress gradient terms in the mean momentum equation balance each other [17]. It is not unreasonable to expect that a similar balance will hold in the case of wall jets also, especially, near the beginning of the overlap layer (see Fig. 14c of [9]); details of the balance further out in the overlap region could be different. Therefore, one may expect the overlap layer in wall jets to begin around $z_+ \sim \sqrt{\mathrm{Re}_\tau}$. Since $z_+B \sim \mathcal{O}(1)$ in the overlap layer, it follows that $B \sim 1/\sqrt{\mathrm{Re}_\tau}$, i.e., $m = -1/2$ and therefore $n = 1/2$, i.e., $C \sim \sqrt{\mathrm{Re}_\tau}$.

Since $z_+B \sim \eta C$ and A is a universal constant; (6) and (7) imply $\left(\tilde{f}/D\right) \sim (\tilde{g}/E)$. Physically, functions D and E could be interpreted the Re_τ-dependent modulation of the inner and outer velocity scales, respectively. For wall jets without an external stream, the jet momentum is expected to overwhelm the drag at the surface as $Re_\tau \to \infty$ [12, 14] with the asymptotic state being the half free jet. Therefore, the effect of the wall (Re_τ dependence) in the overlap region on the outer velocity scale may be expected to be very weak and can possibly only enter in the length scale. Therefore, to the lowest order in Re_τ, it is reasonable to assume $E(\mathrm{Re}_\tau) \approx$ constant. With this, $D(\mathrm{Re}_\tau) \sim \tilde{f}/\tilde{g} \sim U_{\max}/U_\tau$ and the velocity profile (6 and 7) in the overlap layer scales according to either of the two equivalent relations

$$\tilde{f}/D = K_i\left(z_+/\sqrt{\mathrm{Re}_\tau}\right)^A, \tag{8}$$

$$\tilde{g} = K_o \left(\eta \sqrt{\mathrm{Re}_\tau} \right)^A, \quad (9)$$

where K_i, K_o and A are expected to be universal constants. Note that $\tilde{g} = g + (\beta / U_{\max})$ where β is constant for a given profile. Therefore, $\beta / U_{\max}$ could in principle be a function of Re_τ. However, the same arguments that led to $E(\mathrm{Re}_\tau) \approx$ constant, indicate that to the lowest order in Re_τ, one may expect $\beta / U_{\max} \approx$ constant. With this, (9) simplifies to

$$U/U_{\max} = K_o \left(\eta \sqrt{\mathrm{Re}_\tau} \right)^A - \frac{\beta}{U_{\max}}, \quad (10)$$

indicating that the mean velocity profiles in the overlap region should scale in coordinates $U/U_{\max}$ and $\eta \sqrt{\mathrm{Re}_\tau}$.

To test the validity of the hypothesis of *scale-aware overlap* discussed above, Fig. 3 plots all the data (of Figs. 1 and 2) in the coordinates suggested by Eq. (10). Indeed all data remarkably collapse on to a single curve (solid line in Fig. 3) given by (10) over the range $0.6 \le \eta \sqrt{\mathrm{Re}_\tau} \le 6$, i.e., a decade in $\eta \sqrt{\mathrm{Re}_\tau}$. The universal constants in Eq. (10) are found to be $K_o = -0.2858$, $A = -0.5328$ and $\beta / U_{\max} = -1.1121$.

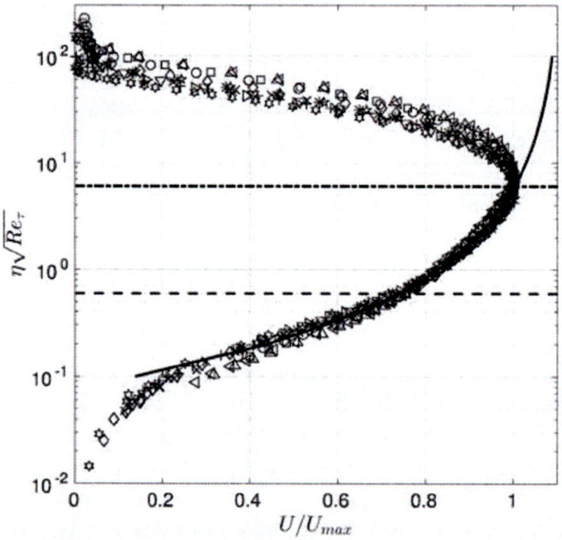

Fig. 3 Overlap layer scaling in terms of the intermediate variable $\eta \sqrt{\mathrm{Re}_\tau}$. Solid line shows the least-squares curve fit of Eq. (10) extending from the dashed line to the dashed–dotted line ($0.6 \le \eta \sqrt{\mathrm{Re}_\tau} \le 6$). The curve is extended beyond these limits for visual aid

4 Conclusions

We have presented a unique approach to the scaling of mean velocity profile in the two-dimensional, fully turbulent wall jets developing on flat surfaces. The conclusions may be summarized as follows.

1. Data support a two-layer description for mean velocity; the inner (wall) layer scaling purely on wall variables U_τ and ν, and the outer (jet) layer scaling purely on outer variables $U_{\max}$ and z_T. These scalings are universal, i.e., independent of Re_τ.
2. Analysis shows that these universal scalings could overlap in a *scale-aware* manner and yield an Re_τ-dependent (non-universal) power-law velocity profile in the overlap layer.
3. Describing the overlap layer in terms of an intermediate variable absorbs the Re_τ-dependence and leads to a universal power-law description given by Eq. (10). All the data collapse remarkably well on to this universal power-law for a decade in the intermediate variable.
4. Physical implications of these findings are revealing. Figure 2a shows that the outer free-jet scaling describes the velocity profile above the maximum as a well as a small portion below the maximum. Similarly, Fig. 2b shows that the inner scaling well-described the region very close to the wall. The region between these two is the overlap seen in Fig. 3. The Reynolds number-dependent overlap implies that the inner–outer interaction is quite strong in wall jets. This interaction occurs between the highly non-linear outer free-jet structure and the viscous-dominated inner wall-flow structure. It is very likely that this interaction leads to the counter-gradient momentum diffusion known to occur below the velocity maximum in wall jets [5] possibly in the same region as the overlap layer in our analysis. We plan to explore this possibility in the near future.

Acknowledgements We thank ERCOFTAC Classic Collection Database for the archived data of EKP experiments. AG, HC, TP and SAD gratefully acknowledge the support by the Director, IITM and the Ministry of Earth Sciences (MoES), Government of India. SAD is thankful to Prof. R. Narasimha (JNCASR, Bengaluru) and Prof. O. N. Ramesh (IISc, Bengaluru) for insightful discussions and valuable suggestions.

References

1. Glauert MB (1956) The wall jet. J Fluid Mech 1(6):625–643
2. Schwarz W, Cosart WP (1961) The two-dimensional turbulent wall-jet. J Fluid Mech 10(4):481–495
3. Bradshaw P, Gee MT (1962) Turbulent wall jets with and without an external stream. ARC R. & M. 3252, Her Majesty's Stationary Office
4. Tailland AC, Mathieu J (1967) Jet pariétal. J Méc 6(1)

5. Narasimha R, Narayan KY, Parthasarathy SP (1973) Parametric analysis of turbulent wall jets in still air. Aero J 77(751):355–359
6. Launder BE, Rodi W (1983) The turbulent wall jet measurements and modeling. Annu Rev Fluid Mech 15(1):429–459
7. Wygnanski I, Katz Y, Horev E (1992) On the applicability of various scaling laws to the turbulent wall jet. J Fluid Mech 234:669–690
8. Schneider ME, Goldstein RJ (1994) Laser doppler measurement of turbulence parameters in a two-dimensional plane wall jet. Phys Fluids 6(9):3116–3129
9. Eriksson JG, Karlsson RI, Persson J (1998) An experimental study of a twodimensional plane turbulent wall jet. Exp Fluids 25(1):50–60
10. Tachie M, Balachandar R, Bergstrom D (2002) Scaling the inner region of turbulent plane wall jets. Exp Fluids 33(2):351–354
11. George WK, Abrahamsson H, Eriksson J, Karlsson RI, Lofdahl L, Wosnik M (2000) A similarity theory for the turbulent plane wall jet without external stream. J Fluid Mech 425:367–411
12. Afzal N (2005) Analysis of power law and log law velocity profiles in the overlap region of a turbulent wall jet. Proc R Soc A 461(2058):1889–1910
13. Barenblatt GI, Chorin AJ, Prostokishin VM (2005) The turbulent wall jet: A triple-layered structure and incomplete similarity. Proc Natl Acad Sci 102(25):8850–8853
14. Gersten K (2015) The asymptotic downstream flow of plane turbulent wall jets without external stream. J Fluid Mech 779:351–370
15. Launder BE, Rodi W (1979) The turbulent wall jet. Prog Aerosp Sci 19:81–128
16. Chauhan K, Henry CH, Marusic I (2010) Empirical mode decomposition and hilbert transforms for analysis of oil-film interferograms. Meas Sci Technol 21(10):105405
17. Wei T, Fife P, Klewicki J, McMurtry P (2005) Properties of the mean momentum balance in turbulent boundary layer, pipe and channel flows. J Fluid Mech 522:303–332

Experimental Investigation on the Role of the Streaky-Like Structures in a Pressure-Induced Separation Bubble

S. Mohamed Aniffa and Alakesh Chandra Mandal

Abstract The role of streaky-like structures present in the separated shear layer has been investigated experimentally using both the conventional and the time-resolved particle image velocimetry (TR-PIV) techniques. Forward jet, backward jet, laminar-like flow structures are observed in the forepart of bubble using the fluctuating velocity vector field. Averaging the PIV realizations associated with each of these structures separately, we study their mean flow characteristics. The maximum values of the root-mean-squared streamwise fluctuating velocity ($u_{\mathrm{rms,max}}$) associated with the forward and backward jet-like structure are found to grow almost linearly, whereas the same for the laminar-like structure grows exponentially. The growth of $u_{\mathrm{rms,max}}$ of the backward jet-like structure is found to be less than the forward jet-like structure. Utilizing the shape factor (H) and the momentum thickness (θ) variations, the transition locations associated with the forward and backward jet-like structures are found to be different compared to the laminar-like flow case.

Keywords Pressure-induced separation bubble · Forward and backward jet-like structure · Particle image velocimetry

1 Introduction

When an attached shear layer experiences a strong adverse pressure gradient, it gets separated from the surface; this type of separation called pressure-induced separation. This separated shear layer undergoes transition-to-turbulent and eventually reattaches with the wall and forms an enclosed region, which is called a separation bubble. This type of separation bubble is encountered in many practical engineering devices, for example, in wings of unmanned aerial and micro-air vehicles, in turbine blades, etc. It affects the aerodynamic performance of the body by increasing the drag and reducing the lift. Hence, from the flow control point of view, the separation bubble is

S. Mohamed Aniffa (✉) · A. C. Mandal
Department of Aerospace Engineering, Indian Institute of Technology Kanpur, Kanpur 208016, India
e-mail: aniffa@iitk.ac.in

L. Venkatakrishnan et al. (eds.), *Proceedings of 16th Asian Congress of Fluid Mechanics*, Lecture Notes in Mechanical Engineering,
https://doi.org/10.1007/978-981-15-5183-3_29

still an active research area in fluid mechanics. The study of stability characteristics of the separated shear layer is important to control the separation bubble.

The Kelvin–Helmholtz (*KH*) instability mechanism is believed to be a primary instability mechanism in the separated shear layer. When the disturbance amplitude (u_{rms}) reaches a threshold value, the shear layer rolls up and forms two-dimensional (2D) spanwise vortices. These 2D vortices break down into small three-dimensional (3D) vortices due to elliptic (spanwise deformation of vortex core) and hyperbolic (instability in braid region) instabilities [1]. It is well known that the formation of a 2D vortex in the separated shear layer is due to the inviscid instability (*KH* instability). The $u_{rms,max}$ associated with the inviscid instability is found to grow exponentially in the separation bubble. Recent work by Simoni et al. [2] shows that the streamwise variation of $u_{rms,max}$ shows transient growth followed by an exponential growth. Moreover, they speculated that transient growth might be due to streaky-like structure. A recent computational study by Hosseinverdi and Fasel [3] confirms that the presence of streaky-like structure (Klebanoff mode, or K-mode) and 2D vortex structures (*KH* mode) in the pressure-induced separation bubble (PISB). Moreover, they reported that even at a low level of FST (0.1%), K-mode emerges in the separated shear layer. Also, they found that transition occurs due to either *KH* mode/K-mode or both in the separated shear layer depends on the level of FST. The K-mode comprises of both low-speed and high-speed regions in the flow [4]. These, low-speed and high-speed regions may affect the transition characteristics of the separated shear layer differently. Hence, in the present study, we aim to investigate experimentally the role of high-speed and low-speed regions/structures in the separated shear layer.

2 Experimental Set-up

Experiments have been carried out in an open-circuit suction-type low-speed wind tunnel. A 12 mm thick (h) flat plate was mounted horizontally at the centre of the tunnel test section. The leading edge of the flat plate had an asymmetrical modified super elliptic leading edge. An adverse pressure gradient in the test section was imposed by a wall-contour, as shown in Fig. 1. The dimensions of the contoured wall

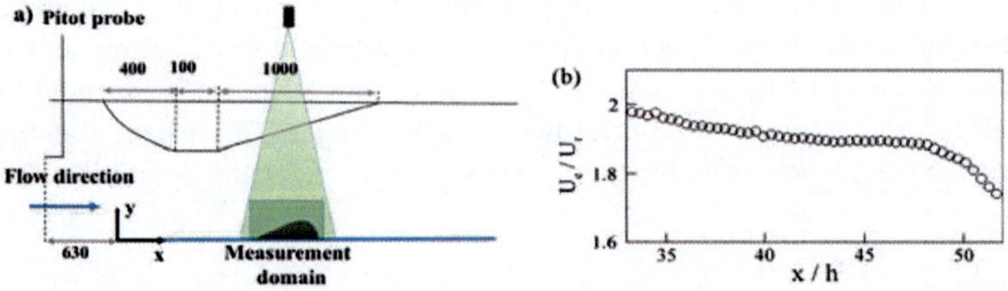

Fig. 1 **a** Schematic of the experimental set-up. All dimensions are in mm. **b** The boundary layer edge velocity (U_e) variation along x

and the location of the Pitot probe (for measuring the reference velocity) are shown in Fig. 1. A 10 mm wide slot along the centreline of the contoured wall was made to pass the laser sheet.

The conventional PIV measurements, i.e. low-frequency measurements, have been carried out in the wall-normal plane (x–y). A double pulsed Nd-Yag laser (InnoLas, 180 mJ/pulse) and a CCD camera (4912 × 3280 pixels) were used to cover the 294 mm × 196 mm measurement domain, as schematically shown in Fig. 1. A total of 600 image pairs was acquired at a rate of 1 Hz. The acquired images were processed using the proVISION-XS package (IDT Vision, USA).

The TR-PIV measurements have been carried out around the maximum height of the bubble in the wall-normal plane. This was done due to the limitation of the spatial resolution of the CMOS camera (Motion Pro, 2560 × 1920 pixels) used here. A double pulsed Nd-YLF laser (Photonics, energy 30 mJ/pulse) was used to illuminate a region of 102 mm × 43 mm. A sequence of 1250 image pairs was acquired at a rate of 365 Hz.

3 Results

Pressure variation along the streamwise direction is important for a separation bubble, as it serves as a boundary condition for the bubble [5]. Instead of the surface pressure variation, we have shown here a variation of the boundary layer edge velocity in Fig. 1b, following Häggmark et al. [6]. In the separation bubble region, the velocity variation is found to be similar to the one reported by Häggmark et al. [6].

3.1 Mean Flow Measurements

Three repetitive events are found in the flow by visual inspection of the fluctuating velocity vectors of all the PIV realizations. These different events are shown in Fig. 2, along with the outline of the mean separation bubble; here, the separation bubble is identified by the trace of zero streamwise mean velocity. Figure 2a shows laminar-like flow in the forepart of the bubble. On the other hand, backward and forward jet-like structures are also observed in the forepart of the bubble, which are shown in Fig. 2b, c, respectively. However, in all these events, one can clearly see the shedding structure in the aft part of the bubble, even though streaky-like structure is present in the flow (Fig. 2b, c). Similar structures are also reported in the case of flat plate bypass boundary layer transition studies [e.g. 7, 8]. To analyse the mean flow characteristics of each event separately, the PIV realizations (frames) are separated according to the laminar-like, backward jet-like and forward jet-like structures. The procedure for separating PIV realizations is as follows: an absolute maximum of the streamwise fluctuating velocity component (u) and its sign (positive/negative) at each streamwise (x) location are determined based on the data at all y locations.

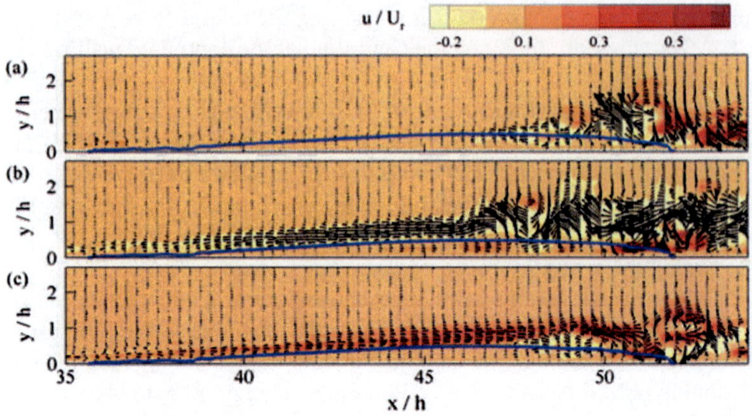

Fig. 2 Fluctuating velocity contour superimposed with fluctuating velocity vectors and the mean separation bubble. **a** Laminar-like structure, **b** backward jet-like structure, **c** forward jet-like structure

Since the shedding structure present in all events, the data points up to the maximum height of the bubble in the streamwise direction are considered to separate the frames. The conditions for each structure given in Table 1 are imposed at each streamwise location while separating the PIV realizations. If 60% of u velocity satisfies the given condition in a single PIV realization till the streamwise distance corresponding to the maximum height of the bubble, the realization is considered for further analysis. The total number of conditionally separated PIV realizations associated with each event is detailed in Table 1. Similar conditional separation of the PIV realizations is reported in the study of an attached boundary layer flow subjected to an elevated level of freestream disturbances [8].

The conditionally separated PIV realizations are also manually verified to check whether these frames correspond to the individual structure separated using MATLAB code. It is also very important to check whether the conditionally separated frames, all together, represent the actual flow characteristics or not. Hence, all

Table 1 Conditions used for separating PIV realizations

S. no.	Events present in the forepart of the bubble	Condition applied to extract	Total number of realizations satisfy the condition (out of 600)	Symbols for representation
1	Laminar	$-0.14 < u/U_r < 0.14$	107	○●
2	Backward jet	$u/U_r < -0.16$	114	○●
3	Forward jet	$u/U_r > 0.16$	119	○●

conditionally separated 340 PIV realizations are ensemble averaged, and the results are compared with the total 600 PIV realizations results, as shown in Fig. 3a and b. One can notice that although the number of conditionally separated realizations is about 56% of the total realizations, the average of these conditionally separated frames closely represents the overall mean flow characteristics.

Mean flow profile corresponding to each event is also shown in Fig. 4. It shows that, due to the backward jet-like structures in the separated shear layer, the mean velocity profile is moved outward from the wall compared to the one associated with the laminar-like event, as shown in Fig. 4a. On the other hand, the mean velocity profile is moved towards the wall due to the forward jet-like structures. A similar variation in the mean flow profile is reported in the case of PISB under different level of forcing amplitude [9]. Hence, it can be inferred that forward and backward

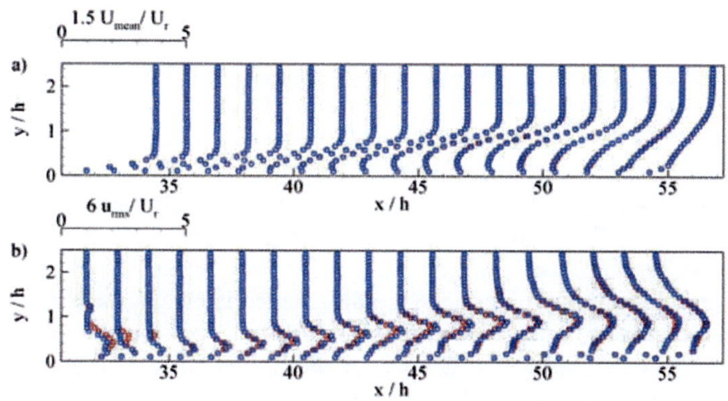

Fig. 3 Comparision of averaged results of total 600 PIV realizations (blue coloured square symbol) with conditionally separated and averaged 340 PIV realizations (red coloured square symbol). **a** Streamwise variation of the mean velocity profile, **b** streamwise variation of u_{rms} profile

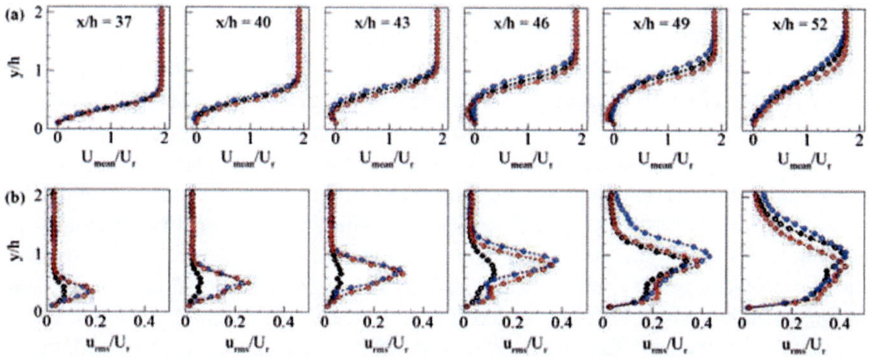

Fig. 4 Mean flow profiles of conditionally separated frames. **a** Streamwise variation of the mean velocity profile. **b** Streamwise variation of the u_{rms} profile. Description of symbols is shown in Table 1

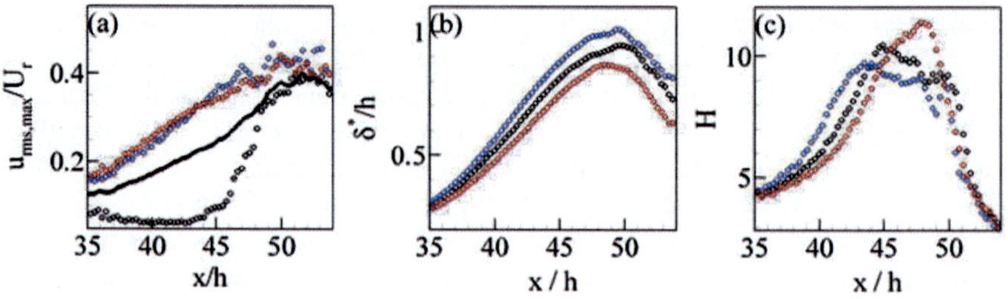

Fig. 5 **a** Streamwise variation of $u_{rms,max}$. The solid line represents the streamwise variation of $u_{rms,max}$ estimated from the total number of PIV realizations, **b** streamwise variation of displacement thickness, **c** streamwise variation of shape factor. Description of symbols is shown in Table 1

jet-like structures impose different forcing amplitude on the separated shear layer compared to the laminar-like flow. The comparison of u_{rms} profile of all three events is shown in Fig. 4b. It shows the wall-normal location of the maximum value of the u_{rms} ($y_{u_{rms,max}}$) is different in each event. This is due to the shift of the inflection point in the velocity profile. In the initial part of the bubble, a shift in the $y_{u_{rms,max}}$ is not clear, whereas it is clear as the streamwise distance increases, as seen in Fig. 4b. In the case of backward jet-like structure, $y_{u_{rms,max}}$ lies above the one corresponding to the laminar case at all the streamwise locations. For the forward jet-like structure, it lies below the laminar case.

Moreover, in the initial part of the bubble, the value of u_{rms} associated with the laminar-like frames is not considerable, compared to the values associated with the forward and backward jet-like structures. However, the value of the u_{rms} profile of laminar-like flow increases rapidly after $x/h = 43$, as further discussed below.

To find the spatial growth, the maximum value of u_{rms} is calculated at each streamwise location. The variations of $u_{rms,max}$ for all three events are shown in Fig. 5a, along with the one corresponding to the total number of 600 PIV realizations (solid line). The variation of $u_{rms,max}$ obtained from total frames found to be similar to the previous studies [2, 5]. The $u_{rms,max}$ values, estimated from the three different events separately, elucidate different types of spatial growth in the flow. The variation of $u_{rms,max}$ due to the forward and the backward jet-like structures show linear growth, whereas the laminar-like structure shows exponential growth in the flow. It indicates the presence of both transient (or algebraic) and exponential growth in the separated shear layer. Types of structure embedded in the shear layer determine the nature of growth (transient/exponential). Initially, the value of $u_{rms,max}$ associated with the forward jet-like structure case is found to be higher, but further downstream, it is overtaken by the $u_{rms,max}$ associated with the backward jet-like structure. It indicates that the growth rate of the backward jet-like structure is higher than the forward jet-like structure. Since the laminar-like structure shows exponential growth, it is not compared with the growth associated with the forward/backward jet-like structure.

In order to understand further the growth of each structure, the displacement thickness is estimated at each streamwise location, as the displacement thickness indicates the location of inflection point from the wall [10]. This is shown in Fig. 5b. One can

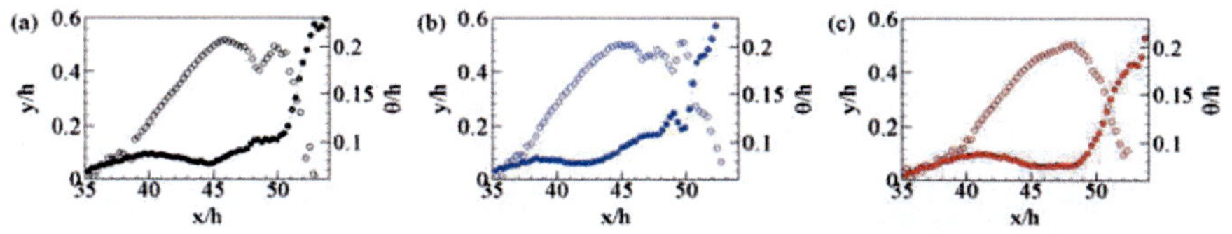

Fig. 6 Separation bubble outline, along with the momentum thickness variation for all three cases. **a** Laminar-like structure. **b** Backward jet-like structure. **c** Forward jet-like structure. The unfilled and filled symbols in (**a**–**c**) indicate the separation bubble outlines and the momentum thicknesses, respectively

see that, at any given streamwise location, value of the displacement thickness associated with the backward jet-like structure is higher than that of the forward jet-like structure. It indicates that the inflection point in the velocity profile corresponding to the backward jet-like structure is also located at a wall-normal distance which is higher than the point of inflection of the forward jet-like structure. Hence, the viscous effect can dampen the growth of the forward jet-like structure, compared to the backward jet-like structure. These different growth mechanisms may alter the transition location in the flow. Since the shape factor is a suitable parameter to find the onset of transition [11], it has been estimated for these three cases, as shown in Fig. 5c. One may notice that the shape factor initially increases to a certain streamwise extent, then it gets saturated/levelled off, followed by a rapid drop further downstream. As the maximum value of the shape factor location is the indicator for the onset of transition [11], the starting location of saturation can be considered as the onset of transition. It reveals that the backward jet-like structure advances the transition process compared to the forward and the laminar-like structures.

For further investigation of the point of transition, separation bubble outline along with the variation of the momentum thickness is shown in Fig. 6a–c, for all the cases. In the laminar-like and the backward jet-like cases, two types of growth in the momentum thickness variation can be noticed in Fig. 6a–b. The first onset of growth is observed near the levelled-off location of the bubble height. One may also note that the shape factor value also gets levelled off at the same location (see Fig. 5c). However, the onset of growth indicates that the transition is initiated in the flow. Otherwise, the bubble height will further grow in the downstream direction. In the first growth region of the momentum thickness, bubble height is not found vary much. However, a sudden drop in the bubble height is observed while the momentum thickness grows abruptly. Hence, it is speculated that the levelled-off region is associated with the transition length of the separated shear layer. Further investigations are necessary to confirm this speculation. However, in the case of the forward jet-like, although the onset of growth of the momentum thickness is observed nearly at the maximum height of the bubble, the growth in the momentum thickness does not seem to include two slopes, as seen in Fig. 6a, b. This may be due to the fact that transition length might be small, as the levelled-off region is found to be absent in this case. However, it is interesting to investigate the unsteady flow features associated with these the two different types of growth (Fig. 6a and b) using the TR-PIV technique, as well.

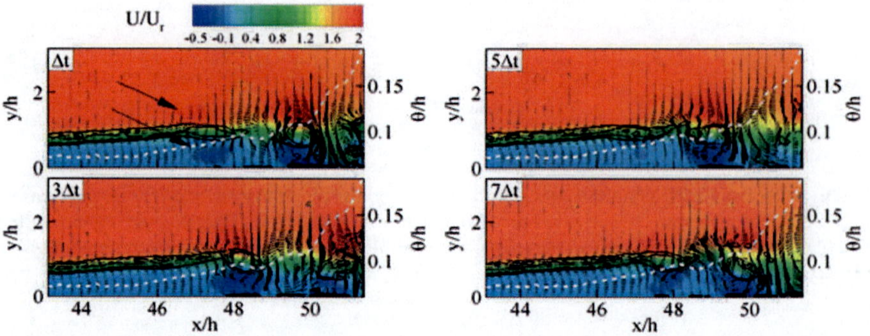

Fig. 7 Instantaneous velocity contour superimposed with fluctuation velocity vector and spanwise vorticity. Contour lines indicate the spanwise vorticity. A packet of high- and low-speed region is shown by a black coloured arrow. The variation of momentum thickness for the backward jet-like structure case is shown by a white coloured dashed line

3.2 TR-PIV Measurements

The TR-PIV measurement has been carried out near the maximum height of the bubble. The time sequence of the backward jet-like structure is shown in Fig. 7. The dashed line shows the variation of momentum thickness, and the line contour shows the spanwise vorticity. We may note that the momentum thickness is calculated here only for the backward jet-like structure from the TR-PIV measurements using the same procedure outlined above. One can also notice in Fig. 7 two growth regions in the momentum thickness variation in the measurement region, consistent with the one seen in Fig. 6b. The concentrated vorticity region is found to be stable to a certain streamwise extent, and then an instability develops in the concentrated vorticity region leading to a high- and low-speed packet, as shown by longest arrows in Fig. 7a. The size of the high- and low-speed packet is found to increase as it travels downstream. However, it cannot be ascertained here whether the instability that arises is due to the interaction of the freestream disturbance with the backward jet-like structure, or is due to the KH instability of the backward jet-like structure. In this region, the value of momentum thickness increases, perhaps due to the entrainment of a packet of a high-velocity region inside the concentrated vorticity region. Moreover, the concentrated vorticity region breaks down to small scales as it moves further downstream, as seen in Fig. 7. At the location of the breakdown of the concentrated vorticity region, an abrupt increase in the momentum thickness indicates the enhancement in the momentum transfer which eventually leads to the reattachment of the separated shear layer with the wall.

4 Conclusions

Both the conventional and the TR-PIV techniques were used to study the role of jet-like structures in a pressure inducted separation bubble. Three different repetitive events, i.e. laminar, backward and forward jet-like flow structures, similar to those seen in the bypass boundary layer transition, are found to exist even in a separated flow. The mean flow characteristics of these structures have been studied individually. It is found that the shear layers for the forward and backward jet-like structures move inward and outward from the wall, respectively, compared to the laminar-like flow case. The $u_{\mathrm{rms,max}}$ associated with the forward and backward jet-like structure found to grow nearly linear. However, the laminar-like flow shows exponential growth in the separated shear layer. Compared to the laminar-like flow case, the backward jet-like structure advances the transition, whereas the forward jet-like structure delays the transition in the separated shear layer. Moreover, two regimes in the growth of the momentum thickness variation are found to exist along the streamwise direction for the cases of the laminar-like structure and the backward jet-like structure. The first regime of growth is found to be associated with the transition onset, whereas the second regime of growth indicates the breakdown of the concentrated vorticity region.

Acknowledgements Financial support for the PIV system from IIT-Kanpur is gratefully acknowledged. We sincerely thank Prof. Sanjay Mittal and Kamal Poddar for providing the CCD camera and its accessories. We thank Mr. Yasar, Balamurugan, Rengaswamy, and lab technicians in low-speed aerodynamics lab for their help during the experiment.

References

1. Marxen O, Lang M, Rist U (2013) Vortex formation and vortex breakup in a laminar separation bubble. J Fluid Mech 728:58–90
2. Simoni D, Lengani D, Ubaldi M, Zunino P, Dellacasagrande M (2017) Inspection of the dynamic properties of laminar separation bubbles: free-stream turbulence intensity effects for different Reynolds numbers. Exp Fluids 58(6):66
3. Hosseinverdi S, Fasel HF (2019) Numerical investigation of laminar–turbulent transition in laminar separation bubbles: the effect of free-stream turbulence. J Fluid Mech 858:714–759
4. Klebanoff PS (1971, January) Effect of free-stream turbulence on a laminar boundary layer. Bull Am Phys Soc 16(11):1323–1305
5. Balzer W, Fasel HF (2016) Numerical investigation of the role of free-stream turbulence in boundary-layer separation. J Fluid Mech 801:289–321
6. Häggmark CP, Bakchinov AA, Alfredsson PH (2000) Experiments on a two–dimensional laminar separation bubble. Philos Trans R Soc Lond Ser A Math Phys Eng Sci 358(1777):3193–3205
7. Jacobs RG, Durbin PA (2001) Simulations of bypass transition. J Fluid Mech 428:185–212
8. Mandal AC, Venkatakrishnan L, Dey J (2010) A study on boundary-layer transition induced by free stream turbulence. J Fluid Mech 660:114–146
9. Marxen O, Rist U (2010) Mean flow deformation in a laminar separation bubble: separation and stability characteristics. J Fluid Mech 660:37–54

10. Boutilier MS, Yarusevych S (2012) Separated shear layer transition over an airfoil at a low Reynolds number. Phys Fluids 24(8):084105
11. Ellsworth RH, Mueller TJ (1991) Airfoil boundary layer measurements at low Re in an accelerating flow from a nonzero velocity. Exp Fluids 11(6):368–374

Importance of Refractive Index Matching of Fluids for PIV and PLIF Measurements in Buoyant Jets

Himanshu Mishra and Jimmy Philip

Abstract From violent volcanic eruptions to the rising clouds, buoyant jets and plumes are omnipresent in nature at wide range of scales (Woods in Annu Rev Fluid Mech 42:391–412, 2010 [15]). The mixing and growth dynamics of a plume is important in fundamental understanding of turbulence as well as in modeling this complex fluid mechanics phenomenon. This necessitates simultaneous measurements of velocity and density of the mixing fluids. Particle image velocimetry (PIV) and planar laser-induced fluorescence (PLIF) are the common methods employed for velocity and density measurement. The usual laboratory fluids, salt solution and pure water, however, introduce optical distortions caused by refractive index mismatch in the two fluids. Although widely known, issues caused by refractive index mismatch are rarely measured and quantified in the literature. Here, we present the results of simultaneous time-resolved measurement of velocity and density in an axisymmetric turbulent plume using 2D-2C PIV and PLIF, respectively. The experiments are performed with and without refractive index matched fluids to compare the uncertainties due to optical inhomogeneities caused by the mixing of two different density fluids. Both mean and turbulence quantities are compared. It is observed that refractive index mismatch adds noise to the measurement, resulting in over prediction of the turbulence quantities. As the distance from jet exit increases the effect of the mismatch decreases.

Keywords Buoyant jets · Refractive index matching · PIV · PLIF

1 Introduction

Waste discharge into environment, such as gaseous releases via chimney stacks or the disposal of wastewater via ocean outfalls, often leads to the formation of turbulent buoyant jets in the initial dilution stage. During this stage, turbulent mixing leads

H. Mishra (✉) · J. Philip
Department of Mechanical Engineering, The University of Melbourne, Melbourne, VIC 3010, Australia
e-mail: h.mishra@student.unimelb.edu.au

L. Venkatakrishnan et al. (eds.), *Proceedings of 16th Asian Congress of Fluid Mechanics*, Lecture Notes in Mechanical Engineering,
https://doi.org/10.1007/978-981-15-5183-3_30

to rapid dilution of pollutants in the ambient fluid, which is typically accompanied by a reduction in the density difference between the released and the environment fluid. Large-scale experiments in configurations with a density difference are challenging. Apart from the significant effort that frequently goes into realizing stratified flows in the laboratory, e.g., the seminal experiments by Kato and Phillips [8], the experimentalist is faced with the challenge of measuring density and velocity information simultaneously to capture the full flow dynamics. Early investigations such as that of Ellison and Turner [6] took fluid samples to obtain density readings and measured local velocities using thermistors. With the emergence of digital cameras and powerful illumination through laser light, it became possible to access density information in a plane using laser-induced fluorescence (LIF) [13, 12] without the need to place devices into the flow. Non-intrusive techniques were also developed to measure flow velocities pointwise (e.g., using laser Doppler anemometry) and in a plane through particle image velocimetry (PIV) [1]. A major challenge in application of optical measurement techniques in buoyancy-driven flows is the distortion caused by refractive index mismatch between by mixing of different density fluids that are employed. A common example is saltwater solution and pure water which have different refractive indices. The problem gets more severe in large-scale flows where negligibly small variation in refractive index may result in large errors due to large distances laser beam has to travel in the medium.

One way to avoid this issue is to ensure uniform refractive index throughout the flow. Refractive index matching (RIM) of the two fluids at all mixing ratios was first suggested by McDougall [10] for laser Doppler velocimetry measurements. After testing many combinations, sugar and Epsom salt combination was suggested. Later Hannoun [7] suggested a combination of ethanol and common salt (NaCl), where the major advantage is the cost because ethanol reduces the solution density as compared to sugar–Epsom salt combination where both solutes raise the density. This combination was also suggested by Daviero et al. [4], for large-scale experiments largely owing to the cost again. Alahayari and Longmire [2] proposed a combination of glycerin and monopotassium phosphate, which offers several advantages over ethanol and salt combination. Although the distortions caused by refractive index mismatch is widely known issue in experimental fluid mechanics research especially in buoyancy driven flows, it has hardly been quantified in any sense. Many studies use refractive index matching for accurate measurements [5, 9], however there are recent studies involving buoyant flows that do not use refractive index matching and obtain reasonable results [3]. This raises important questions like: is refractive index matching important? If yes, when and where? Are there cases where experimentalists can get away without refractive index matching? In other words, what is the error in turbulence statistics that is caused by non-RI matched fluids. Furthermore, it is intuitive that as the buoyant jet develops downstream and mixes, the RI mismatch would reduce, and at some distance the error induced by non-RI matched fluids should be negligible. Unfortunately, we do not know at which downstream location this would happen. As refractive index matching is a tedious and expensive process (especially for large-scale experiments), the answers to these questions are important. In the following, we will present simultaneous measurement of time-resolved velocity and density fields through a combined PIV and PLIF approach, both with and without RI matched fluids in turbulent buoyant jet. We compare the mean and turbulence statistics from these two datasets, which will allow us to obtain the error associated with the optical distortions.

2 Experiments

A schematic of the experimental setup is shown in Fig. 1. The experiments are conducted in a glass test tank (1.2 m high and 0.75 × 0.75 m^2 square base). The sidewalls are made of 15 mm thick clear glass, and bottom wall constructed from marine grade aluminum is removable, allowing us to change the inlet conditions. For the non-refractive index matched experiment, the tank is filled with a solution of water and salt (heavy fluid) to a depth of 1 m. Water (lighter fluid) is injected vertically upward, driven by a constant head tank through a 3D printed nozzle with an exit diameter $d = 1$ cm. For refractive index matched experiment, the heavy fluid is replaced by a solution of water and monopotassium phosphate and the lighter fluid is replaced by solution of water and glycerin. The water–glycerin–monopotassium phosphate solution allows us to obtain a constant refractive index and varying density throughout the mixing process. For reference, we note that refractive index of water is 1.3326 whereas it increases linearly with increasing salt concentration. Salt solution used for current experiment has a refractive index of 1.3348 with a density difference

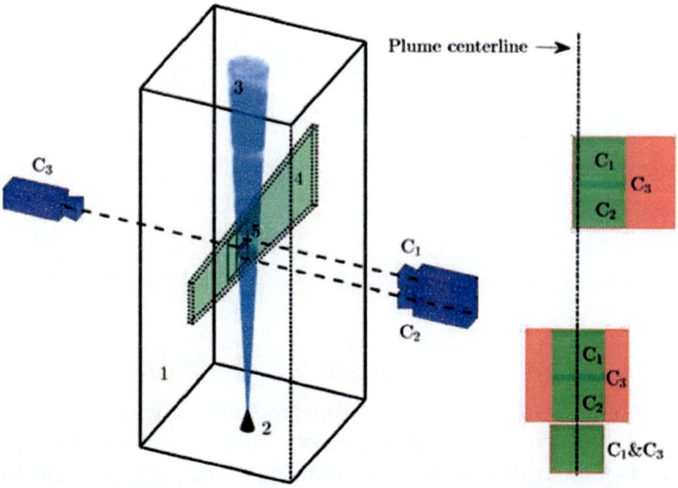

Fig. 1 Experimental setup schematic (1) main glass tank (2) nozzle (3) plume (4) laser sheet (5) field of view. C_1, C_2 represent the two PIV cameras and C_3 is the PLIF camera. Right-hand side of the figure shows the three locations chosen for measurements in the present experiments. The green color is PIV field of view with overlap between C_1 and C_2, red color is the PLIF field of view of C_3

Table 1 Experiment parameters

Q_0 (cm^3/s)	M_0 (cm^4/s^2)	F_0 (cm^4/s^3)	Re_0	Ri_0	z/d
39.25	1963	1155	5000	0.021	0–7; 8–21; 39–52

of 1% at the jet inlet. The RI matched fluids have refractive index of 1.3365 with same initial density difference.

Three PCO-dimax HS4 12-bit high-speed cameras with 32 GB of RAM and maximum frame rate of 2277 frames per second at full resolution were used for the measurements. As PIV provides lower spatial resolution compared to PLIF, two cameras are stacked over each other for PIV, and one was placed on the opposite side of the tank for PLIF. PIV cameras are attached with 105 mm lens and 532 nm notch filters, whereas PLIF camera is fitted with 60 mm lens and 550 nm bandpass filter with half width of 25 nm. The filters keep the frequencies of scattered laser light out of the PLIF camera and dye fluorescence from the PIV cameras. The same camera configuration is moved vertically for measurements at three different axial locations. The field of view of the cameras and the axial measurement locations are shown in Fig. 1. All three cameras are synchronized and operate at the full resolution of 2000 × 2000 pixels. Flow is illuminated by Spectra Physics 25 W and 532 nm diode pumped solid-state continuous laser. The data processing techniques for both PIV and PLIF are same as described in [11]. Experimental parameters such as volume flux (Q), momentum flux (M), integral buoyancy (B), buoyancy flux (F), initial Reynolds number (Re_0), and Richardson number (Ri_0) are defined in (1). The values are shown in Table 1.

$$Q = 2\int_0^\infty \overline{w} r \mathrm{d}r,\ M = \int_0^\infty \overline{w}^2 r \mathrm{d}r,\ B = 2\int_0^\infty \overline{b} r \mathrm{d}r,$$

$$F = 2\int_0^\infty \overline{w}\overline{b} r \mathrm{d}r,\ \mathrm{Re} = \frac{\mathrm{d}Q_0}{v\left(w^2/4\right)},\ \mathrm{Ri} = \frac{B_0 Q_0}{M_0^{3/2}} \quad (1)$$

3 Results

Figure 2a, b shows the raw PIV and PLIF images with and without refractive index matching at all three axial measurement locations. It can be seen in the PIV image that the refractive index mismatch causes blurring in the jet region. The blurring is high close to the nozzle exit, and it reduces as the fluids mix along the downstream direction. The distortion seems to be mostly gone at the furthest axial measurement location. Similarly, non-refractive index matched PLIF images are diffused and blurry, getting better with axial distance.

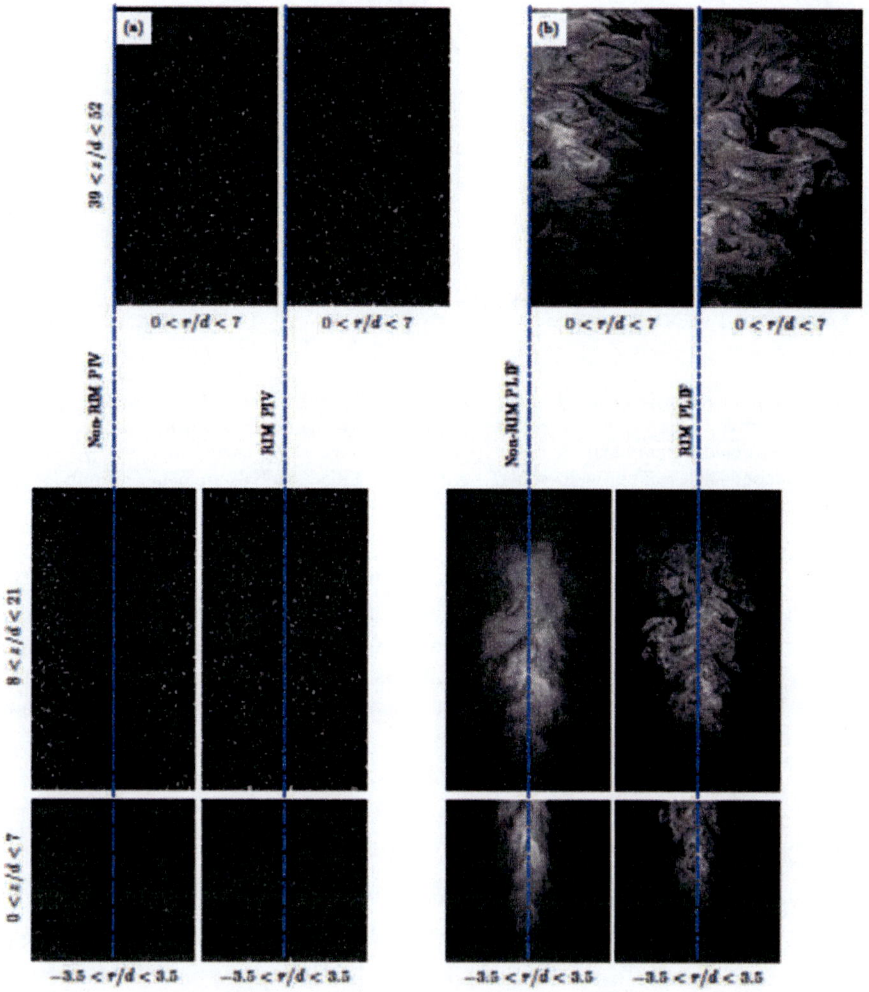

Fig. 2 **a** Raw PIV and **b** PLIF images at different axial locations with and without refractive index matching. The blue line represents the centerline of the plume

Figure 3a shows the comparison of mean axial velocity in both the cases. The mean does not seem to be affected much by the distortions barring the plume edges where non-refractive index matched profile shows slight oscillations. Figure 3b shows the axial rms velocity, and it can be clearly seen that non-refractive index matched case over predicts the rms. This can be attributed to loss of PIV cross-correlation caused by the blur seen in Fig. 2a. The blur adds noise to the measurement which can be seen as causing pseudoturbulence in the flow. Far from the nozzle exit both the cases show the similar values, suggesting a point in the flow where refractive index matching

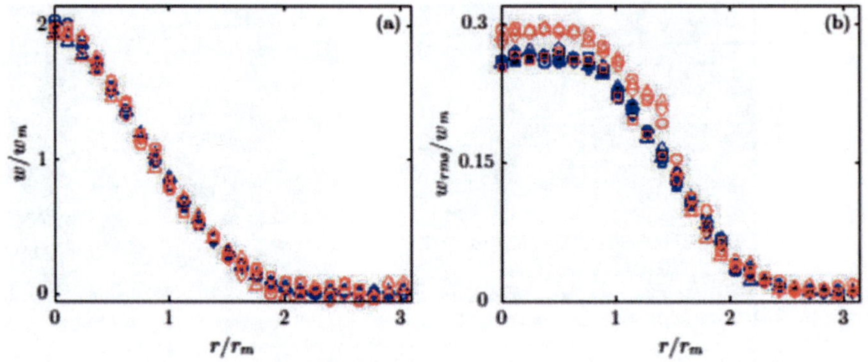

Fig. 3 **a** Normalized mean axial velocity **b** axial rms velocity at different z/d locations. Blue symbols are for refractive index matched experiment. Red is non-refractive index matched experiments. Symbols are for different axial locations [(white circle—$z/d = 6$), (white triangle—$z/d = 12$), (white diamond—$z/d = 19$), (white square—$z/d = 41$)]. Filled symbols are obtained from [14] at an axial distance $z/d = 45$

can be neglected. In Fig. 3b we also show results from Wang and Law [14] at axial location of $z/d = 45$. Their experiments were carried out without RIM.

Figure 4a shows the radial velocity rms and has a behavior similar to Fig. 3b. Again, the data falls over each other for furthest axial measurement location. Figure 4b shows the rms buoyancy obtained using PLIF. It shows that the density measurements are subject to noise as well and far from the nozzle the measurement improves.

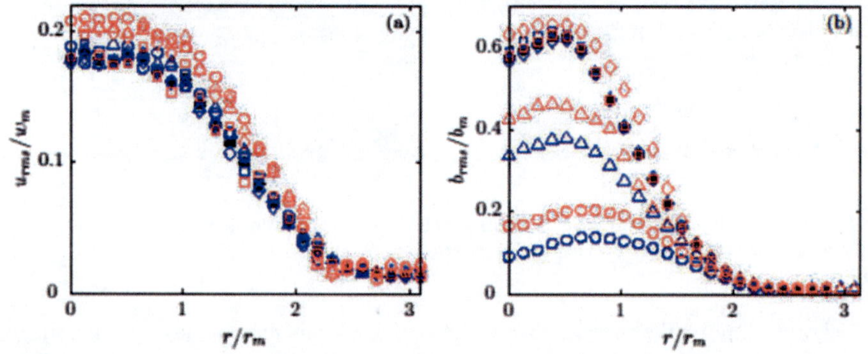

Fig. 4 **a** Normalized radial rms velocity and **b** Buoyancy rms at different z/d locations. Blue symbols are for refractive index matched experiment. Red is non-refractive index matched experiments. Symbols are for different axial locations [(white circle—$z/d = 6$), (white triangle—$z/d = 12$), (white diamond—$z/d = 19$), (white square—$z/d = 41$)]. Filled symbols are obtained from [14]

4 Conclusions

Simultaneous measurements of velocity and density in a turbulent buoyant plume are carried out using PIV and PLIF with and without refractive index matching. It is observed that not matching the refractive index has less effect on the mean quantities. The turbulence, however, is over predicted in both velocity and density fluctuations especially closer to the jet exit. The errors start to reduce as the measurement location moves downstream and optical distortion seem to have minimal effect after a certain axial location. In our case, we observe that by $z/d \approx 40$ the turbulence statistics in RIM case are almost indistinguishable from the non-RIM case. As the refractive index of salt solution is known and actual dilution of plume is measured with the refractive index matched experiment, this can be used to find a generalized value of acceptable refractive index mismatch. Such an estimate would be helpful in any future investigation involving buoyancy.

Acknowledgements The authors gratefully acknowledge the Australian Research Council for the financial support of this work.

References

1. Adrian L, Adrian RJ, Westerweel J (2011) Particle image velocimetry. Cambridge University Press, Cambridge
2. Alahyari A, Longmire EK (1994) Particle image velocimetry in a variable density flow: application to a dynamically evolving microburst. Exp Fluids 17(6):434–440
3. Burridge HC, Parker DA, Kruger ES, Partridge JL, Linden PF (2017) Conditional sampling of a high Peclet number turbulent plume and the implications for entrainment. J Fluid Mech 823:26–56
4. Daviero GJ, Roberts PJW, Maile K (2001) Refractive index matching in large-scale stratified experiments. Exp Fluids 31(2):119–126
5. Earl TA, Paetzold J, Cochard S (2013) Tomographic PIV measurements of turbulent fountains with refraction index matching. J Flow Vis Image Process 20(3)
6. Ellison TH, Turner JS (1959) Turbulent entrainment in stratified flows. J Fluid Mech 6(3):423–448
7. Hannoun I. Matching the refractive index in density stratified flows
8. Kato H, Phillips OM (1969) On the penetration of a turbulent layer into stratified fluid. J Fluid Mech 37(4):643–655
9. Krug D, Holzner M, Luthi B, Wolf M, Kinzelbach W, Tsinober A (2015) The turbulent/non-turbulent interface in an inclined dense gravity current. J Fluid Mech 765:303–324
10. McDougall TJ (1979) Measurements of turbulence in a zero-mean-shear mixed layer. J Fluid Mech 94(3):409–431
11. Mishra H, Philip J (2018) Simultaneous velocity and density measurements using PIV and PLIF in turbulent axisymmetric buoyant plumes. In: Proceedings of the 21st Australasian fluid mechanics conference, Adelaide, Australia
12. Papanicolaou PN, List EJ (1988) Investigations of round vertical turbulent buoyant jets. J Fluid Mech 195:341–391

13. Walker DA (1987) A fluorescence technique for measurement of concentration in mixing liquids. J Phys E Sci Instrum 20(2):217
14. Wang H, Law AW (2002) Second-order integral model for a round turbulent buoyant jet. J Fluid Mech 459:397–428
15. Woods AW (2010) Turbulent plumes in nature. Annu Rev Fluid Mech 42:391–412

A Staggered Update Procedure (SUP) for Higher-Order Cell-Centre Finite Volume Method

Shubhashree Subudhi and N. Balakrishnan

Abstract A Staggered Update Procedure (SUP) wherein cell-centre finite volume update happens in conjunction with a generalized finite difference-based cell vertex update is introduced. This is in variance with the classical finite volume method, where the vertex values are interpolated to a required accuracy. The implementation of SUP involves computation of inviscid interfacial fluxes, both along face normal and tangential directions. A recursive defect corrected Green-Gauss procedure-based gradient finding is proposed for determining second-order accurate gradients at face centroid, cell centroid and vertices. The gradients computed at cell centres and vertices are used to represent the linear polynomial required for these higher-order upwind flux computations at volume interface. The second-order gradients also facilitate the estimation of viscous fluxes to the same order of accuracy as that of the convective fluxes, when used with a linear reconstruction procedure. Implicit time integration scheme is used for both cell and vertex update. All demonstrations pertain to linear convection diffusion equation. The benefits of the proposed procedure are brought out, particularly by solving diffusion dominated problem, with implications in accurately resolving turbulent wall boundary layer and diffusion dominated wake flows. Also, the residual convergence and computational cost of the proposed methodology are addressed.

Keywords Higher-order finite volume · Staggered update procedure · Defect correction

1 Introduction

Developing an accurate, efficient and robust numerical scheme for industrial computations involving complex geometric configurations is one of the active areas of research in CFD. The unstructured mesh-based solvers offer flexibility in terms of

S. Subudhi · N. Balakrishnan (✉)
Computational Aerodynamics Laboratory, Department of Aerospace Engineering, Indian Institute of Science, Bangalore 560012, India
e-mail: nbalak@iisc.ac.in

L. Venkatakrishnan et al. (eds.), *Proceedings of 16th Asian Congress of Fluid Mechanics*, Lecture Notes in Mechanical Engineering,
https://doi.org/10.1007/978-981-15-5183-3_31

handling complicated geometries. The present-day unstructured data-based finite volume solvers typically employing linear solution reconstruction offer formally first-order accuracy on irregular meshes. Although these methods have been greatly successful in the last two decades in solving several problems of industrial relevance, there has always been a demand for greater accuracy. The need for such methods has further increased with the significant advancements in the HPC technology and the use of higher fidelity CFD methods such as DES becoming more common, in the recent years. On the other hand, attempts to achieve higher-order accuracy severely hamper the robustness of the unstructured grid-based solvers and make these methods unsuitable for routine industrial solution. Therefore, the search for a robust higher-order unstructured grid-based procedure for industrial computations has been elusive and the present work on Staggered Update Procedure (SUP) is an attempt in that direction. The essential idea of SUP, proposed in reference [1], is to update solution variables not just at the cell centres, but also at the vertices. This work highlights the fact that the solutions obtained using a rotated difference scheme to determine fluxes on volume interface can show significant dependence on the way the solution is interpolated at the vertices, clearly underlining the importance of accurate vertex value interpolation. The stability issue with SUP exploiting a rotated difference scheme was carried out by Pinto [2]. The present work differs from both these earlier works in the sense that SUP framework is rather used to achieve higher-order accuracy and per se the rotated difference scheme proposed earlier is not of interest. It is in this context SUP, where a generalized finite difference update is employed at the vertices or in other words, using the governing pde itself as an interpolant was proposed. A recursive defect correction strategy in combination with Green-Gauss procedure is proposed for determining second-order accurate gradients at face centroid, cell centroid and vertices. The gradients thus obtained are used in linear solution reconstruction for estimating the upwind inviscid fluxes to second order and also the viscous fluxes on the finite volume interface to the same order of accuracy. It is worth noting that a mere use of Green-Gauss procedure for determination of viscous fluxes can result in degeneration of order of accuracy of the finite volume update. Here, we have investigated the procedure for simple linear scalar convection–diffusion problem.

2 Solution Methodology

The 2D scalar convection–diffusion equation is given as follows:

$$\frac{\partial \varphi}{\partial t} + \nabla \cdot \left(\vec{f}^{\text{con}} - \vec{f}^{\text{vis}} \right) = 0 \tag{1}$$

where $\vec{f}^{\text{con}} = (u\varphi \; v\varphi)^T$ and $\vec{f}^{\text{vis}} = \left(\alpha \frac{\partial \varphi}{\partial x} \alpha \frac{\partial \varphi}{\partial y} \right)^T$ are convective and diffusive flux vectors and φ is the scalar being transported. (u, v) are convective speeds and α is the diffusion coefficient. The formulation of SUP consists of following steps. The details can be found in reference [4].

Calculation of convective fluxes at volume interface J (Fig. 1):

$$F_{\perp J}^{\text{con}} = F_{\perp}^{\text{con}^+}\left(\varphi_i^J\right) + F_{\perp}^{\text{con}^-}\left(\varphi_j^J\right)$$
$$F_{\uparrow J}^{\text{con}} = F_{\uparrow}^{\text{con}^+}\left(\varphi_{J_1}^J\right) + F_{\uparrow}^{\text{con}^-}\left(\varphi_{J_2}^J\right) \tag{2}$$

where $\varphi_{(.)}^J = \varphi_{(.)} + \nabla\varphi_{(.)}\vec{r}_{J(.)}$.

The normal inviscid flux directly goes into finite volume state update at cell centres.

Volume Update:

The finite volume state update at cell i reads as:

$$\frac{\varphi_i^{n+1} - \varphi_i^n}{\Delta t} = -\frac{1}{\Omega_i}\sum_J F_{\perp J}\Delta S_J \tag{3}$$

Vertex update (Fig. 2):

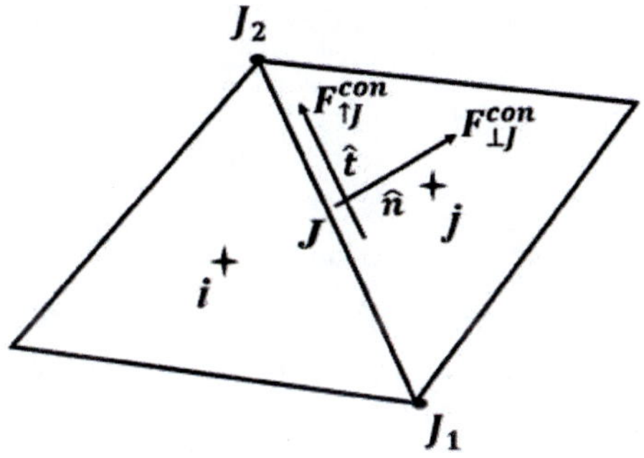

Fig. 1 Upwind fluxes at an interface

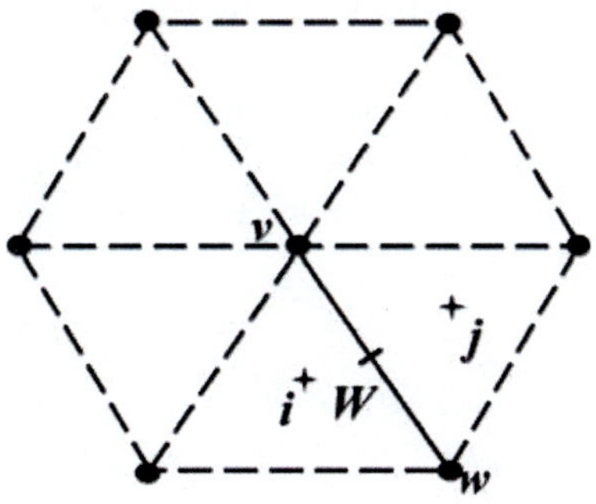

Fig. 2 2D cluster for LSFD-U

The state update at the vertex v using upwind least-squares finite difference (LSFD-U) procedure is given as:

$$\frac{\varphi_v^{n+1} - \varphi_v^n}{\Delta t} = -\left(f_{x_v} + g_{y_v}\right) \tag{4}$$

The conventional LSFD-U presented in [3] requires flux along the ray v-w. However, in SUP framework, both the normal and directional upwind inviscid fluxes available at interface are used for obtaining upwind fluxes along coordinate directions (denoted as f, g) at W as:

$$\begin{aligned} f_W^{\text{con}} &= F_{\perp J}^{\text{con}} n_x + F_{\uparrow J}^{\text{con}} t_x \\ g_W^{\text{con}} &= F_{\perp J}^{\text{con}} n_y + F_{\uparrow J}^{\text{con}} t_y \end{aligned} \tag{5}$$

The conventional linear least-squares procedure is employed for evaluating the flux derivatives as given below:

$$\begin{aligned} f_{x_v} &= \frac{\sum\limits_W \Delta y_W^2 \sum\limits_W \Delta f_W \Delta x_W - \sum\limits_W \Delta x_W \Delta y_W \sum\limits_W \Delta f_W \Delta y_W}{\sum\limits_W \Delta x_W^2 \sum\limits_W \Delta y_W^2 - \left(\sum\limits_W \Delta x_W \Delta y_W\right)^2}; \\ g_{y_v} &= \frac{-\sum\limits_W \Delta x_W \Delta y_W \sum\limits_W \Delta g_W \Delta x_W + \sum\limits_W \Delta x_W^2 \sum\limits_W \Delta g_W \Delta y_W}{\sum\limits_W \Delta x_W^2 \sum\limits_W \Delta y_W^2 - \left(\sum\limits_W \Delta x_W \Delta y_W\right)^2} \end{aligned} \tag{6}$$

Gradient Computations:

For linear solution reconstruction, the left and right states are obtained by extrapolating the cell averaged states (attributed to the cell centroids in case of linear reconstruction) from the left and right cells, respectively, for evaluation of normal flux. Likewise the states at the vertices forming the face are extrapolated to evaluate the directional flux. The gradients computed at cell centres and vertices are used to represent the linear polynomial required for these higher-order flux computations (Fig. 3).

A defect correction strategy is developed in relation to Green-Gauss procedure for estimating gradients. The second-order gradients at face centre obtained through defect correction are given as:

$$\nabla \varphi_J = \frac{1}{\Omega_J} \sum_k \tilde{\varphi}_k \hat{n}_k \Delta S_k \tag{7}$$

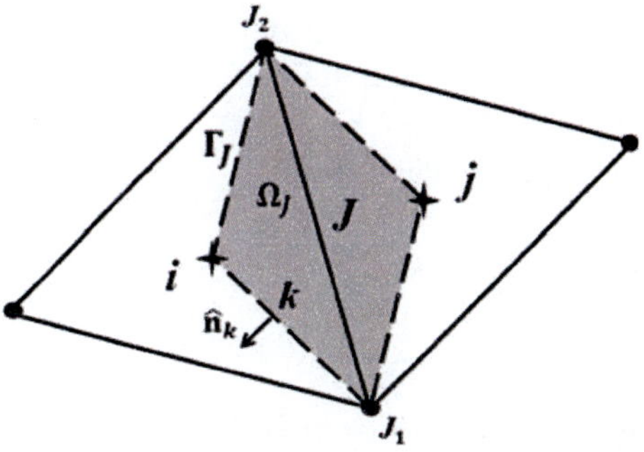

Fig. 3 Diamond path covolume

with $\tilde{\varphi}_k = \varphi_k^{\text{corr}} - \frac{1}{2}\left[\Delta\varphi_{x_k}(x_k - x_J) + \Delta\varphi_{y_k}(y_k - y_J)\right]$ where $\Delta\varphi_{x_k} = \varphi_{x_k} - \varphi_{x_J}$; $\Delta\varphi_{y_k} = \varphi_{y_k} - \varphi_{y_J}$ and $\varphi_k^{\text{corr}} = \frac{\varphi_1+\varphi_2}{2} - \left[\left.\frac{\partial\varphi}{\partial s}\right|_2 - \left.\frac{\partial\varphi}{\partial s}\right|_1\right]\frac{\Delta S_k}{8}$; $\left.\frac{\partial\varphi}{\partial s}\right|_{1,2} = \nabla\varphi_{1,2}.\hat{s}$.

It is worth noting that the above correction is valid iff the gradients are recovered to second order both at volume centroid and vertex. In order to accomplish this, closed paths are identified around cell centre and vertex as shown in Fig. 4. The figures suggest that the paths involved in obtaining these gradients coincide with the path employed for face gradients. This immediately suggests a three-layered recursive defect correction strategy which recovers second-order gradients not only on volume face, but also at centroids and vertices.

The second-order solution gradient obtained at face centre goes into computation of normal viscous flux for volume update. Regarding vertex update, the viscous flux derivatives are estimated modifying the linear least-squares expressions given in Eq. 6.

In the current work, the presence of two independent updates makes the formulation of implicit procedure tricky. It can be noticed that the finite volume state update at cell centres is loosely coupled to the LSFD-U update at vertices, whereas the latter is strongly coupled to finite volume update. Thus, the implicit formulation for

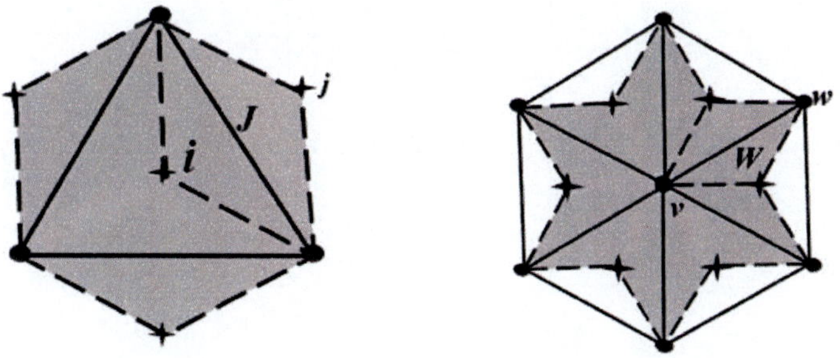

(a) Diamond path for cell centre

(b) Petal path for vertex

Fig. 4 Closed path for surface integration

finite volume state update remains unchanged. An implicit scheme for vertex update in SUP framework is designed by coupling with implicit finite volume update. An inconsistent linearization wherein the first-order implementation is implicitized is proposed for volume and vertex update in [4].

3 Results

The scalar convection diffusion equation in 2D has been solved on a square domain of unit length with constant u velocity. A sine wave is given as inlet profile on the left boundary. On the right boundary corresponding to the exit, a zero normal solution gradient is enforced. On the top and bottom boundaries, a zero φ condition is implemented. The results are obtained for low Peclet number ($Pe = 1$) on a sequence of distorted mesh with and without gradient correction within SUP framework (denoted as SUP (W/Correction) and SUP (W/O Correction) in the legend, respectively) and compared against exact solution. A diffusive problem corresponding to a low Pe has been chosen for demonstration, as the primary objective is to bring out the utility of the proposed methodology in accurately capturing the turbulent wall boundary layer and other viscous dominated regions of the flow. For error analysis, standard finite volume procedure employing a pseudo-Laplacian-based vertex value interpolation (indicated as FVM (PL) in the legend) is also included.

Comparison of solution contours on coarse distorted mesh for $Pe = 1$ is made in Fig. 5. Interestingly, the SUP solution with gradient correction becomes indistinguishable from the exact solution even on coarse grid as against the SUP solution without correction. Grid convergence results presented in Fig. 6 clearly demonstrate the superiority of defect corrected SUP, in the sense that it exhibits not only higher fall rates (~2) but also an error level almost two orders lesser as compared to the standard finite volume procedure.

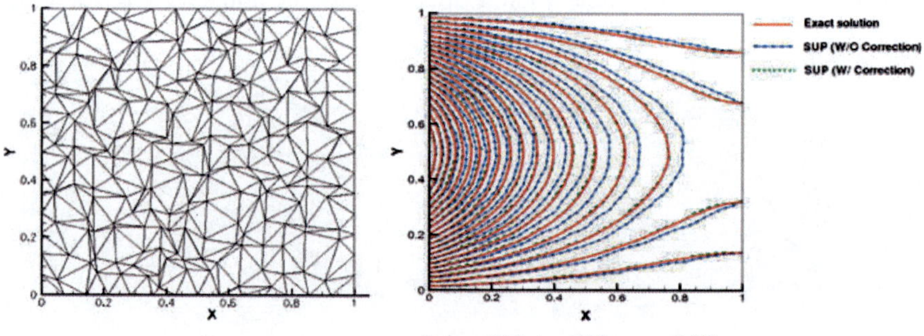

[min = 0.05, Δ = 0.05, max = 0.95]

Fig. 5 **a** Distorted mesh (424 volumes); **b** Comparison of solution contours for $Pe = 1$

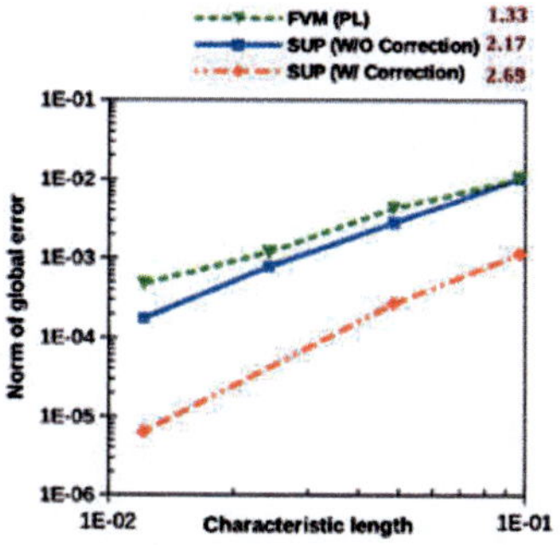

(a) Cell centre: Solution

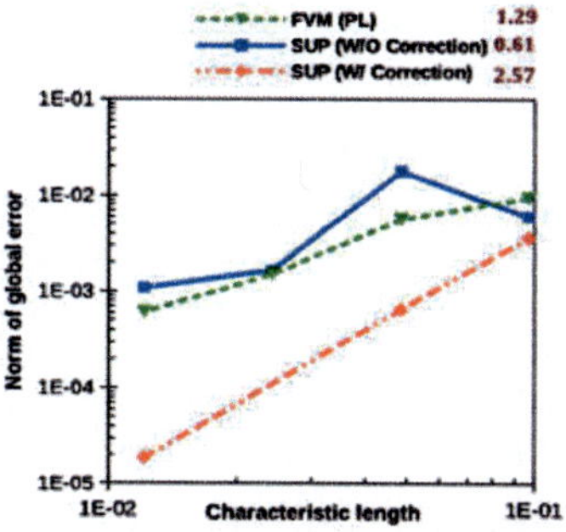

(b) Vertex: Solution

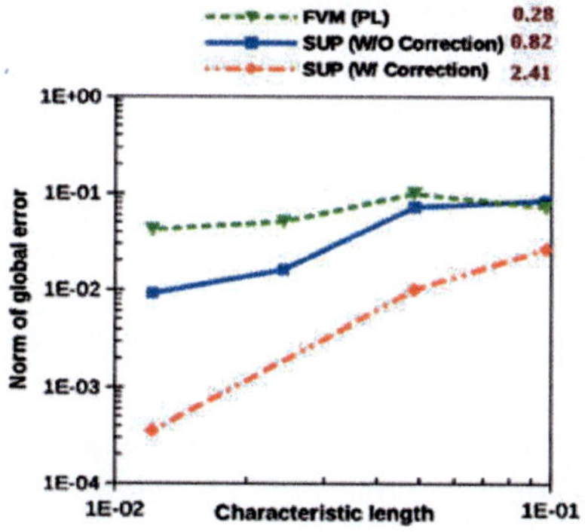

(c) Face centre: Solution gradient in x

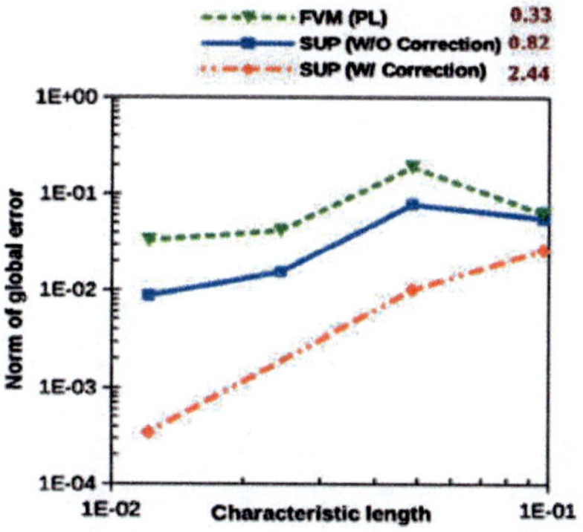

(d) Face centre: Solution gradient in y

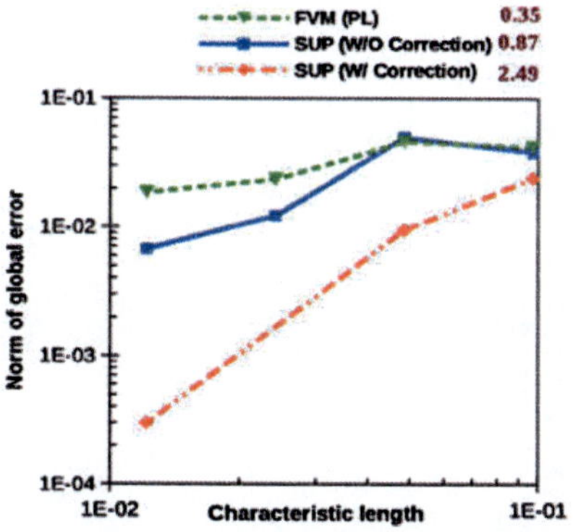

(e) Cell centre: Solution gradient in x

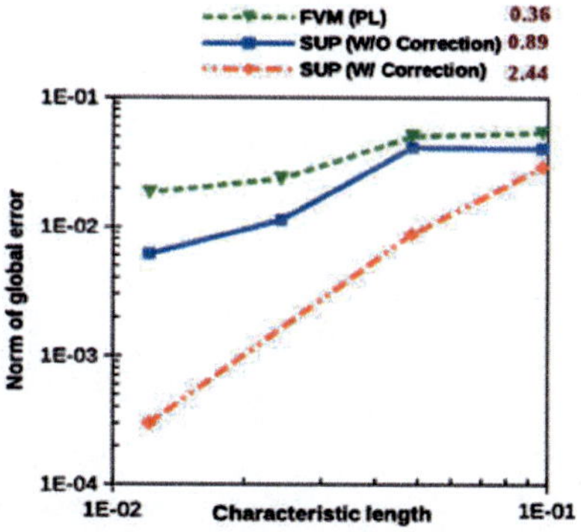

(f) Cell centre: Solution gradient in y

Fig. 6 Comparison of fall rates of global errors for $Pe = 1$ for solution values and x and y gradients on distorted mesh

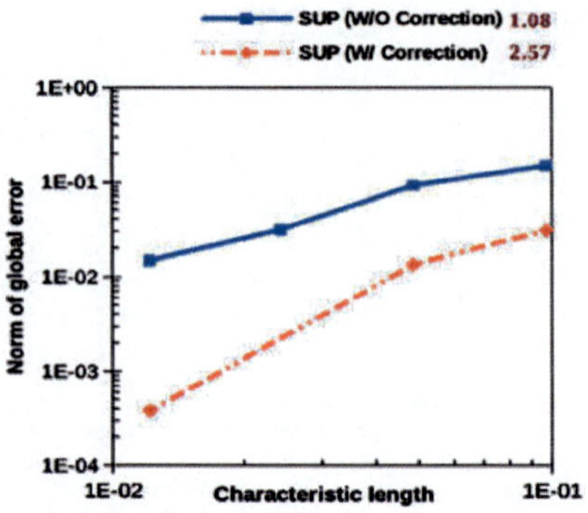

(g) Vertex: Solution gradient in x

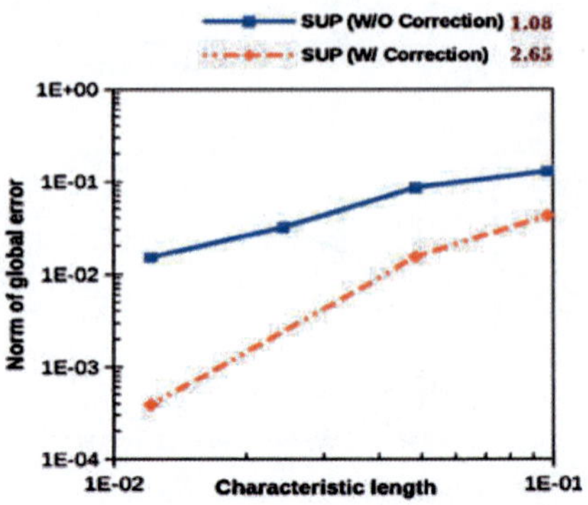

(h) Vertex: Solution gradient in y

Fig. 6 (continued)

Figure 7 shows that SUP (W/Correction) is about 30 times computationally more expensive than classical finite volume method and is about 5 times more expensive than SUP (W/O Correction) in terms of computational time. The time per iteration analysis shows that 75–80% of total time per one implicit iteration is taken by explicit calculations which consists of the recursive correction strategy for gradient evaluation. An important aspect of this algorithm is that for the same memory fetched, more computational effort is expended, for achieving higher accuracy, without significantly losing out on the robustness. This feature, in fact, is ideally suited for next-generation computer hardware supporting massive data parallelism.

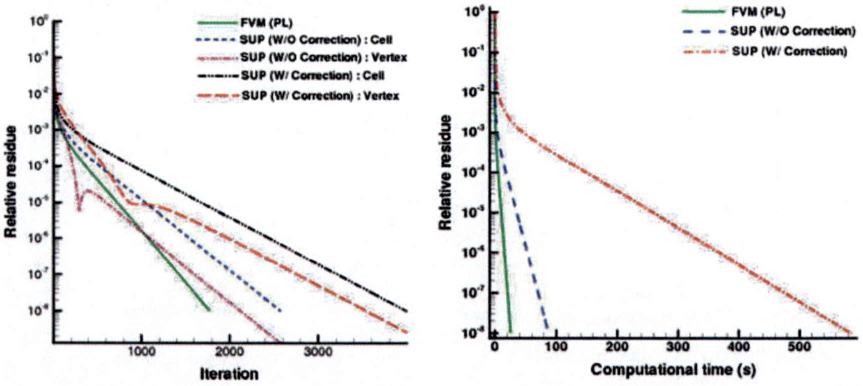

Fig. 7 Comparison of convergence history for implicit time scheme on distorted mesh (6784 volumes)

4 Conclusions

In the work presented, superior accuracy of the defect corrected SUP as compared to the conventional finite volume method is established, particularly for diffusion dominated flows. Though computationally expensive, the inherent data parallelism of the proposed method makes it suitable for next-generation computer hardware, which would make the additional computational effort imperceptible to the end user.

References

1. Shende N, Balakrishnan N (2001) Towards a staggered update procedure for cell centre finite volume scheme. Technical report, IISc ARDB Fluid Mechanics Report FM02
2. Pinto Brijesh S (2012) Towards high fidelity cfd solvers. Master's thesis, Indian Institute of Science, Bangalore
3. Sridar D, Balakrishnan N (2003) An upwind finite difference scheme for meshless solvers. J Comput Phys 189(1):1–29
4. Subudhi S, Balakrishnan N (2018) Towards a staggered update procedure for higher order cell centre finite volume method: I. Technical Report, CAd Lab Technical Note: CAd TN 2018:01

Turbulent Flow Computations on a Hybrid Unstructured Point Distribution Using Meshless Solver LSFD-U

Amol Khillare, Mohamed Yousuf, N. Munikrishna, and N. Balakrishnan

Abstract Upwind—Least Square Finite Difference (LSFD-U) is a generalized finite difference method capable of operating on an arbitrary distribution of points. The method is known to work well on an isotropic distribution of points typically needed for inviscid computations. However, on highly anisotrpoic distribution of points as encountered in the viscous padding region associated with the RANS computations, (i.e. high aspect ratio volumes in finite volumes parlance) these methods tend to fail because of the ill-conditioning of the associated geometric matrix. This necessitates the use of rotated coordinates for update in this region. Yet another difficulty associated with such point distributions is the sharp turning the grid lines experience at certain locations (for example, near the trailing edge of airfoil). These are identified as regions of grid folding. The work proposes two different ways for mitigating this difficulty. In the first approach points are added around points identified with grid folding and the second approach simply involves placing the fictitious interface close to these points, in contrast with the earlier convention of placing these fictitious interfaces mid-way along a ray connecting two points under consideration. Both these methods attempt to recover an isotropic distribution of points in the region of grid folding. The method is successfully demonstrated for simulating 2D high lift flows.

Keywords Meshless solver · Least squares · Generalized finite difference

1 Introduction

For the last few decades there has been considerable research in the area of meshless solvers, particularly for solving fluid flow problems [1–4]. The primary interest in such algorithms stems from the fact that these methods are expected to operate on any arbitrary distribution of points, not requiring classical discretization of the computational domain into non-overlapping volumes. This feature is expected to

A. Khillare · M. Yousuf · N. Munikrishna · N. Balakrishnan (✉)
Computational Aerodynamics Laboratory, Department of Aerospace Engineering, Indian Institute of Science, Bangalore 560012, India
e-mail: nbalak@iisc.ac.in

L. Venkatakrishnan et al. (eds.), *Proceedings of 16th Asian Congress of Fluid Mechanics*, Lecture Notes in Mechanical Engineering,
https://doi.org/10.1007/978-981-15-5183-3_32

considerably reduce the cost and time involved in grid generation and also, further enable CFD process automation. These methods can be broadly classified as generalized finite difference procedure and invariably employ method of least-squares in some form or other. The Upwind Least Squares Finite Difference method christened as LSFD-U [3] is one such method which has been successfully used for solving problems of industrial relevance [5]. These methods [6, 7] are known to work well on isotropic distribution of points typically needed for inviscid computations. However, on highly anisotrpoic distribution of points as encountered in the viscous padding region of hybrid-unstructured or hybrid-Cartesian point distribution needed for RANS computations, (i.e. high aspect ratio volumes in finite volumes parlance) these methods tend to fail because of the ill-conditioning of the associated geometric matrix. This necessitates the use of rotated coordinates for update in this region [6, 7]. Yet another difficulty associated with such point distributions is the sharp turning the grid lines experience at certain locations (for example, at the trailing edge of airfoil). The discretization strategy employed in the viscous padding fails in these regions associated with grid folding. The focus of this work is to evolve strategies to mitigate this difficulty.

2 Methodology

The compressible Reynolds-Averaged Navier–Stokes equation in the conservative form is expressed as,

$$\frac{\partial W}{\partial t} + \frac{\partial (f + F)}{\partial x} + \frac{\partial (g + G)}{\partial y} = 0 \tag{1}$$

where $W = (\rho\ \rho u\ \rho v\ e)^{\mathrm{T}}$ is the vector of conserved variables, f and g are inviscid flux vector, and F and G are viscous flux vectors.

The flow solver employs LSFD-U methodology [3, 4] which is basically upwind generalized finite difference procedure involving method of least squares. While an upwind discretization is used for inviscid flux derivatives calculation, a robust viscous discretization procedure is selected based on positivity analysis [6]. Linear solution reconstruction procedure is employed for higher order accuracy along with Venkatakrishnan limiter [8] in order to preserve solution monotonicity. The Spalart–Allmaras [9] turbulence model is employed for the transport of eddy viscosity. LSFD-U flow solver exploits the directionality exhibited by local cloud for discretization of inviscid fluxes. The details of discretization procedure can be obtained from [6].

The strategy of exploiting directionality exhibited by local cloud of points faces difficulty when such point distributions experience sharp turning of the grid lines. These regions are identified as grid folding regions. This work proposes two different methods for mitigating this difficulty. In the first approach, new points are added around points identified with grid folding and in second approach, fictitious interfaces are placed close to the points identified with grid folding. The second approach differs

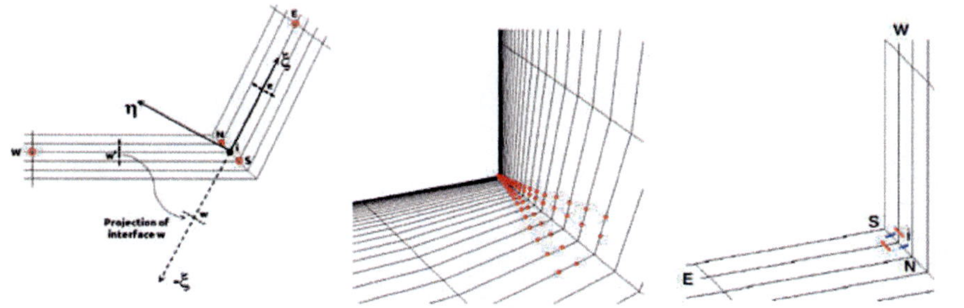

(a) Rotated LSFD – U (b) Point addition (c) Modified placement offictitious interfaces

Fig. 1 Grid folding region

from the earlier approach [3, 4] where ficititous interfaces are placed at location midway along the ray joining two points under consideration. These three different approaches are shown pictorially in Fig. 1.

The conventional LSFD-U [4] approach when applied to point involving grid folding, the associated fictitious interfaces may fall in non physical region because of thin boundary layer. This problem to some extent can be solved by the way of limiting the solution variable at these interfaces. The newly proposed approaches eliminate the possibility offictitious interfaces lying in the non physical region.

3 Results

The efficacy of solver is demonstrated by solving turbulent flow past MDA3, a multi-element airfoil. Flow conditions used in simulation corresponds to free stream Reynolds number of 9 million, Mach number of 0.2 and angle of attack of 16.21°. The results are compared against an industry standard finite volume solver, HiFUN [10].

In order to capture boundary layer, hybrid unstructured mesh with 38 prism layers is generated around the geometry, Fig. 2. Total number of points present in the grid are 237,048 and maximum y^+ exhibited by grid is 2.5. Iterative convergence of lift and drag coefficient is shown in Fig. 5 and is tabulated in Table 1. Density residual, pressure coefficient distribution and Mach contours are plotted in Figs. 3, 4 and 6 respectively. Residual convergence is better with modified interfaces approach as seen from Fig. 3 and it is easy to implement in 3D framework.

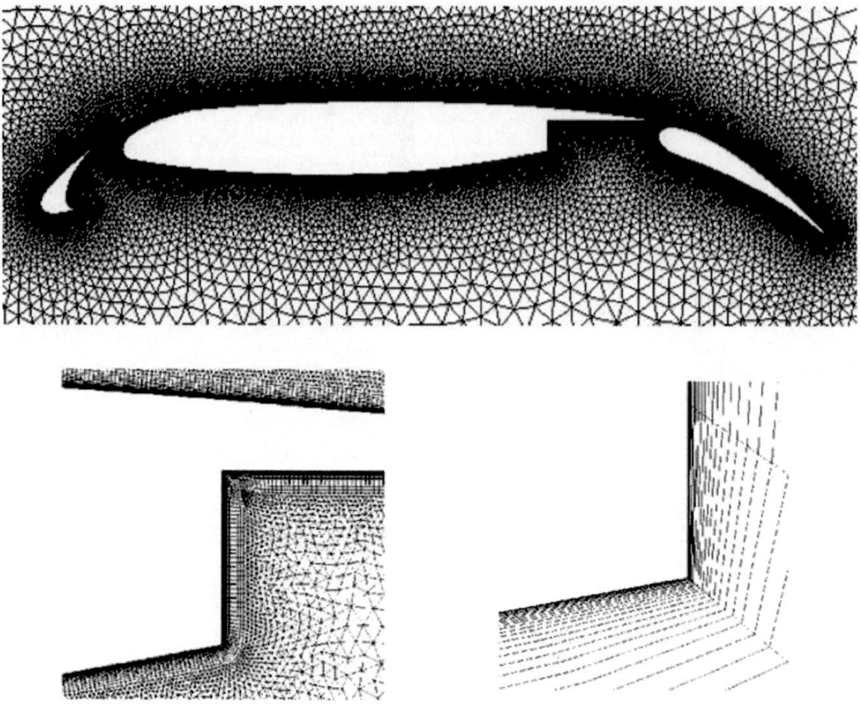

Fig. 2 Mesh

Table 1 Lift and drag coefficients

	Lift coefficient	Drag coefficient
HiFUN	4.12	684×10^{-4}
LSFD-U: midpoint interfaces	4.09	921×10^{-4}
LSFD-U: point addition	4.11	642×10^{-4}
LSFD-U: modified interfaces	4.08	952×10^{-4}

4 Conclusions

The work presented can be considered as a further step in the realization of meshless LSFD-U based RANS solver. While the issues pertaining to the development of meshless RANS solver involving highly anisotropic point distribution was addressed in our previous work, its extension to more realistic point distributions involving grid turning has been successfully addressed in this work. While both the proposed methodologies show promise, extending the proposed methods to 3D can be more easily realized using the modified interface approach. The solutions presented show a good comparison with those obtained using industry standard CFD solver HiFUN.

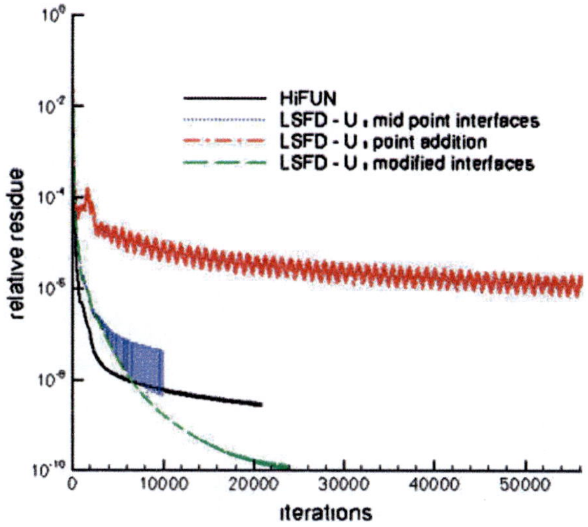

Fig. 3 Density residual

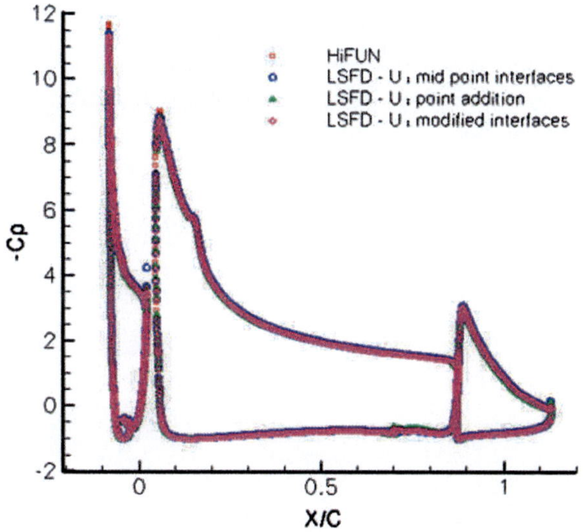

Fig. 4 Pressure coefficient distribution

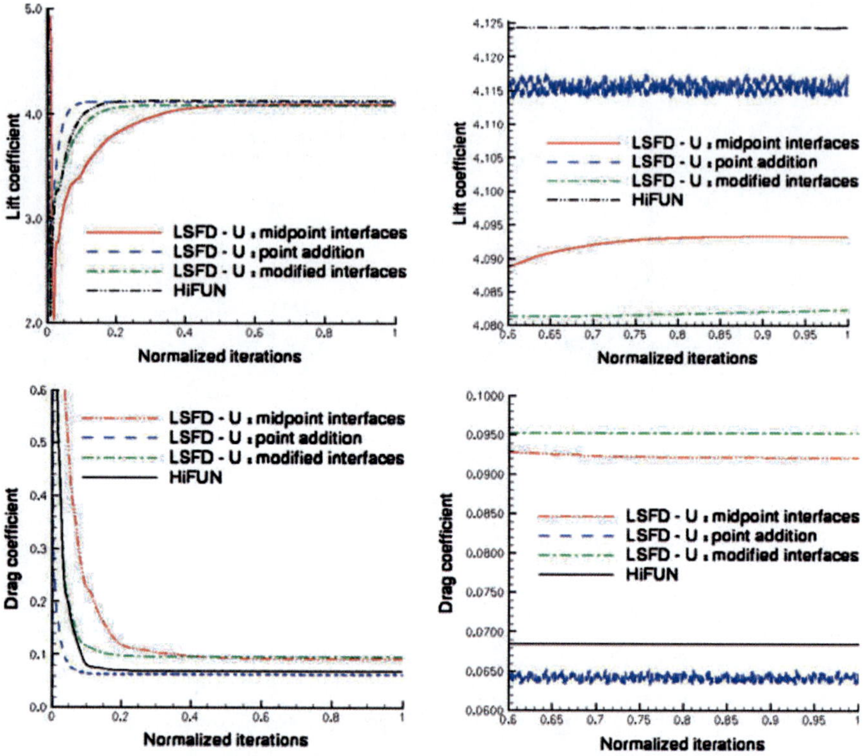

Fig. 5 Lift and drag coefficient iterative convergence

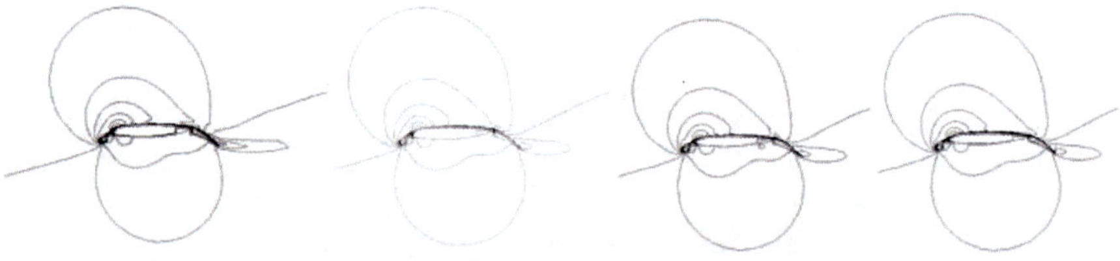

(a) HiFUN (b) Midpoint interfaces (c) Point addition (d) Modified interfaces

Fig. 6 Mach Contour, min = 0.05, max = 0.65, $\Delta = 0.05$

References

1. Ghosh AK, Deshpande SM (1995) Least squares kinetic upwind method for inviscid compressible flow. AIAA-95-1735-CP
2. Katz A, Jameson A (2010) Meshless scheme based on alignment constraints. AIAA J 48(11):2501–2511
3. Balakrishnan N (1999) New least squares based finite difference method. Report No. 99 FM 9, De- partment of Aerospace Engineering, Indian Institute of Science, Bangalore, India

4. Sridar D, Balakrishnan N (2003) An upwind finite difference scheme for meshless solvers. J Comput Phys 189:1–23
5. Mohamed Y, Munikrishna N, Parthiban A, Balakrishnan N (2014) Demonstration of automated CFD process using meshless technology. In: 6th European Conference on Computational Fluid Dynamics (ECFD VI), Barcelona, Spain, July 20–25, 2014
6. Munikrishna N (2007) On viscous flux discretization procedures for finite volume and meshless solvers. PhD Thesis, Department of Aerospace Engineering, Indian Institute of Science, Bangalore, India
7. Munikrishna N, Balakrishnan N (2010) Turbulentflow computations on a hybrid Cartesian point distribution using meshless solver LSFD-U. Comput Fluids 23:118–138
8. Venkatakrishnan V (1995) Convergence to steady state solutions of the Euler equations on unstructured grids with limiters. J Comput Phys 118:120–130
9. Spalart PR, Allmaras SR (1992) A one-equation turbulence model for aerodynamic flows. AIAA-92-0439
10. https://www.sandi.co.in/

Computational Hydraulics of an Underground Wastewater Drainage System

Navaneeth Krishnan, Janki Shinde, Madhukar M. Rao, and M. Damodaran

Abstract A high-fidelity computational method for modeling multi-phase wastewater flow is developed to understand the unsteady flow structures so as to facilitate the design of components of a typical underground wastewater drainage system consisting of a tangential vortex drop shaft, de-aeration chamber, and an adit. Two-phase flow consisting of air and water is addressed in this study. It is found that the flow structure is profoundly affected by hydraulic structures such as the de-aeration chamber and the air-vent system. The commonly accepted practice is to consider only the approach flow, vortex intake, and the drop shaft for analysis. The present study shows that the full hydraulic system should be modeled in an integrated manner to ensure a safe design.

Keywords Computational multi-phase hydraulics · Vortex drop shaft · De-aeration chamber · Adit

1 Introduction

The impact of climate change on water resources and the challenges of high population densities and scarcity of land in urban centers have resulted in innovative public policy for the development of innovative underground systems for the conveyance of wastewater by gravity to facilitate water collection, treatment, storage, and disposal such as Singapore's Deep Tunnel Sewerage System [1] and London's Thames Tideway Tunnel [2]. Such systems consist of a network of underground conveyance tunnels and hydraulic shafts which typically consist of a link sewer, a vortex drop structures, drop shafts, de-aeration chambers, air-vent pipes, and adits as shown in Fig. 1 which is considered in the present study. As empirical design assessment of

N. Krishnan (✉) · J. Shinde · M. M. Rao
ACRi Infotech Pvt. Ltd, No. 611 (2nd Floor), AECS Layout Main Road, Bangalore 560037, India
e-mail: navaneeth.krishnan@acricfd.com

M. Damodaran
Temasek Laboratories, National University of Singapore, 5A Engineering Drive 1, Singapore 117411, Singapore

L. Venkatakrishnan et al. (eds.), *Proceedings of 16th Asian Congress of Fluid Mechanics*, Lecture Notes in Mechanical Engineering,
https://doi.org/10.1007/978-981-15-5183-3_33

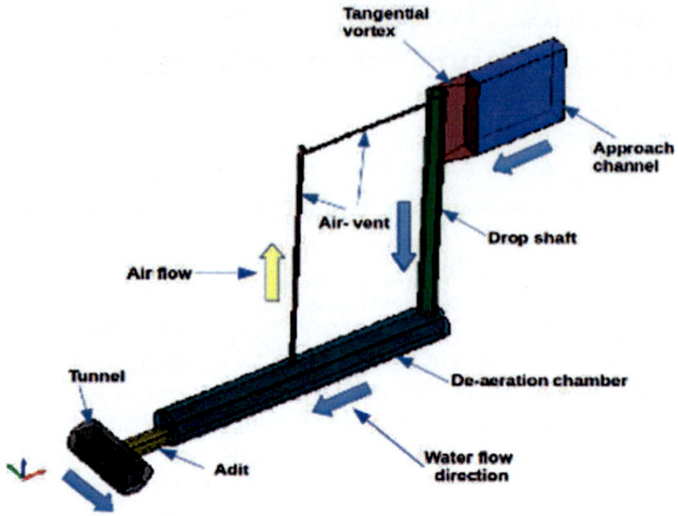

Fig. 1 Components of a typical underground hydraulic system

such structures are usually available and can be generated using expensive experimental assessment as in Yu and Lee [3], multi-phase computational fluid dynamics (CFD) simulations are fast becoming the norm for the design assessment of such structures such as Chan et al. [4] in which a two-phase flow in a system consisting of the tangential vortex the approach channel, tangential vortex, and the drop shaft only referred to as Configuration 1 in this study.

The focus of this study is to access the impact of adding a de-aeration chamber, air vent, and an adit to the same tangential vortex drop shaft configuration considered in Chan et al. [4] which is referred to as Configuration 2 in this study. The dropshaft diameter D of 0.0124 m is also the width of the approach channel. The length and the depth of the approach channel excluding the tapered channel are 5D, the drop shaft length is 10D (excluding the portion of the dropshaft that is flush with the top of the approach channel); the diameters of the de-aeration chamber and adit are 2D and D, respectively, and the lengths of the de-aeration chamber and adit are 10D and 2D, respectively. The air vent is a straight circular pipe of diameter 0.25D which starts at the midpoint of the de-aeration chamber and reaching beyond the top of the approach channel and open to the atmosphere at the ground level with a horizontal portion entering near the top of the approach channel above dropshaft. As the flow is gravity driven, unless otherwise stated the gradient in the direction of the flow is 1:1000 for all the components downstream of the drop shaft, i.e., the de-aeration chamber, the adit, and the deep drainage tunnel. In accordance with the experimental setup in Chan et al. [4], the approach channel has no slope.

2 Computational Hydraulics Model

The governing equations of incompressible multiphase volume-of-fluid (VOF) in integral form for a control volume V with a surface area S are given by the continuity and momentum equations, i.e.,

$$\int_S \vec{U} \cdot \mathrm{d}S \tag{1}$$

$$\rho\left(\frac{\partial}{\partial t}\int_V \vec{U}\mathrm{d}V + \int_S \left(\vec{U} \oplus \vec{U}\right) \cdot \vec{n}\mathrm{d}S\right) = -\int_S p\vec{n}\mathrm{d}S + \int_S \vec{\vec{\tau}}\vec{n}\mathrm{d}s + \int_V f\mathrm{d}V \tag{2}$$

where $\vec{U}$ is the vector of flow variables which in this case consist of the flow velocity in the three coordinate directions, $\vec{\vec{\tau}}$ is the deviatoric viscous stress tensor, ρ is the density of the air–water mixture which will be estimated after solving the transport equation for volume fraction of water, p is the pressure, and $\vec{f}$ is the specific body force (inclusive of gravity). Additionally, the hydrostatic component of the pressure can be removed from the governing equations, so that the dynamic part of the pressure can be used to satisfy the continuity equation. The dynamic part of the pressure is given by

$$p_{\mathrm{rgh}} = p - \rho(\vec{g} \cdot \vec{r}) \tag{3}$$

where $\vec{g}$ is the acceleration due to gravity (vector) and $\vec{r}$ is the position vector of the finite volume cell centroid with reference to any suitable point in the domain, so that the formulation is invariant with respect to translations and rotations of the coordinate system. In such a situation, the body force vector will exclude the gravity effects, and the pressure p appearing in Eq. (2) will be replaced with p_{rgh} as defined in Eq. (3). The integral equation for the transport of volume fraction of water (α) takes the following form:

$$\frac{\partial}{\partial t}\int_V \alpha\mathrm{d}V + \int_S \left(\vec{U}\alpha + \vec{U}_r\alpha(1-\alpha)\right) \cdot \vec{n}\mathrm{d}S = 0 \tag{4}$$

where $\vec{U}_r$ is the compression velocity needed to prevent the smearing of the air–water interface. The fluid density is computed as a mixture density as outlined in Hirt et al. [6] as follows:

$$\alpha \cdot \rho_{\mathrm{water}} + (1-\alpha) \cdot \rho_{\mathrm{air}} \tag{5}$$

Equations (1–5) are discretized using the finite volume method and solved simultaneously using the *interFoam* solver for multi-phase flow in the *OpenFOAM*™ [5]

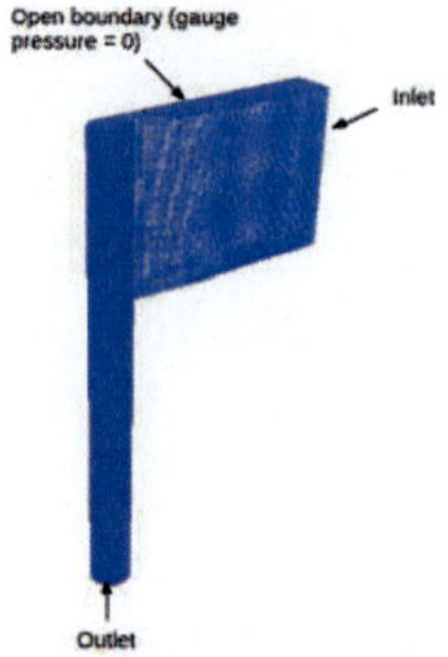

(a) Configuration 1

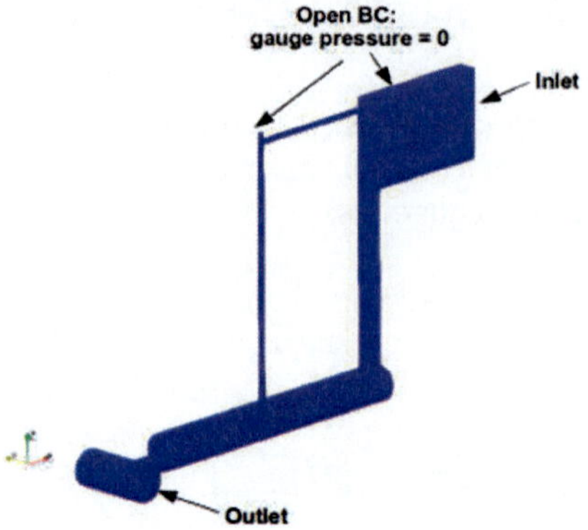

(b) Configuration 2

Fig. 2 Mesh for **a** Configuration 1: Standalone tangential vortex intake with approach channel, vortex intake, and drop shaft **b** Configuration 2: Full system integrating Configuration 1 with a de-aeration chamber, air vent, adit, and part of an exit tunnel

library. The SST k-turbulence model outlined in Menter [6] are additional equations which are numerically solved in conjunction with the flow Eqs. (1) and (2) to model the turbulent viscosity in the context of Reynolds-averaged Navier–Stokes equations (RANS). Various mesh generation utilities in *OpenFOAM*™ library such as *blockMesh*, *snappyHexMesh*, and *mergeMesh*, have been used to generate hexahedral meshes, cluster and merge meshes in independent component shown in Fig. 1 to form the overall computational meshes shown in Fig. 2. The mesh integrity is checked using the *checkMesh* utility. A sequence of meshes ranging from coarse to fine have been generated for the purpose of assessing numerical accuracy. Computations have been carried out for various incoming flow rates of ranging from 2 to 8 L/s and the corresponding water height and velocity at the inlet are estimated from steady normal flow considerations for a hydraulic gradient of 1:1000. The initial condition consisted of the entire system being filled with air at zero velocity.

The top of the approach channel, vortex intake, and air vent were considered as open boundaries, as shown in Fig. 2, with gauge pressure set to zero. At all walls no-slip conditions are set. The outlet is located at one end of the exit tunnel as shown in Fig. 2. The upstream end of the exit tunnel is closed off and is modeled as a wall. At the downstream end of the exit tunnel, all other dependent variables including volume fractions are set to zero gradient.

3 Mesh Convergence and Numerical Accuracy

Computations have been carried out on a sequence of meshes, a coarse mesh of 29,316 elements, a medium mesh of 232,683 elements, and a fine mesh of 1,017,286 elements for Configuration 1. The air–core area ratio (fraction of area occupied by air at various locations in the dropshaft have been compared with the results of Chan et al. [4], and computations presented have been carried out on the fine mesh. Similar studies for Configuration 2 extend the medium mesh by 129,292 elements and the fine mesh by 820,575 elements. Results are presented on the fine mesh.

4 Results and Discussion

Computational assessment are presented and discussed for both configurations. Configuration 1 is targeted primarily at assessing the accuracy of the present computations against the benchmark numerical and experimental results of Chan et al. [4] and Yu and Lee [3]. Configuration 2 is aimed at understanding the hydraulics of the overall wastewater conveyancing system and to assess the impact of adding other components downstream of the drop shaft.

(a) **Configuration 1**: Figure 3 shows the computed minimum air–core area ratio for various flow rates in the drop shaft corresponding to inflow flow rates at the inlet of the approach channel of 2, 4, 8, and 10 L/s. Comparing Fig. 3a–d which show the computed water phase and air core at selected Z-locations along the drop shaft and from the present study with Fig. 3e–h which are the corresponding results from Chan et al. [4] show good agreement with the benchmark results and experimental data. The air–core area is ratio is substantially greater than 25% in all cases, which represents the requirement for stable flow. The vortex structure is also well defined, with the water moving close to the periphery of the

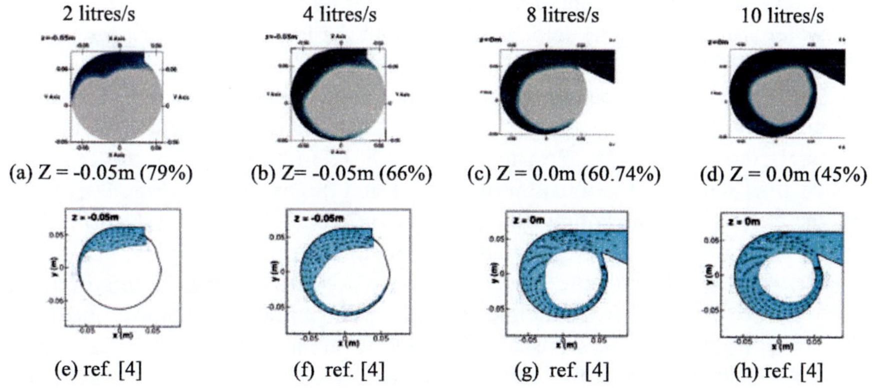

Fig. 3 **a–d** Computed minimum air core area ratio and flow structure in the drop shaft of configuration corresponding to 2, 4, 8 and 10 L/s approach channel inlet flow rate at indicated Z-locations along the drop shaft. **e–h** corresponding results taken from Chan et. al. [4]. Note that the dark blue areas in **a–d** indicate the water phase and the light grey areas indicate air core. The percentages in brackets next to the Z-locations refer to the fraction of area occupied by the air phase

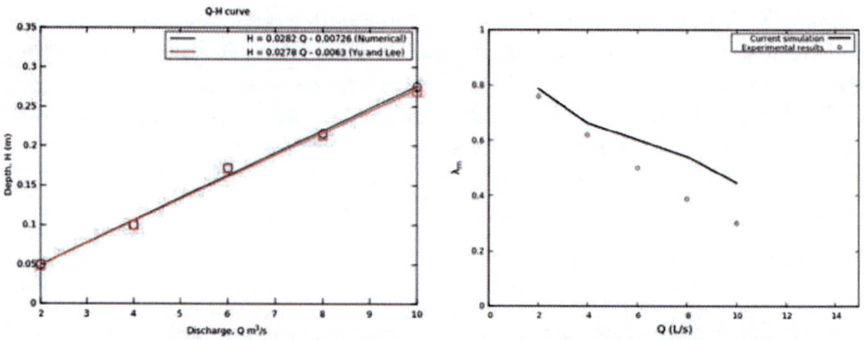

(a) ratingcurve

(b) minimum air core area ratio

Fig. 4 Computed **a** rating curve for approach channel and the variation of computed **b** minimum air–core area ratio in the drop shaft for various intake flow rates compared with Ref [3, 4]

drop shaft cross section, due to centripetal acceleration which forces the air to remain at the center, as is desired for stable operation of the hydraulic structure. Figure 4a shows the rating curve for the approach channel showing water heights in the channel for different flow rates showing good agreement with that from Chan et al. [4] and Yu and Lee [3]. Figure 4b compares the computed minimum air–core area ratio in the drop shaft against that of Chan et al. [4]. The overall results are apparently satisfactory from the view point of assessing the impact of adding the de-aeration chamber, air vent, and adit to Configuration 1.

(b) **Configuration 2**: Selected computed results for the complete hydraulic system for a flow rate of 8 L/s in the approach channel are interpreted here. Figure 5 shows the computed contours of the air–water interface (i.e., iso-surface of 0.5 water volume fraction) colored according to the local value of the flow speed. Figure 5a shows the computed air–water interface corresponding to configuration 1 at fully developed state. Figures 5b–d show the evolution of the computed contours of the air–water interface at selected instants t of 4 s, 6 s, and 8 s, respectively during which the flow has completed the full circuit from the approach channel intake to the junction of the adit and the exit tunnel. Figure 6 shows the evolution of the computed air core at various Z-locations in the drop shaft of Configuration 2 at various instants of time. Figure 5a clearly shows that the water surface is stable and smooth over the entire domain, starting from the approach channel, via the inclined part of the tangential vortex intake, and down the drop shaft. The spiraling of the flow to form the vortex structure is evident and is consistent with flow structures shown in Fig. 3. Figure 5b shows the water surface at time instant $t = 4$ s indicating that the water has crossed the air-vent location in the de-aeration chamber. Figure 6a shows a clear air core structure in the drop shaft clearly indicating a good vortex structure; however, disturbances and/or wave-like structures can be observed on the water surface in the approach channel as well as in the de-aeration chamber and the air vent is clear. Figure 5c

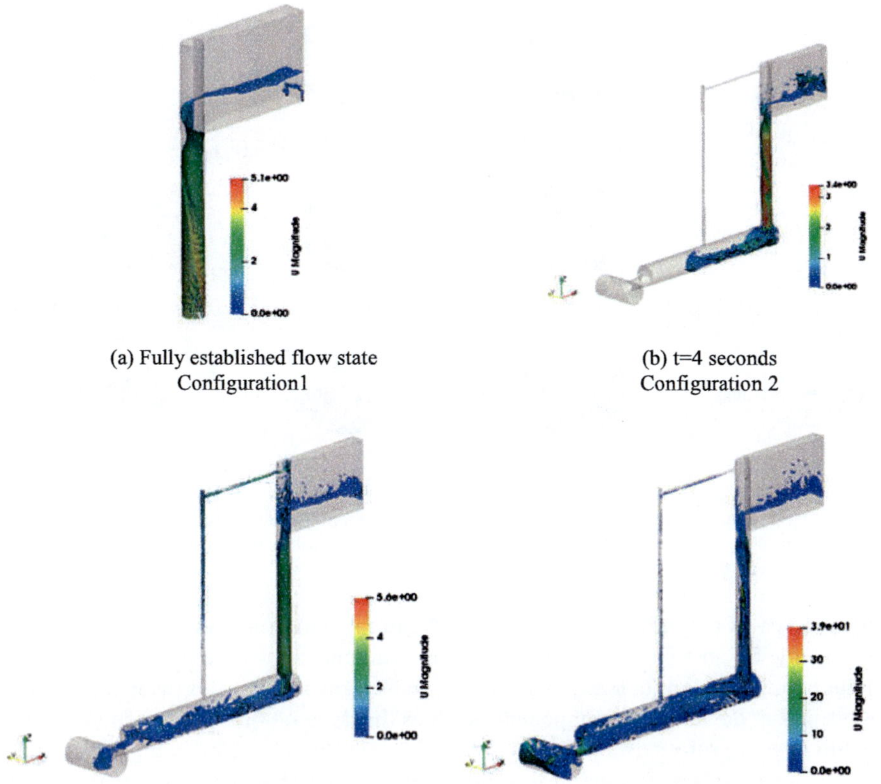

(a) Fully established flow state
Configuration1

(b) t=4 seconds
Configuration 2

(c) t=6 seconds
Configuration 2

(d) t=8 seconds
Configuration 2

Fig. 5 Computed contours of the air–water interface corresponding to an approach channel flow rate of 8 L/s for **a** Fully established flow state in Configuration 1 and for Configuration 2 at **b** $t = 4$ s, **(c)** $t = 6$ s, (d) $t = 8$ s as flow fills the entire system

shows the air–water interface at time $t = 6$ s and the corresponding air core and vortex structure can be seen in Fig. 6b. It can also be seen that water has crossed the adit and has entered the deep sewage tunnel and also some water has entered the air vent. This causes substantial disruption to the entire flow structure throughout the domain, including the upstream components such as the approach channel and the drop shaft.

Some disruption of the air core and vortex structure evident in Fig. 6b is probably due to injection of water from the air vent, which is a significant breach of the usual design norms. Figure 5d shows that water has crossed into the deep sewage tunnel and is flowing out of the outlet. Water entry into the air vent is evident, and the flow in the de-aeration chamber and the adit is extremely unsteady and unstable, with spray-like structures in the de-aeration chamber. Figure 6c shows a complete

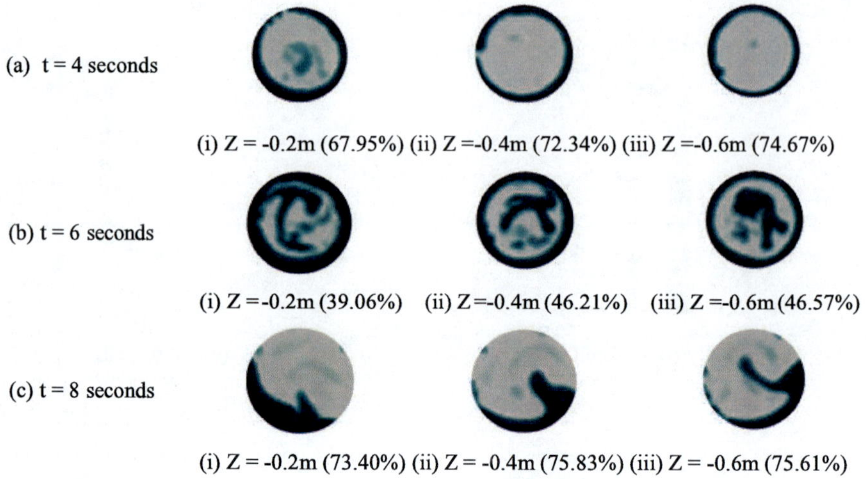

Fig. 6 Evolution of the computed air core (indicated as percentage of the local cross-sectional area) in Configuration 2 at various Z-locations in the drop shaft at various time instants

breakdown of the air core and vortex in the drop shaft. Such a flow is completely unsuitable for an efficient functioning of the hydraulic conveyancing system, and the failure of the design (at least, in terms of a stable flow structure) is evident. It is also evident that downstream components such as the de-aeration chamber and air vent. could have a profound impact on the flow even in the upstream components such as the vortex intake and the drop shaft. Therefore, CFD analysis for the design of hydraulics system must consider the total hydraulic conveyancing system to assess the performance of the total system instead of predicting design based on individual components of the system in isolation.

5 Conclusions

The computational study of hydraulic conveyancing system shows that while stable flow is apparently reached in Configuration 1 which only consists of the approach channel, tangential vortex intake, and the drop shaft, the addition of the downstream components such as de-aeration chamber, air vent, and adit exiting into a deep tunnel can completely alter the flow field to the point of making the design invalid. The ingress of water into the air vent causes a feedback loop in which the vortex air core in the drop shaft is progressively destroyed in time. The flow in the de-aeration chamber being very unsteady and chaotic calls for the incorporation of additional energy dissipation devices at the downstream end of the drop shaft.

References

1. Public Utilities Board. Singapore, Deep tunnel sewerage system. https://www.pub.gov.sg/dtss/about
2. Thames tideway scheme. https://en.wikipedia.org/wiki/Thames_Tideway_Scheme
3. Yu D, Lee JHW (2009) Hydraulics of Tangential vortex intake for urban drainage. J Hydraulic Eng 135:164–174
4. Chan SN, Qiao QS, Lee JHW (2018) On the three-dimensional flow of a stable vortex intake. J Hydro-Environ Res 21:29–42
5. OpenFOAM^{TM} v.18.06 (2018) The open source CFD tool box user guide. https://www.openfoam.com/. OpenCFD, ESI Group
6. Menter FR (1994) Two-equation eddy-viscosity turbulence models for engineering applications. AIAA J 32(8):1598–1605

Exploring Physics-Informed Neural Networks for Compressible Flow Prediction

M. Chaudhari, I. Kulkarni, and M. Damodaran

Abstract This work aims to understand the impact of hyperparameters of neural networks on the capability of physics-informed neural network (PINN) surrogates for compressible flow predictions and how they compare with traditional neural network (NN) surrogates by considering steady inviscid compressible flow in a 1-D converging–diverging nozzle subjected to different back pressures. Both NN and PINN are trained with known random flow characteristics estimated using analytical and computational methods corresponding to a set of back pressures and used to predict flow characteristics corresponding to arbitrary back pressures which are not in the training dataset.

Keywords Physics-informed neural networks · Compressible nozzle flows · Highly nonlinear flow prediction

1 Introduction

Recent advances in machine learning have shown the capability of physics-informed neural network (PINN) surrogates as outlined in Raissi et al. [1] in making fast, economic and reliable predictions in thermofluids while preserving their physics. Such networks are trained to learn the spatial and temporal distribution of flow variables while complying with the governing equations. Artificial neural networks

M. Chaudhari (✉) · I. Kulkarni
Department of Mechanical Engineering, Indian Institute of Technology Madras, Chennai, TN 600036, India
e-mail: mohilc@smail.iitm.ac.in

I. Kulkarni
e-mail: ishank@smail.iitm.ac.in

M. Damodaran
Temasek Laboratories, National University of Singapore, 5A Engineering Drive 1, Singapore 117411, Singapore
e-mail: tslmura@nus.edu.sg

L. Venkatakrishnan et al. (eds.), *Proceedings of 16th Asian Congress of Fluid Mechanics*, Lecture Notes in Mechanical Engineering,
https://doi.org/10.1007/978-981-15-5183-3_34

(NNs) can be trained with known solutions of a problem of interest for specific conditions (be it analytical, experimental or computational), and then, post-training is used to predict the outcome of the problem for arbitrary conditions at insignificant computational power. In this work, an NN is implemented for predicting flow variables using scarce training data. Additionally, the outputs from NN are forced to benefit from the knowledge of governing physics to create a PINN, and their accuracy is assessed. An inviscid one-dimensional compressible flow in a converging–diverging nozzle is considered for this comparative study. The back pressure drives the fluid flow in the nozzle connected to a reservoir at the inlet. Initially, there is no flow in the nozzle as the back pressure and the inlet stagnation pressures are the same. As the back pressure is decreased to initiate the flow and reaches maximum flow rate resulting in choked conditions at the throat and normal shocks form in diverging part of the nozzle thereafter as the back pressure is decreased further. Flow prediction for the range of back pressures due to the presence of a shock in the diverging part of the nozzle is of interest in this study. The compressible governing equations are given by

$$\mathbf{w}_t + \mathbf{F}_x + \mathbf{S} = 0 \tag{1}$$

where $\mathbf{w} = \left(\rho S \ \rho u S \ \rho E S\right)^T$, $\mathbf{F} = \left(\rho u S \ \rho\left(u^2 + p\right)S \ (\rho E + p)u S\right)^T$ and $\mathbf{S} = \left(0 \ -p S_x \ 0\right)^T$, ρ is density, u is speed, E is the specific energy, $p = (\gamma - 1)\left(E - 0.5\rho u^2\right)$ is the pressure assuming ideal (calorically perfect) gas law and $S(x) = 1 + 2.2(3x - 1.5)^2, x \in [0, 1]$ is the cross-sectional area of the nozzle considered in this study. Equation (1) can be approximately solved analytically for steady-state flows (for which the temporal derivatives in Eq. (1) can be dropped) using isentropic and normal shock relations or computationally using well-known numerical schemes by time marching the solution from an initial flow condition to steady state. The aim of this study is to use a NN with minimal training data and network depth to predict the shock strength and location in the nozzle for a given back pressure and to enhance the quality of the predictions by using the governing physical laws of the problem to form the PINN,

2 Methodology

Figure 1 shows the PINN architecture considered for this study to ascertain the impact of network hyperparameters on the prediction capability. The network in the blue box corresponds to the NN architecture.

The parameters in the NN and PINN are determined by optimizing a loss function defined as Loss $=$ MSE $+ \lambda\overline{w}$(RES) $\equiv \frac{1}{Ns}\left(\sum_{j=1}^{4}\sum_{i=1}^{Ns}(yj^{\text{pred}} - yj^{\text{truth}})^2 + \lambda\overline{w}\sum_{i=1}^{4} ej^2\right)$ where N_s is the number of points in the dataset, with subscript j referring to an output variable, i.e., (ρ, u, p, E),

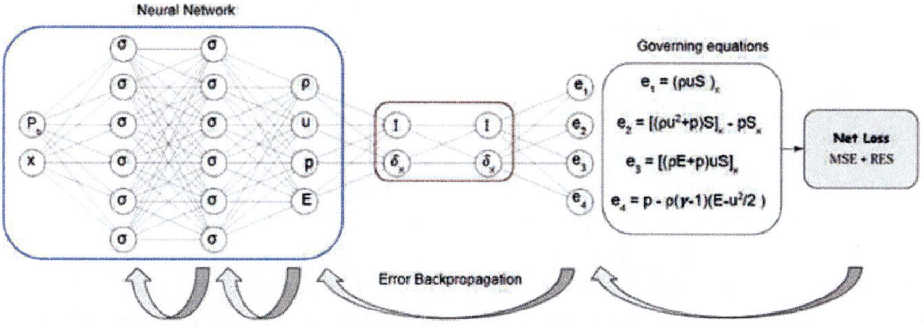

Fig. 1 Physics-informed neural network architecture

y_j^{pred} is the output value as predicted by the network and y_j^{truth} is the actual ground truth, $e_1 = (\rho uS)_x$, $e_2 = ((\rho u^2 + p)S)_x - pS_x$, $e_3 = ((\rho E + p)uS)_x$ and $e_4 = p - \rho(\gamma - 1)(E - u^2/2)$ are the steady-state compressible flow equations and equation of state governing the physics of the problem,λ is a selector switch to activate PINN if $\lambda = 1$ and NN if $\lambda = 0$, $\overline{w}$ is a parameter to bias the optimizer to RES. Ideally, *Loss* should be zero to ensure that the output values predicted by the network are exactly equal to the true output values, and all the predicted values satisfy the underlying governing equations. The derivatives of the flow variables appearing in RSE are evaluated using *Automatic Differentiation* as outlined in Hoffmann [2] in *Tensorflow* of Abadi et al. [3]. The data structure for the given problem consists of two independent (input) variables, namely back pressure P_b and location inside the nozzle measured from nozzle inlet x. The dependent (output) variables are pressure p, density ρ, speed u and specific energy E associated with x in the nozzle for a particular P_b. Compressible Nozzle Flow Simulator of Devenport [4] and a numerical scheme of Toro [5] are used to generate datasets for this study. The nozzle length is divided into 100 nodes. First, a mother dataset corresponding to 27 different back pressures for all x values is generated. The training dataset is created by randomly selecting a fraction f of the points from the mother dataset. Due to the random selection, the data points in the training set do not necessarily have any continuous relation with respect to P_b and x among them. A validation dataset is generated for a particular P_b for all values of x to study the effect of the NN and PINN hyperparameters on the quality of the prediction of shock strength and its location from the predicted pressure distribution along the nozzle length. A test dataset corresponding to six different back pressures for all values of x is generated to compare the accuracy of NN and PINN predictions. The accuracy of NN and PINN trained using analytical data is compared by varying three different hyperparameters of the networks, namely f, the size of the network, and $\overline{w}$. The study of the effect of each hyperparameter on the quality of the prediction is assessed, while keeping all the other hyperparameters fixed. The accuracy of the PINN is explored further by varying two additional parameters, i.e., inclusion of state error (e_4) and nonlinear activation functions.

3 Results and Discussion

3.1 *Effect of Training Data Size*

The first study demonstrates the variation of f on the accuracy of NN and PINN. A network architecture consisting of input, output and four hidden layers is represented by the vector of length 6, i.e.,[2, 15, 25, 25, 15, 4], and the value of each element in this vector represents the number of neurons in a particular layer, and $\overline{w}$ is set to be 1. Figures 2 and 3 compare the pressure distribution along the nozzle length predicted by NN and PINN with the ground truth and the corresponding components of the loss function, i.e., MSE and RES for three values of f. The discontinuity in the MSE plots can be attributed to the presence of sharp discontinuity near the shock location. It can be seen from Fig. 3a–b that the accuracy of both PINN and NN is of the same order with *MSE* and *RES* tending to zero everywhere except at the shock location for $f = 20\%$ and 10%. As f is decreased further to 1%, the order of *Loss* increases indicating that the network depth is too large for the training data size resulting in overfitting described in Nusrat and Jang [6] in the predictions as can be seen from Fig. 2c. Figure 3c shows that while the PINN drives *RES* to zero, the accuracy of PINN relative to NN decreases, implying that network depth should be decreased in order for the network to benefit the characteristics of PINN.

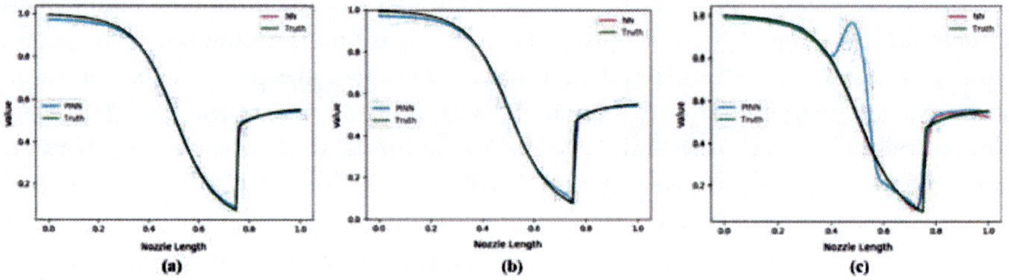

Fig. 2 Pressure distribution along the nozzle length corresponding to **a** $f = 20\%$ **b** $f = 10\%$ **c** $f = 1\%$

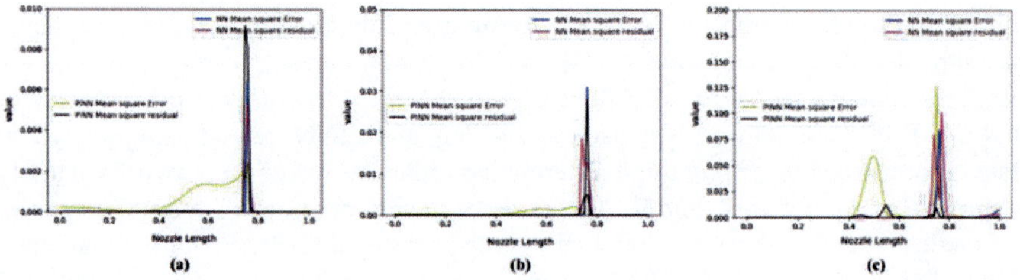

Fig. 3 MSE and RES distribution along the nozzle length corresponding to **a** $f = 20\%$ **b** $f = 10\%$ **c** $f = 1\%$

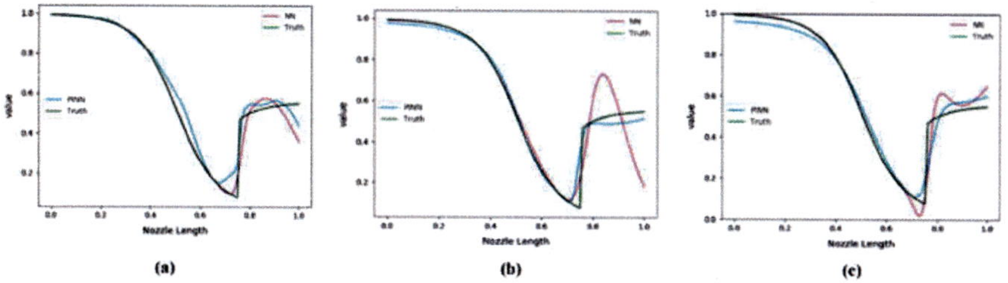

Fig. 4 Pressure distribution along the nozzle length corresponding to **a** [2, 15, 25 15, 4], **b** [2, 10, 10, 10, 4], **c** [2, 5, 5, 4]

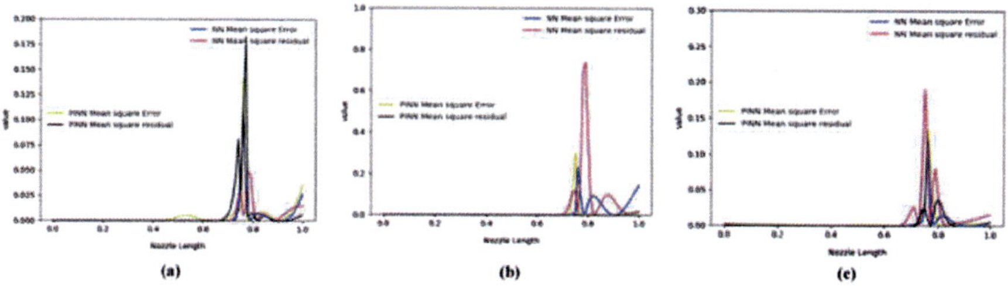

Fig. 5 MSE and RES distribution along the nozzle length corresponding to **a** [2, 15, 25 15, 4], **b** [2, 10, 10, 10, 4], **c** [2, 5, 5, 4]

3.2 Effect of Network Size

The second study demonstrates the effects of the variation of network size on the accuracy of NN and PINN. Here, f and $\overline{w}$ are set to be 1% and 1, respectively. Figures 4 and 5, respectively, compare the pressure distribution along the nozzle length predicted by NN and PINN with the ground truth and the corresponding components of *Loss*, i.e., MSE and RES for three network architectures, i.e., [2, 15, 25 15, 4], [2, 10, 10, 10, 4] and [2, 5, 5, 4]. As the depth of the network is reduced, the accuracy of the prediction increases initially to a peak value followed by a decrement which can be attributed to underfitting described in Nusrat and Jang [6], implying that the network may be too shallow to capture the trend in the data.

3.3 Effect of $\overline{w}$

The third study demonstrates the variation of $\overline{w}$ on the accuracy of NN and PINN. Here, f and network architecture are 1% and [2, 10, 10, 10, 4], respectively. In order to make the network robust against underfitting and overfitting described in Nusrat and Jang [6], w is increased in *Loss*. Figures 6 and 7, respectively, compare the pressure distribution along the nozzle length predicted by NN and PINN with the

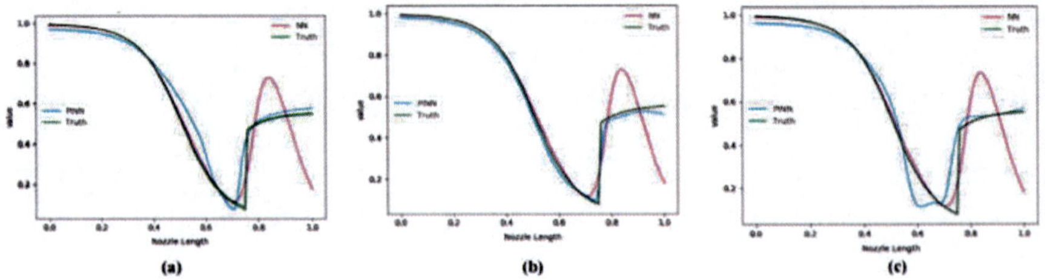

Fig. 6 Pressure distribution along the nozzle length corresponding to **a** $w = 4$, **b** $w = 40$, **c** $w = 400$

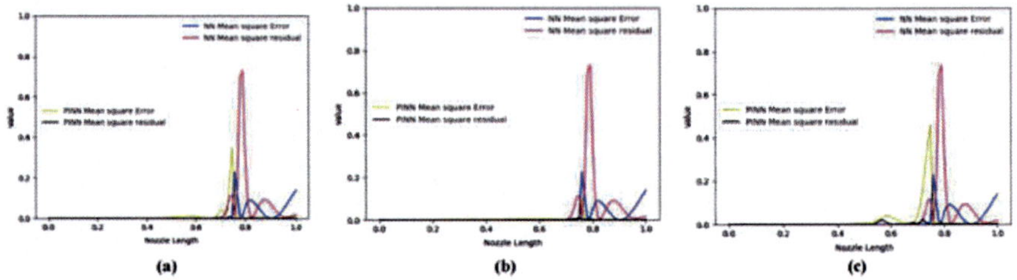

Fig. 7 MSE and RES distribution along the nozzle length corresponding to **a** $w = 4$, **b** $w = 40$, **c** $w = 400$

ground truth and the corresponding components of *Loss*, i.e., MSE and RES for three different values of $\overline{w}$. It can be seen from Fig. 6 that as $\overline{w}$ increases, the accuracy of the prediction shows a trend similar to that in Fig. 4 implying that for large values of $\overline{w}$, RES is overpowering the *Loss*,, thereby not capturing the discontinuity within an acceptable range. The PINN prediction using the *optimal hyperparameter* settings, i.e., f being 1%, network architecture structure of [2, 10, 10, 10, 4] and $\overline{w}$ of 40 shown in Fig. 6b appears to comply with ground truth.

3.4 Effect of Including State Equation Error

Figures 8a, b compares the pressure distribution along the nozzle length with and without including e_4 in the *RES* in the PINN with *optimal hyperparameter* settings defined Sect. 3.3. The quality of the predictions is increased by imposing tighter constraints on the flow variables by forcing them to satisfy the state equation in addition to the steady compressible flow equations.

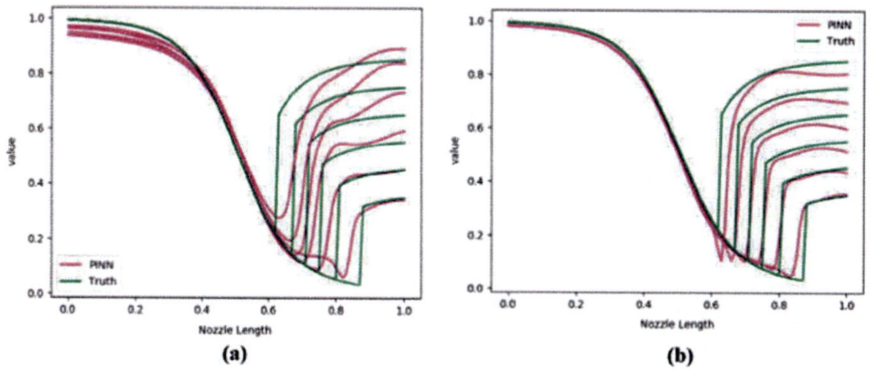

Fig. 8 Pressure distribution along the nozzle length corresponding to six different back pressures for PINN **a** without e4 and **b** with e4

3.5 *Effect of Activation Function*

Figures 9a–c compares the pressure distribution using PINN along the nozzle length with the same *optimal hyperparameter* settings for three different types of activation functions.

Figure 9b shows the highest accuracy due to closest resemblance of *tanh* to a *step function* (which can approximate any function using a weighted linear combination) compared to *sigmoid* and *arctangent*. The same *optimal parameters* obtained from the validation dataset are tested against the test dataset using ADAM optimizer of Kingma and Ba [7]. Figure 10 compares the predicted pressure, density, speed, specific energy and the normalized test error of the predictions by PINN and NN against the ground truth. Although PINN predicts the shock location accurately, in Fig. 10k, the shock strength prediction is not due to the scarce training data. Figures 10c, f, i, l show the effect of increasing f on the test accuracy.

The robustness of PINN is validated by training both the networks using numerical data. The inherent error in the training data due to the numerical scheme is compensated by increasing f to 2%, network size to [2, 10, 15, 10, 4] and w to 4,

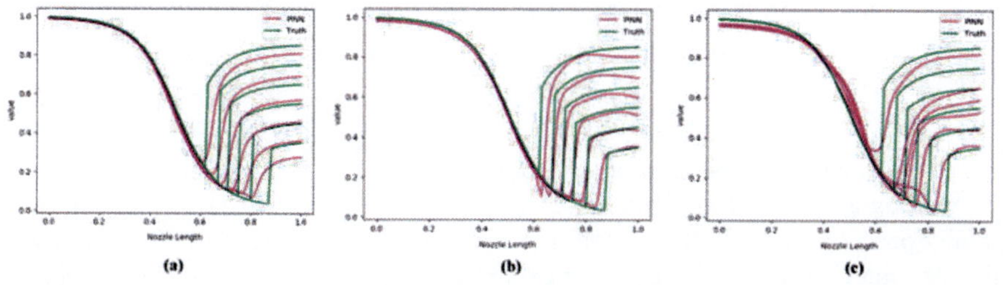

Fig. 9 Pressure distribution along the nozzle length corresponding to 6 different back pressures for **a** atan **b** tanh **c** sigmoid activation function

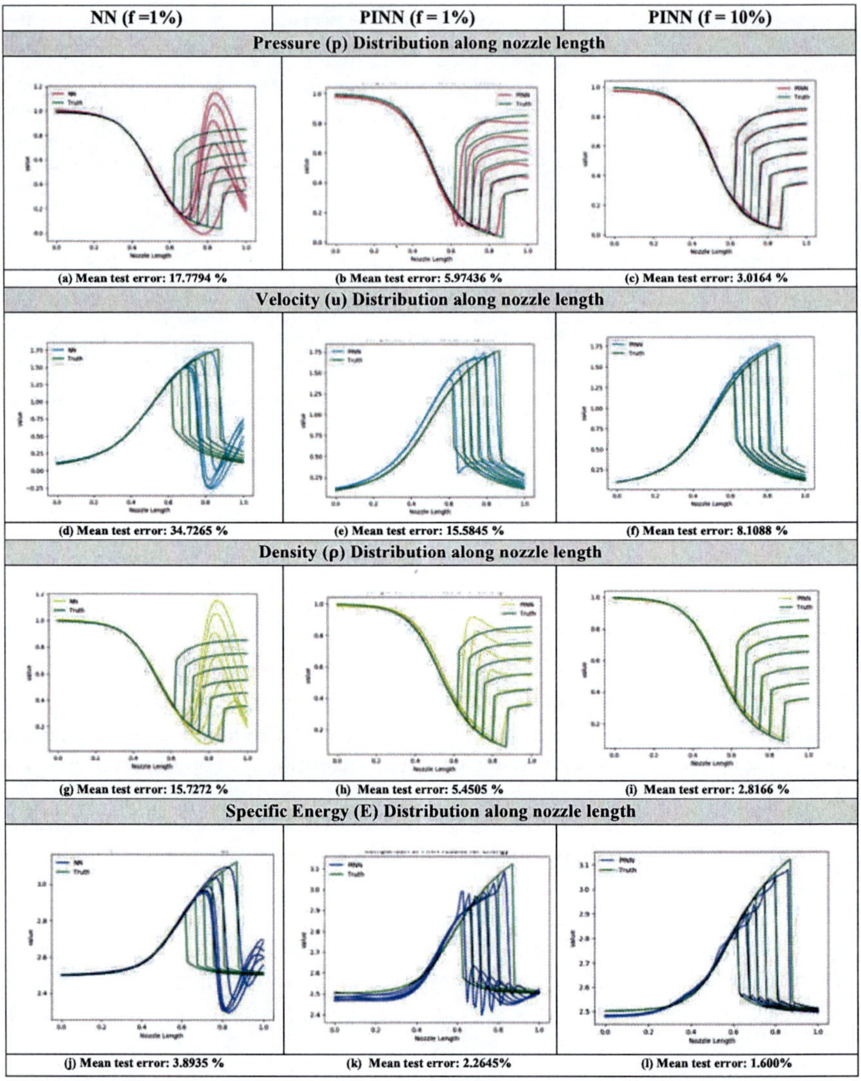

Fig. 10 Predictions of pressure, speed, density, specific energy distribution along nozzle length for PINN and NN

respectively, and testing against a particular value of P_b. From Fig. 11, it is observed that compared to NN (test error of 16.7677%), the PINN (test error of 10.635%) prediction is observed to follow the trend closely with well-defined shock strength and location making it well suited for capturing high-resolution nonlinearity even in the presence of noisy training data.

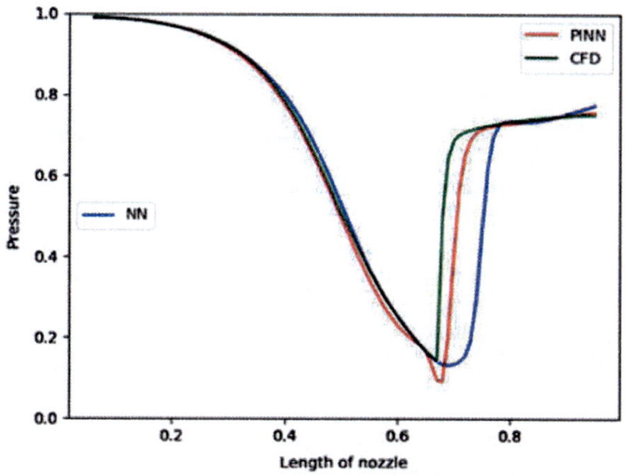

Fig. 11 Pressure distribution predictions trained on numerical data for NN and PINN

4 Conclusion

PINN is computationally more expensive than NN for similar hyperparameters (parameters of the network). However, PINN requires fewer training data for capturing the overall physics of the problem. Due to the presence of governing physics information, PINN can capture the general trend with fewer network architecture size than NN for the same training data. That means PINN obtains similar results with less computational power and time compared to NN. Additionally, PINN can work very well with numerical and experimental data having some inherent noise to obtain predictions complying with the physics which means PINN requires fewer hyperparameter conditioning compared to NN for avoiding the problems of underfitting and overfitting. The key point here is that PINN essentially does not require any training data to predict, i.e., given the governing equations, and PINN has the capability of predicting the physics with high accuracy with just the knowledge of initial or the boundary conditions (specific to the problem in-hand) given sufficient hyperparameter tuning. In addition to super-sparse data, the inclusion of high number of collocation points (marching with time and space) where the residual error is driven to zero is analogous to increasing the number of nodes and decreasing time step in a typical numerical scheme. With clear comparison to NN, the question of how computationally expensive and time taking compared to a CFD solver scheme for replicating similar results still remains unexplored, enabling the potential of PINN in partially or fully substituting all the CFD numerical schemes used for simulating flows of interest.

Acknowledgements The undergraduate summer internship of the first two authors from May-July 2019 at Temasek Laboratories, NUS, was supported from seed project funds of TLSP1903-2019.

References

1. Raissi M, Perdikaris P, Karniadakis G (2019) Physics-informed neural networks: a deep learning framework for solving forward and inverse problems involving nonlinear partial differential equations. J Comput Phys 378:686–707
2. Hoffmann P (2015) A Hitchhiker's guide to automatic differentiation. Numer Algor 72:775–811 (2016)
3. Abadi M et al (2016) Tensorflow: A system for large-scale machine learning. In: Proceedings of 12th USENIX Conference on Operating Systems Design and Implementation (OSDI'16), Nov 02–04, pp 265–283, Savannah, GA, USA; ISBN 978-1-931971-33-1
4. Devenport W (2014) Converging-diverging nozzle matlab simulator. Virginia Polytechnic Institute and State University, USA. https://www.dept.aoe.vt.edu/~devenpor/aoe3114/CD%20Nozzle%20Sim/index.html
5. Toro EF (2019) Riemann solvers and numerical methods for fluid dynamics: a practical introduction. Springer, Heidelberg. 978-3-642-06438-8
6. Nusrat, Jang S (2018) A comparison of regularization techniques in deep neural networks. Symmetry 10(11):648. https://doi.org/10.3390/sym10110648
7. Kingma D, Ba J (2015) ADAM: a method for stochastic optimization. In: Proceedings of the 3rd International Conference on Learning Representations (ICLR 2015), May 7–9, San Diego, USA

Flow Separation Control on a NACA-4415 Airfoil at Low Reynolds Number

S. Sudhakar and N. Karthikeyan

Abstract An experimental study using 3C-PIV was carried out for investigating the effect of adapting a passive flow control device similar to that found on the humpback whale flippers (tubercles) to the leading edge of a NACA-4415 airfoil at a Reynolds number of 120,000. The evolution of the mean velocity field on the NACA-4415 airfoil and with leading-edge tubercles are measured using stereo-particle image velocimetry at an angle of attack of 18°. The leading-edge tubercles exhibited attached flow up to 50% of the airfoil and reduced the height of the separated region over modified airfoil as compared to the baseline. This results in an improved aerodynamics performance by the modified airfoil.

Keywords Tubercles · Low Reynolds number · Separation · Flow control

1 Introduction

The characterization and understanding of flow features over the airfoil at low Reynolds number are very important in the context of the design of efficient wings for MAV/UAV and rotor blades for the wind turbine. The flight Reynolds numbers of MAV/UAV and operating conditions of wind turbine blades fall in the low Reynolds number flow regime. Prominent flow feature at low Reynolds number on airfoil/wing is the laminar separation bubble (LSB). It can be either a short or long bubble. The short LSB does not affect the airfoil performance at intermediate angles of attack before the stall. However, it produces an abrupt stall (leading-edge stall) due to bubble burst and exhibits hysteresis in the aerodynamic characteristics. Whereas, the long LSB exhibits adverse airfoil performance (lower lift and higher drag) at intermediate angles and exhibits thin airfoil stall [1–4]. UAV/MAVs perform several aerobatic maneuvers during their flight. They also encounter sudden gust [5]. The

S. Sudhakar (✉) · N. Karthikeyan
Experimental Aerodynamic Division, CSIR-National Aerospace Laboratories, Bangalore 560017, India
e-mail: ssudha@nal.res.in

L. Venkatakrishnan et al. (eds.), *Proceedings of 16th Asian Congress of Fluid Mechanics*, Lecture Notes in Mechanical Engineering,
https://doi.org/10.1007/978-981-15-5183-3_35

wind turbine blades with a high aspect ratio, operating in the low Re regime, experience tip stall [6]. Under these extreme conditions, it is necessary to control the flow separation for the improved aerodynamic performance of UAVs and wind turbines.

Flow separation control on the airfoils at low Reynolds number is essential for enhancing the aerodynamic performance of the MAV/UAV and wind turbine rotor blades. Flow control methods can be classified as active (suction/blowing, synthetic, dimple, and plasma actuators) or passive (roughness, riblets, vortex generators). Active control methods are very complex to integrate into the system, as they require additional power for their functioning [7]. Further, active control techniques are not cost-effective. Passive control techniques are cost-effective, easier to implement, and highly reliable for implementing in wide engineering applications [7].

Over the past two decades, several studies report the benefits of the application of a novel passive flow control device called 'tubercles,' to the leading edge of airfoils originally inspired by flippers of humpback whales (Fig. 1). The leading-edge tubercles enabled the whale flippers to maintain the lift at higher angles of incidence and aiding them to make sharp turning maneuvers for catching the prey [8, 9]. Studies [10, 11] show that the incorporation of leading-edge tubercles results in inferior (lower lift with drag penalty) or superior (higher lift with no drag penalty) performances depending upon the angle of attack either in the pre-stall and post-stall angles, respectively, as compared to an airfoil without tubercle. On the other hand, a wing with leading-edge tubercles exhibits delayed stall (40%) and increased maximum lift (6%) as compared to the baseline [12, 13] at all angles of attack. However, the effect of tubercles on airfoils and wings performance depends on a variety of parameters such as airfoil geometry (thickness and camber), wing geometry (aspect ratio, and sweep angle), and Reynolds number.

The mechanism behind the effectiveness of the tubercles has been investigated in the past [14, 15]. The streamwise vortices generated from the leading-edge tubercles energize the boundary layer, controlling the flow separation at the higher angles

Fig. 1 Humpback whale flipper

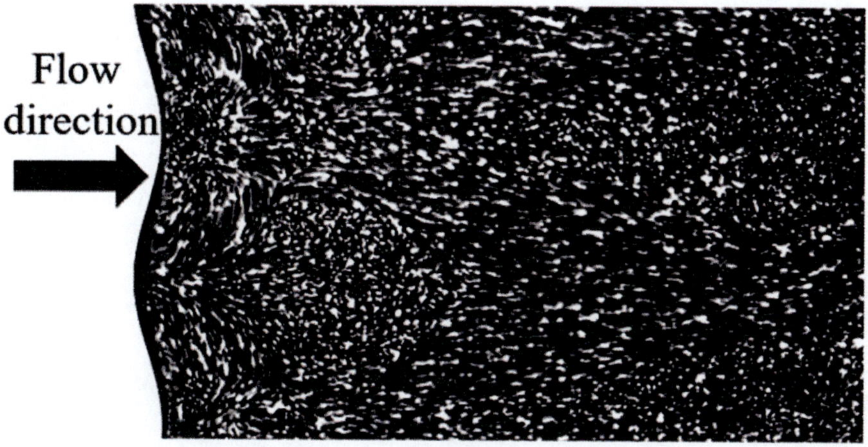

Fig. 2 Oil flow pattern on NACA-4415 airfoil with tubercles [16]

attack resulting in an improved aerodynamic performance on wings/airfoil. The flow field near the tubercles at high angles of attack is highly three-dimensional (Fig. 2) [16], which requires a technique like stereo-particle image velocimetry (SPIV) or tomographic-PIV to characterize.

In the present study, the flow field over a NACA-4415 airfoil with and without leading-edge tubercles have been investigated using SPIV measurements, and the results are presented.

2 Experimental Setup

Experiments were carried out in the low-speed wind tunnel facility experimental aerodynamic division at NAL. The tunnel is having a square test section with a cross-section area of 0.55 m × 0.55 m and a length of 1.5 m. The freestream turbulence in the test section is 0.15% for the freestream velocity of up to 20 m/s. Two airfoil models were used in the present study. The first airfoil model is the baseline case with the NACA-4415 profile and is shown in Fig. 3a. The chord (c) and span of the airfoil are 250 mm and 545 mm, respectively. The second model is the modified version of the baseline with tubercles incorporated at the leading edge. The tubercles had a wavelength of 0.25c and an amplitude of 0.025c (Fig. 3b). The model is mounted in the tunnel with the help of support rods at both ends of the model. The one end of the model is supported at the one side of the tunnel wall, and the other end is mounted with the angle of attack adjustment mechanism at another side of the tunnel wall.

SPIV measurement was carried out on both the airfoil models at a post-stall angle of attack of 18°. Figure 4 shows a schematic of the stereo-PIV measurement setup in the tunnel. SPIV measurements were carried out using two MotionPro® Y5 cameras

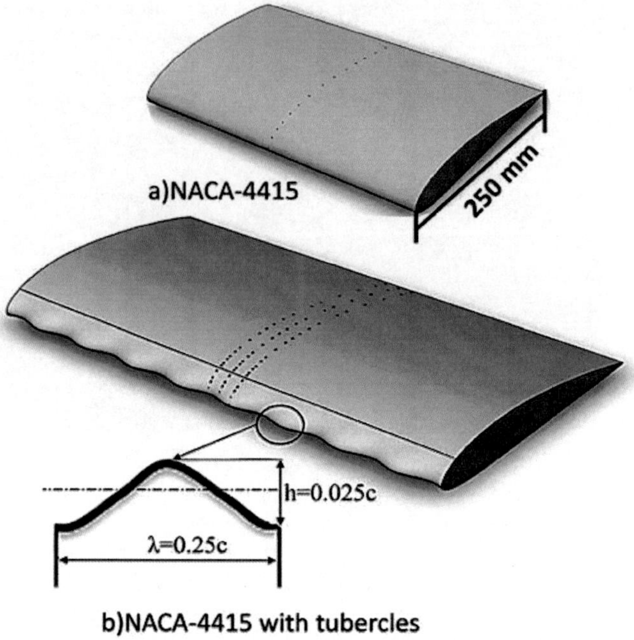

Fig. 3 Airfoil models

of 2314(H) pixels × 1728(V) pixels. The cameras were mounted on either side of the tunnel, fitted with Nikkor® 85 mm lens with Scheimpflug mounts. The cameras were positioned in such a way that they satisfy the Scheimpflug condition for off-axis imaging to obtain a stereo-view of the measurement plane. The cameras and the light sheet optics were mounted on a traverse made out of LINOS X-RAIL®, and the entire setup along with the light sheet is moved to the required measurement location. Such an arrangement eliminates the calibration at each measurement location. The flow in the spanwise plane (y–z plane) was imaged at five chordwise locations (x/c = 0.18, 0.27, 0.36, 0.52 and 0.66). The flow was seeded with particles of 10 microns' diameter and illuminated using a dual cavity 400 mJ Nd: YAG pulsed laser with help of light sheet optics. An area of 80 mm × 120 mm in the plane perpendicular to the streamwise direction was imaged. The synchronization of the laser and the camera image acquisition was achieved through IDT MotionPro® timing hub. The laser pulse separation time was fixed at 50 μs to allow the particles to move at least six pixels corresponding to the maximum velocity in the measurement region. All measurements and analyses were made using the IDT ProVision XS® software. The velocity vectors were obtained by processing the images with a 24 × 24 pixel interrogation window and 50% overlap. The spatial resolution obtained for this case was 0.75 mm (magnification factor (M) = 0.063 mm/pixel). Mean velocity vector fields were obtained from an ensemble average of 500 image pairs.

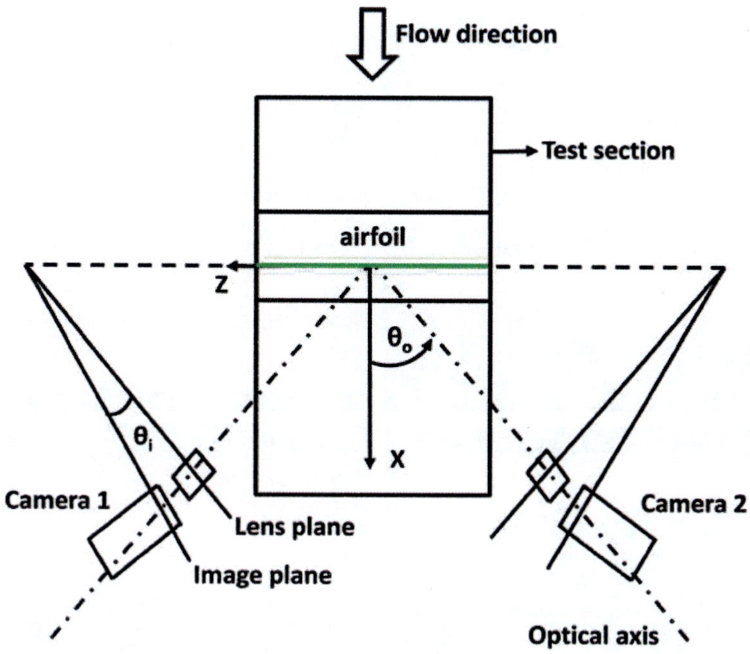

Fig. 4 Stereo-PIV setup

3 Results

A comparison of mean streamwise velocity contour (U_{mean}) for the NACA-4415 airfoil and with the leading-edge tubercles at various chord locations are shown in Fig. 5a, b. The negative streamwise velocity values in the flow field near the airfoil surface (Fig. 5a) indicate the flow separation along the entire span of the measurement region at $x/c = 0.18$. The extent of a separated region normal to the airfoil surface grows in the downstream direction toward the trailing edge ($x/c = 0.18$–0.66). At $x/c = 0.66$, the streamwise velocity contour in the entire measurement plane is negative (Fig. 5a), indicating a presence of massive re-circulation region on the airfoil surface. This results in a significant drop in the lift and an increase in the drag.

In contrast to the baseline, at $x/c = 0.18$, except the small low-velocity region behind every trough section, the attached flow is seen on the entire measurement area (Fig. 5b). The velocity magnitude is positive in the low-velocity region behind the trough section as compared to that seen in the baseline (negative). The streamwise velocity contour also shows a spanwise variation in the boundary layer thickness from the peak (thinner) to the trough (thicker) section. This spanwise variation imparts a sinusoidal pattern to the boundary layer thickness. The size of the low-velocity region increases in the downstream direction ($x/c = 0.27$). The presence of a negative streamwise velocity close to the airfoil surface indicates the onset of flow separation

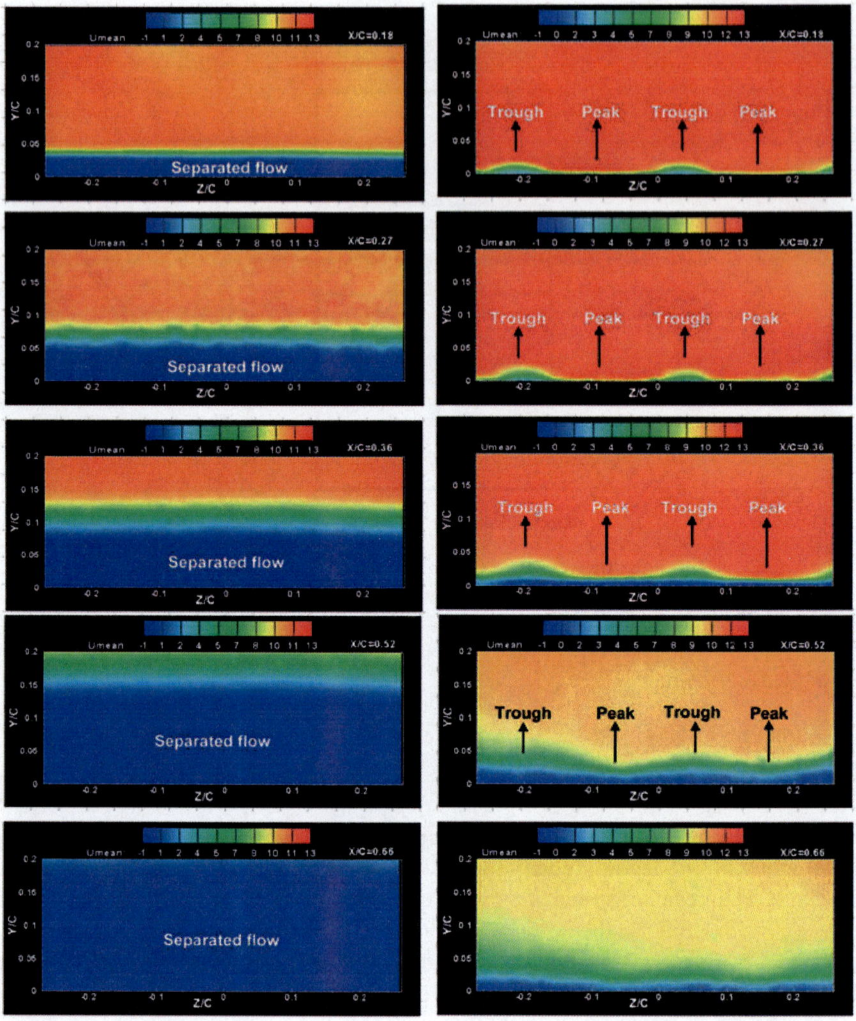

a) NACA-4415 airfoil b) NACA-4415 airfoil with tubercles

Fig. 5 Comparison of mean streamwise velocity contour at different chordwise locations

at $x/c = 0.36$. The extent of this negative velocity region increases in the downstream direction, as seen in other measurement planes beyond $x/c = 0.36$. The extent of the separated region for the modified airfoil is significantly reduced as compared to the baseline. This results in improved post-stall performance of the modified airfoil (greater lift and smaller drag), as reported in the literature [9, 10].

Figure 6 shows a comparison of vertical velocity contours (V) at various chordwise positions ($x/c = 0.18, 0.27, 0.36,$ and 0.66) for the baseline and modified airfoil,

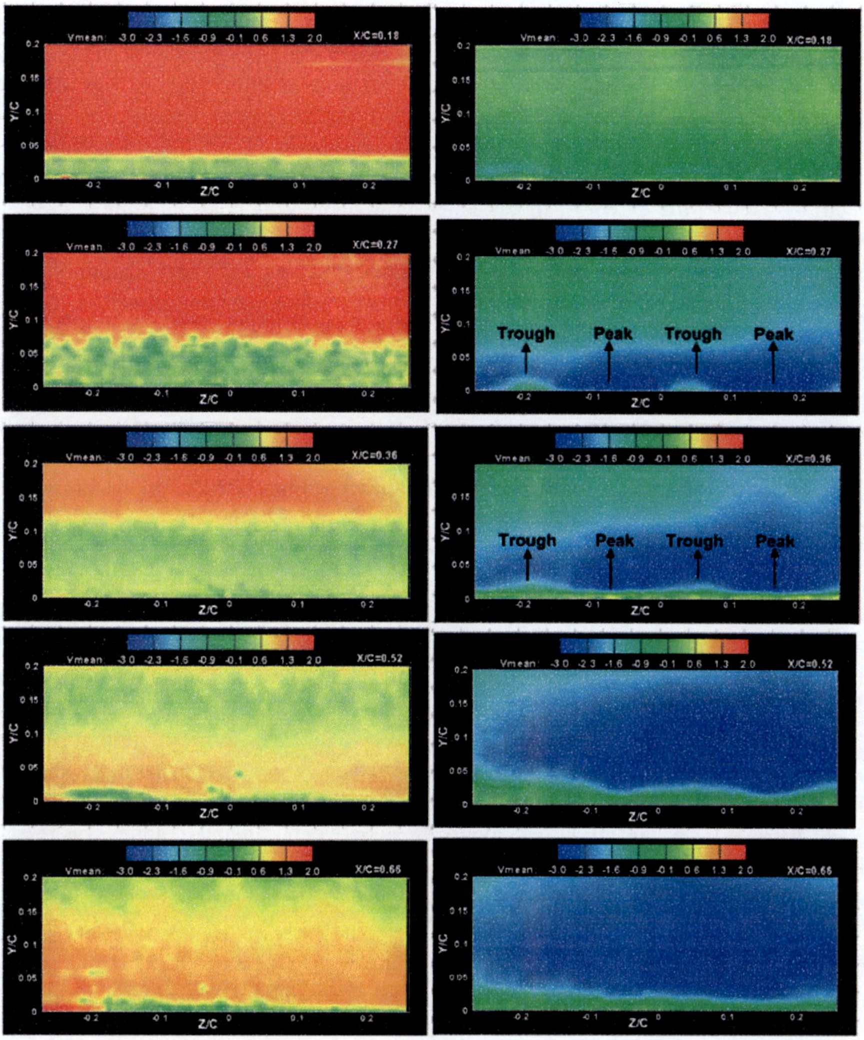

a) NACA-4415 airfoil b) NACA-4415 airfoil with tubercles

Fig. 6 Comparison of mean vertical velocity contour at different chordwise locations

respectively. The positive vertical velocity magnitude in the velocity contour indicates the flow moving away from the airfoil and reveals the separated flow on the airfoil surface (Fig. 6a) at all the positions. These regions have negative streamwise velocities in the corresponding measurement planes, as observed in Fig. 5a. Comparatively, negative values dominate the vertical velocity contour over the modified airfoil (Fig. 6b), at all measurement locations. The negative vertical velocities are an indication of the flow moving toward the airfoil surface. Similar to the streamwise velocity contour, a sinusoidal wavy pattern in the spanwise direction is also seen in the vertical velocity contour for the modified airfoil.

An examination of streamwise and vertical velocity contour plots in Figs. 5b and 6b show regions exhibiting both positive streamwise and negative vertical velocities of varying extents behind the peak and trough regions of the modified aerofoil. The extent of such regions is greater behind the peak sections as compared to the trough sections. Hence, the extent of the attached flow behind the peak sections is greater than that in the trough sections. The possible mechanism for the greater extent of the attached flow observed behind the peak region could be due to the secondary flow induced by the spanwise pressure gradient across the peak to trough sections of the tubercles [17, 18].

Using the 3C-PIV data obtained at different chordwise locations, it is attempted to reconstruct a volume of streamwise velocity. Data between the measurement locations were interpolated using a cubic b-spline technique. This allows for an overall view, providing an understanding of the influence of the tubercles on the flow field. The volumetric streamwise velocity field is shown in Fig. 7a, b for the baseline and modified airfoils, respectively. The streamwise contour shows the separated flow throughout the volume (Fig. 7a) for the baseline. Further, the height of the separated region (y-direction) increases in the streamwise direction. Whereas in the modified airfoil, the flow is attached over most of the airfoil, and the height of the separated region is significantly reduced as compared to the baseline (Fig. 7b). The greater attached flow region over the modified airfoil results in a better post-stall lift and drag characteristics as compared to the baseline.

Further, the vertical coordinates of the velocity field in the measurement planes ($x/c = 0.27$, 0.36, and 0.66) from the airfoil surface ($y = 0$) were manually shifted to match with the vertical coordinate of airfoil surface ($y = 0$) for the measurement plane at $x/c = 0.18$. The streamwise velocity in between translated planes was interpolated using a cubic b-spline technique, and a streamwise velocity volume was built (Fig. 8). For a qualitative understanding of flow parallel to the airfoil surface,

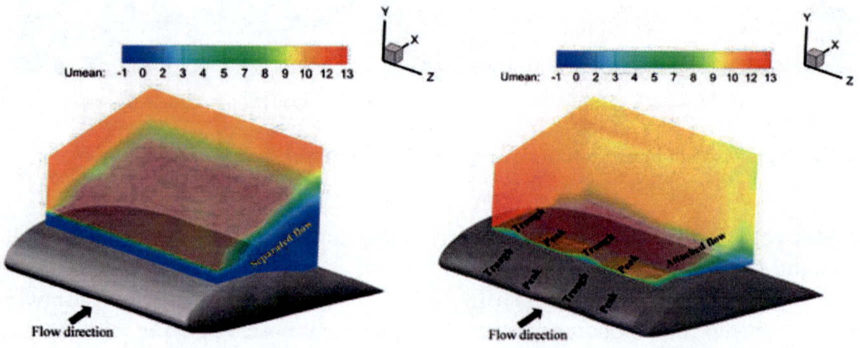

a) NACA-4415 airfoil b) NACA-4415 airfoil with tubercles

Fig. 7 Streamwise velocity contour volume obtained from interpolation of measurement planes

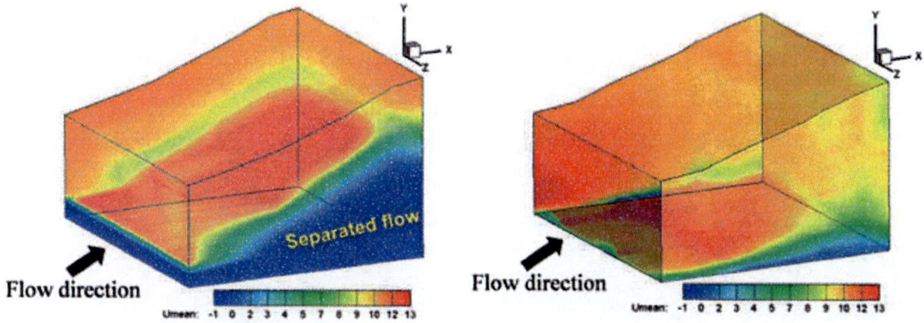

a) NACA-4415 airfoil b) NACA-4415 airfoil with tubercles

Fig. 8 Streamwise velocity contour volume obtained from interpolation of translated measurement planes

streamwise velocity contour slices (x–z plane) extracted from the volume (Fig. 8) in the vertical direction ($y = 1, 3, 4, 6, 8, 10$ mm) are shown in Figs. 9 and 10 for the baseline and modified airfoil, respectively. The streamwise velocity contour has negative values in the entire area of the slices up to the height of 6 mm, which indicates a completely separated flow on the aerofoil for the baseline case (Fig. 9). For the modified case, at $y = 1$ mm, the streamwise velocity contour slices have positive streamwise velocities up to 35% of the chord for the modified airfoil (Fig. 10). However, the region behind the peak has a higher streamwise velocity magnitude than the trough region. The alternate low and high-velocity islands in velocity contour supports the flow compartmentalization hypothesis reported in the literature [17]. Additionally, the streamwise velocity contour slice at $y = 1$ mm is fairly consistent with the surface oil flow pattern over the modified airfoil for the angle of attack of 18° (Fig. 2). As the vertical distance is increased to 3 mm, the regions behind the peaks reached the local boundary layer edge velocity. Whereas behind the trough section, though the streamwise velocity magnitude has increased, it does not reach boundary layer edge velocity. At $y = 4$ mm, the streamwise extent of a high-velocity region (boundary layer edge) has increased in the peak section. However, streamwise velocities behind the trough sections are still lower than the boundary layer edge velocity. At $y = 6$ mm, the streamwise velocity behind the trough sections has increased to boundary layer edge velocity, while in the peak section, the streamwise extent of the high-velocity region has further increased. The compartments of alternate high-velocity and low-velocity regions behind the peak and trough section, respectively, observed at $y < 6$ mm, have vanished, since the spanwise variation of the streamwise velocities has ceased to exist, near the leading edge. However, the extent of the high streamwise velocity region behind the peak to trough section differs. While behind the peak the extent is up to 50% of the chord, behind the trough, the extent is close to 30% of the chord only. Hence, the velocity contours show the greater chordwise extent of the attached flow in the peak region as compared to the trough section. Also, the separated flow regions indicated by negative or zero streamwise velocities

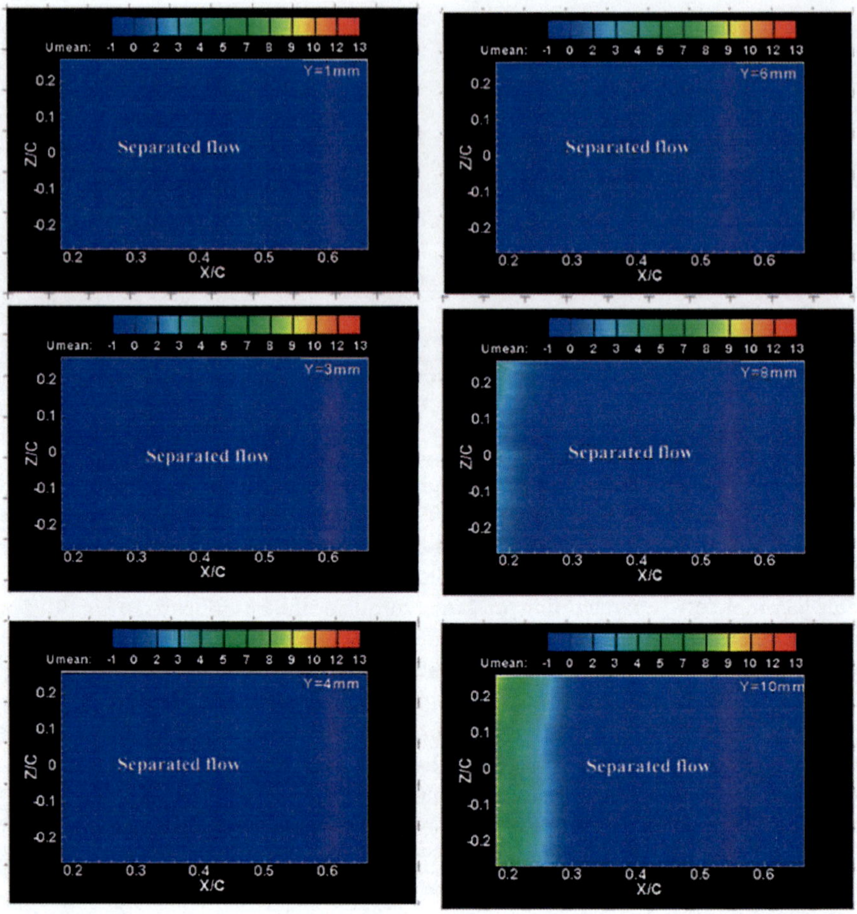

Fig. 9 Streamwise velocity contour slices extracted from the volume in the vertical direction for NACA-4415

are not seen at $y = 6$ mm. A similar pattern is seen at slices of higher y values (8 and 10 mm).

The results indicate that with the inclusion of the tubercles and exhibit variation of streamwise velocity in the direction normal to the surface of the airfoil in addition to the spanwise and chordwise directions behind the peak and trough section. The physical mechanism behind such variation should be investigated further.

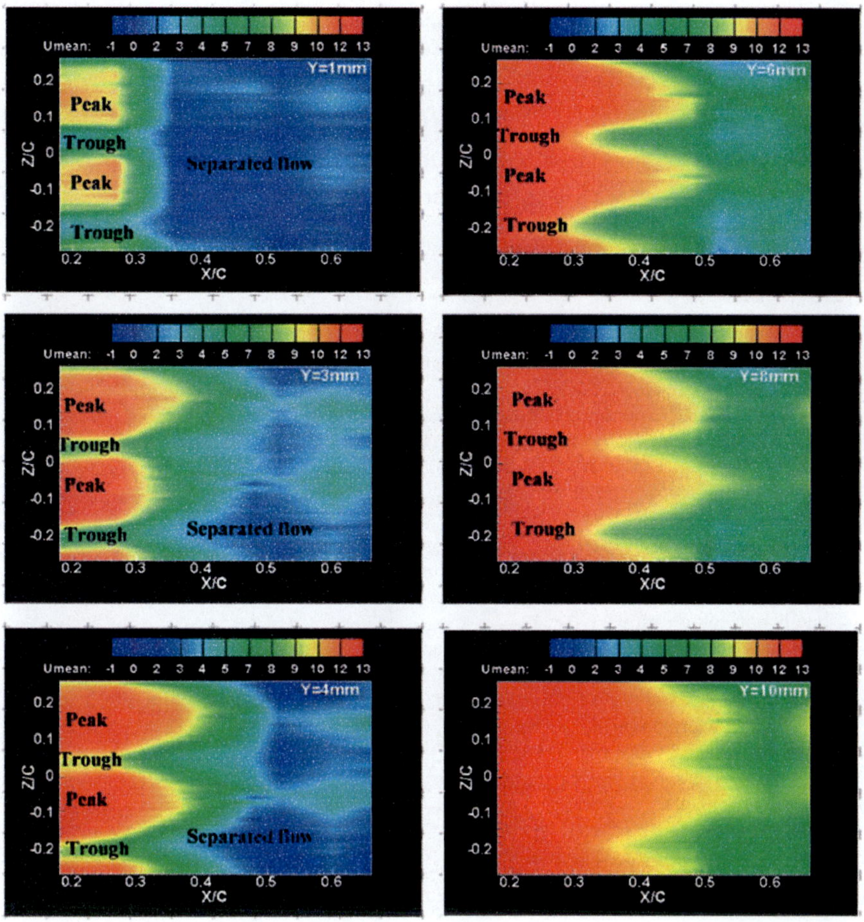

Fig. 10 Streamwise velocity contour slices extracted from the volume in the vertical direction for modified airfoil

4 Conclusions

Stereo-particle image velocimetry was carried out on a NACA-4415 airfoil and one with the leading-edge tubercles for an angle of attack of 18° and Reynolds number of 120,000. The flow is completely separated over the baseline airfoil. Leading-edge tubercles increased the chordwise extent of the attached flow and reduced the height of the separated flow over the modified airfoil as compared to the baseline airfoil. The chordwise extent of the attached flow is greater behind the peak region as compared to the trough region. Similarly, the height of the separated region is smaller behind the peak region as compared to the trough region. This could be the reason for the improved post-stall aerodynamic performance of the modified airfoil with reference to the baseline.

Acknowledgements The authors would like to acknowledge the efforts of Mr.Nitin Prakash Pawar in carrying out the experiments in the 0.55m large low-speed wind tunnel. The technical support of Mr.Ramachandra during model fabrication is gratefully acknowledged.

References

1. Tani I (1984) Low speed flows Involving bubble separations. Progress in aeronautical sciences, vol 5. Macmillian, New York
2. Liebeck R (1992) Laminar separation bubbles and airfoil design at low Reynolds numbers. In: 10th AIAA applied aerodynamics conference, Palo Alto, CA, USA
3. Yang Z, Igarashi H, Martin M, Hu H (2008) An experimental investigation on aerodynamic hysteresis of a low-Reynolds number airfoil. In: 46th AIAA aerospace sciences meeting and exhibit, Reno, Nevada, USA
4. Ellsworth R, Mueller T (1991) Airfoil boundary layer measurements at low Reynolds number in an accelerating flow from a nonzero velocity. Exp Fluids 11(6):368–374
5. Bulka E, Nahon M. Automatic control for aerobatic maneuvering of agile fixed-wing UAVs. J Intelligent Robot Syst 93(1–2):213
6. Schubel PJ, Crossley RJ (2012) Wind turbine blade design. Energies 5(9):3425–3449
7. Gad-el-Hak M (2001) Flow control: the future. J Aircraft 38(3):402–418
8. Fish FE, Battle JM (1995) Hydrodynamic design of the humpback whale flipper. J Morphol 225(1):51–60 (cit. on pp. 4, 18, 35)
9. Fish FE, Weber PW, Murray MW, Howle LE (2011) The tubercles on humpback whales' flippers: application of bio-inspired technology (cit. on pp. 6, 18)
10. Johari H, Henoch C, Custodio D, Levshin A (2007) Effects of leading-edge protuberances on airfoil performance. AIAA J 45(11):2634–2642
11. Hansen KL, Kelso RM, Dally BB. Performance of leading-edge tubercles for distinct airfoil profiles. AIAA J 49(2011019):85–100
12. Miklosovic DS, Murray MM (2007) Experimental evaluation of sinusoidal leading edges. J Aircraft 44:1404–1407
13. Mikilosovic DS, Murray MM, Howle LE, Fish FE (2004) Leading-edge tubercles delay stall on humpback whale flippers. Phys Fluids 16(5)
14. Rostamzadeh N, Hansen K, Kelso R, Dally B (2014) The formation mechanism and impact of streamwise vortices on NACA 0021 airfoil's performance with undulating leading edge modification. Phys Fluids 26(10)
15. Hansen KL, Rostamzadeh N, Kelso RM, Dally BB (2016) Evolution of the streamwise vortices generated between leading-edge tubercles. J Fluid Mech 788:730–766
16. Karthikeyan N, Sudhakar S, Suriyanarayanan P (2014) Experimental studies on the effect of leading-edge tubercles on laminar separation bubble. In: 52nd aerospace sciences meeting
17. Watts P, Fish FE (2001) The influence of passive, leading-edge tubercles on wing performance. In: Proc Twelfth Intl Symp Unmanned Untethered Submers Technol Durham New Hampshire: Auton Undersea Syst Inst
18. Skillen A et al (2013) Investigation of wing stall delay effect due to an undulating leading edge: an LES study. TSFP Digital Library Online. Begel House Inc

Assessment of Flow Field Behind the Mechanical Vortex Generators at Mach 2.0

C. Manisankar and S. B. Verma

Abstract Experiments were carried out at $M\infty = 2.05$ to understand the flow development downstream of the mechanical vortex generators. Four control devices were tested, (i) rectangular vane, *MVG1* (ii) ramp vane, *MVG2* (iii) Anderson vane, *MVG3* (iv) split-Anderson vane, *MVG4*. The total pressure values were obtained through two rakes (each consisting of 12 pitot tubes) placed along three streamwise locations of $z = 5\delta$, 10δ and 15δ (where $\delta = 12.5$ mm is the local boundary layer thickness for no-control). The velocity contour shows well-defined wake region behind that of all the control configurations. The wake appears to lift off from the plate surface with the increase in the streamwise distance. The relative velocity contour captures the momentum re-distribution occurring between the outer edge of boundary layer and the near-wall region. The oil flow visualization for control highlights the presence of vortex trails, which for *MVG1* appears to be larger than other control configurations. The boundary layer profile along the vortex trails showed fullness to the profile for *MVG1*, whereas for *MVG3-4*, there is minimal change relative to that of the no-control.

Keywords Shock waves · Boundary layer · Vortex generator

1 Introduction

The separation of the flow over an aerodynamically designed body is undesirable, but often encountered at places like intake of engine [1], nozzles while operating at low nozzle pressure ratio [2], along control surface and fins in supersonic flights, etc. Additionally, in transonic [3, 4] and supersonic flow, the shockwave boundary interaction (SWBLI) induces unsteadiness in the flow field. Over several decades, researchers [5] are working to understand the source behind the low frequency and high amplitude unsteadiness associated with the SWBLI. It is clear from the literature survey [6–15] that the flow control device in the form of active or passive is proven

C. Manisankar (✉) · S. B. Verma
Experimental Aerodynamics Division, CSIR-NAL, Bangalore, India
e-mail: manisankar@nal.res.in

L. Venkatakrishnan et al. (eds.), *Proceedings of 16th Asian Congress of Fluid Mechanics*, Lecture Notes in Mechanical Engineering,
https://doi.org/10.1007/978-981-15-5183-3_36

effective in modifying the separation characteristics induced by the SWBLI. Most of the studies pertaining to SWBLI are performed at the transonic flow regime [3, 4], not much have been reported for the purely supersonic flow cases [10–12]. Moreover, majority of the studies were focused in understanding the effectiveness brought in by these devices, with very few computational studies [15] have attempted to describe the flow field development downstream of such devices. From literatures [6–15], it is clear that the geometrical parameters of the passive devices like height, spanwise spacing and thickness have active role in deciding the effectiveness of the device. However, only few computational and experimental studies have been reported, with variation in the device height and shapes of the passive devices. The aim of the present study is to map the flow field development at different streamwise distance (5δ, 10δ and 15δ) behind that of vortex generators with different shapes like (i) rectangular vane, *MVG1* (ii) ramp vane, *MVG2* (iii) Anderson vane, *MVG3* (iv) split-Anderson vane, *MVG4*.

2 Experimental Set-up

Experiments were performed in the 0.46 m $\times$ 0.3 m blow-down tri-sonic wind tunnel at CSIR-National Aerospace Laboratories (NAL). The wind tunnel is a capable of achieving Mach number in the range 0.4–4 with maximum Reynolds number of 50 million per meter. For the current work, Mach number of 2.05 $\pm$ 0.02 (U_∞ = 523 ms^{-1}) was maintained with blow-down total pressure of 208.5 kPa $\pm$ 2% (absolute) and total temperature of 298 K $\pm$ 0.4%. The control devices were mounted on the sidewall of the tunnel wall as shown in Fig. 1. The mechanical vortex generator configurations used were in the form of (i) rectangular vane (*MVG1*), (ii) ramp vane (*MVG2*), (iii) Anderson type (*MVG3*), (iv) Split-Anderson (*MVG4*) as in Fig. 2. The vortex generator height ($h/\delta = 0.5$) was determined based on the un-disturbed boundary layer thickness. The surface oil flow visualization was carried out to capture vortex trails on the plate surface. The mean total pressure values were obtained using electronic pressure scanners with a operational range of 45psid. The in situ method was adopted for calibrating the scanner using Druck calibrator of model DPI-610. The analog signal from the ESP scanner was acquired using eight-channel signal conditioner module from National Instruments of model SCXI-1520. For the current experimental work, the data was sampled at the rate of 1000 S/s with 1000 samples per port leading to sampling interval of 1 s.

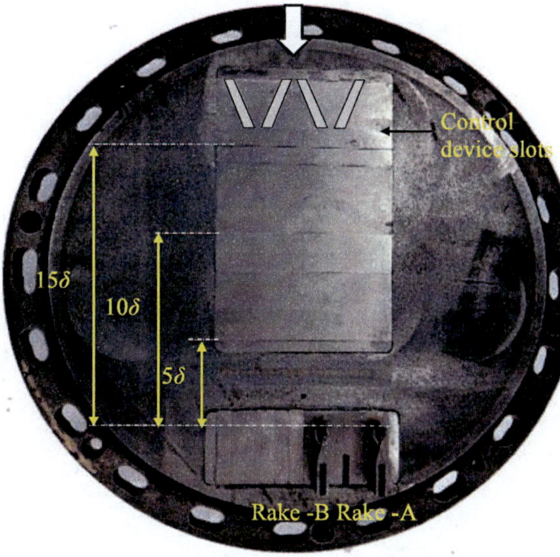

Fig. 1 Rake mounted in the circular flange

3 Results

3.1 Velocity Profile

Figure 3 shows the boundary layer profile at two different spanwise location (i) along the vortex trail, $x = -12.5$ mm (ii) along the central axis between two vane, $x = -22.5$ mm for the spanwise location of $z = 5\delta$. From the obtained pitot pressure values and the corresponding static pressure values through Rayleigh supersonic pitot formula, the values of local Mach number and subsequently velocity values inside the boundary layer are obtained across the boundary layer. It is clear that the boundary layer profile for the no-control has velocity deficit nearer to the wall. With the application of the control *MVG1*, the profile gets modified and appears to be fuller as in Fig. 3a. This may be due to the momentum exchange occurring between the outer layer of the boundary layer toward the near-wall region through vortices. At the same instance for *MVG2*, the velocity distribution is not as full as *MVG1*, which indicates that *MVG2* possess relatively lesser vortex strength than *MVG2*. On the other hand, for the control *MVG3* and *MVG4*, the profile almost overlaps with that of the no-control. Moving downstream to $z = 10\delta$, the velocity profile remains similar to that of $z = 5\delta$ as in Fig. 3b. However, along the central axis between two vanes, the profile shows reduction in the velocity deficiency at $z = 10\delta$ as in Fig. 3d, indicating the wake lift off.

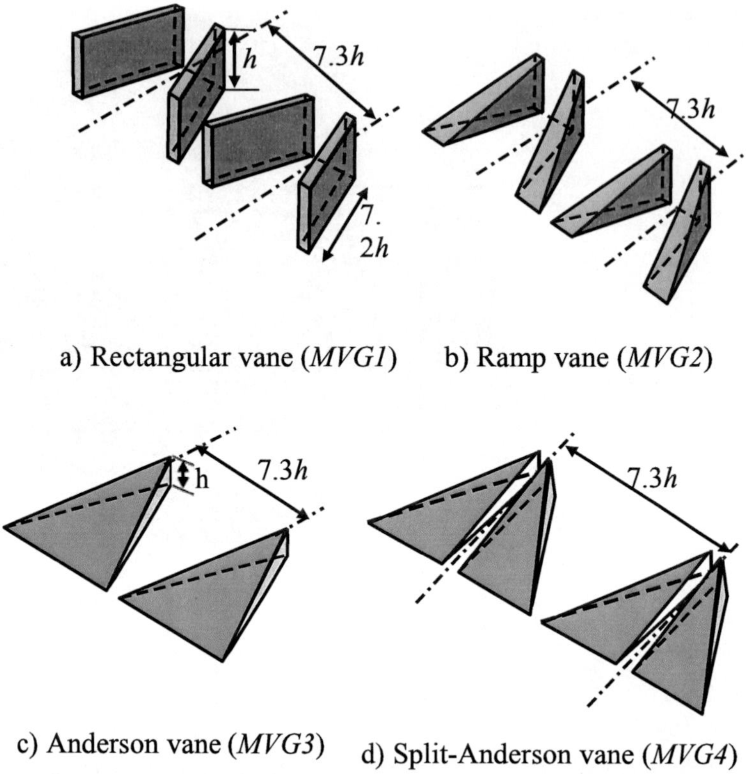

Fig. 2 Geometric details of micro-vortex generator

3.2 *Velocity Contour*

Figure 4a shows the velocity contour for the control configurations at $z = 5\delta$; at this location, the pitot rake is traversed in spanwise direction covering $x = \pm 52.5$ mm with a span of 10 mm for all the control configurations. Wherein the core of the wake is observed to be aligned with that of the central axis between two vanes. Moreover, inside the wake region are clearly visible. Out of all the configurations, *MVG1* is observed to have larger wake spread, with about only 18 mm gap between two wake structures in the spanwise direction. With the increase of the streamwise distance to $z = 10\delta$, the wake appears to be lift off from the surface. Further, at larger streamwise distance, the wake width is significantly higher, causing the merger of the two wakes. It is important to note that wake region occupies significant portion of the boundary layer resulting in low momentum along the wake region at this streamwise location. Further moving downstream at $z = 15\delta$ for the rectangular vane, *MVG1* the wake region has larger lift off added to the increased wake width. Among all the configurations the Anderson vane, *MVG3* appears to have smaller wake region.

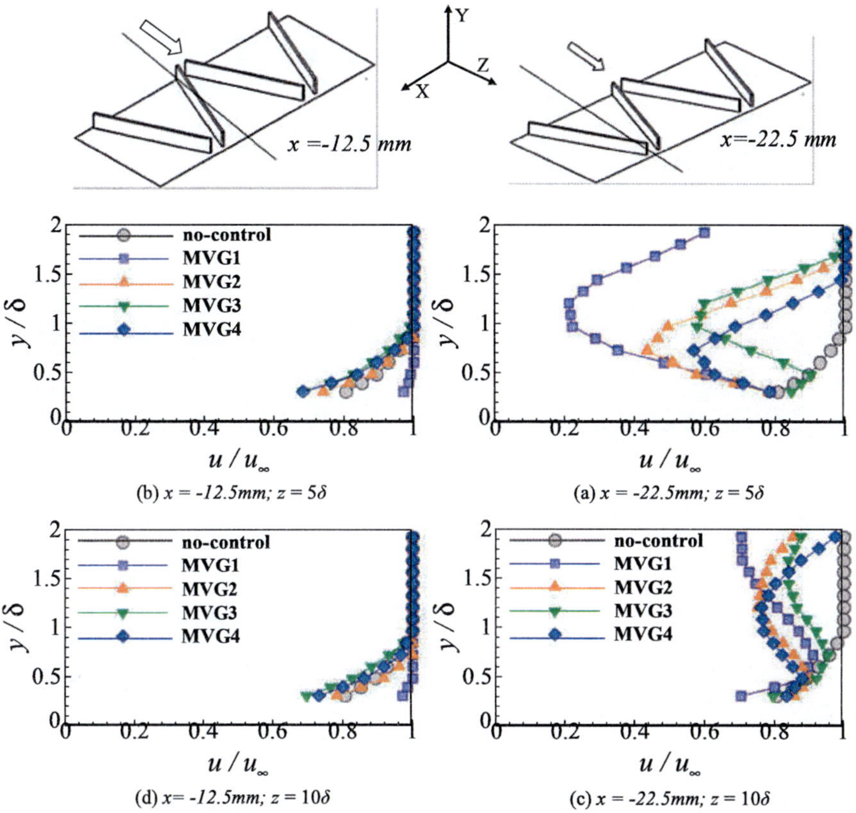

Fig. 3 Velocity profile with and without control at $x = -12.5$ mm, -22.5 mm and $z = 5\delta$, 10δ

Moreover, the effect of introducing trailing edge gap *MVG4* resulted in increase of the wake width.

As observed in *MVG1* the *MVG2*, at $z = 5\delta$, indicates wake region consisting of two structures inside the wake as in Fig. 4b. The vertical extent of the wake region is felt only until $y = 20$ mm which is almost 50% less than the *MVG1*. With the increase in the streamwise distance $z = 10\delta$ for *MVG2*, the two wakes seem to lift off from the surface with increase in the width of the wake. For the further increase in the streamwise distance to z = 15δ for *MVG2*, the wake shows more lift off and increase in the wake width.

In the case of *MVG3*, at $z = 5\delta$, as in Fig. 4c, the wake region reflects that of the ramp vane; however, the two vortices present for other control device are not visible for *MVG3*. The gap between the two wake regions is about 35 mm which is relative larger than other two control devices discussed above. For the increase in the streamwise distance to $z = 10\delta$, and $z = 15\delta$ for *MVG3*, the wake shows lift off from the surface accompanied with 31 and 28 mm gap between two wake regions.

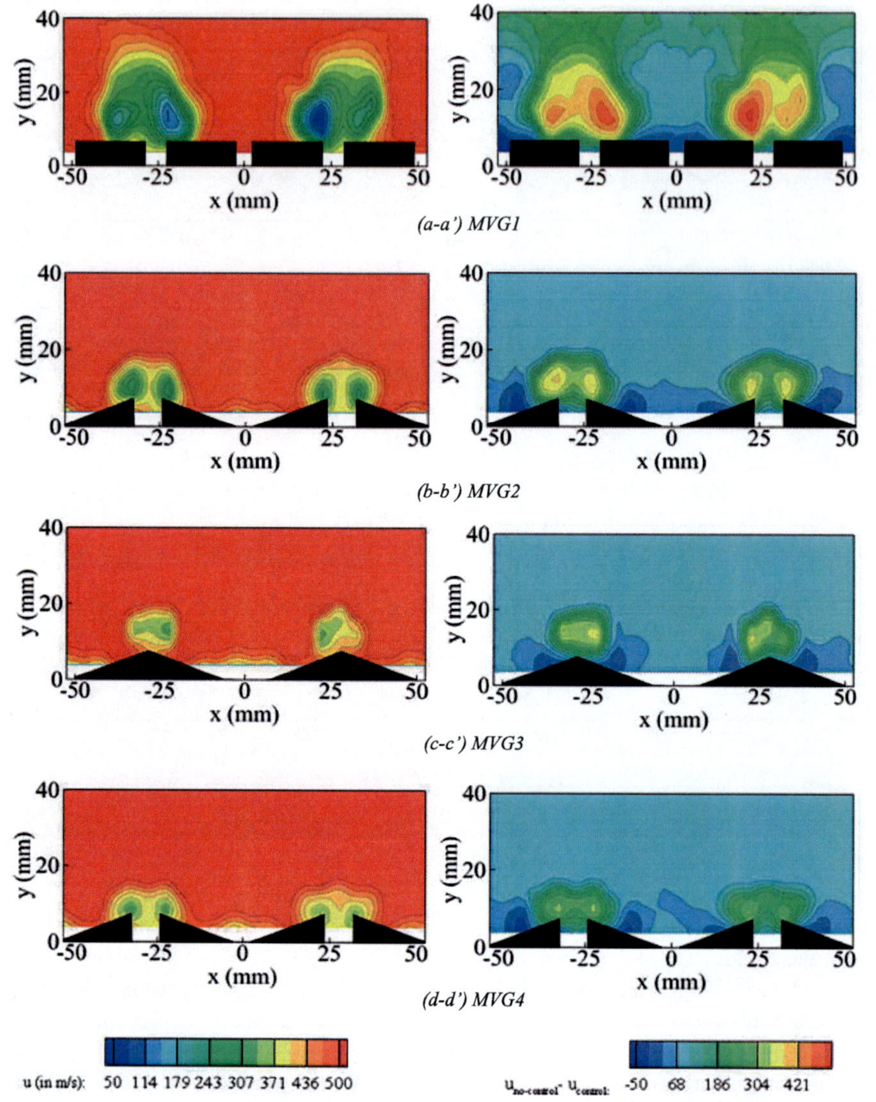

Fig. 4 Velocity contour (**a–d**); relative velocity contour (**a'–d'**); at $z = 5\delta$ for control configurations *MVG1-4*

For *MVG4*, as in Fig. 4d, the wake region shows no major change with the trailing edge gap. At all streamwise locations, the vertical extent of the wake remains like that of the *MVG2* and *MVG3*. However, inside the wake region, two distinct vortices are visible in streamwise location of $z = 5\delta$.

3.3 Relative Velocity Contour

The relative velocity values are calculated by subtracting the undisturbed velocity value with that one obtained behind the control device as discussed earlier. Figure 4a' shows the relative velocity contour plot for the *MVG1* at $z = 5\delta$; the positive values are more concentrated in the wake region, which is expected, as the wake possesses mostly low momentum fluid. There are negative values closer to the wall indicating the high momentum being brought in by the primary vortex. With the increase in the streamwise distance to $z = 10\delta$, the positive value associated with the wake is shifted away from the wall indicating wake lift off. Moving downstream in the streamwise direction at $z = 15\delta$, the positive values show shift in the direction normal to the wall, indicating wake lift off from the wall surface. The negative values also show a vertical shift strongly suggesting upward movement of the vortex.

For *MVG2*, as in Fig. 4b' at $z = 5\delta$, the positive values lie very close to wall surface, and the negative values is sparsely distributed. The reduced level of the relative velocity for the ramp vane suggests that the vortex size is small relative to the rectangular vane. At $z = 10\delta$, the spread of the positive value is expected as the velocity contour showed increase in the wake width with the increase of the streamwise distance. Further downstream at $z = 15\delta$, the positive values moved significantly upward and its impact on the boundary layer is felt only in the upper layer of the boundary layer region. In the case of *MVG3* as in Fig. 4c' at $z = 5\delta$, the positive values pertaining to the wake are very close to the wall surface. At $z = 10\delta$ and 15δ, the positive values move away from the wall surface as expected as the velocity contour showed movement of wake away from the wall surface. Almost similar observations were evident for *MVG4* as in Fig. 4d' at all streamwise locations.

Figure 5 gives isometric view of the velocity contour at three streamwise distance $z = 5\delta$, 10δ and 15δ. As discussed earlier with the increase in the streamwise distance, the wake appears to lift away from the wall surface. The vertical distance with which

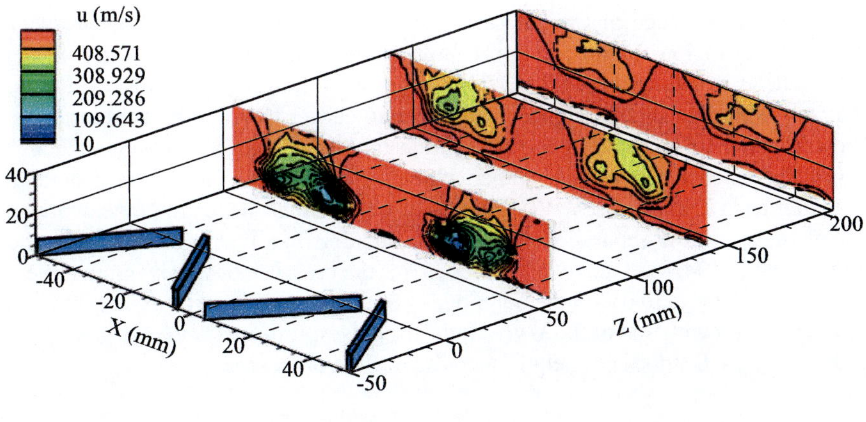

Fig. 5 Isometric view of velocity contour showing wake lift off for *MVG1*

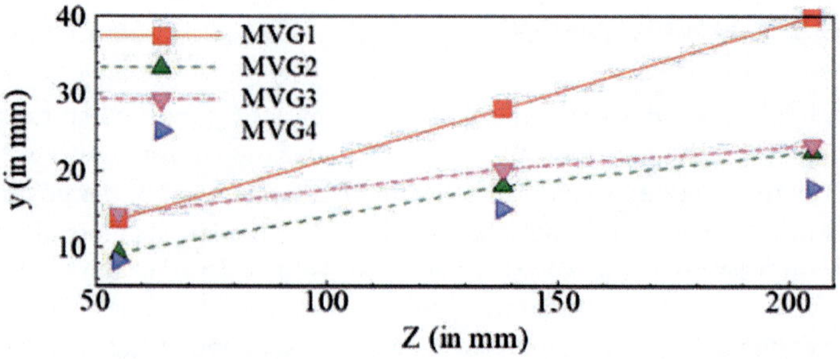

Fig. 6 Variation in the wake lift off along vertical direction from the plate surface along streamwise direction at $x = -22.5$ mm

the core of the wake has shifted is calculated for all the control configurations as in Fig. 6. It is clear that for all the control configurations, larger the streamwise distance more is the wake lift off value. Among all controls, for the increase in streamwise distance, the *MVG1* has larger vertical shift in the core of the wake. On the other hand, *MVG4* shows least vertical push in the wake away from the wall surface.

3.4 Surface Oil Flow Visualization for Control Device

Figure 7a shows the surface oil flow visualization and their corresponding flow pattern (Fig. 7a') for *MVG1*, at the outer edge of the rectangular vanes lines representing the vortex impingement (marked as *AL*) and vortex lift off (marked as *SL*) are clearly visible in Fig. 7a'. The trails of the vortex are also clearly visible, which increases in size while moving downstream. In the gap between the two vortex generators, the direction of oil flow as seen through trails suggests presence of counter rotating vortices.

At the trailing edge of the device, the oil flow is stagnant indicating presence of separation region. Figure 7b is for *MVG2*; the width of the vortex trail in the ramp vane is less than that of the *MVG1*. The separation region is observed to be relatively less than the rectangular vane as seen earlier through the contour plots. The counter rotating vortex generated in the middle of the ramp vanes maintains significant distance between each other, which shows the size of the vortex generated is smaller as in Fig. 7b'. For *MVG3*, there is a well-defined vortex impingement and lift off lines and the separation is relatively smaller than that of both *MVG1* and *MVG2*. For *MVG4*, as in the oil pattern confirms the presence of vortex impingement and vortex lift off activity on the outer edges of the device. The vortex streak width almost matches with that of the *MVG3* indicating the splitting of *MVG3* to give space in the trailing edge does not help in increase of the vortex size.

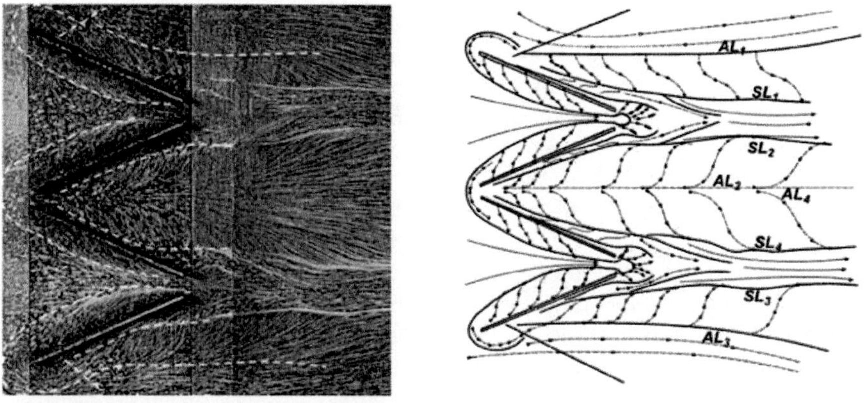

(a-a') MVG1

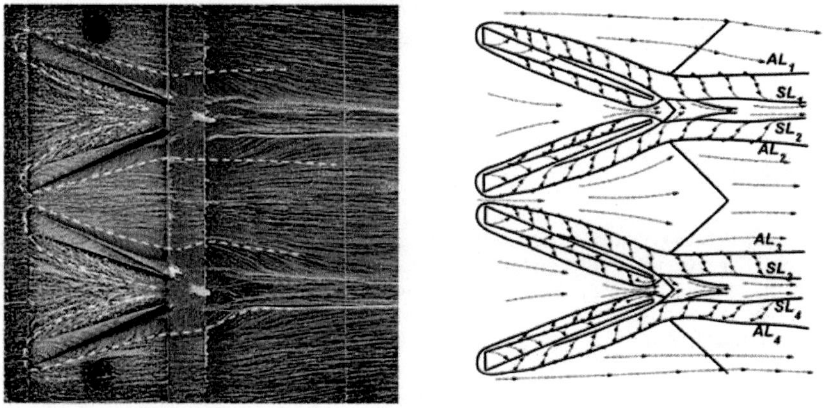

(b-b') MVG2

Fig. 7 Oil flow visualization image (**a–b**) and corresponding flow pattern (**a**'–**b**') for *MVG1* and *MVG2*

4 Conclusions

Experiments were carried out to understand the flow field development behind the mechanical vortex generators. Four control configurations were used for this study with height maintained at $h = 0.5\delta$ for all the control configurations. The velocity contour showed presence of large wake structures behind control devices, with *MVG1* having larger wake structures relative other devices. The relative velocity contour captured the momentum distribution from the outer region of the boundary layer toward the near-wall region. With the increase in the streamwise distance there is an increase in the wake lifts off, which allows uniform distribution of the near-wall momentum. Out of all the control configuration tested, *MVG1* exhibits larger near-wall momentum exchange all along the spanwise direction. The surface oil flow visualization revealed the vortex impingement line and the vortex lift off line along with the wake region. The control *MVG1* showed larger vortex trail width which almost covered the spanwise gap between the *MVG1* devices. The wake region as visualized through the oil flow showed larger wake region for the *MVG1*. For *MVG3* and *MVG4*, the vortex trace was observed to be very less in size. The velocity profile along the vortex trail showed fullness in the profile with control *MVG1* with almost overlap in profile for the *MVG3-4*. Among all the control configurations, tested *MVG1* is observed to be optimal device in terms of generation of the vortices and momentum exchange nearer to the wall region which gives fullness to the boundary layer resulting in delay in the separation.

Acknowledgements The authors wish to thank the Aeronautical Research and Development Board of India for supporting this project. The technical support of Narayana during the model design and fabrication as well as of Janardhan and staff of the 0.3 m wind tunnel facility at National Aerospace Laboratories (NAL) during the test campaigns is gratefully acknowledged.

References

1. Babinsky, Ogawa H, Hideaki (2008) SBLI control for wings and inlets. Shock Waves 2:89–96
2. Frey M, Hagemann G (1999) Flow separation and side-loads in rocket nozzles. AIAA 99:2815
3. Rybalko M, Babinsky H, Loth E (2010) VGs for a normal SBLI with a downstream diffuser. AIAA Paper 2010-4464
4. Bur R, Corbel B, Delery J (1998) Study of passive control in a transonic shock wave/boundary-layer interaction. AIAA J 36(3):394–400
5. Green JE (1970) Interactions between shock waves and turbulent boundary layers. Progr Aerosp Sci 11:235–340
6. Lin JC, Howard FG, Bushnell DM (1990) Investigation of several passive and active methods for turbulent flow separation control. AIAA paper 90-1598
7. Raghunathan S (1988) Passive control of shock boundary layer interaction. Progr Aerosp Sci 25(3):271–296
8. Taylor HD (1947) The elimination of diffuser separation by vortex generators. United Aircraft Corporation Report No. R-4012-3, June 1947
9. McCormick D (1993) Shock/boundary-layer interaction control with vortex generators and passive cavity. AIAA J 31(1):91–96
10. Verma SB, Manisankar C (2012) Shockwave/boundary-layer interaction control on a compression ramp using steady micro jets. AIAA J 50:2753–2764
11. Verma SB, Chidambaranathan M (2011) Transition control of Mach to regular reflection induced interaction using an array of micro ramp vane-type vortex generators. Phys Fluids 27:107102 1-23
12. Verma SB, Manisankar C (2017) Assessment of various low-profile mechanical vortex generators in controlling a shock-induced separation. AIAA J 50:2228–2240
13. Delery J (1985) Shock wave/turbulent boundary layer interaction and its control. Prog Aerosp Sci 22:209–280

14. Dussauge JP, Dupont P, Debiève JF (2006) Unsteadiness in shock wave boundary layer interactions with separation. Aerosp Sci Technol 10:85–91
15. Lee S (2009) Large eddy simulation of supersonic boundary layer interaction control using micro-vortex generators. Ph.D. dissertation, University of Illinois at Urbana-Champaign

Effect of Forward Splitter Plate Leading Edge Shape on the Cylinder Flow

Kiran Chutkey, P. Suriyanarayanan, and L. Venkatakrishnan

Abstract Experiments have been conducted on a cylinder with a forward splitter plate at Reynolds number of 5.33×10^4. Ten different splitter plate configurations are considered wherein the leading edge of the splitter plate is modified. The flow development on the splitter plate and on the cylinder has been examined using PIV. Static pressure on the cylinder surface and hot wire measurements in the cylinder wake have been carried out. A maximum of 57% increase in base pressure is observed with the use of splitter plate leading edge modification as compared to the simple cylinder case.

Keywords Cylinder flow · Shedding frequency · Splitter plate

1 Introduction

The flow past a circular cylinder is one of the oldest problems in fluid mechanics. It still remains a relevant problem even today because of its wide practical applicability as well as for its theoretical interest. From an engineering point of view, the interest has been toward reducing the drag and unsteady forces on the cylinder using flow control devices. The most popular among the control devices are the splitter plates

K. Chutkey (✉) · P. Suriyanarayanan · L. Venkatakrishnan
Experimental Aerodynamics Division, CSIR National Aerospace Laboratories, Bangalore 560017, India
e-mail: kiranc@nal.res.in

L. Venkatakrishnan et al. (eds.), *Proceedings of 16th Asian Congress of Fluid Mechanics*, Lecture Notes in Mechanical Engineering,
https://doi.org/10.1007/978-981-15-5183-3_37

placed at the rear end of the cylinder which are referred in the literature [1, 2] as the backward splitter plate (BSP). The BSP reduces the communication between the top and bottom of the shear layer and has been found to be effective in reducing the unsteadiness of the wake [2] for plate length more than five times the cylinder diameter. On the contrary for a splitter plate placed at the front end of the cylinder, referred to as the forward splitter plate (FSP), the flow field at the front end of the cylinder is modified. Because of the presence of FSP, a separation bubble is formed at the splitter plate cylinder junction. This has been referred to as nose streamlining by Viswanath et al. [3]. The optimum plate length was reported to be equal to the cylinder diameter. They have also reported that the shedding frequency increases because of the introduction of a FSP. The authors in their previous work [4] have conducted experiments on this configuration and have discussed the differences in the wake of the cylinder between a simple cylinder and an FSP mounted cylinder. The results have shown an 13% reduced in drag in the presence of FSP. The present work is in continuation of this. In the present work, we have tried investigating the effect of leading edge shapes of the FSP on cylinder flow.

2 Experimental Setup

Experiments were carried out in the 0.55×0.55 m low speed wind tunnel at NAL. The circular cylinder model along with the splitter plate was mounted horizontally in a removable test section with a glass window on one side which provided viewing access. Measurements were made at a free-stream velocity (U_∞) of 20 m/s; the corresponding Reynolds number Re_D (based on the cylinder diameter) was 5.33×10^4. Figure 1 shows the schematic of cylinder model inside the wind tunnel.

The cylinder model was made out of glass fiber reinforced plastic and was polished to a smooth surface finish. The model diameter and span were equal to 40 mm and 545 mm, respectively. A slot of 2 mm was cut across the span of cylinder to insert a splitter plate. The end plates at the span-wise sides of the cylinder model had extended shafts to accommodate the splitter plates. The model has 35 pressure ports in the mid span region. The pressure ports were of 1 mm diameter and were offset

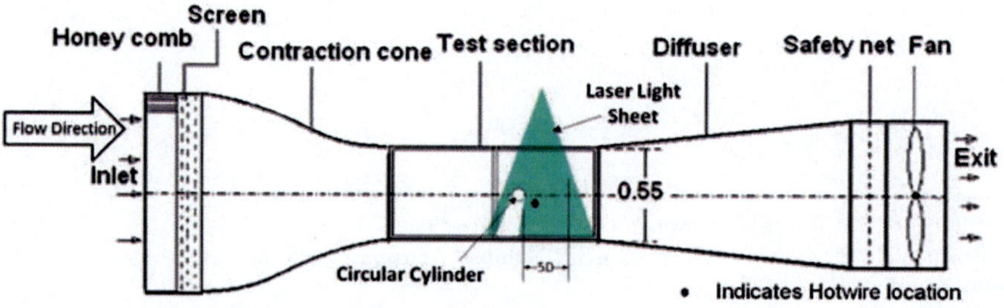

Fig. 1 Schematic of cylinder model in tunnel

from each other by an angle of 10°. The surface pressure measurements were carried out using two 32 port ESP scanners.

2.1 Hot Wire Setup

The shedding frequency of the circular cylinder is measured using the single component Dantec constant temperature hot wire system. The hot wire probe was placed at a diameter distance vertically and two diameters distance horizontally from the center of cylinder. The analog signal from the hot wire was acquired at a sampling rate of 1 kHz.

2.2 PIV Setup

Particle image velocimetry (PIV) technique has been used to obtain the velocity field. A double pulsed Nd: YAG laser with an energy pulse of 400 mJ is used which operates at a frequency of 15 Hz. The beam from the laser unit was steered using an optical arrangement to obtain a laser sheet of thickness approximately one millimeter.

A Nikon 105 mm lens is used as the imaging optics for the camera. The pixel density of the images is 2330 × 1728. DAVIS software is used to acquire and process 1325 pair of images. The interrogation size is 24 × 24 pixels with an overlap of 50%. The time between the pulses is 6.2 microseconds and is chosen based on a minimum of two-pixel displacement of a particle.

2.3 Model Configurations

The models considered in the present study are presented in Table 1. These models represent inline tandem bluff body configurations. The leading edge of FSP for configuration 1–3 is modified with a circular shape. The splitter plate for these configurations acts as a BSP for the front end small cylinder. The diameter of this leading edge circle is chosen such that it is one fifth of the splitter plate length, which corresponds to an optimum BSP configuration.

For configurations 5–7, a perpendicular plate is placed ahead of the cylinder. For configurations 8–10, the plate is placed at an angle of 45° to the streamwise direction. For these cases, three different heights of leading edge plate are chosen. The dimensions for all these cases are listed in Table 1. For all the cases the main cylinder will be either partly or completely in the wake of the leading edge shapes.

Table 1 Description of the model configurations

Configuration	Leading edge (LE) modification	LE modified height	Splitter plate width
Round end LE			
Configuration 1	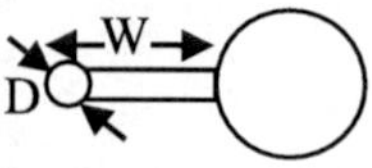	LE diameter (D) 3.8 mm	20 mm
Configuration 2		LE diameter (D) 6.6 mm	34 mm
Configuration 3		LE diameter (D) 2 mm	20 mm
Configuration 4		LE diameter (D)infinity, blunt edge	40 mm
Perpendicular plate at LE			
Configuration 5	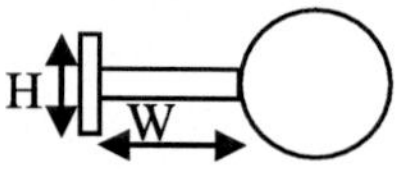	LE height (H) 8 mm	40 mm
Configuration 6		LE height (H) 16 mm	40 mm
Configuration 7		LE height (H) 32 mm	40 mm
Angle plate at LE			
Configuration 8	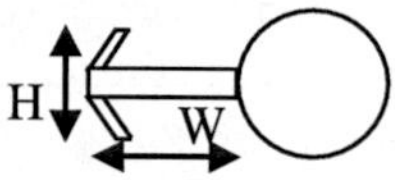	LE height (H) 7 mm	40 mm
Configuration 9		LE height (H) 10.8 mm	40 mm
Configuration 10		LE height (H) 22.8 mm	40 mm

3 Results

The pressure distribution around the cylinder for different configurations is presented in Fig. 2. Because of the splitter plate leading edge shape, the pressure on the cylinder close to 0° azimuthal angle is reduced as compared to a simple cylinder configuration. The base pressure is increased for all the cases as compared to a simple cylinder case. The change in base pressure is listed in Table 2 for all cases. For configuration 3 case, the base pressure has reduced. For this particular configuration, 0.25 D length splitter plate length is used which is off from an optimum configuration of 1D splitter plate length.

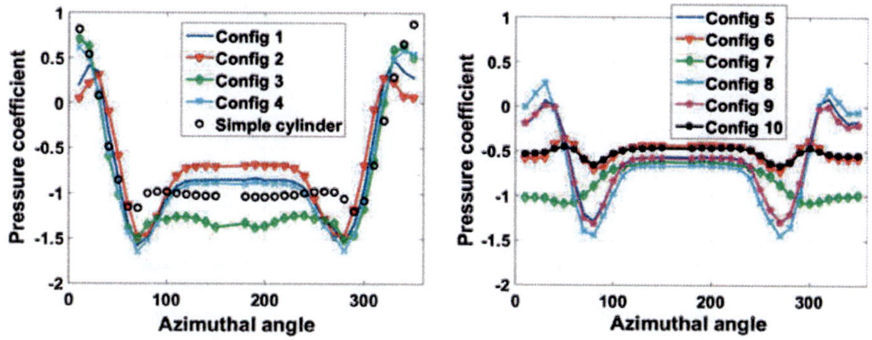

Fig. 2 Pressure distribution around cylinder for different configurations

Table 2 Pressure coefficient and wake shedding Strouhal number for different configurations

	Cpb	% change in Cpb	Strouhal number (St)	% change in St	Drag coefficient (Cd)	% change in Cd
Simple cylinder	−1.03		0.20		1.076	
Configuration 1	−0.85	−17.48	0.22	12.30	0.847	−21.262
Configuration 2	−0.70	−32.04	0.26	30.30	0.809	−24.777
Configuration 3	−1.33	29.13	0.21	4.60	0.999	−7.150
Configuration 4	−0.90	−12.62	0.24	21.75	0.910	−15.382
Configuration 5	−0.58	−43.69	0.23	15.35	0.847	−21.268
Configuration 6	−0.44	−57.28	0.26	32.05	1.132	5.213
Configuration 7	−0.63	−38.83	0.09	−55.60	1.788	66.135
Configuration 8	−0.67	−34.95	0.26	29.90	0.760	−29.320
Configuration 9	−0.59	−42.72	0.26	32.25	0.818	−23.998
Configuration 10	−0.47	−54.37	0.23	12.85	1.184	10.065

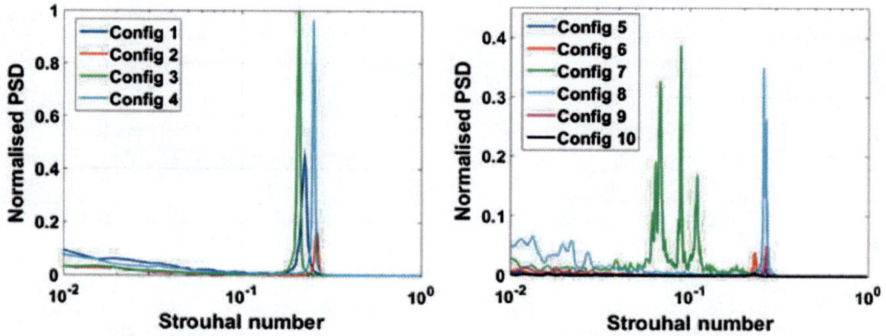

Fig. 3 Power spectra measured using hot wire placed in wake of cylinder for different configurations

The leading edge disturbances considerably alter the cylinder base wake. The shedding frequency of the cylinder wake is shown in Fig. 3 for all the cases. For all the configurations except configuration 7, the shedding frequency is increased. For the configuration 7 case, the LE perpendicular plate height ahead of cylinder is almost close to cylinder diameter. The main cylinder is completely in the wake of this leading edge perpendicular plate. Hence, a difference is noticed in the shedding frequency for this particular case. The Strouhal number for all the cases is listed in Table 2. Also, the drag coefficient, calculated from the wake defect profile at x/D of 2.0, for all the cases, is listed in Table 2.

The streamwise direction velocity contours obtained from PIV measurements for all the configurations are presented in Fig. 4. From the contours, it can be noticed that a separation bubble is present on the splitter plate, which represents the wake of the splitter plate leading edge shape. For different cases, the shear layer coming from the leading edge impinges at different locations on the cylinder surface. This alters the initial flow development on the main cylinder, and hence, considerable changes are seen in the near wake development of the cylinder flow.

4 Conclusions

The changes in the flow field for different forward splitter plate configurations have been brought out through pressure, hot wire and PIV measurements. The front end splitter plate configurations manipulate the initial conditions for the cylinder wake flow. The base pressure increases for all the cases except configuration 3. A maximum increase of 57% in base pressure is noticed for a configuration 6 case as compared to simple cylinder case.

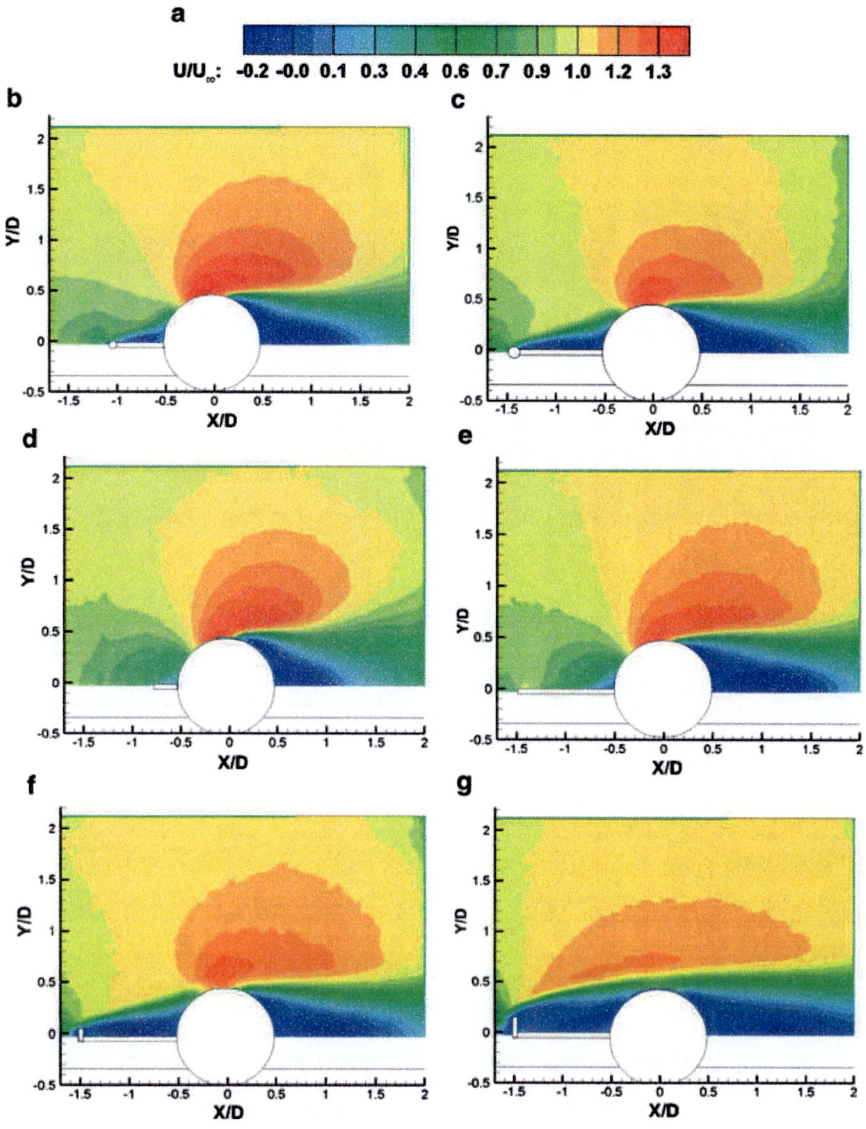

Fig. 4 **a** Contour legend. **b** Configuration 1, Streamwise velocity contours. **c** Configuration 2, Streamwise velocity contours. **d** Configuration 3, Streamwise velocity contours. **e** Configuration 4, Streamwise velocity contours. **f** Configuration 5, Streamwise velocity contours. **g** Configuration 6, Streamwise velocity contours. **h** Configuration 7, Streamwise velocity contours. **i** Configuration 8, Streamwise velocity contours. **j** Configuration 9, Streamwise velocity contours. **k** Configuration 10, Streamwise velocity contours

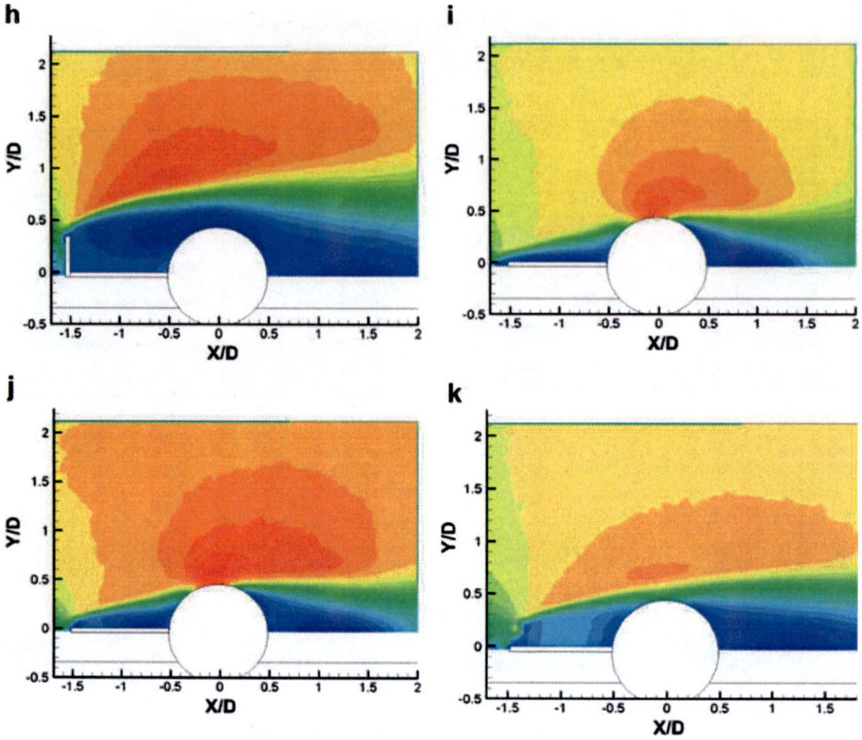

Fig. 4 (continued)

References

1. Roshko A (1954) On the drag and shedding frequency of two-dimensional bluff bodies. NACA TN 3169
2. Zradkovich MM (1981) Review and classification of various aerodynamic and hydrodynamic means of suppressing vortex shedding. J Wind Eng Ind Aerodyn 7:145
3. Viswanath PR, Prabhu A, Srinivas T (1991) Drag reduction of circular cylinder with a forward splitter plate. In: Zhuang FG (ed) Recent advances in experimental fluid mechanics, Proceedings of the 1st international conference on experimental fluid mechanics, Chengdu, China, p 293
4. Chutkey K, Suriyanarayanan P, Venkatakrishnan L (2018) Near wake field of circular cylinder with a forward splitter plate. J Wind Eng Ind Aerodyn 173:28–38

Turbulent/Non-turbulent Interface of a Transient Diabatic Plume

Samrat Rao, G. R. Vybhav, P. Prasanth, S. M. Deshpande, and R. Narasimha

Abstract In order to define the edge of a plume subject to off-source heating, referred to as a transient diabatic plume (TDP), it is necessary to define a boundary of such a plume. The vorticity magnitude ($|\boldsymbol{\omega}|$), appropriately scaled by suitable local length and velocity scales, has been used to determine the boundary turbulent/non-turbulent interface (TNTI) of the TDP. This enables us to use this value of ($|\boldsymbol{\omega}_t|$), to compute the true turbulent mass flux (MF) which is a measure of fluid entrained into the TDP.

Keywords Transient diabatic plume · Turbulent/non-turbulent interface · Vorticity magnitude · Cumulus cloud

1 Introduction

The clouds are an extremely important component in determining the weather and climate and play a major role in climate change (reports of Intergovernmental Panel on Climate Change; [3]). Fundamental observational data on clouds are not easily available, and it is unlikely that such measurements will yield a meaningful understanding of the fluid dynamics of an individual cloud. In the absence of such data, in early work on the subject, the self-similar hypothesis of [4] has been adopted by [9] to model the behaviour of clouds. However, this model could not capture the fact that through most of the height of a cumulus cloud tower, the radius was relatively constant [8]. Thereafter, it has been shown [1, 2] that a plume subjected to off-source diabatic heating has more resemblance to the dynamic behaviour of the clouds.

S. Rao (✉)
Department of Mechanical Engineering, IIT Jammu, Jammu, India
e-mail: samrat.rao@gmail.com

G. R. Vybhav · S. M. Deshpande · R. Narasimha
Engineering Mechanics Unit, JNCASR, Bengaluru, India

P. Prasanth
Department of Physics, Michigan Technological University, Michigan, USA

L. Venkatakrishnan et al. (eds.), *Proceedings of 16th Asian Congress of Fluid Mechanics*, Lecture Notes in Mechanical Engineering,
https://doi.org/10.1007/978-981-15-5183-3_38

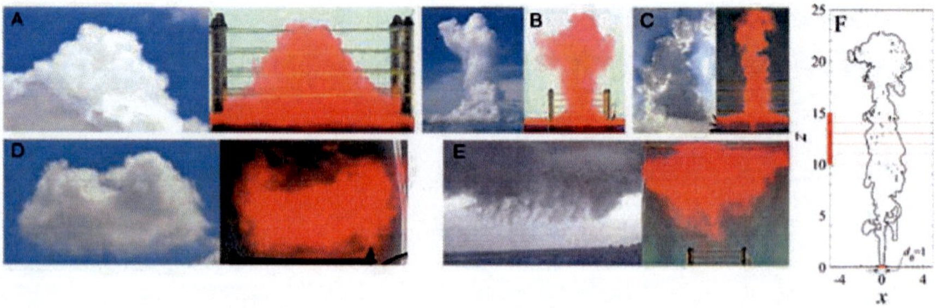

Fig. 1 **a** In each of A, B, C, D, E, the left panel is a natural cloud, and the right panel is the laboratory 'cloud' (reproduced from [5]). **b** F shows schematic of the flow. The dotted red colour lines represent the extent of the off-source heat addition zone. The outline contour is drawn using vorticity magnitude, $|\omega| = 0.5$ as the criteria

Clouds are represented in atmospheric general circulation models (models that simulate global atmospheric circulation of the Earth) by using various cloud parameterization schemes (such as Arakawa–Schubert scheme and Zhang Mac Farlane schemes) which are not always satisfactory rational approximations of actual physical phenomena of clouds. This leads to significant biases in climate predictions.

A cloud may be considered as a transient diabatic plume (TDP). Recent laboratory experiments [5] have successfully reproduced a variety of shapes closely resembling naturally occurring clouds (Fig. 1a). The model of a cloud used by them is a TDP. Hence, it was concluded that a TDP is a promising model for a cumulus cloud. The schematic of a TDP is shown in Fig. 1b.

Though extremely useful for demonstrating the usefulness of a TDP as a model of a cumulus cloud, a shortcoming of laboratory experiments is that detailed 3D data of the various flow parameters like velocity, temperature, scalar concentration and in particular the vorticity vector cannot be easily obtained. This precludes a detailed understanding of the dynamics of the flow. This shortcoming can be addressed by doing direct numerical simulations (DNS) of a TDP. At JNCASR, efforts towards DNS of a TDP are an ongoing process. In this paper, we discuss one such DNS computation and show how choosing an appropriate local length and velocity scale gives more insight into characterising a TDP. In Sect. 2, we briefly describe the numerical code used. In Sect. 3, we present visuals of the DNS data showing the impact of the scaling proposed. Finally, in Sect. 4, we present the concluding remarks.

2 Governing Equations and Computational Details

A code, Megha 3, has been used for DNS of a TDP. Megha 3 solves the 3D-Boussinesq approximations of the continuity, momentum and energy equations. The equations solved are ([1]):

$$\text{Continuity: } \nabla \cdot \boldsymbol{u} = 0 \tag{1a}$$

$$\text{Momentum: } \frac{\partial \boldsymbol{u}}{\partial t} + (\boldsymbol{u} \cdot \nabla)u = -\frac{1}{\rho_0}\nabla P + \upsilon \nabla^2 \boldsymbol{u} + \alpha(\Delta T)\boldsymbol{g} \tag{1b}$$

$$\text{Energy: } \frac{\partial(\Delta T)}{\partial t} + (\boldsymbol{u} \cdot \nabla)\Delta T = \kappa \nabla^2(\Delta T) + \frac{\boldsymbol{J}(t)}{\rho_0 C_P} H(x,t) \tag{1c}$$

where $\boldsymbol{u}$ is the velocity vector, $\Delta T_0 = T(\boldsymbol{x}, t) - T_0$ is the temperature differential with respect to ambient at T_0, P is the pressure, ρ_0 is the ambient density, v is the kinematic viscosity, κ is the thermal diffusivity, α the coefficient of the thermal expansion, $\boldsymbol{g}$ the acceleration due to gravity ($\boldsymbol{g} = -g\hat{z}$, $\hat{z}$ is the unit vector along the vertical axis), $\boldsymbol{J}(t)$ is the heat added per unit volume per unit time, and H $(\boldsymbol{x},t)$ is the distribution function which determines the special region within which heat is released at time t. The density of plume is $\rho(x,t) = \rho_0(1 - \alpha(\Delta T))$.

Figure 1b shows schematic of TDP. A hot patch of diameter d_0 at $z = 0$ is the buoyancy source. The scales used for non-dimensionalising the above equations are: d_0 for length, $(\Delta T)_0$ for temperature, $W_0 = \sqrt{g\alpha d_0 \Delta T_0}$ for velocity. This gives a non-dimensional time t using d_0/W_0 as the scale, called a flow unit (*FU*). Henceforth, all quantities discussed are non-dimensional.

A finite-volume staggered-grid formulation is used in a domain of size 100 × 100 × 40 in the x, y and z directions. The grids used are Cartesian with highest density in the central part of the domain which is where the TDP is formed. The total number of grid points is ≈1.6 billion. The boundary conditions for temperature and pressure are Neumann except for the hot patch at the bottom, and no-slip for the velocities except for Neumann at the top boundary. Since the TDP must have off-source diabatic heating, we add heat from $z = 10$ to $z = 15$ (called the heat injection zone (HIZ)) from $t \approx 42.5$ *FU* to $t = 63$ *FU*. The total timespan of the simulation is 90 *FU*. The solution procedure involves computing the required non-divergent velocity by solving the pressure-Poisson equation. Additional details are in [6].

3 Results

We focus on analysing the bulk parameters of the TDP. To visually identify the TDP, we use the vorticity magnitude $|\omega|$ ($|\omega| = (\omega_x^2 + \omega_y^2 + \omega_z^2)^{1/2}$), since all turbulent flows are characterised by (stochastic) vorticity fields. Hence, if we wish to separate the core of the TDP from the ambient outside, it is appropriate to use a vorticity threshold to define the turbulent/non-turbulent interface (TNTI).

Figure 2 shows axial sections of TDP. We choose time $t = 72$, where the flow, in Fig. 2a, has a tall tower (stem) with a cauliflower head (cap). In Fig. 2a, b, the quantity plotted is $|\omega|$. There is an increase in $|\omega|$ by almost two orders of magnitude post heat injection and beyond the HIZ. The baroclinic torque [5] is responsible

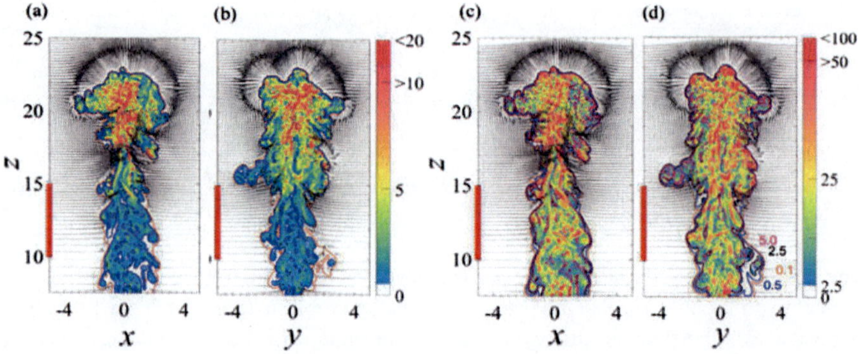

Fig. 2 Velocity vectors and vorticity magnitude in axial sections of the simulated TDP at $t = 72$: **a**, **b** vorticity magnitude using hot patch scaling ($|\omega|$); **c**, **d** vorticity magnitude using local scaling ($|\omega_l|$). The contours in all figures are as in (**d**). Region in red from $z = 10$ to $z = 15$ is the HIZ

for this enormous increase. The other important observations are that the centreline velocity increases (not shown) with height and the width does not linearly increase as in a classical plume. In Fig. 2a, an inrush of ambient fluid, visualised by velocity vectors, is seen around $(x, z) \approx (1, 18.5)$. Other regions where fluid is being engulfed and expelled are also observed. This entrainment and detrainment changes the mass flux in the TDP.

3.1 Choice of Appropriate Length and Velocity Scales

The vorticity plotted in Fig. 2a, b is based on d_0 and W_0. This is not entirely appropriate for the purpose of separating the TNTI of the TDP since the nature of the flow changes due to heat injection. It is preferable to choose local scales. Accordingly, the following local scales are defined

$$\text{Local volume}: \; Q_l(z,t) = 2\pi \int_0^{R_0} \langle w(r,z,t)\rangle_\theta r\,\mathrm{d}r \tag{2a}$$

$$\text{Local kinematic momentum:} \; M_l(z,t) = 2\pi \int_0^{R_0} \langle w(r,z,t)\rangle_\theta^2 r\,\mathrm{d}r \tag{2b}$$

$$\text{Local length scale:} \; R_l(z,t) = \frac{Q_l}{\sqrt{M_l}} \tag{2c}$$

$$\text{Local velocity scale:} \; W_l(z,t) = \frac{M_l}{Q_l} \tag{2d}$$

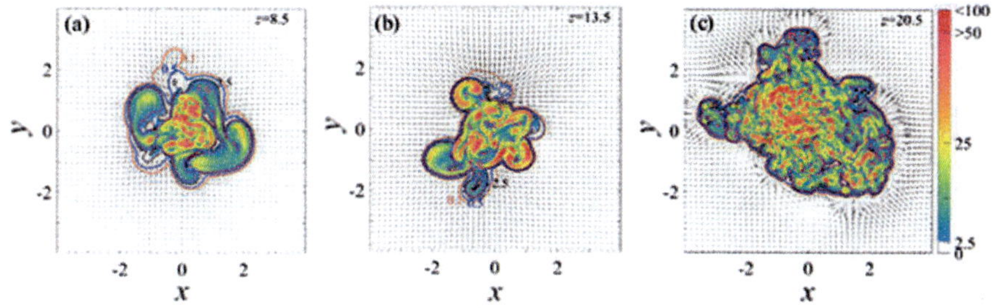

Fig. 3 Velocity vectors and $|\omega|$ in diametral sections of the simulated TDP at $t = 72$. The contours in all figures are as in Fig. 2c

Here, $\langle w(r, z, t)\rangle_\theta$ denotes the azimuthally averaged vertical velocity $w(r, z, t)$, and R_0 is the radial location where the mass flux (discussed in Sect. 3.2) reaches the maximum value.

The locally scaled $|\omega|$ is thus defined as $|\omega_l| = |\omega|\ (R_l/W_l)$. Comparing Fig. 2a, b and Fig. 2c, d, it can be seen that with the above-defined local scales, the turbulence in the flow becomes comparable at all levels. This is not the true picture of vorticity magnitude ($|\omega|$) variation as in Fig. 2a, b, but the scale transformation enables us to more easily choose an appropriate universal $|\omega|$ threshold for defining the TNTI. A detailed study of the boundary using $|\omega_l| = 0.1, 0.25$ up to 5.0 in steps of 0.25 has been carried out. This showed that the boundaries corresponding to $|\omega| = 2.0$ to 3.0 were closest to each other. We have therefore adopted $|\omega_l| = 2.5$ as the TNTI.

An interesting observation is that when $|\omega_l|$ is chosen, the fluctuations in $|\omega_l|$ are of the same order before and after the HIZ, whereas the fluctuations in $|\omega|$ are much larger. This is clearly seen in the diametral sections shown in Fig. 3 and the line plots shown in Fig. 4. This shows the appropriateness of the chosen local scales.

3.2 *Mass Flux*

The rate of variation of the mass flux (MF) in z is a measure of the entrainment into the TDP. To determine the MF, one needs to fix a boundary of the TDP. There have been multiple definitions of what is considered as the boundary. Due to the lack of data, researchers have chosen various terms such as half-width and *e*-folding width. Such measures are misleading since it is known that the TNTI forms the boundary between the turbulent core and the ambient, by definition of turbulence [7]. Thus, the vorticity magnitude can be used as a true boundary of the TDP. We compute MF using the following formula:

$$\text{MF}\ (z, t) = \int_0^{2\pi}\int_0^{R} w(r, \theta, z, t) r \mathrm{d}r \mathrm{d}\theta \tag{3}$$

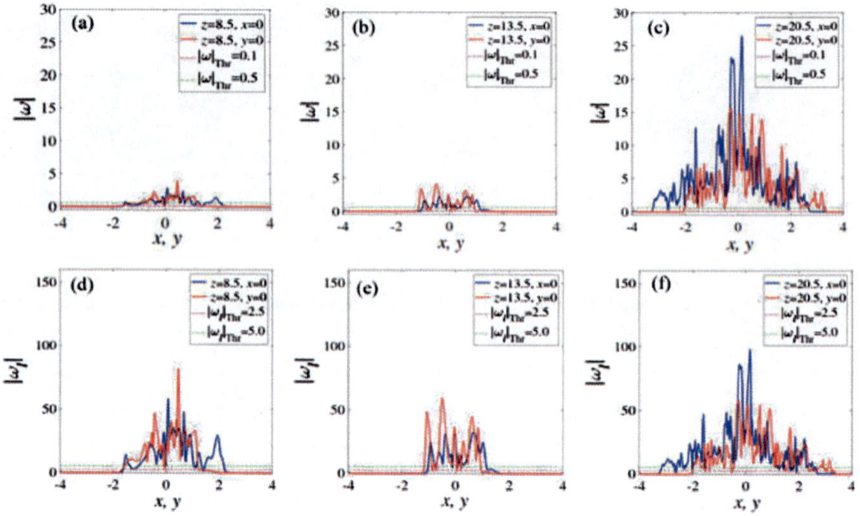

Fig. 4 Horizontal variation of $|\omega|$ and $|\omega_l|$ at various vertical levels. **a** and **d** below HIZ. **b** and **e** in HIZ. **c** and **f** above HIZ

where R could be the radius, or a chosen width of the TDP as in Eq. 2c. As mentioned previously, we have chosen $|\omega_l| = 2.5$ as marking the TNTI. The corresponding width is defined as $R = 2.5$. Figure 5 shows the cumulative MF (CMF) curve at various heights. CMF is the MF calculated using Eq. 3, and hence, it changes with radius R. Various measures of MF based on the chosen R are shown. The value of R corresponding to $|\omega|$ thresholds of 0.5 and 2.5 implies that grid points where $|\omega_l|$ is greater than the threshold value have been chosen for calculating MF. The location where d(MF)/d$r = 0$ may also be used to define the width R (Eq. 2c) of the TDP. From the discussion in Sect. 3.1, it is evident that the choice of $R_{2.5}$ gives the best possible estimate of the true turbulent MF (Fig. 5).

4 Conclusions

In this paper, we have shown that it is necessary to use appropriate scales to separate the turbulent core of the TDP from the non-turbulent ambient. Using local scales yields a value of the local vorticity magnitude whose range is uniform across the entire length of the TDP. This is used to determine the true turbulent mass flux which is necessary to compute entrainment into the TDP.

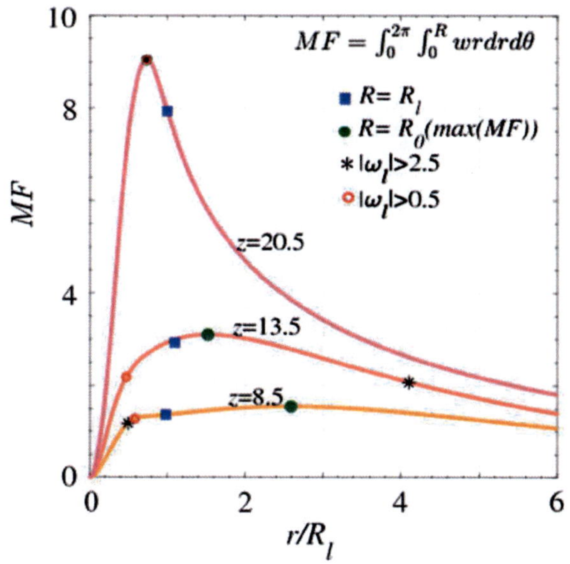

Fig. 5 Cumulative mass flux (CMF) at three vertical levels. Measures of MF based on the different criteria (R) of the TDP are shown in the legend. The turbulent MF is shown by star (*)

References

1. Basu A, Narasimha R (1999) Direct numerical simulation of turbulent flows with cloud-like off-source heating. J Fluid Mech 385:199–228. https://doi.org/10.1017/S0022112099004280
2. Bhat GS, Narasimha R (1996) A volumetrically heated jet: large-eddy structure and entrainment characteristics. J Fluid Mech 325:303–330. https://doi.org/10.1017/S0022112096008130
3. Houze RA (2014) Cloud dynamics, 2nd edn. Academic Press
4. Morton BR, Taylor GI, Turner JS (1956) Turbulent gravitational convection from maintained and instantaneous sources. Proc R Soc Lond Ser A Math Phys Sci 234(1196):1–23. https://doi.org/10.1098/rspa.1956.0011
5. Narasimha R, Diwan SS, Duvvuri S, Sreenivas KR, Bhat GS (2011) Laboratory simulations show diabatic heating drives cumulus-cloud evolution and entrainment. Proc Natl Acad Sci USA 108(39):16164–16169. https://doi.org/10.1073/pnas.1112281108
6. Prasanth P (2014) Direct numerical simulation of transient cumulus cloud-flow. Master's thesis, Jawaharlal Nehru Centre for Advanced Scientific Research, Bangalore
7. Shinde SR, Prasanth P, Narasimha R, to be submitted
8. Telford JW (1966) The convective mechanism in clear air. J Atmos Sci 23(6):652–666. https://doi.org/10.1175/1520-0469(1966)023%3c0652:TCMICA%3e2.0.CO;2
9. Turner JS (1986) Turbulent entrainment: the development of the entrainment assumption, and its application to geo-physical flows. J Fluid Mech 173:431–471. https://doi.org/10.1017/S0022112086001222

Noise Source Characteristics of Compressible Jet from Non-axisymmetric Nozzles

Shailesh R. Nikam, Siddhesh Bhikole, and Malay Suvagiya

Abstract Present experimental investigation reports effect of change in nozzle exit geometry on acoustic characteristics of compressible jet in the near and far field at jet exit Mach number 0.8. Three different nozzle configurations used in this investigation are circular, elliptical, and rectangular. Aspect ratio of elliptical and rectangular nozzle is 3. Equivalent nozzle exit diameter of all nozzle configurations is maintained constant as 40 mm. Results are presented by obtaining cross-correlation between near field and far field simultaneously measured pressure fluctuations. Acoustic characteristic along major and minor axis is different for elliptical and rectangular nozzle. Cross-correlation obtained indicates that noise source shifts closer to nozzle exit for elliptical as well as rectangular nozzle.

Keywords Jet noise · Acoustic characteristics · Mach number · Cross-correlation

1 Introduction

Understanding of jet noise and its control remained a topic of investigation from last six decades. Lighthill initiated jet noise theory [1, 2] which is further investigated by many researchers [3–8]. The two noise contributors in the jet flow are known to be the small or fine-scale structures (FSSs) and the large-scale structures (LSSs). The fine-scale structures are developed in the near region of the nozzle and are distributed throughout the mixing layer of the jet, whereas spatially coherent large-scale structures are developed after the end of the potential core. With increase in frequency of aircraft operation, noise regulations are becoming more stringent. Various active and passive noise control techniques are used by many researchers [9–17] to address this issue. Change in nozzle exit geometry is one of the passive techniques to alter acoustic

S. R. Nikam (✉) · S. Bhikole
K. J. Somaiya College of Engineering, Vidyavihar, Mumbai 400077, India
e-mail: shailesh.n@somaiya.edu

M. Suvagiya
Aerospace Engineering Department, IIT Bombay, Mumbai 400076, India

L. Venkatakrishnan et al. (eds.), *Proceedings of 16th Asian Congress of Fluid Mechanics*, Lecture Notes in Mechanical Engineering,
https://doi.org/10.1007/978-981-15-5183-3_39

Table 1 Details of test nozzles

S. No.	Nozzle geometry	Major axis (mm)		Minor axis (mm)	
1	Circular	40		40	
2	Elliptical	69.382		23.08	
3	Rectangular	61.384		20.463	

characteristic of jet flow. Attempts are made earlier by many researchers [18–21] to investigate effect of elliptical and rectangular nozzle on aero-acoustic characteristics of jet flow. However, most of the work is carried out for supersonic flow condition and effect of change in nozzle exit geometry on noise source characteristics is not discussed in detail. The present experimental investigation is aimed to gain better understanding of how a change in nozzle exit geometry from circular to elliptical and rectangular affects the noise source characteristics and its far-field radiation characteristics at a high subsonic Mach number of 0.8. The simultaneous measurement of near-field and far-field pressure fluctuations using a microphone is used to estimate the radiated component of near-field pressure fluctuations propagating to the far field using a cross-correlation technique.

2 Experimental Setup and Test Nozzle

The experiments are conducted in free jet acoustic facility at Aerospace department of IIT Bombay, Mumbai. Test nozzle of jet facility is mounted inside the anechoic chamber. Three convergent nozzles with different nozzle exit geometry are used in the present investigation. Details of test nozzles are given in Table 1. Circular nozzle has nozzle exit diameter of 40 mm and 150 mm length. For elliptical and rectangular nozzle, equivalent nozzle exit diameter (D_{eq}) maintained is same as 40 mm. Aspect ratio 3 is used for elliptical and rectangular nozzle. Photograph of test nozzles inside anechoic chamber is shown in Fig. 1.

3 Experimental Method

Experiments are conducted at a plenum chamber temperature of about 301 $\pm$ 2 K and a pressure of 1.524 times the ambient pressure to produce a free jet at $M = 0.8$. Pressure in plenum chamber is continuously monitored using pressure scanner and maintained using pressure regulating valve. Simultaneous near-field and far-field pressure fluctuations were recorded using microphones. Figure 2 shows the schematic

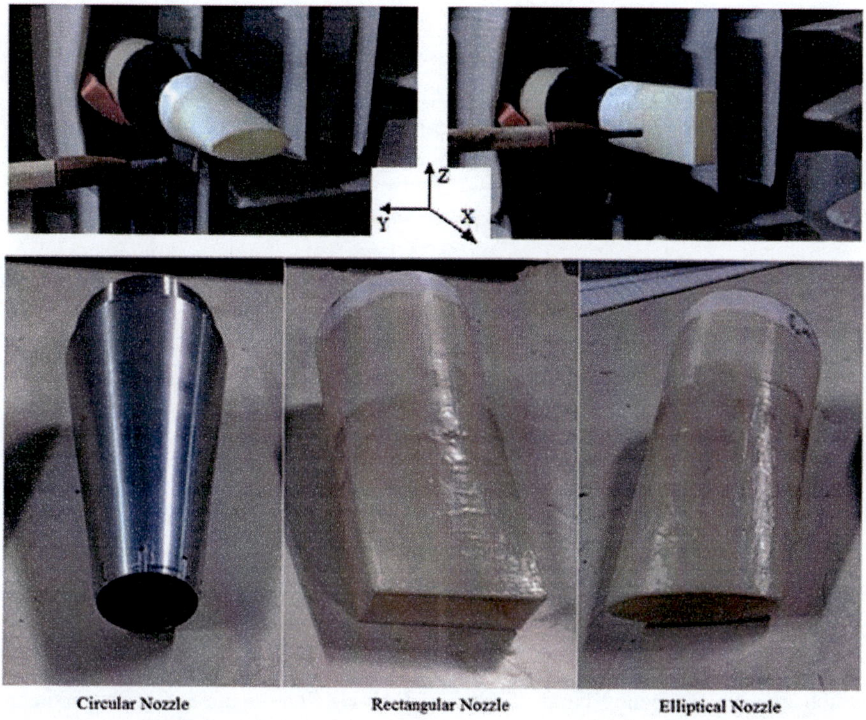

Fig. 1 Photograph of test nozzles

arrangement of the microphones in the near field and far field of the jet. The near-field, microphone is traversed parallel to the jet edge from $x/D_{eq} = 0.5$ to $x/D_{eq} = 14.5$ in a step of $2D$. First position of microphone from jet exit is $1.9D_{eq}$ in y-direction and $0.5D_{eq}$ in x-direction. Jet divergence angle is considered to be about 9° obtained from aerodynamic measurement (not reported in this paper). Computer controlled traverse is used for traversing near-field microphone. Far-field microphones are positioned at polar angle 30°–90° in a step of 10° (excluding at 70°) using linear array of microphone as shown in Fig. 2. All far-field measurement data is scaled to constant radial distance of $50D_{eq}$ in the post processing.

B&K 4939 free-field condenser microphones with a 6.53 mm (1/4′) diameter were used in conjunction with B&K 2970 pre-amplifiers for the near-field and far-field acoustic measurements. The microphones were powered by four channel NEXUS model 2690-0S4 signal conditional amplifiers. Each microphone has an open circuit sensitivity of 4.5 mV Pa^{-1} and a flat frequency response from 4–100 kHz. The microphones were calibrated using B&K 4226 multifunction acoustic calibrator before commencing the measurements. A National Instrument PCI-4462 card-based DAQ and LabVIEW software were used for the acoustic data acquisition. One hundred blocks of narrow band data were acquired at a rate of 200 kHz with 8192 samples,

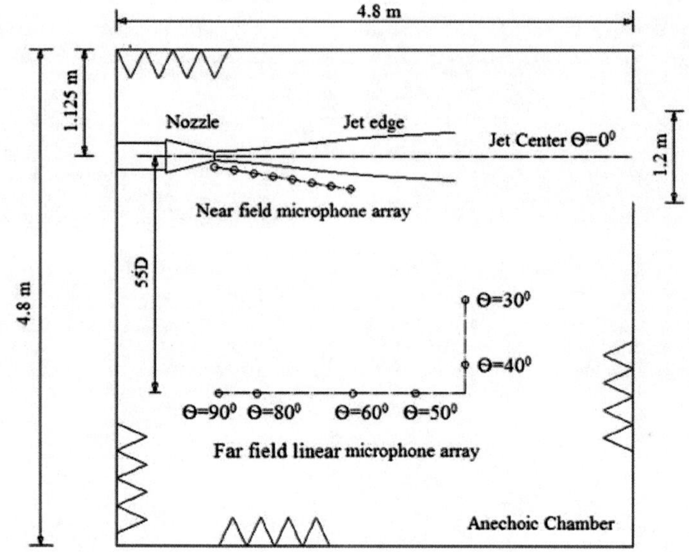

Fig. 2 Schematic of test setup showing microphone placement in the near field (along jet edge) and far field (polar angle 30°–90°)

which gives a frequency band width of 24.41 Hz. The accuracy of the acoustic measurement as per the manufacturer's specification is within ± 0.3 dB. The acquired data is post-processed using the MATLAB function. Cross-correlation function is obtained from two simultaneously measured near-field and far-field pressure fluctuations. The normalized cross-correlation is calculated by using MATLAB programs. The normalized cross-correlation is given by Eq. (1), where m is reference microphone and n is correlated microphone. The normalized cross-correlation function is a non-dimensional parameter which gives information about degree of similarity of two signals with respect to set time lag.

$$R_{mn}(\tau) = \frac{P_m(t)P_n(t+\tau)}{P_m^2(t)^{\frac{1}{2}}\langle P_n^2(t)\rangle^{\frac{1}{2}}} \tag{1}$$

4 Result

4.1 Near-Field Sound Pressure Level (SPL) Spectra

Pressure fluctuations are measured along the jet boundary at different nozzle downstream locations (x/D_{eq}) for the circular, elliptical and rectangular nozzle at jet exit

Mach numbers 0.8. Measurements are taken in both major and minor axis plane for elliptical and rectangular nozzle. Jet boundary is determined from aerodynamic measurement at different longitudinal location in the transverse direction. Microphone is traversed at different streamwise position starting from $x/D_{\text{eq}} = 0.5$ till 14.5 in a step of 2D keeping radial distance from jet boundary same for all the longitudinal position. Figure 3 shows the sound pressure level (SPL) spectra obtained from these measurements for the circular nozzle against non-dimensional frequency, Strouhal number $\left(St = \frac{f D_{\text{eq}}}{U_{\text{exit}}}\right)$. At $x/D_{\text{eq}} = 0.5$, spectra is broadband in nature for all the nozzle configuration. Surprisingly, tonal noise is obtained at about $St = 0.1$ for rectangular nozzle along minor axis, reason for this tonal noise is not very clear. Elliptical nozzle shows almost similar spectra along major and minor axis except variation at around $St = 1$. At this frequency, noise level is higher along major axis. Similar observations are seen for rectangular nozzle. At $x/D_{\text{eq}} = 14.5$, after $St = 0.1$, spectral level for major axis of elliptical and rectangular nozzle shows lower level than their respective minor axis. Also spectral levels along major axis are less than circular nozzle after $St = 0.1$.

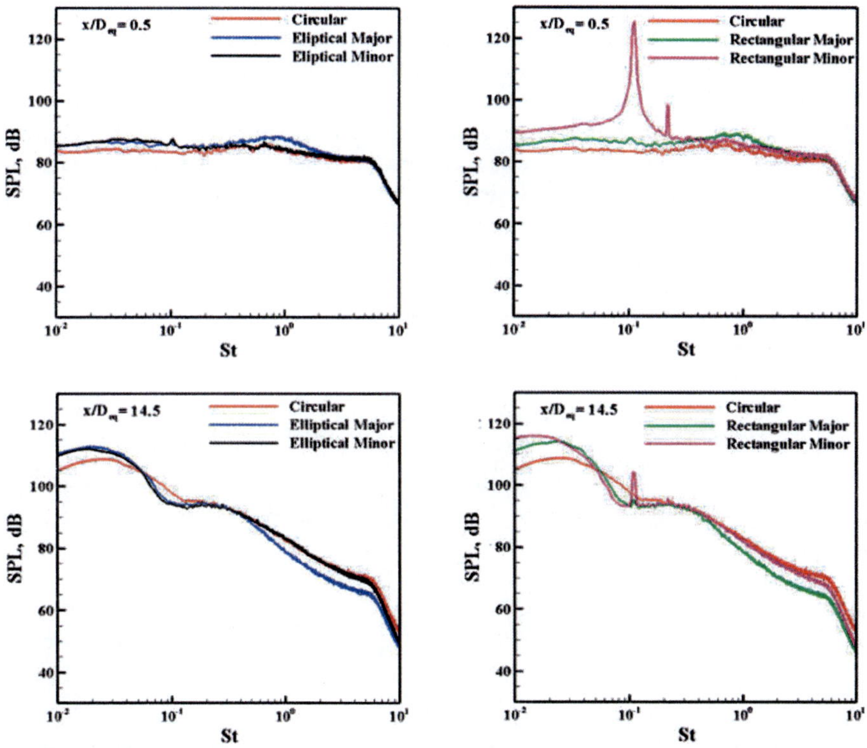

Fig. 3 Comparison of SPL spectra for circular, elliptical, and rectangular nozzle at different streamwise location in the near field

4.2 Far-Field SPL Spectra

Pressure fluctuations recorded in the far field using linear array of microphones is post-processed to obtain SPL spectra. Scaling law is used to get spectra at constant radial distance of $50D_{eq}$. SPL spectra changes from narrowband at lower polar angle (30°) to broadband at higher polar angle (90°). Peak noise level decreases and peak frequency increases from lower polar angle to higher polar angle. This is expected as noise level closer to the jet at lower polar angle is mainly due to small-scale turbulent structures produced downstream of the potential core, whereas noise in the sideline direction (higher polar angle) is due to small-scale turbulent structure produced in the mixing layer of the jet. Noise level signifies that large-scale structures produce more acoustic energy as compared to small-scale structures. Peak noise level at polar angle 30° is about 82 dB obtained at about Strouhal frequency 0.2 which matches very well with dominating jet frequency reported earlier by many researchers [8, 22, 23]. Near-field spectra along the jet edge (Fig. 3) shows its peak at low frequency. Spectra in the near field at $x/D_{eq} = 14.5$ shows its peak at Strouhal frequency 0.026. This peak frequency in the near field is far from the peak frequency ($St = 0.2$) obtained in the far field. This is due to the fact that SPL obtained in the near field closer to the jet edge is highly influenced by hydrodynamic pressure rather than pure acoustic. Figure 4 shows comparison of SPL spectra for circular, elliptical, and rectangular nozzle. As like in the near-field spectra, tonal noise is seen at far-field spectra for rectangular nozzle along minor axis at same Strouhal frequency 0.1. At polar angle 30°, peak level is same for all nozzle configurations; however, there is difference in the level along major and minor axis for elliptical and rectangular nozzle. At this polar angle (30°) elliptical nozzle shows lower level along major axis as compared to circular nozzle at higher frequency.

Figure 5 shows variation of overall sound pressure level (OASPL) at different far-field polar angles. OASPL is shown in two different frequency band, low frequency (0.01–10 kHz) and high frequency (10–100 kHz). OASPL shows directivity of noise

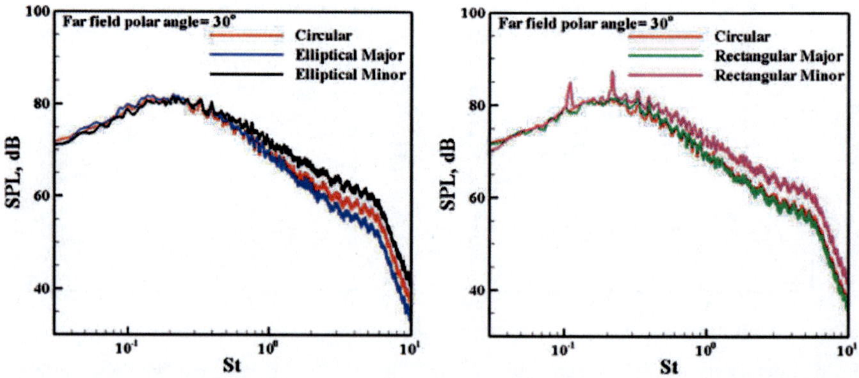

Fig. 4 Comparison of far-field noise spectra (SPL) for all nozzle configurations at polar angle 30°

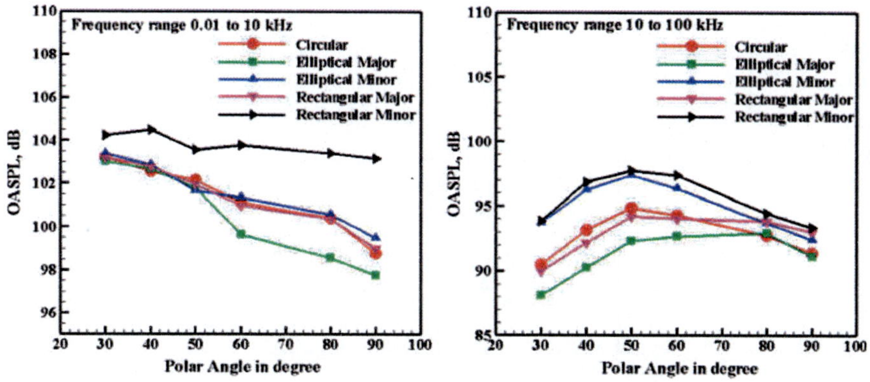

Fig. 5 Comparison of OASPL for all the nozzle configurations in different frequency band

radiation in the far field in both the frequency band. OASPL in low frequency band shows its peak at lower polar angle 30°, i.e., closer to the jet axis, and it reduces as polar angle increases from 30° to 90°. This trend is observed for all nozzle configurations. In high frequency band, peak noise level is obtained at polar angle 50° for all nozzle configurations. Rectangular nozzle along minor axis shows maximum OASPL at all polar angles as compared to other nozzle configuration in low frequency band, whereas elliptical nozzle along major axis shows lowest OASPL at polar angle 60°, 80°, and 90°. At (10–100 kHz) higher frequency, both elliptical and rectangular nozzle along minor axis shows lower OASPL as compared to circular nozzle at lower polar angles. However, along minor axis, level is consistently high at all the polar angles as compared to circular nozzle. This variation in OASPL in low and high frequency band is consistent with SPL spectra shown in Fig. 4.

4.3 Near-Field Far-Field Cross-Correlation

Cross-correlation function characterizes the propagation of coherent part of the signal embedded within a random field measured simultaneously using a pair of microphone; thus, cross-correlation coefficient gives information about common energy between the two signals. Pressure recorded by microphone in the near field composed of two components: hydrodynamic pressure and acoustic pressure as stated by Arndt et al. [24]. To understand effect of change in nozzle exit geometry on the near-field pressure that radiates to the far field, cross-correlation is obtained between near-field and far-field pressure fluctuations from simultaneous measurements of the near-field pressure fluctuations along the jet edge from $x/D_{eq} = 0.5$ to 14.5 in a step of 2D and the far-field pressure fluctuations at polar angles of $\theta = 30°$ to 90° in steps of 10° except at 70°. Similar method is used earlier by Viswanathan et al. [25] and Nikam et al. [26].

Figure 6 shows the cross-correlation obtained between signals from the near-field reference microphone, traversed along the jet edge from $x/D_{\text{eq}} = 0.5$ to 14.5 in steps of 2D, and the far-field microphone positioned at the polar angle $\theta = 30°$

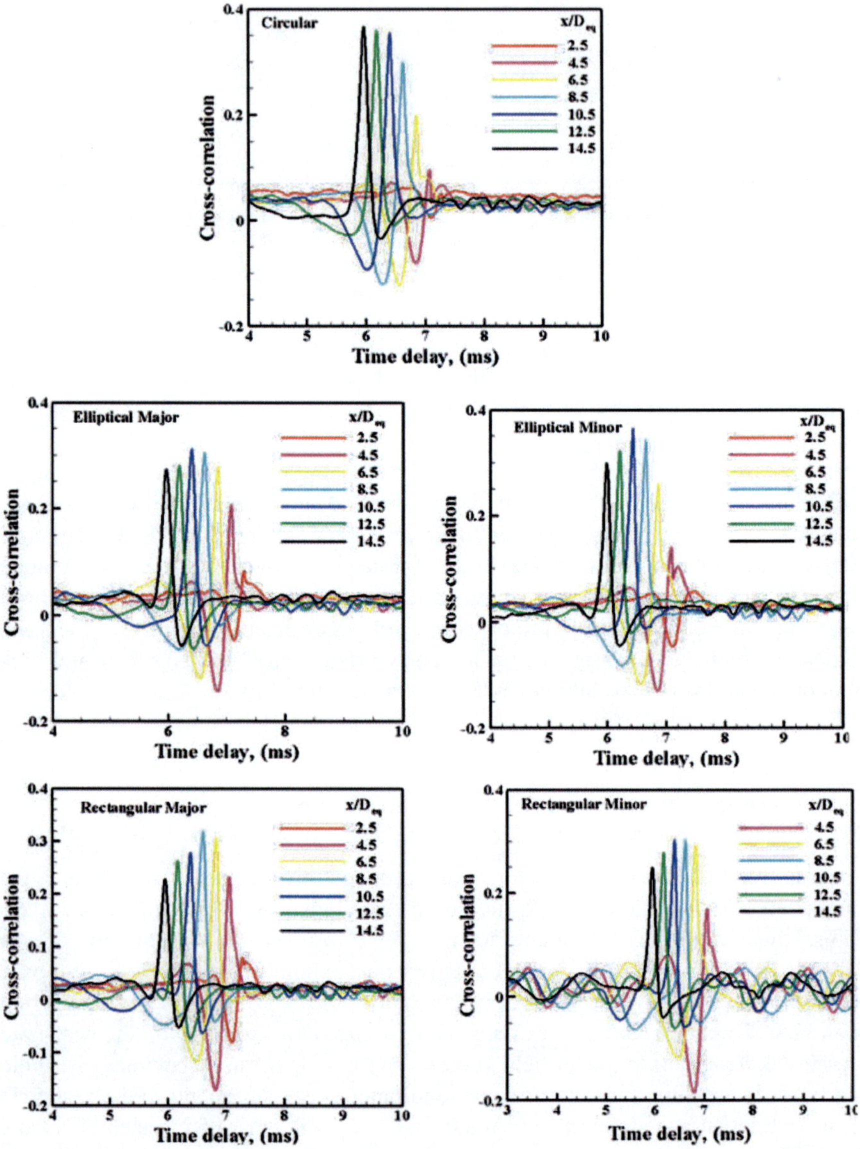

Fig. 6 Plots of near-field/far-field cross-correlation for all nozzle configurations. Near field along jet edge and the far-field microphone positioned at the polar angle $\theta = 30°$

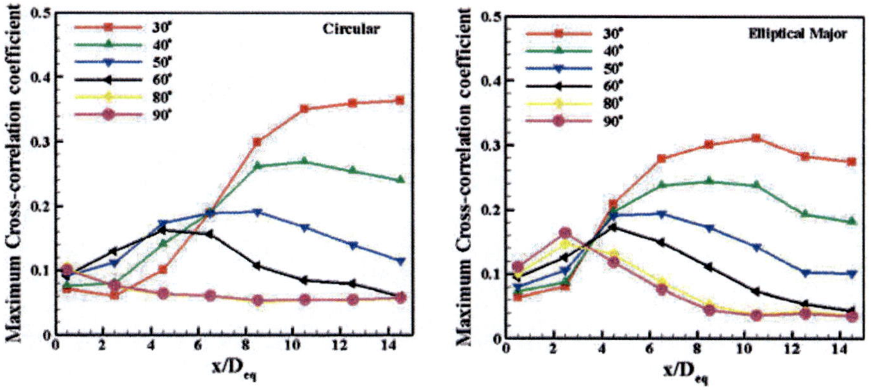

Fig. 7 Maximum of the near-field/far-field cross-correlation coefficient for circular and elliptical (major axis) nozzle

for all nozzle configurations. First noticeable peak is seen from $x/D_{\mathrm{eq}} = 2.5$, for all nozzle configuration except rectangular nozzle with minor axis where noticeable peak is seen from $x/D_{\mathrm{eq}} = 4.5$. With increase in axial distance, positive peak increases, reaches its peak, and then decreases. The time delay reduces with a reduction in distance between the two microphones as the near-field microphone is moved downward. Geometric distance between near-field far-field microphone position and time delay obtained between these two microphones gives acoustic wave speed. Magnitude and position of peak cross-correlation coefficient changes with change in nozzle configuration, which indicates that noise source location in the near field, radiating noise more effectively in the far field at 30°, is different for different nozzle configuration. Thus, change in nozzle configuration changes noise source location. Cross-correlation is obtained between all the near-field positions and far-field polar angles, and maximum value of cross-correlation coefficient is obtained for the entire nozzle configuration. However, Fig. 7 shows plots of the maximum of the near-field/far-field cross-correlation coefficients only for the base and the elliptical nozzle along major axis at all the polar angles from $\theta = 30°$ to 90° when the near-field reference microphone is traversed along the jet edge in steps of 2D. For each far-field angle, peak attains at different near-field position which indicates that noise sources radiating noise in the far field at different polar angle is from different near-field position. Elliptical nozzle shows difference in the magnitude of peak coefficient and corresponding near-field position for different polar angle as compared to circular nozzle, suggesting shift in the noise sources due to change in nozzle geometry. Figure 8 shows comparison of maximum of the near-field/far-field cross-correlation coefficients for all nozzle configurations at 30° and 40°. It is clearly seen that peak value shifts closer to nozzle exit for elliptical and rectangular nozzle as compared to circular. Thus, change in nozzle geometry from circular to elliptical/rectangular shifts noise source toward nozzle exit.

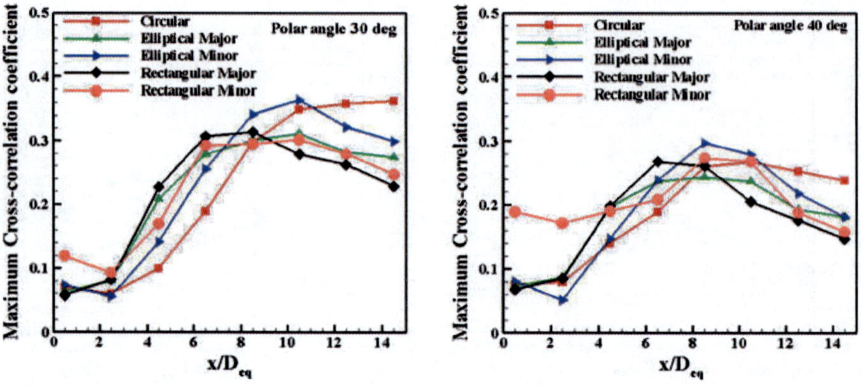

Fig. 8 Comparison of maximum of the near-field/far-field cross-correlation coefficients for all nozzle configurations at 30° and 40°

5 Conclusion

Present experimental investigation confirms that change in nozzle exit geometry from circular to elliptical or rectangular, changes the acoustic characteristics of jet flow. Noise level along major and minor axis of elliptical and rectangular nozzle is different. Noise level along major axis is less than minor axis for both elliptical and rectangular nozzle. Noise level of jet from elliptical nozzle along major axis is less than circular nozzle at frequency more than $St = 0.4$ at lower polar angle. Near-field, far-field cross-correlation confirms that change in nozzle geometry shifts noise source closer to the nozzle exit.

Acknowledgements The authors wish to thanks Dr. S. D. Sharma, Professor, Aerospace Engineering Department, IIT Bombay, Mumbai, for his guidance and permission to conduct experiment in Jet acoustic facility at IIT Bombay.

References

1. Lighthill MJ (1952) On sound generated aerodynamically I. General Theory. Proc R Soc Lond Ser A 211:564–587
2. Lighthill MJ (1954) On Sound generated aerodynamically II. Turbulence as a Source of Sound. Proc R Soc Lond Ser A 222:1–32
3. Mollo-Christensen E, Kolpin MA, Martuccelli JR (1964) Experiments on jet flows and jet noise far-field spectra and directivity patterns. J Fluid Mech 18:285–301
4. Lush PA (1971) Measurements of subsonic jet noise and comparison with theory. J Fluid Mech 46:477–500
5. Tam CKW (1998) Jet noise: since 1952. Theoret Comput Fluid Dyn 10:393–405
6. Tanna HK (1977) An experimental study of jet noise part I: turbulent mixing noise. J Sound Vib 50:405–428

7. Zaman KBMQ (1986) Flow field and near and far sound field of a subsonic jet. J Sound Vib 106:1–16
8. Tam CKW, Viswanathan K, Ahuja KK, Panda J (2008) The sources of jet noise: experimental evidence. J Fluid Mech 615:253–292
9. Alkislar MB, Krothapalli A, Butler GW (2007) The effect of streamwise vortices on the aeroacoustics of a Mach 0.9 jet. J Fluid Mech 578:139
10. Arakeri VH, Krothapalli A, Siddavaram V, Alkislar MB, Laurenko LM (2003) On the use of microjets to suppress turbulence in a Mach 0.9 axisymmetric jet. J Fluid Mech 490:75–98
11. Castelain T, Béra J, Sunyach M (2006) Noise reduction of a Mach 0.7–0.9 jet by impinging microjets. Comptes Rendus Mécanique 334:98–104
12. Bridges J, Wernet M, Brown C (2003) Control of jet noise through mixing enhancement
13. Callender B, Gutmark EJ, Martens S (2005) Far-field acoustic investigation into chevron nozzle mechanisms and trends. AIAA J 43:87–95
14. Tide PS, Babu V (2009) Numerical predictions of noise due to subsonic jets from nozzles with and without chevrons. Appl Acoust 70:321–332
15. Mohan NKD (2017) Active chevrons for jet noise reduction. In: 24th International Congress on Sound and Vibration, London, July 2017, pp 1–8
16. Hileman J, Samimy M (2003) Effects of vortex generating tabs on noise sources in an ideally expanded mach 1.3 jet. Int. J. Aeroacoust 2:35–63
17. Zaman KBMQ, Reeder MF, Samimy M (1994) Control of an axisymmetric jet using vortex generators. Phys Fluids 6:778
18. Ahuja K, Manes JP, Massey KC (1990) An evaluation of various concepts of reducing supersonic jet noise. In: 13th Aeroacoustics conference, reston, Virigina: American Institute of Aeronautics and Astronautics, pp AIAA-90-3982
19. Verma SB, Rathakrishnan E (2003) Influence of aspect-ratio on the mixing and acoustic characteristics of plain and modified elliptic slot jets. Aerosp Sci Technol 7:451–464
20. Kim J-H, Samimy M (1999) Mixing enhancement via nozzle trailing edge modifications in a high speed rectangular jet. Phys Fluids 11:2731–2742
21. Kerechanin CW, Samimy M, Kim J-H (2001) Effects of nozzle trailing edges on acoustic field of supersonic rectangular jet. AIAA Journal 39:1065–1070
22. Bogey C, Barré S, Fleury V, Bailly C, Juvé D (2007) Experimental study of the spectral properties of near-field and far-field jet noise. Int J Aeroacoust 6:73–92
23. Nikam SR, Sharma SD (2017) Effect of chevron nozzle penetration on aero-acoustic characteristics of jet at M = 0.8. Fluid Dyn Res 49
24. Arndt REA, Long DF, Glauser MN (1997) The proper orthogonal decomposition of pressure fluctuations surrounding a turbulent jet. J Fluid Mech 340:1–33
25. Viswanathan K, Underbrink JR, Brusniak L (2011) Space-time correlation measurements in nearfields of Jets. AIAA J 49:1577–1599
26. Nikam SR, Sharma SD (2018) Effect of a chevron nozzle on noise radiation from a compressible jet. AIAA J., 1–18

Effect of Reynolds Number on the Non-reacting Turbulent Flow Structures of a Double Swirler Burner

Dhanalakshmi Sellan, Raju Murugan, and Saravanan Balusamy

Abstract A deeper understanding of the effect of the swirl flow mechanism in gas turbine combustion is essential to design optimized combustors under various operating conditions. The double swirl flow arrangement has an advantage over the single swirl flow by proving better mixing and flame anchoring due to the increased shear action between the two swirl flows. The non-reacting flow characteristics of the unconfined double swirler burner are studied for various Reynolds numbers. 2D particle image velocimetry (PIV) technique with proper spatial calibration was used to carry out the experiments, and the velocity field is calculated using Insight4G software (TSI). The velocity field and the turbulence parameters such as Reynolds stress and turbulent intensity under different Reynolds numbers are analyzed, which helps to enhance our understanding of the turbulence mixing. The results show that the increase in outer swirl Reynolds number increases the radial component of the velocity; therefore, the swirl flow spread in the radial direction is increasing. Similarly, the outer swirl Reynolds number affects the turbulent parameters such as Reynolds stress and turbulence intensity. Because of this phenomenon, the spread of the flame front may occur and further to the local quenching in the worst case.

Keywords Particle image velocimetry · Double swirler burner · Reynolds number · Reynolds stress · Turbulent intensity

1 Introduction

The flow field characteristics studies under non-reacting conditions always play a role in understanding the effect of swirl flow mechanism of gas turbine combustion, which is essential to design optimized combustors under various operating conditions. The swirl has a significant impact on the development of the combustor flow field because introducing swirl creates a central toroidal recirculation zone, vortex breakdown, which enhances the combustion performance [1]. Other experimental studies also

D. Sellan (✉) · R. Murugan · S. Balusamy
Indian Institute of Technology Hyderabad, Kandi 502285, India
e-mail: me16resch11003@iith.ac.in

L. Venkatakrishnan et al. (eds.), *Proceedings of 16th Asian Congress of Fluid Mechanics*, Lecture Notes in Mechanical Engineering,
https://doi.org/10.1007/978-981-15-5183-3_40

show that swirl has large-scale effects on flow fields: jet evolution, entrainment, decay and flame size, shape, stability, and combustion intensity (for reacting flows) are affected by the degree of swirl imparted to the flow field [2]. The outer flow has a strong effect on the formation of recirculation zones and on mixing characteristics [3], and the strength of the recirculation region is dependent on the swirler design.

In the present study, the co-axial double swirler burner is used to study the non-reacting flow structure under various outer Reynolds numbers (Re_o). The double swirl structure in the burner configuration has more advantages than the single swirl burner in terms of mixing and flame stabilization [4]. The flow characteristics of the swirling array are compared with a single swirl, and it is shown that the co-rotating swirler array has shorter recirculation zones and stronger turbulence intensity [5]. Various techniques help to understand the flow field characteristics, and one among them is particle image velocimetry (PIV), which is used by the research community effectively [6–9]. The present study used the PIV technique to capture the two-component velocity field of non-reacting flows.

2 Experimental Setup and Operating Conditions

A swirl-stabilized axisymmetric annular burner is designed and fabricated for the present experimental study. The design philosophy of the co-axial double swirler burner is similar to the Cambridge/Sandia stratified swirl burner (CSB), which has proven to be capable of stabilizing a flame at arrange of flow rates, stratification levels, and swirl numbers [10–13]. As shown in Fig. 1a, the burner consists of three concentric annuli and a bluff body, which serves flame stabilizations. The inner and outer annulus flows may be run with the axial flow or on various swirl numbers. For

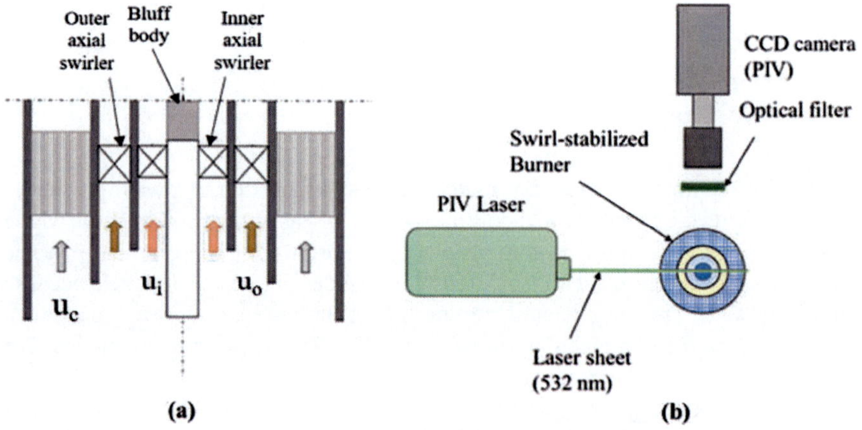

Fig. 1 Schematic of **a** cross-sectional view of the burner and **b** experimental setup

Table 1 Test matrix

Case No.	Bulk velocity (m/s)			Reynolds number		
	Inner (u_i)	Outer (u_o)	Co-flow (u_c)	Inner (Re_i)	Outer (Re_o)	Co-flow (Re_c)
1	7.5	4	0.4	5309	4402	2826
2	7.5	6	0.4	5309	6581	2826
3	7.5	8	0.4	5309	8804	2826

the present study, a moderate swirler with geometric swirl number (S) 0.8 is used. The geometric swirl number can be calculated based on Eq. (1) where D_h is the hub diameter and D_s is the swirl diameter [14–16]. Unlike CSB, this burner configuration will allow us to vary the swirl number and Reynolds number independently, which aids a wide range of operating conditions.

$$S = \frac{2}{3}\frac{\left[1-\left(\frac{D_h}{D_s}\right)^3\right]}{\left[1-\left(\frac{D_h}{D_s}\right)^2\right]} \tag{1}$$

The experimental setup of a double swirler burner along with the PIV arrangement is shown in Fig. 1b. Mass flow controllers (MFC-Alicat MCR-D series) are used to regulate the flow rates of the inner and outer annulus air and accuracy of ±1% of full scale. The airflow through three annular is seeded with micrometric vegetable oil particles using seed particle generators that are in-house fabricated. In these particle generators, Laskin nozzle design is used to produce oil particles at the micron level, efficiently. The PIV is the non-intrusive technique used to measure the two-component instantaneous velocities of the flow field.

The micron-level particles are illuminated by laser sheet, and the laser sheet is generated using 532 nm wavelength, 135 mJ Litron double pulsed Nd:YAG laser system. A CCD camera, which is synchronized with the laser system, acquires the image pairs at a given time interval Δt. The captured images are post-processed using 4G Insight software (TSI), which uses cross-correlation to determine the displacement. The present study was carried out for three experimental cases, which are given in Table 1. The inner Reynolds number (Re_i) and co-flow Reynolds number (Re_c) are maintained constant for three different outer Reynolds number (Re_o).

3 Results

To understand the flow structure and velocity field, radial and axial velocity contours are plotted and presented here for various outer Reynolds number (Re_o). 'x' indicates the axial co-ordinate and 'r' indicates the radial co-ordinates for all the results

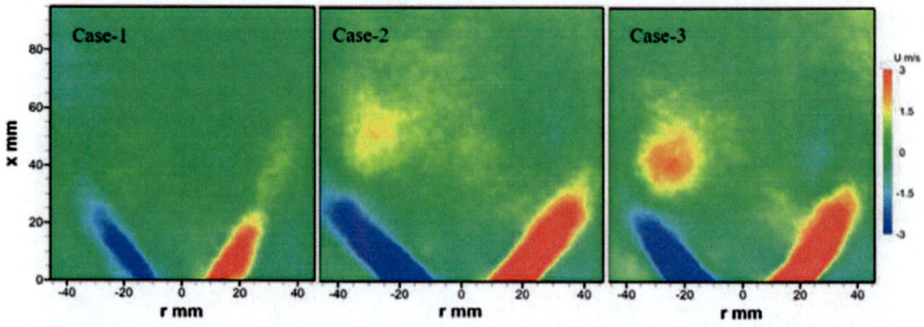

Fig. 2 Mean radial velocity field for three different cases

provided in this section. The mean of 200 post-processed data is used to get the plots presented here. Figure 2 shows the mean radial velocity field for three different outer Reynolds numbers (Re_o). From case-1 to case-3, as increasing Re_o the radial velocity increases, also strong vortex break down was observed.

The effect of outer swirl flow plays an essential role in the shear action between the inner swirl flow and co-flow conditions, which shows the increase in the radial flow field than the axial flow field, which is shown in Fig. 3. But the radial spread of axial and radial velocities shows a similar configuration. In case-1, the inner axial bulk velocity dominates the outer flow; therefore, the outer spreading is less, and the magnitude of axial velocity is less due to less overall bulk velocity (Inner bulk velocity + Outer bulk velocity). But the flow structure becomes wider for the case-2 than case-1 and case-3. In case-3, due to an increase in outer axial bulk velocity, the mean axial velocity starts dominating the outward pull. Therefore, in case-3, the flow structure has a lesser radial spread compared to case-2 but more than case-1. Due to these flow structures, the flame length will be affected in reacting conditions. When considering the same flow condition (Case-2 and case-3) for reacting case, the flame length may reduce, because there will be an effect of thermal expansion radially.

The streamline plots are very useful in identifying the recirculation zones in the flow field. Figure 4 shows the streamline plot superimposed with the axial

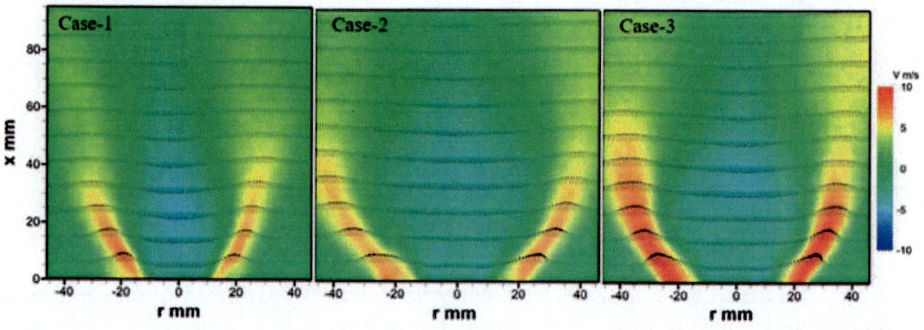

Fig. 3 Mean axial velocity field for three different cases

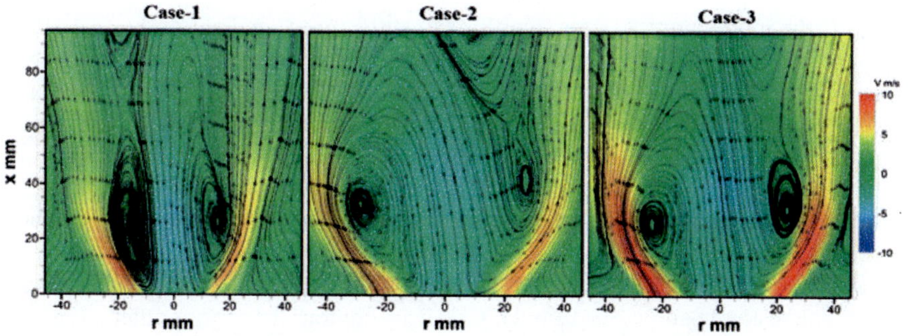

Fig. 4 Streamline plot superimposed with axial velocity contour

velocity contour, where inner recirculation zones (IRZ) are identified. When the inner Reynolds number (Re_i) dominates over the outer (Re_o) (Case-1), the IRZ is very strong and close to the centreline of the burner exit. In case-2, when increasing the outer axial bulk velocity, the inner recirculation zones are moved outward and axially upward. On the other hand, when Re_o dominates on the Re_i, IRZ becomes weak and moves radially outward. Also, case-2 shows compact inner recirculation zones, whereas case-1 has elongated IRZ. The radial outward spreading and the shifting of inner recirculation zone (IRZ) downstream might lead to the spreading of the flame front due to the thermal expansion of gas, and it varies at different swirl conditions [11] which we have not reported here.

3.1 Reynolds Stress and Turbulent Intensity

The mean Reynolds stress (τ_R) contour for various outer swirl flow conditions with constant inner swirl flow and co-flow are shown in Fig. 5. The Reynolds stress expressed as '$-u_i u_j$' by assuming incompressible for the present study. Reynolds

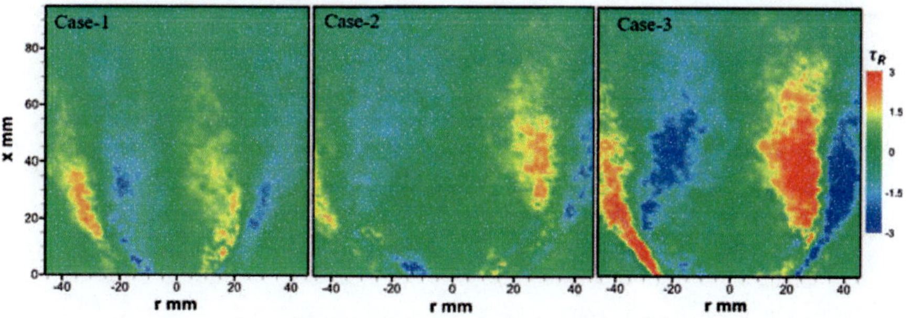

Fig. 5 Mean Reynolds stress contour plots

stress is the momentum flux that works against the mean gradient of the main flow, and this results in the production of turbulence. This is evident from the production term in Eq. (2) (turbulent kinetic energy equation). The positive sign in turbulence production term shows that there is energy transfer from the main flow to turbulence. The dissipation is almost negligible due to assumption of inviscid flow condition. Therefore, there is more energy transfer than dissipation. In the present study, an increase in outer axial bulk velocity contributing to Reynolds stress increment at radially outward as well as axially upward. When the Reynolds stress is high, the energy transfer to the turbulence will also be high, and thus, it promotes the fuel/air mixing for complete combustion. The turbulent intensity contour for various outer swirl flow conditions with constant inner Reynolds number (Re_{i}) and co-flow Reynolds number (Re_{c}) is shown in Fig. 6.

$$\underbrace{\left[\frac{\partial}{\partial t}+U_j\frac{\partial}{\partial x_j}\right]K}_{\text{Variation of TKE with time}} = \underbrace{\frac{\partial}{\partial x_j}\left\{-\frac{1}{\rho}PU_j\delta_{ij}-\frac{1}{2}u_iu_jU_j+2\nu S_{ij}U_i\right\}}_{\substack{\text{Redistribution term}\\ \text{(Pressure diffusion+Turbulent transport+Molecular viscous term)}}} - \underbrace{u_iu_j\frac{\partial U_i}{\partial x_j}}_{\text{Production term}} - \underbrace{2\nu S_{ij}S_{ij}}_{\text{Dissipation term}} \tag{2}$$

The turbulent intensity is one of the major parameters to characterize the swirl flow under both non-reacting and reacting conditions, and the energy transfer between the flow to turbulence can be understood by Eq. (2). With an increase in outer flow, turbulent intensity also increases in both the axial and radial direction. As the Reynolds stress transfers the energy to the turbulence, the increase in Reynolds stress leads to an increase in the turbulence level. However, increasing the swirl Reynolds number at the flow field is not always favorable to enhance flame stabilization like an increase in turbulence level at lean fuel conditions may lead to blow-off. Therefore, it is essential to study the flow characteristics under various operating conditions such as different

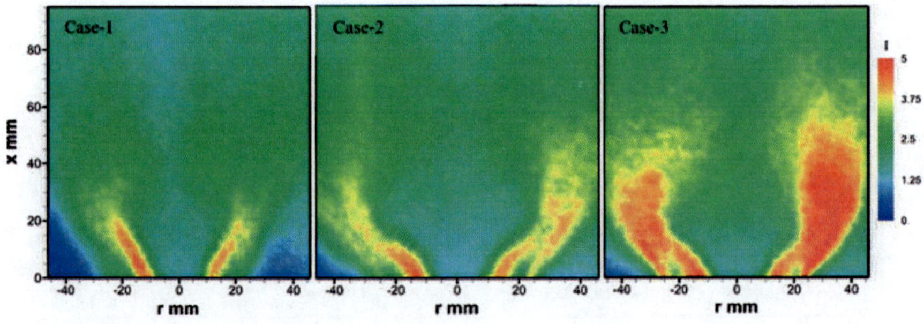

Fig. 6 Mean turbulent intensity contour for three different cases

swirl number, mass flow rate, and equivalence ratio in order to identify the optimized operating conditions.

4 Conclusions

The flow field characteristics and flow structures for various outer flow Reynolds numbers under non-reacting conditions are observed using the 2D PIV technique. Two-component velocity fields are plotted to understand the flow structure. It is observed that the increase in outer flow Reynolds number will increase the radial velocity and has less effect on the increase in axial velocity. Vortex breakdown was observed from the radial velocity field while increasing the outer flow Reynolds stress. From the axial velocity field, it is noted that the inner recirculation zones shift toward downstream with an increase of Reynolds number. Also, turbulent parameters such as Reynolds stress and turbulent intensity are plotted to understand the turbulent swirl flow characteristics. The higher the Reynolds number at outer flow, the higher the Reynolds stress and turbulence intensity. The turbulent parameters also increase radially since the increase in outer flow Reynolds number affects the flow field. Because of this phenomenon, the flame size may vary, and flame may not be a compact one under-reacting flow condition.

Acknowledgements This work is supported by the Department of Science and Technology (DST), India (Grand no ECR/2015/000343). The authors are grateful to the Ministry of Human Research and Development (MHRD), India, for research fellowship.

References

1. Ahmed SA, Nejad AS (1992) Velocity measurements in a research combustor part 1: Isothermal swirling flow. Exp Therm Fluid Sci 5(2):162–174
2. Lilley DG (1977) Swirl flows in combustion: a review. AIAA J 15(8):1063–1078
3. Vu BT, Gouldin FC (1982) Flow measurements in a model Swirl combustor. AIAA J 20(5):642–651
4. Tung C, Shiung S, Hochgreb S (2015) Combustion performance of a counter-rotating Double Swirl flame burner under stratified burning condition. Chem Eng Trans 45(December):193–198
5. Cai J, Jeng SM, Tacina R (2001) Mult-Swirler aerodynamics: experimental measurements. AIAA J 2001-3574(c)
6. Bhatia D, Atkinson C, Kitsios V, Mazellier N, Soria J (2017) Estimation of turbulent dissipation rate using 2D data in channel flows. In: Proceeding of the 3rd World Congress on mechanical, chemical, and material engineering, no 1, pp 1–8
7. Watson KA, Lyons KM, Donbar JM, Carter CD (2000) Visualization of multiple scalar and velocity fields in a lifted jet flame. J Vis 3(3):275–285
8. Takeuchi J, Miraghaie R, Yuki K, Satake S, Kunugi T, Morley NB (2006) Turbulent velocity profile measurement in circular pipe flow using particle image velocimetry technique, pp 140–143

9. Schabacker J, Bölcs A (1996) Investigation of turbulent flow by means of the PIV method. In: Proceedings of the 13th symposium measuring techniques for transonic supersonic flows cascades turbomachines, p 9
10. Sweeney MS, Hochgreb S, Dunn MJ, Barlow RS (2011) A comparative analysis of flame surface density metrics inpremixed and stratified flames. Proc Combust Inst 33(1):1419–1427
11. Zhou R, Balusamy S, Sweeney MS, Barlow RS, Hochgreb S (2013) Flow field measurements of a series of turbulent premixed and stratified methane/air flames. Combust Flame 160(10):2017–2028
12. Sweeney MS, Hochgreb S, Dunn MJ, Barlow RS (2013) Multiply conditioned analyses of stratification in highly swirling methane/air flames. Combust Flame 160(2):322–334
13. Sweeney MS, Hochgreb S, Dunn MJ, Barlow RS (2012) The structure of turbulent stratified and premixed methane/air flames II: Swirling flows. Combust Flame 159(9):2912–2929
14. Chong CT, Hochgreb S (2014) Spray flame structure of rapeseed biodiesel and Jet-A1 fuel. Fuel 115:1–8
15. Murugan R, Balusamy S, Kolhe PS (2018) Experimental investigation of the spray structure of novel flow blurring twin-fluid atomizer 1, pp 1–8
16. Vanierschot M, Van den Bulck E (2008) Influence of swirl on the initial merging zone of a turbulent annular jet. Phys Fluids 20(10)

Wake Interaction of Two Rotationally Oscillating Cylinders Placed in a Side-By-Side Configuration

Izhar Hussain Khan, Rahul Yadav, and Sanjay Kumar

Abstract Experimental investigation of flow past two rotationally oscillating cylinders placed in a side-by-side configuration in a crossflow has been carried out at Reynolds number 150. The results on wake structure are presented for normalized cylinder spacing (center-to-center spacing normalized by cylinder diameter) of 1.4, 1.8, 2.5, 4.0 and 7.5. Forced sinusoidal oscillation amplitudes about the cylinder axes varied from $\pi/8$ to π radians and normalized forcing frequency varied from 0.25 to 5.0. Both in-phase and out-of-phase configurations are considered. Hot wire anemometry and planar laser-induced fluorescence technique are used to characterize the wake structure. The interaction between the two wakes is seen to be strongly dependent on the spacing ratio and phase of oscillation between the cylinders in addition to being subjected to the effects of oscillation frequency and amplitudes. Some of the distinct wake structures that have been observed are the synchronized anti-phase vortex wake structure, bi-stable-biased wake structure with superimposed small-scale vortices, single bluff body large wavelength vortex wake structure, asymmetric P + S and P + 3S vortex wake structures, etc. An interesting observation of symmetry breaking was observed at $T/D = 2.5$ and 4.0 in the in-phase oscillation mode. Wake structure becomes independent at spacing ratio of 7.5, and each cylinder wake structure is similar to a single cylinder performing rotational oscillation. Some of the wake structures are also seen to be chaotic and highly irregular in nature. A wake boundary diagram is also shown for $T/D = 4.0$ case.

Keywords In-phase and anti-phase shedding · Constructive and destructive interference · Lock-on

I. H. Khan (✉) · R. Yadav · S. Kumar
Department of Aerospace Engineering, Indian Institute of Technology Kanpur, Kanpur, Uttar Pradesh 208016, India
e-mail: izhar@iitk.ac.in

L. Venkatakrishnan et al. (eds.), *Proceedings of 16th Asian Congress of Fluid Mechanics*, Lecture Notes in Mechanical Engineering,
https://doi.org/10.1007/978-981-15-5183-3_41

1 Introduction

Multiple bluff bodies occur in many practical engineering applications such as cables of a bridge, transmission wires and tubes in a heat exchanger. Significant work can be found in the literature regarding the mutual interference effects of two or more bluff bodies and their resulting wake patterns. Bluff body wakes and their modifications have been primarily studied for circular cylinder geometry, partly because of their simplicity in both experimentation and numerical/theoretical studies. Studies have been made for different spacing between the cylinders and in various configurations such as side-by-side, tandem or a general staggered configuration. Spacing is characterized by T/D, the ratio of distance between cylinder axes and cylinder diameter.

Williamson [1] showed that a stable wake structure, which is either in the in-phase or in the anti-phase mode, can be seen if the spacing was between $T/D = 2$ and 6. Sumner et al. [2] and Wang et al. [3] have shown that the wake behaves as a single bluff body wake if the spacing is very less which is T/D is < 1.2. Therefore, in order to observe and understand the wake structure in various regimes that depend greatly on the spacing, five different spacings were chosen in the present investigation, two spacings below $T/D = 2$ (1.4 and 1.8), two between $2 < T/D < 6$ (2.5 and 4.0) and one above $T/D = 6$ (7.5). Kumar et al. [4] conducted experiments on a single rotationally oscillating cylinder at various forcing amplitudes and frequencies. The phenomenon of lock-on was observed in which the dominant wake frequency in the flow is same as the applied forced oscillation frequency. It was shown that the dependence of circulation values of the shed vortices on the forcing frequency follows a decay character independent of forcing amplitude beyond forcing frequency of ~1.0 and a scaling behavior with forcing amplitude at forcing frequencies ≤1.0. Thiria et al. [5] also performed similar experiments and studied the effect of oscillation amplitude and frequency on the wake and also measured the drag force. They showed that the phase lag between the vortex shedding and the rotary motion of the cylinder gives either a constructive contribution to the wake which is responsible for a global increase in the fluctuations in the wake or a destructive contribution which gives the opposite behavior.

Rotationally oscillating the cylinder offers the opportunity to control the flow structures and alter the drag force on the body. It is very likely that such an arrangement will be used in coordination with others in any practical application related to drag reduction or fluid mixing. This offers the opportunity to study two rotationally oscillating cylinders in a side-by-side configuration at various forcing parameters. The forced oscillations on the two cylinders are of the nature, $\theta = \theta_0 \sin(2\pi ft)$, where θ_0 is the oscillation amplitude and f is the oscillation frequency about the axis of the cylinders. The forced oscillation frequency is normalized by the vortex shedding frequency from a single stationary cylinder at the given Reynolds number and is denoted by FR. The schematic of the problem studied is shown in Fig. 1.

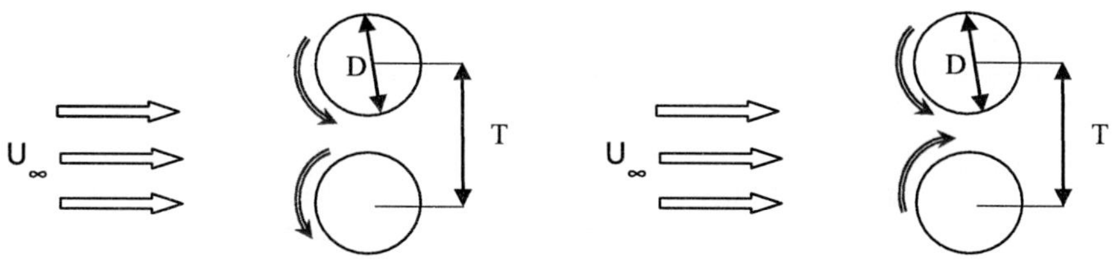

Inphase Oscillation, $\theta = \theta_o \sin(2\ \ ft)$ Antiphase Oscillation, $\theta = \theta_o \sin(2\ \ ft)$

Fig. 1 Schematic of the oscillating motion

2 Experimental Setup

The experiments were performed in a water tunnel at the Fluid Dynamics Laboratory of the Department. The water tunnel (Model: 0710, Rolling Hills Inc.) had a 0.25 m deep, 0.46 m long and 0.18 m wide test section. The sides of the test section are made of 6 mm thick tempered glass which allow the laser sheet to illuminate the required region of visualization. The top portion of the test section is a transparent plate which is made of plexiglas (acrylic) material allowing the camera to capture the required wake pattern. It was 180 mm wide, 500 mm long and 12 mm thick and contained a wide slot at one end to house the bearings, which were fitted to the two cylinders, at any given spacing. The water surface was touching this top plate at all times, and the velocity of water in the test section needed to achieve Re 150 was 0.018 m/s. The tunnel water temperature was maintained at 25 ± 2 °C. An anodized aluminum plate was used at the bottom portion of the test section which also contained a wide slot, corresponding to the top plate, to house the bearings that were fitted to the ends of the cylinders.

Simultaneous rotational motion of both the cylinders was obtained using timing belts and pulleys. Two timing belts with pulleys were fitted to the cylinders and connected to an adapter fitted to the motor shaft. The cylinders used in the experiments were stainless steel rods of length 270 mm and diameter 8 mm. The cylinder had an aspect ratio of 31.25 based on the wetted length. The blockage resulting due to the cylinders in the tunnel test section was 8.88%.

The motor used to impart sinusoidal rotary oscillations at various amplitudes and frequencies was an AC servomotor (Model: SM0602AE4-KCD-NNV, SHANGHAI MOONS' Electric Co. Ltd) connected to a servomotor drive which was computer controlled using the corresponding software (M Servo Suite). The oscillation amplitude was controlled through this software by varying the position count of the motor. In addition, the motor was also coupled to a waveform signal generator (Model: DG 1000, RIGOL) which provided the required analog sinusoidal signals of known frequency. These signals controlled the oscillation frequency of the cylinders.

Flow visualization was done by planar laser-induced fluorescence technique using Rhodamine-B as the dye substance. The plane of visualization was illuminated using a 3 W continuous laser, and the wake structure was captured with a digital camera

(Model: D810, Nikon) at 60fps. The camera which was mounted on a gearhead and a traverse mechanism allowed viewing the wake structure illuminated in the plane of visualization.

Dye injection system consists of a pressurized apparatus in order to pump the dye liquid into the tunnel test section. Air was pumped to three 500 ml HDPE bottles containing the dye, and the dye was piped to three corresponding high-precision needle valves that controlled the flow rate. Plastic tubes were then used to transport the dye to the L-shaped SS hollow tubes of diameter 0.9 mm from where the dye is finally released into the flow. The dye releasing point was 12 cm (15D) upstream of the cylinders. The flow rate of dye is adjusted in such a way that the exit velocity of the dye from the tube is approximately equal to the flow velocity. This condition was evident from the mixing characteristics of the dye.

3 Results

This section describes the wake structure as obtained using laser-induced fluorescence (LIF) technique. An extensive study was made to observe the effects of all the parameters as separate sets of experiments. For a particular T/D, first, the phase was set to either in-phase or anti-phase mode; then, for every amplitude, the forcing frequency was varied and the resulting wake structures were observed. A total of 450 cases are studied.

3.1 *Effect of Oscillation Frequency*

The effect of forcing frequency on the wake structure is presented at an amplitude of π rad at $T/D = 4.0$. Both in-phase and anti-phase cases are presented. The wake structure for the in-phase mode is shown in Fig. 2.

(i) **In-phase mode**

It is observed that at low FR of 0.25, the mode of shedding switches between in-phase and anti-phase modes although predominantly anti-phase shedding mode is observed. This behavior is similar to the unforced case [1]. At FR 0.50, the shear layers from all four surfaces seem to be relatively stretched before they curl up into vortices as we can clearly see the large wavelength structure. Shedding, however, appears to be in the in-phase mode. The two vortex pairs shed from the cylinders merge downstream to create larger vortex pair.

Increasing the FR to 0.80 renders the in-phase shedding mode as more stable and locked to the cylinder forcing. In every half cycle, the vortices behind each cylinder are shed as two single rows. The fluid between any two subsequent vortices in a row is displaced inward between the two vortices of the other cylinder leading

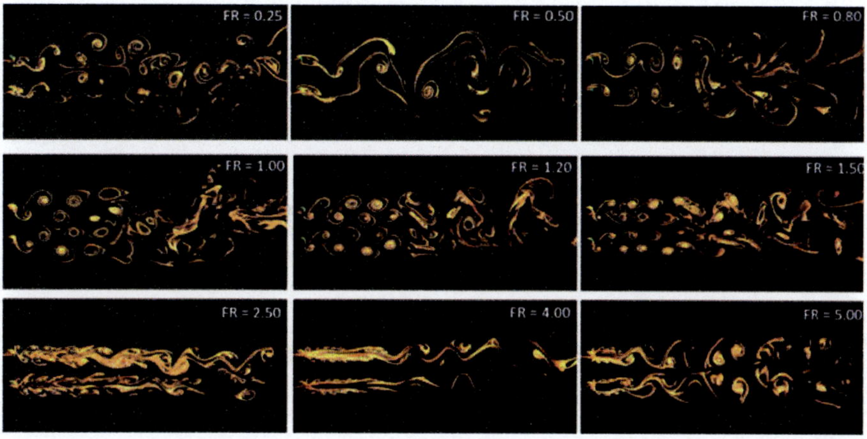

Fig. 2 Effect of forcing frequency on the wake structure, in-phase mode, $T/D = 4.0$, Amplitude $= \pi$ rad

to a structure that can be called 2S + 2S (intertwine). As these vortices advect downstream, the merger of the same sign vortices takes place. At FR = 1.0, 1.2 and 1.5, the wake structure changes to the double row mode. It is seen that the size of the shed vortices from the cylinder also reduces as forced frequency is increased leading to less interaction between the two wakes. This is very evident at FR = 2.5, 4.0 and 5.0. The small-scale vortices shed at higher frequencies merge to form two independent vortex streets which interact with each other and produce symmetrical wake patterns in the far wake similar to anti-phase shedding mode of the stationary cylinders as can be seen in the case of FR = 5.0.

(ii) **Anti-phase mode**

The effect of changing the oscillation mode to anti-phase is examined in this section. The wake structure for the anti-phase mode of forcing for similar spacing and forcing frequencies as the in-phase mode is shown in Fig. 3.

At a low FR of 0.25, both in-phase and anti-phase shedding modes are observed with no observable periodicity. The two vortex streets merge downstream leading to a more mixed wake.

Increasing the frequency to FR = 0.50, we can clearly see a large wavelength pattern where in every half cycle of the cylinder oscillation, the outer shear layers of both the cylinders are subjected to constructive interference and they roll up to form vortices; however, during the end of this half cycle, a small portion of the inner shear layers also tends to form vortices which actually undergoes destructive interference. This leads to a pattern where a total of six vortices are shed for every cycle of oscillation forming a 2P + 2S mode.

This shedding mode transitions to a very symmetric anti-phase shedding of two rows of vortex streets at FR = 0.8 which is also locked-on to the cylinder forcing. In every half cycle, a vortex is shed each from the outer shear layers and in the next

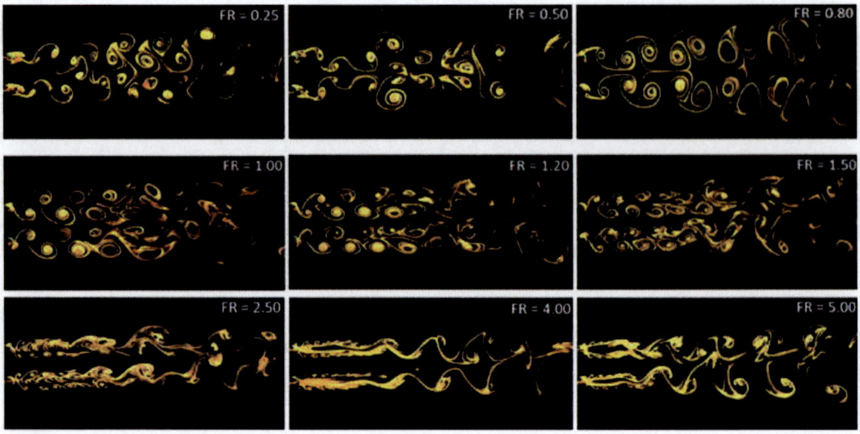

Fig. 3 Effect of forcing frequency on the wake structure, anti-phase mode, $T/D = 4.0$, Amplitude = π rad

half cycle, from the inner shear layers. This entire process is repetitive, giving rise to a fine pattern of vortices of opposite sign adjacent to each other and all advecting at the same speed relative to each other. It can be called P + P mode. Due to the opposite sign of the vortices, there is no coalescence observed.

As the FR is increased to 1.0, the spacing between vortices reduces and also the wake width. At a certain distance downstream, the vortices smear out leading to randomness and a lot of mixing in the far wake downstream. As FR is increased to 1.20 and 1.50, the wake width reduces further and the distance between the two vortex rows reduces. The far wake here slightly resembles an anti-phase shedding mode of the vortex wake.

Increasing the FR to 2.5 and higher results in two independent vortex streets near the cylinders comprising of very small-scale vortices locked to cylinder forcing. These merge, at a distance depending on forcing frequency, to form a vortex street with larger vortices. The two large scale vortex streets further interact to produce somewhat symmetric wake structure as visible in FR = 4.0 and 5.0 cases.

3.2 *Effect of Oscillation Amplitude*

To observe the effect of oscillation amplitude on the wake structure, the frequency ratio was fixed at FR = 1, the spacing between the cylinders was the same $T/D = 4.0$, and only, the antiphase mode is considered.

At a low amplitude of **π/8**, even though the motion of the cylinder is comparatively smaller, it seems enough to promote constructive interference between the cylinders and the forming vortices. The shedding is clearly in the antiphase mode. The wake seems to be relatively unstable since the mechanism of interference is weaker and

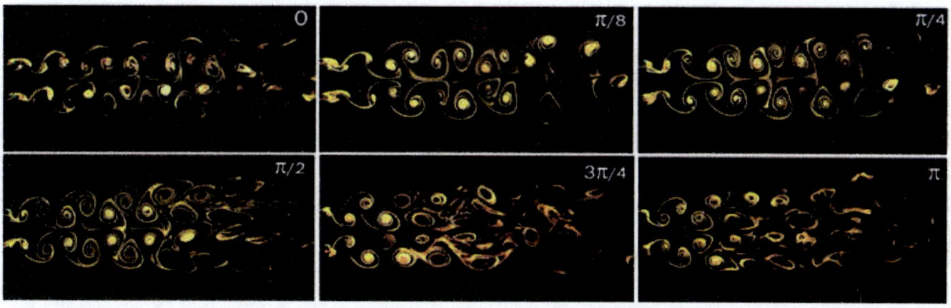

Fig. 4 Effect of forcing amplitude on the wake structure, anti-phase mode, $T/D = 4.0$, FR = 1

the vortices do not seem to pair up readily. As the amplitude is increased to **π/4**, the inter-vortex spacing becomes more compact. At higher amplitudes of **π/2**, **3π/4** and **π**, the vortex size becomes relatively smaller, the spacing becomes much more compact, and the vortices seem to merge and coalesce moving upstream closer to the cylinders. The wake structure for the antiphase mode is shown in Fig. 4.

3.3 Wake Modes

Wake modes and their boundaries are shown in the amplitude–frequency plane for both in-phase and antiphase oscillations at $T/D = 4.0$. Unlike the case of two side-by-side stationary cylinders, the influence of rotational oscillation gives rise to a whole new set of wake patterns. Many distinctive wake patterns are identified in the present parameter space while a lot of other irregular or unidentifiable patterns also exist. The mode shapes correspond to one complete cycle of cylinder oscillation (Table 1).

As shown in the table, there exist distinctive wake structures or mode shapes based on the forcing amplitude and frequency. At lower FRs and amplitudes, the wake structure is similar to the unforced case where the two vortex streets are synchronized in anti-phase or in-phase manner. A peculiar behavior of the wake occurs for the FR = 0.5 case where a transition from unforced-like wake structure to locked-on wake structure takes place. Major difference occurs near the resonant frequency as there is complete constructive interference between the cylinders and the forming vortices. These are the 2P and the 2S + 2S modes as seen in Figs. 2 and 3. In the higher FR end, the 2P double row mode dominates where no significant interaction between the two vortex wakes can be seen in both the in-phase and anti-phase modes and the far wakes again tend to behave as synchronized wakes. In the in-phase case, an asymmetric P + 2S mode is also observed between the transition region from the more stable 2S + 2S mode and the 2P(d) mode. The boundaries are shown in Fig. 5.

Table 1 Wake modes for $T/D = 4.0$. Syn: synchronized wake, FW Syn: far wake synchronized, ✷: 2S + 2S intertwine, **O**: P(double row) + 2S, ■: Large wavelength structure, ★: 2P(double row), □: 3/2(2P + 2S), ⅄: 2P(antiphase), ---: unidentifiable/irregular wake pattern

	θ_o\FR	0.25	0.5	0.8	1.0	1.2	1.5	2.5	4.0	5.0
In-phase	π/8	Syn	---	---	✷	✷	---	---	FW Syn	FW Syn
	π/4	Syn	---	---	✷	O	---	---	FW Syn	FW Syn
	π/2	Syn	■/---	✷	✷	★	★	★	★	★
	3π/4	---	■	✷	O	★	★	★	★	★
	π	---	■	✷	O	★	★	★	★	★
Anti-phase	π/8	Syn	Syn	---	⅄	⅄	---	---	FW Syn	FW Syn
	π/4	Syn	Syn	⅄	⅄	⅄	---	---	FW Syn	FW Syn
	π/2	---	□/---	⅄	⅄	⅄	★	★	★	★
	3π/4	---	□	⅄	⅄	⅄	★	★	★	★
	π	---	□	⅄	⅄	★	★	★	★	★

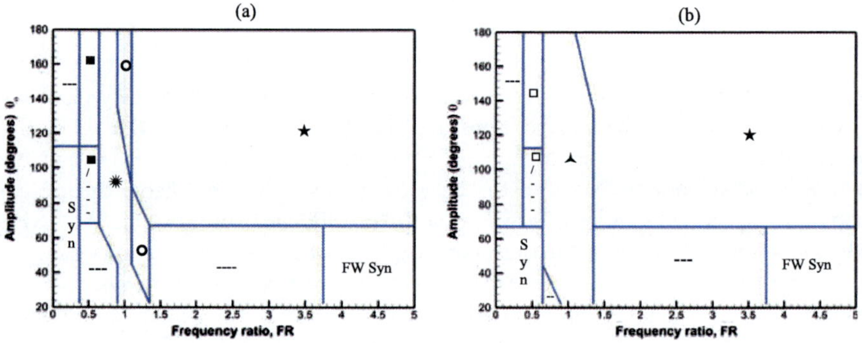

Fig. 5 Wake boundaries in the amplitude-frequency plane. **a** In-phase. **b** Anti-phase Syn: synchronized wake, FW Syn: far wake synchronized, ✷: 2 S + 2S intertwine, **O**: P(double row) + 2S, ■: Large wavelength structure, ★: 2P (double row), □: 3/2(2P + 2S), ⅄: 2P (antiphase), ---: unidentifiable/irregular wake pattern

4 Conclusions

Extensive flow visualization study of the wake of two rotationally oscillating cylinders placed in a side-by-side configuration is conducted. Dependence of the wake structure on the forcing frequency, amplitude, and phase of oscillation and spacing between the cylinders is observed. Many distinct flow patterns have been noticed such as the single bluff body biased wake, synchronized wake structure for both in-phase and anti-phase modes: 2P pattern which is seen in the anti-phase mode for FR ranging from 0.8 at higher amplitudes and 1.2 at lower amplitudes. This is observed for T/D 2.5 and 4.0 and in few cases of 1.8 as well. A different form of 2P pattern, which is named 2S + 2S, is observed for $T/D = 4.0$ where the vortex pairs are same in number but shedding is in the in-phase mode. Another wake structure where the vortices advect as pairs, with shedding in the in-phase mode, but the spacing equal to T/D 2.5 was observed to be significantly different. Some wake structures have been seen to be unique to a particular parameter such as the (2P + 2S) pattern seen in the antiphase mode for FR 0.5, amplitude of $\pi/2$ and above and for spacing ratio of $T/D = 4$ only. Similarly, (P + 3S) pattern is seen for FR = 1 and amplitude = π at $T/D = 1.4$ in the anti-phase mode. A diagram showing wake boundaries for $T/D = 4.0$ is shown.

Acknowledgements We would like to thank Dr Kamal Poddar for providing us the laser facility for our flow visualization experiments. We are also grateful to the personnel in the department workshop who helped to fabricate the setup and also our laboratory mates for their support and fruitful discussions.

References

1. Williamson CHK (1985) Evolution of a single wake behind a pair of bluff bodies. J Fluid Mech 159:1–18
2. Sumner D (2010) Two circular cylinders in cross-flow: a review. J Fluids Struct 26:849–899
3. Wang ZJ, Zhou Y (2005) Vortex interactions in a two side-by-side cylinder near-wake. Int J Heat Fluid Flow 26(3):362–377
4. Kumar S, Lopez C, Probst O, Francisco G, Askari D, Yang Y (2013) Flow past a rotationally oscillating cylinder. J Fluid Mech 735:307–346
5. Thiria B, Goujon-Durand S, Wesfreid JE (2006) The wake of a cylinder performing rotary oscillations. J Fluid Mech 560:123–147

Growth of Disturbances in a Pre-transitional Boundary Layer Downstream of Distributed Surface Roughness

Robin Joseph and Sourabh S. Diwan

Abstract The growth of disturbances in the pre-transitional region of a boundary layer in presence of surface roughness is of relevance to many engineering situations and has been a topic of research for several studies in the past. In this work, we present velocity measurements in a boundary layer developing on a flat plate (mounted in a wind tunnel) downstream of a 2-D strip of distributed surface roughness of limited streamwise extent. Two different grades of sandpaper roughness are used toward this purpose. The power spectra of the streamwise velocity fluctuations show that the roughness introduces distinct frequencies as well as a broad range of disturbances into the boundary layer, with the latter being primarily responsible for the onset of transition. The disturbances in the pre-transitional region exhibit amplification which can be attributed to the transient growth mechanism. For the finer grade of roughness, the disturbance kinetic energy grows approximately linearly with streamwise distance from the mean roughness position, which is broadly consistent with the linearized disturbance theories. The wall-normal location of the peak disturbance energy shifts away from the wall with increasing streamwise distance, and the fluctuating velocity profiles exhibit a non-similar distribution in the streamwise direction. These results are qualitatively similar to the theoretical predictions of Fransson et al. (Phys Fluids 16:3627–3638, 2004) but in contrast to the optimal-disturbance theory of Luchini (J Fluid Mech 404:289–309, 2000), both of which deal with steady disturbances. The results for the coarser roughness, however, depart considerably from these theoretical trends.

Keywords Distributed roughness · Pre-transitional boundary layer · Hot-wire anemometry · Transient growth

R. Joseph · S. S. Diwan (✉)
Department of Aerospace Engineering, Indian Institute of Science, Bangalore 560012, India
e-mail: sdiwan@iisc.ac.in

L. Venkatakrishnan et al. (eds.), *Proceedings of 16th Asian Congress of Fluid Mechanics*, Lecture Notes in Mechanical Engineering,
https://doi.org/10.1007/978-981-15-5183-3_42

1 Introduction

It is well known that under elevated levels of external disturbances, boundary layers bypass the *natural* route to transition via Tollmein–Schlichting wave growth, and this phenomenon is known as "bypass transition" [1]. The disturbances are typically introduced in the form of free stream turbulence (FST) or surface roughness. There have been many studies on the FST-induced transition, which have identified the key stages in the transition process [1]. The roughness-induced transition is particularly diverse due to a variety of types and configurations possible like isolated roughness elements, two dimensional steps, array of roughness, distributed roughness, etc. Despite being representative of practical engineering surfaces, distributed roughness has not received much attention in comparison with other types of roughness [1]. In an early study, von Doenhoff and Horton [2] determined the critical Reynolds number for transition based on the roughness height from measurements downstream of an airfoil with a strip of sandpaper roughness pasted on it. Reshotko and Levanthal [3] conducted measurements on a Blasius boundary layer with two grades of roughness covering the whole plate; they found that the coarse roughness causes a significant growth in the disturbance amplitudes. Other studies have either used an array of isolated roughness elements (e.g., [4, 5]) or an array of local patches of quasi-random distributed roughness [6].

There have also been efforts, in the last two decades, to understand the theoretical aspects of bypass transition, especially using transient growth ideas. Luchini [7] introduced disturbances outside a boundary layer (akin to FST-induced transition studies) and calculated the initial "optimal" disturbances that would result in a maximum transient growth. A key outcome of Luchini's theory is the near universality of streamwise fluctuation velocity profiles irrespective of whether the input disturbance is optimal or not. These results were experimentally validated by Matsubara and Alfredsson [8] for FST-induced transition. Fransson et al. [5] carried out a theoretical analysis by introducing disturbances downstream of the plate leading edge and found that at certain downstream locations, such disturbances can produce higher growth than optimal perturbations at the leading edge. They reported streamwise fluctuation velocity profiles that are not self-similar in contrast to the Luchini's [7] theory. Moreover, by adjusting the sub-optimality of the input disturbances, they were able to match their theoretical predictions with the experimental results on an array of cylindrical roughness elements. Downs et al. [6] reported the presence of transient growth of steady disturbances downstream of patches of quasi-random roughness. On the other hand, White and Reshotko [9] did not find a clear indication of transient growth for a distributed sandpaper roughness.

The aim of the present work is to study the growth of pre-transitional disturbances in a boundary layer in the presence of a 2-D strip of distributed (grit) roughness. Two types of roughness are used to study the similarities and differences in the disturbance growth. We observe amplification of disturbances downstream of roughness for subcritical Reynolds numbers, implying the presence of transient growth. The finer roughness grade exhibits linear growth of disturbance energy and non-similar fluctuating velocity profiles consistent with the theory of Fransson et al. [5]. The coarser roughness shows significant departures from the theoretical predictions.

2 Experimental Setup

The experiments are conducted in an open-circuit wind tunnel at the Department of Aerospace Engineering, IISc, Bangalore. The test section dimensions are 0.5 m × 0.5 m in cross section and 3 m in length. The free-stream turbulence intensity is about 0.1%. A flat plate of length 2.1 m with a super elliptic leading edge is mounted in the test section. At 100 mm from the leading edge, a sandpaper strip of 40 mm streamwise length is pasted that covers the whole span (Fig. 1). Two sets of measurement are conducted, one with extra coarse emery cloth (Grade 24) of overall height ~1.5 mm and the other with silicon carbide waterproof paper (Grade 80) of height ~0.6 mm (referred to as R24 and R80, respectively). The roughness height (k) includes the thickness of the cloth/paper base on which the roughness particles are glued. Streamwise velocity measurements are carried out using a hot-wire probe connected to the Streamline Pro CTA from Dantec Dynamics. Hot-wire data is acquired using a National Instruments DAQx board and is corrected for ambient temperature variation. The data is sampled at 20 kHz and is low-pass filtered at 10 kHz as per the Nyquist criterion to remove aliasing. The streamwise, wall-normal and spanwise co-ordinates are represented by X, y and z, respectively. U and u represent the streamwise mean and fluctuating velocity respectively.

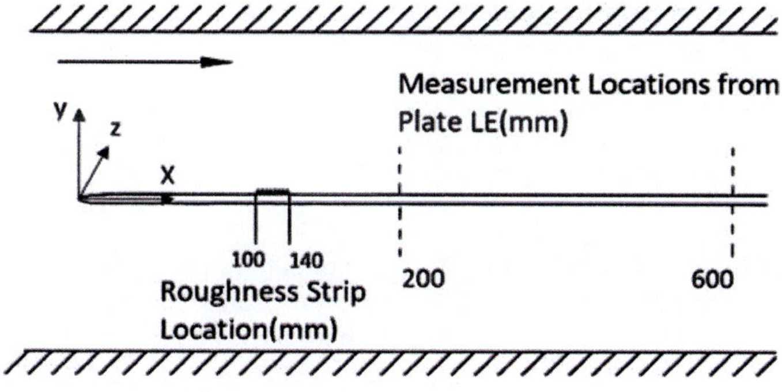

Fig. 1 Schematic of the experimental setup

3 Results

First, we determine the precise streamwise location where the boundary layer becomes transitional. A perusal of the literature shows that there is no commonly accepted definition of the start of transition zone. Some studies define the onset of transition based on the location where the turbulent spots just start appearing (e.g., Westin et al. [10]). Others define it based on crossing a certain threshold for the intermittency in the velocity signals (e.g., Fransson et a. [11]). For the present purposes we use the former method and consider the first appearance of spots during the acquisition time of 60 s to be the onset of transition. For convenience, $X = 500$ mm is chosen as the location for the start of transition and the free-stream velocity (U_∞) is adjusted accordingly. The corresponding free-stream velocities are 11.5 m/s for R80 and 6.7 m/s for R24. Table 1 shows the relevant parameters associated with each roughness grade. The displacement thickness, δ_k^*, and the 99% boundary layer thickness (δ_{99}) are calculated at the leading edge of the roughness. Roughness Reynolds number is defined as $Re_k = U_k * k/\nu$, where U_k is the velocity at the roughness height and ν is kinematic viscosity. δ_{99}, δ_k^* and U_k are determined from the standard Blasius relations. The critical Reynolds number, $\mathrm{Re}_{\mathrm{k\_crit}}$, is defined as the Reynolds number at which transition starts at the location of roughness. The Reynolds number based on the displacement thickness at the roughness leading edge, $Re_{\delta * k}$, is found be 350 and 457, respectively, for R24 and R80; these values are less than $Re_{\delta *} = 520$ at which the Blasius boundary layer becomes unstable with respect to Tollmein–Schlichting waves. Further downstream, for both R24 and R80, $Re_{\delta *}$ does become higher than 520, suggesting TS-waves may be present. However, the disturbance levels quickly become high (~1%) downstream of roughness and the peaks seen in the spectra decay downstream (Fig. 3). We therefore conclude that TS instability is not the primary mechanism for transition in the present case.

3.1 *Characterization of the Boundary Layer Downstream of Roughness*

Velocity profiles are measured from $X = 200$ to 600 mm in steps of 100 mm. The mean velocity profiles are seen to follow the Blasius distribution quite well in the pre-transitional region, for $X \geq 300$ mm, for both the roughness grades (not shown here).

Table 1 Summary of relevant parameters for the two roughness grades

Roughness	k (mm)	U_∞ (m/s)	U_k(m/s)	δ_k* (mm)	$Re_{\delta * k}$	δ_{99}(mm)	k/δ_{99}	k/δ_k*	Re_k
R24	1.5	6.7	5.74	0.84	350	2.405	0.64	1.7857	535
R80	0.6	11.5	5.98	0.64	457	1.816	0.33	0.9375	223

To quantify the growth of disturbances, profiles of u_{rms}/U_∞ are plotted in Fig. 2 for different X locations; here, u_{rms} is the root mean square of the velocity fluctuations. The scatter in the data for $X = 500$ and 600 mm is typical to that found in early stages of transition in the previous studies [10] and does not imply measurement inaccuracies. In the pre-transitional region, the u_{rms} profiles peak around $y/\delta^* \approx 1$ (except for $X = 200$ mm for R24), which is similar to the other roughness-induced transition cases involving isolated or arrays of roughness [5]. At $X = 500$ mm where spots first appear, a secondary peak, much closer to the wall, is observed for both R24 and R80 roughness (Fig. 2). Beyond this location, there is a rapid increase in the magnitude of u_{rms} and the secondary peak at $y/\delta^* \approx 0.3$ becomes prominent at $X = 600$ mm. This observation supports our definition of the onset of transition (at $X = 500$ mm) based on the appearance of spots.

The power spectral densities of the fluctuating velocity signals are plotted in Fig. 3a, b for R24 and R80, respectively. At each streamwise location, the signal

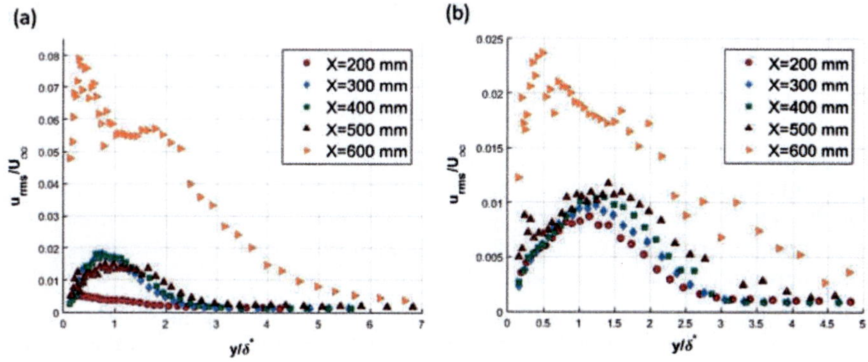

Fig. 2 Profiles of total u_{rms} fluctuation for **a** R24 roughness and $U_\infty = 6.7$ m/s, **b** R80 roughness and $U_\infty = 11.5$ m/s

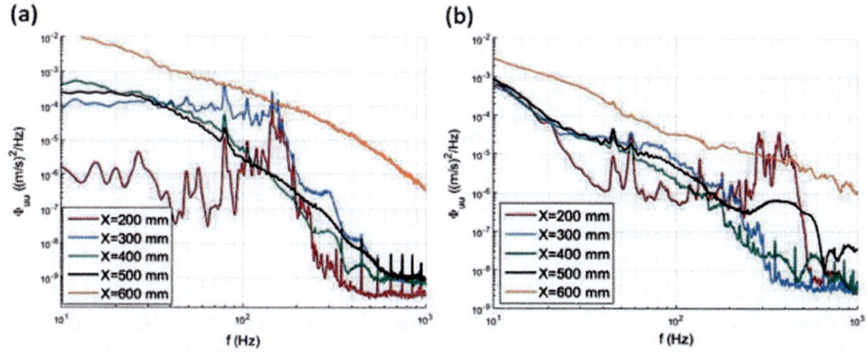

Fig. 3 Power spectral density (Φ_{uu}) versus frequency (f) for **a** R24, $U_\infty = 6.7$ m/s. **b** R80, $U_\infty = 11.5$ m/s

at the wall-normal location of maximum u_{rms} is used to calculate the spectra. At $X = 200$ mm, several peaks are found in the signal along with an elevated broadband energy in the range 100–200 Hz (Fig. 3). This shows that the distributed roughness introduces highly unsteady disturbances in the boundary layer for both the grades of roughness, with the peaks in the spectra possibly representing the vortices shed from the roughness strip. This should be contrasted with the previous studies on arrays of isolated roughness elements which typically result in steady disturbances at comparable values of Re_k; for example, Fransson et al. [5] found steady disturbances up to $Re_k = 340$, while Re_k for R80 is 223 (Table 1). The distinct peaks in the spectra become progressively less prominent as we approach the transition location ($X = 500$ mm), at which point they are barely discernible (Fig. 3). This suggests that the broadband disturbance field introduced by the roughness is primarily responsible for causing the transition, and the shed vortices do not play an important role. This scenario is similar to the FST-induced transition wherein broadband disturbances are introduced in the free stream [8]; in the present case, the disturbances are introduced right inside the boundary layer, close to the wall.

3.2 Comparison with Transient Growth Studies

Luchini [7] predicted the peak in the streamwise fluctuating velocity (u) profile for optimal disturbance to be located at $y/\delta^* = 1.3$, and this was validated experimentally by Matsubara and Alfredsson [8]. Luchini [7] further showed that the u-profiles are mostly invariant of the spanwise wavenumber of the initial disturbances. This explains why the u_{rms} profiles in the FST-induced transition experiments, which largely introduce sub-optimal disturbances, match well with the profile of the optimal disturbance over the lower part of the boundary layer.

In the present experiments, neither R24 nor R80 roughness has a well-defined u_{rms} maximum at $y/\delta^* = 1.3$ (Fig. 4). In the R80 roughness case, the peak starts at $y/\delta^* < 1.3$ and increases to $y/\delta^* > 1.3$ at $X = 500$ mm (Fig. 4b), while the R24 roughness has a peak that is at $y/\delta^* < 1$ for all X locations (Fig. 4a). In both the roughness configurations, the location of the peak in u_{rms} gradually increases with increasing X. Thus, the u_{rms} profiles in the present case do not follow a self-similar distribution in the pre-transitional region. These results are closer in spirit to the experiments of Fransson et al. [5] involving measurement in a boundary layer past an array of cylindrical roughness elements. They also observed a non-self-similar distribution of the u_{rms} profiles and a gradual movement of the wall-normal location of the maximum u_{rms} away from the wall.

The downstream evolution of the maximum u_{rms} (denoted as $(u_{rms})_m$) and the transitional intermittency are plotted in Fig. 5; the intermittency is calculated using the method used in [11]. At each X, the velocity signal at the wall-normal location of maximum u_{rms} is chosen to calculate intermittency. For R80 roughness, the disturbance kinetic energy, i.e., $(u_{rms})_m^2$, is seen to increase linearly with X–X_r in the pre-transitional region, with an increase in growth rate in the early part of transition

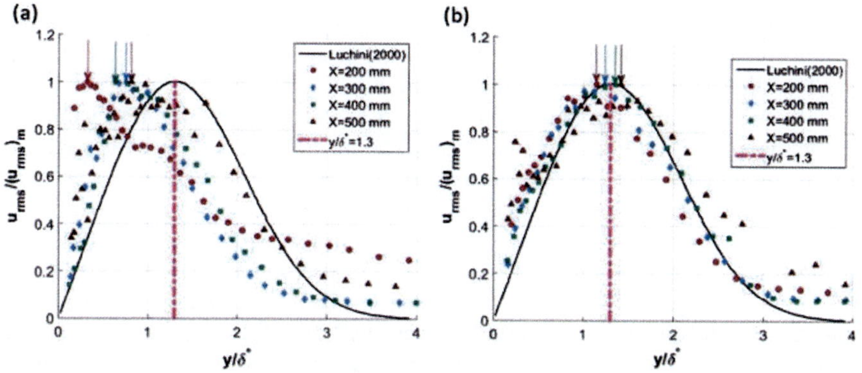

Fig. 4 Comparison of fluctuation velocity profiles with Luchini's optimal perturbation profile. **a** R24 roughness, $U_\infty = 6.7$ m/s. **b** R80 roughness, $U_\infty = 11.5$ m/s. The arrows mark the peak for each profile

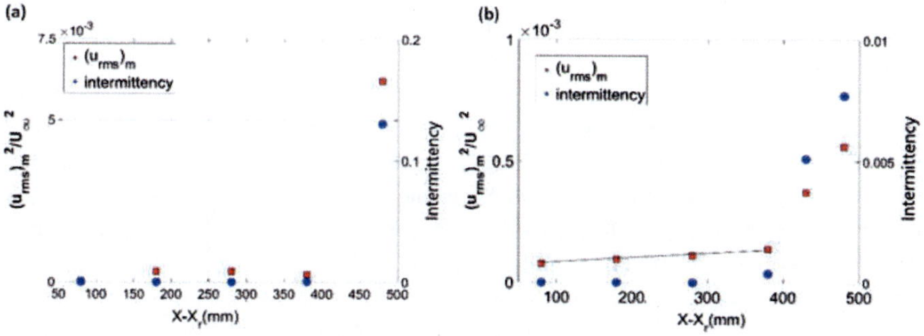

Fig. 5 Variation of $(u_{\text{rms}})_m^2$ and intermittency with X–X_r for **a** R24, $U_\infty = 6.7$ m/s. **b** R80, $U_\infty = 11.5$ m/s

zone ($X > 500$ mm); here, $X_r = 120$ mm is the mean roughness location. The linear growth of the disturbance energy in the pre-transition region was also reported in [8] for the FST-induced transition and is broadly consistent with the outcome of the linearized disturbance theories ([12]). We are not aware of any experiments on roughness-induced transition which have reported such linear disturbance energy growth downstream of distributed roughness. For R24 roughness, there is a marginal growth in $(u_{\text{rms}})_m^2$ from $X = 300$ mm to $X = 400$ mm and in fact a slight drop at $X = 500$ mm (Fig. 2a), after which the disturbance energy shows a sharp rise as the boundary layer becomes transitional (Fig. 5a). Thus, for this case, the disturbances seem to decay somewhat (after the initial amplification), before the onset of transition. We do not have sufficient number of measurement locations upstream of $X = 400$ mm for the R24 case to assess the linearity of the disturbance energy.

The variation of intermittency (Fig. 5) confirms that the transition to turbulence is much quicker in R24 roughness in comparison with R80 roughness. The R24 roughness is able to reach an intermittency value of nearly 0.13 at $Re_X = 2.5 \times 10^5$, whereas for R80 roughness a Reynolds number of around 3.5×10^5 is needed for the first appearance of spots (< 0.01 intermittency).

Next, we compare our results with the predictions of the transient growth theories of Luchini [7] and Fransson et al. [5]. The theory of Fransson et al. [5], who introduced disturbances downstream of the leading edge, is expected to be more directly applicable to the present case wherein disturbances are introduced from $X = 100$ mm to $X = 140$ mm on the plate. Towards this, we compare the wall-normal location (y_{max}) of maximum u_{rms} obtained in our experiment with that predicted by these theories (Fig. 6). Here, X_{peak} is the streamwise location corresponding to the peak value of the disturbances. In the theories of Luchini [7] and Fransson et al. [5], the disturbances initially exhibit a transient growth followed by a decay (i.e., they do not lead to transition) and X_{peak} corresponds to the location of maximum disturbance energy. For the present experiment, the boundary layer does become transitional for $X > 500$ mm, and therefore, some care is needed in choosing X_{peak}. For R24 roughness, $X_{peak} = 400$ mm since the disturbance energy shows a slight decay beyond this point (Fig. 5a). For R80 roughness, disturbances grow monotonically up to the transition location, and we assume that this location corresponds to the peak disturbance energy, i.e., $X_{peak} = 500$ mm.

Figure 6 shows that the increasing trend in y_{max} with X for the R80 case is qualitatively similar to the theoretical result of Fransson et al. [5] and departs considerably from that of Luchini [7]. Note that the conditions in Fransson et al. are not exactly identical to those in the present experiment. Fransson et al. considered *steady* disturbances generated by an array of roughness elements, whereas the distributed roughness introduces a broad range of *unsteady* disturbances (Fig. 3). Furthermore, the sub-optimality parameter in Fransson et al. was tuned to match their experimental results, and the same value of the parameter may not hold for the present conditions. The role of unsteady disturbances in bypass transition for a roughness array has been

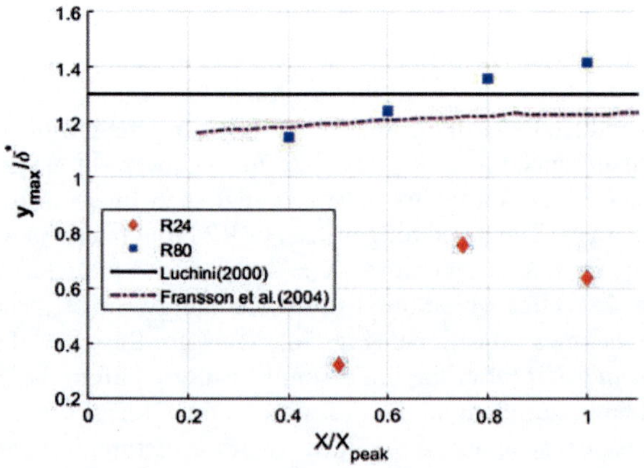

Fig. 6 Comparison of y_{max} between the present experiment and existing theoretical results

studied in some of the earlier studies (e.g., Ergin and White [4]). However, no theoretical comparison of the experimental results for unsteady disturbances has been made, to the best of our knowledge. Finally, the y_{max} locations for the R24 case are much lower than those for the R80 case and therefore much farther from the theory (Fig. 6). The reason for this behavior is not entirely clear but may be attributed to the higher k/δ_{99} for R24 as compared to R80 (table 1). Since k is a larger fraction of the local boundary layer thickness (δ_{99}) for R24 roughness, this could cause a considerable modification to the boundary layer structure downstream of it. This may result in an initial condition for the growth of disturbance for R24 which is qualitatively different from that for R80. This is a matter of ongoing investigation.

4 Conclusions

In this work, we have characterized the pre-transitional region downstream of a 2-D strip of distributed surface roughness, using two roughness grades: 24-grit (R24) and 80-grit (R80). The power spectra of the streamwise velocity fluctuations show that the roughness introduces distinct frequencies as well as a broad range of disturbances into the boundary layer. The former become less prominent as the transition location is approached and the transition is seen to be caused primarily by the broadband disturbance field. The disturbances grow in the pre-transitional boundary layer; for R80, the growth is monotonic, whereas, for R24, the disturbance level reaches a maximum and exhibits a slight decay before the boundary layer becomes transitional. For the R80 roughness, $(u_{rms})_m^2$ grows linearly with streamwise distance in the pre-transitional region, which is broadly consistent with the transient growth and other linearized disturbance theories [12]. Furthermore, for this roughness, the y-location of the $(u_{rms})_m$ increases with streamwise distance and the fluctuating velocity profiles exhibit a non-similar distribution. These results are qualitatively (though not quantitatively) similar to the theoretical predictions of Fransson et al. [5] but in contrast to the optimal disturbance theory of Luchini [7], both of which deal with steady disturbances. The results for the R24 case depart considerably from the theoretical trends, for reasons not entirely clear. The existing transient growth theories may have to be extended to include the effect of unsteady disturbances, to reconcile the differences between theory and the present experimental results.

Acknowledgements The authors thank Dr. P Phani Kumar and Mr Aditya Anand for their help in setting up the experiment.

References

1. Durbin PA (2017) Perspectives on the phenomenology and modeling of boundary layer transition. Flow Turbulence Combust 99:1–23
2. von Doenhoff AE, Horton EA (1958) A low-speed experimental investigation of the effect of a sandpaper type of roughness on boundary-layer transition. NACA Tech. Rep. TR 1349
3. Reshotko E, Leventhal L (1981) Preliminary experimental study of Disturbances in a LAMINAR boundary layer DUE to distributed roughness. AIAA Paper 81-1224
4. Ergin FG, White EB (2006) Unsteady and transitional flows behind roughness elements. AIAA J 44(11):2504–2514
5. Fransson J, Brandt L, Talamelli A, Cossu C (2004) Experimental and theoretical investigation of the non-modal growth of steady streaks in a flat plate boundary layer. Phys Fluids 16:3627–3638
6. Downs SD, White EB, Denissen NA (2008) Transient growth and transition induced by random distributed roughness. AIAA J 46:451–462
7. Luchini P (2000) Reynolds-number-independent instability of the boundary layer over a flat surface: optimal perturbations. J Fluid Mech 404:289–309
8. Matsubara M, Alfredsson PH (2001) Disturbance growth in boundary layers subjected to free-stream turbulence. J Fluid Mech 430:149–168
9. White EB, Reshotko E (2002) Roughness-induced transient growth in a flat-plate boundary layer. AIAA Paper 2002-0138.
10. Westin KJA, Boiko AV, Klingmann BGB, Kozlov VV, Alfredsson PH (1994) Experiments in a boundary layer subjected to free stream turbulence. Part 1. Boundary layer structure and receptivity. J Fluid Mech 281:193–218
11. Fransson J, Matsubara M, Alfredsson PH (2005) Transition induced by free stream turbulence. J Fluid Mech 527:1
12. Brandt L (2014) The lift-up effect: the linear mechanism behind transition and turbulence in shear flows. Eur J Mech (B/Fluids) 47:80–96

Reynolds Stress Gradient and Vorticity Fluxes in Axisymmetric Turbulent Jet and Plume

Rohit Singhal, S. Ravichandran, Sourabh S. Diwan, and Garry L. Brown

Abstract The relation between the Reynolds stress gradient and the vorticity fluxes in a turbulent shear flow has been known since the time of G. I. Taylor. With recent advances in scientific computing, this question has received a renewed attention. In this work, we present results from a well-resolved direct numerical simulation of an axisymmetric turbulent jet and plume. The simulation reproduces the self-preserving features of the two flows reported in the literature and satisfies the identity relating the Reynolds stress derivative to the vorticity fluxes, to a reasonably good degree, establishing the veracity of the simulation. The axial derivative term in the identity is shown to be negligibly small, particularly when the jet/plume is in a self-preserving state. The radial profiles of the two vorticity fluxes for the jet and plume are nearly identical, suggesting a similarity of the underlying structure. The significant result is that the vorticity flux term involving the radial vorticity and azimuthal velocity is non-zero in the core of the jet as well as plume, while it is zero in the outer region of these flows. Furthermore, this term, in the core of the flow, is equal in magnitude (but opposite in sign) to the second flux term relating azimuthal vorticity and radial velocity. Coherent vorticity in the outer region of the jet/plume is in the form of hairpin vortices whereas in the core of these flows it shows a more complex shape.

Keywords Turbulent axisymmetric jet and plume · Vorticity flux · Reynolds stress · Direct numerical simulation

---

R. Singhal · S. S. Diwan (✉)
Department of Aerospace Engineering, Indian Institute of Science, Bengaluru, Bengaluru 560012, India
e-mail: sdiwan@iisc.ac.in

S. Ravichandran
Nordita, KTH Royal Institute of Technology and Stockholm University, SE-106 91 Stockholm, Sweden

G. L. Brown
Department of Mechanical and Aerospace Engineering, Princeton University, Princeton NJ 08544, USA

L. Venkatakrishnan et al. (eds.), *Proceedings of 16th Asian Congress of Fluid Mechanics*, Lecture Notes in Mechanical Engineering,
https://doi.org/10.1007/978-981-15-5183-3_43

1 Introduction

The turbulent jets and plumes play an important role in many natural and engineering situations and have been investigated in several studies in the past, e.g., see the experimental studies of Hussein et al. [1], Wang and Law [2] and the reference therein. More recently, direct numerical simulation (DNS) studies of jets/plumes have been performed (e.g., Plourde et al. [3], Reeuwijk et al. [4]) and this has provided more detailed information on the structure and statistics of these flows. Reeuwijk et al. [4] carried out a well-resolved DNS of a round jet and plume and found that the second-order statistics for the two flows were nearly identical (see also [2]). The coherent structure in a thermal plume was studied by Plourde et al. [3], who observed presence of complex three-dimensional hairpin-like vortices in the fully turbulent regime of the plume. An important advantage of the DNS is that it provides the three-component vorticity field which would be very difficult to obtain in an experimental study.

The gradients of the Reynolds shear stresses in a turbulent flow, which appear in the Navier–Stokes equations, can be related to the "fluxes" of vorticity and gradients of the normal stresses using a well-known vector identity [5]; see Eq. (1) below. Taylor [6] was one of the first to realize this relation and used it to develop a vorticity transport theory for two-dimensional flows. He attempted to extend it for three-dimensional flows [7] but found the expressions too complicated to be of practical use. As a result of this, and in part due to the experimental difficulties in measuring vorticity, Taylor's ideas were not pursued further for a long time. There has been a renewed interest, in the last decade or so, in understanding the connection between the Reynolds stress gradient and vorticity fluxes due to the availability of the numerical data sets. Brown and Roshko [5] used a large-eddy-simulation calculation to examine different components of the vorticity fluxes in a plane turbulent wake. Brown et al. [8] performed a DNS of a turbulent channel flow and investigated the transfer of the viscous stress to Reynolds stress in terms of the vorticity fluxes. The vorticity transport ideas have also been recently examined in the context of relaminarization and retransition of a turbulent boundary layer [9].

In this work, we take a close look at the relation between the Reynolds stress gradient and vorticity fluxes in a statistically axisymmetric jet and plume. To the best of knowledge, such an exercise has not been carried out in the previously reported studies on these flows. We find that the vector identity relating the Reynolds stress gradient and vorticity fluxes is satisfied to a reasonably good degree and that the axial derivative term makes a negligibly small contribution to the identity for both jet and plume. The radial distributions of the two vorticity flux terms are presented and their possible connection to the coherent vorticity in the flow has been examined.

2 Numerical Details and Validation

The governing equations used are the incompressible Navier–Stokes equations for the jet and the Boussinesq equations for the plume [4]. For both cases, a finite-difference DNS code is developed using a uniform structured Cartesian grid with a resolution of 1296 * 972 * 972 in z (axial) * x (lateral) * y (lateral) directions. The grid resolution used here is comparable to that used in [4] and is found sufficient for the present purposes. The values of Reynolds number (Re) are 3200 and 1500 for the jet and plume, respectively, whereas Prandtl number (Pr) for the plume is chosen to be 0.7. Round jet and plume flows are initiated by applying Dirichlet boundary condition for velocity (U) and temperature (T), respectively, on the bottom face of the domain ($z = 0$) as $K_o = K(z = 0, r \leq d/2) = 1$ and $K(z = 0, r > d/2) = 0$, where $K = U$ for the jet and $K = T$ for the plume. Here, r represents the radial coordinate measured from the jet/plume axis and d is the diameter of the orifice/hot-patch. The governing equations are discretized using a second-order scheme for spatial accuracy and a second-order accurate Adams–Bashforth method for time advancement. The flow domain spans $60d$ in z and $45d$ each in x and y directions. For all variables in the simulations, lateral walls are operated on classical non-reflecting open boundary conditions. For the upper surface, outflow open boundary condition defined in [10] is used. Time and azimuthal averaged statistics are indicated by an overbar and are represented in polar coordinate system, with ϕ indicating the azimuthal direction.

Figure 1 shows the variation of the typical length and velocity scales with z/d. In the region $15 < z/d < 45$, the centerline velocity decay ($U_c(z) = \bar{u}_z(r = 0, z)$), the centerline temperature decay for the plume ($T_c = \overline{T}(r = 0, z)$), and the radial width of the flow (b_{ue}, defined as $\bar{u}_z(z, r = b_{ue}) = U_c(z)/e$) follow the self-similar relations for the axisymmetric jet and plume quite well (Fig. 1); see [4]. The characteristic velocity scale for normalizing U_c is the orifice outflow velocity U_o for the jet and the buoyancy velocity $U_o = \sqrt{\beta g T_o d}$ for the plume; β is coefficient of thermal expansion, and g is acceleration due to gravity. The spread rates of the jet and plume, defined as $b_{ue} = s(z - z_o)$, are found to be similar and both $\approx$0.11, consistent with the literature [11]. The axial velocity ($\overline{u}_z(r, z)/U_c(z)$) and temperature ($\overline{T}(r, z)/T_c(z)$) profiles in the radial direction collapse well onto Gaussian curves (not shown here), confirming the existence of self-similarity. The second-order statistics, scaled on the

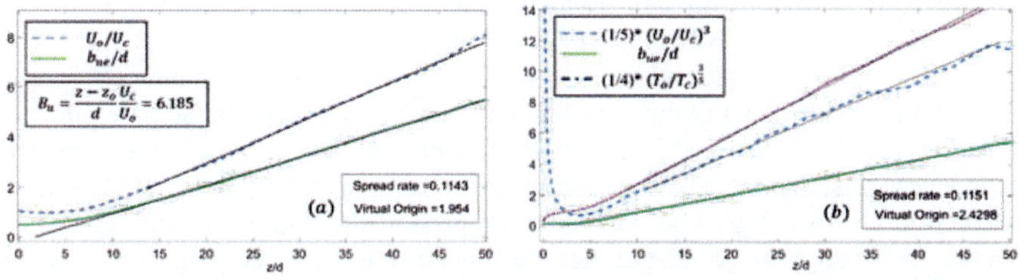

Fig. 1 Variation of centerline quantities (U_c, T_c) and radial width (b_{ue}) with z for **a** jet **b** plume

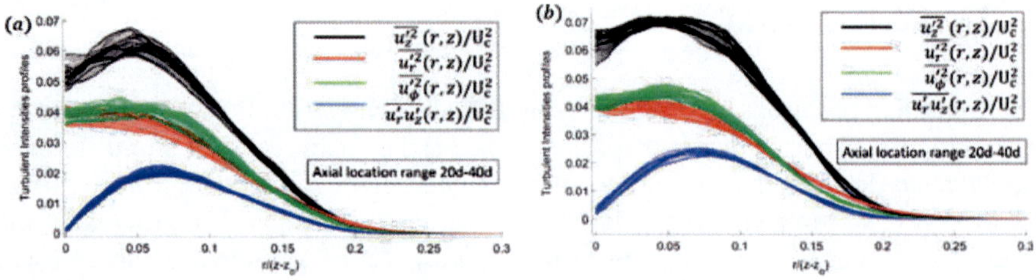

Fig. 2 Radial profiles of the second-order turbulent statistics of **a** jet **b** plume. Here, $u'_{\square}$ is the fluctuating velocity

mean centerline velocity, shows self-similar behavior to a good degree, for both jet and plume (Fig. 2), implying that the flow in both the cases has reached a "self-preserving" state in this region [12]. The distributions in Fig. 2 compare well with those obtained in earlier DNS studies [4].

3 Results

3.1 Reynolds Stress Gradient and Vorticity Fluxes

For the present context of an axisymmetric jet/plume, the gradient of the Reynolds shear stress can be expressed in terms of the vorticity fluxes as in Eq. (1). The terms in Eq. (1) have been multiplied by $(z - z_o)/U_c^2$ to render them non-dimensional by the local variables and these non-dimensional terms are denoted as I, II, and III.

$$\underbrace{\left[\frac{1}{r}\frac{\partial}{\partial r}\left(r\overline{u'_r u'_z}\right)\right]\frac{z-z_o}{U_c^2}}_{\text{I}} = \underbrace{\left(\overline{\omega'_r u'_\phi} - \overline{\omega'_\phi u'_r}\right)\frac{z-z_o}{U_c^2}}_{\text{II}} + \underbrace{\frac{1}{2}\frac{\partial}{\partial z}\left(\overline{u'^2_r} + \overline{u'^2_\phi} - \overline{u'^2_z}\right)\frac{z-z_o}{U_c^2}}_{\text{III}} \tag{1}$$

Here, ω'_r and ω'_ϕ represent the fluctuating vorticity components in r and ϕ directions, respectively. The vorticity components are calculated by post-processing the velocity field obtained from the simulations. Figure 3 shows the radial distributions of terms I, II, and III in Eq. (1). It can be seen that the net vorticity flux (term II) nearly matches with the Reynolds stress gradient (term I), with the axial derivative term (III) making a small contribution for both jet (Fig. 3a) and plume (Fig. 3b). Terms I and II exhibit a relatively larger scatter close to the axis presumably due to the difficulties in calculating the radial derivatives near the axis as a result of the $1/r$ factor. Notwithstanding the scatter, Fig. 3 shows that the identity in Eq. (1) has been satisfied to a reasonably good degree by the present simulation, adding support to the veracity of the simulation. The identity is satisfied to a greater degree of accuracy for the plume than for the jet (Fig. 3). We are presently investigating the reasons for this behavior.

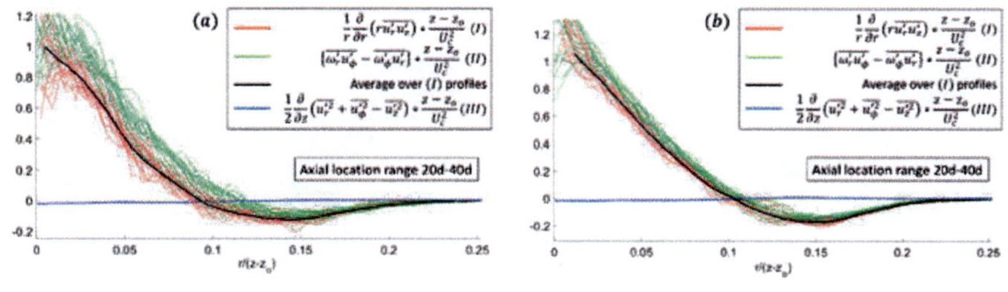

Fig. 3 Radial profiles of the different terms in Eq. (1) for **a** jet **b** plume

Next, we attempt to understand the reasons for the smallness of the axial derivative term (term III) in Eq. (1). The thin-shear-layer approximation implies that the axial derivatives should be small as compared to radial derivatives. However, since $b_{\mathrm{ue}} \propto (z - z_o)$ for the two flows, the ratio of the axial and radial derivatives does not necessarily decrease with increase in z, unlike in a turbulent boundary layer. It is therefore worth looking into this question more closely. Toward this we rearrange term III in Eq. (1) as

$$\frac{(z-z_o)}{U_{\mathrm{c}}^2}\frac{1}{2}\frac{\partial}{\partial z}\left(\overline{u_r'^2}+\overline{u_\phi'^2}-\overline{u_z'^2}\right) = (z-z_o)\frac{1}{2}\frac{\partial}{\partial z}\left\{\frac{\left(\overline{u_r'^2}+\overline{u_\phi'^2}-\overline{u_z'^2}\right)}{U_{\mathrm{c}}^2}\right\} - (z-z_o)\left(\overline{u_r'^2}+\overline{u_\phi'^2}-\overline{u_z'^2}\right)\frac{1}{2}\frac{\partial}{\partial z}\left\{\frac{1}{U_{\mathrm{c}}^2}\right\}. \quad (2)$$

Using the self-similar variations of the centerline axial velocity, i.e., $U_{\mathrm{c}} \propto (z - z_o)^{-1}$ for jet and $U_{\mathrm{c}} \propto (z - z_o)^{-1/3}$ for plume, we get $(z-z_o)\frac{1}{2}\frac{\partial}{\partial z}\left\{\frac{1}{U_{\mathrm{c}}^2}\right\} = C_1\left(\frac{1}{U_{\mathrm{c}}^2}\right)$, where $C_1 = 1$ for jet and $C_1 = 1/3$ for plume. Furthermore, the self-preserving nature of the flow (Fig. 2) implies that, the term $\left(\overline{u_r'^2}+\overline{u_\phi'^2}-\overline{u_z'^2}\right)/U_{\mathrm{c}}^2$ can be replaced by a self-similar profile: $f(r/b_{\mathrm{ue}} = \eta)$. Substituting these into Eq. (2) and with some manipulation, we get,

$$\frac{(z-z_o)}{U_{\mathrm{c}}^2}\frac{1}{2}\frac{\partial}{\partial z}\left(\overline{u_r'^2}+\overline{u_\phi'^2}-\overline{u_z'^2}\right) = \frac{-\eta}{2}\frac{\mathrm{d}f(\eta)}{\mathrm{d}\eta} - C_1 f(\eta). \quad (3)$$

Here, we have used $(\partial/\partial z) = (\mathrm{d}/\mathrm{d}\eta)\partial\eta/\partial z$ and the self-similar relation, $b_{\mathrm{ue}} = s(z - z_o)$ (Fig. 1). Figure 4 shows the variation of $f(\eta)$ as a function of η. Note that the higher scatter of this quantity is due to the fact that we are subtracting two quantities, i.e., $\overline{u_r'^2}+\overline{u_\phi'^2}$ and $\overline{u_z'^2}$, of nearly equal magnitude (Fig. 2b), which amplifies small differences. The representative variation of $f(\eta)$, obtained by averaging over the axial range $z/d = 20$–40, is shown in Fig. 4 by a dashed line. It is now possible to estimate the contribution of the terms on the RHS in Eq. (3), based on the profiles in Fig. 4, by evaluating the maximum values of $f(\eta)$ and $\mathrm{d}f/\mathrm{d}\eta$. This shows that the

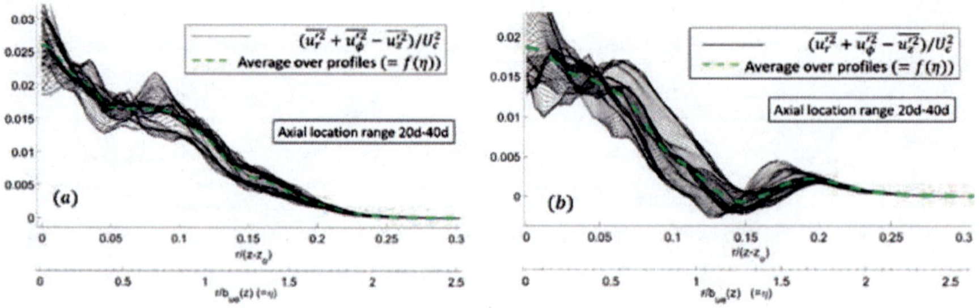

Fig. 4 Radial profile of $\left(\overline{u_r'^2} + \overline{u_\phi'^2} - \overline{u_z'^2}\right)/U_c^2$ and its self-similar representation, $f(\eta)$ for **a** jet **b** plume

maximum value of the axial derivative term (term III) is about 2% of the maximum value of term I, which explains the negligibly small contribution made by term III in satisfying the identity in Eq. (1). This exercise suggests that in addition to the thin-shear-layer approximation, the presence of self-preservation may be responsible for the axial derivative to have a negligible contribution. (This should be contrasted with the fully developed turbulent channel flow in which the streamwise derivative term is identically zero [8].) It would be interesting to evaluate the contribution of the axial derivative term for flows far from self-preservation, such a transitional jet/plume or a re-laminarizing axisymmetric wake. Incidentally, the term III in Fig. 3 is plotted using the expression in Eq. (3), which gives a much smoother distribution than obtained by directly evaluating it, as the axial derivative operator on the second-order statistics amplifies small variations and produces a noisy distribution.

3.2 Vorticity Fluxes and Coherent Structures

Figure 5 shows the contributions of the individual fluctuating vorticity fluxes to the Reynolds stress gradient term. It is interesting to note that the radial distribution of the two fluxes is quite close to each other for the jet and plume. It is known

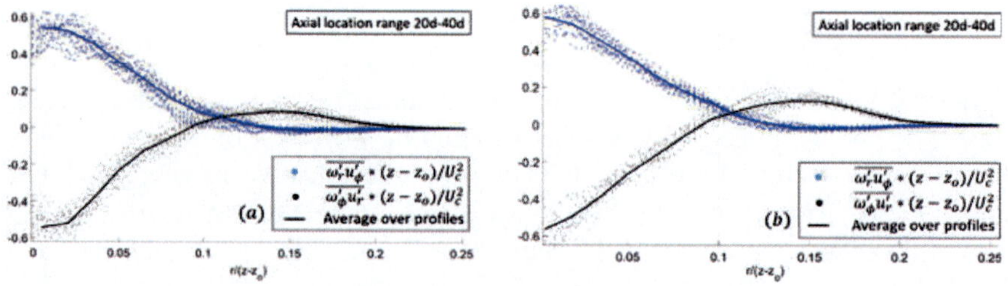

Fig. 5 Radial profiles of individual fluctuating vorticity fluxes for **a** jet **b** plume

that the Reynolds shear stress profiles for an axisymmetric jet and plume are nearly identical to each other (see [4], Fig. 2), although there is an additional buoyancy term in the governing equations for the plume. However, Fig. 5a, b shows that the individual vorticity fluxes for the two flows are also nearly identical, which is an interesting new result. The term $\overline{\omega'_\phi u'_r}$ is positive for $r/z - z_o > 0.12$ and becomes negative in the core of the jet as well as the plume. On the other hand, $\overline{\omega'_r u'_\phi}$ is zero for $r/z - z_o > 0.12$ but acquires large positive values near the axis of the jet/plume (Fig. 5). Furthermore, $\overline{\omega'_\phi u'_r}$ and $\overline{\omega'_r u'_\phi}$ are nearly equal in magnitude in the core of the jet/plume. The present results for the jet (Fig. 5a) are consistent with those obtained by [13]. The presence of large correlation between the relevant velocity and vorticity fluctuations in the core of the flow and their nearly equal magnitudes are important results; these can help us better understand the turbulence structure of the jet/plume. In particular, it is significant that $\overline{\omega'_r u'_\phi}$ has nonzero values in the core of the flow, despite the fact that the individual mean quantities, i.e., $\overline{\omega}_r$ and $\overline{u}_\phi$, are zero due to the axisymmetry condition.

It is of interest to investigate the link between the behavior of the vorticity fluxes and the coherent structures in the flow. Towards this, we use the Q criterion on the instantaneous vorticity field to identify coherent vortices in the simulated jet and plume. The Q criterion defines a vortex as a region with positive second invariant of the velocity-gradient tensor [14]. A suitable numerical value for Q is chosen to reveal the vortex-like structures; see Fig. 6a for jet and 6b for plume. In the outer region of the jet and plume, we observe presence of hairpin-like structures. This implies that in this region the radial vorticity ($\omega_r = \omega'_r$) appears in pairs of positive and negative values, corresponding to the "legs" of the hairpins. The legs of the hairpin vortices are clearly seen in the diametral sections of the flow in which alternate pairs of positive and negative radial vorticity appear in the outer region. This is shown in Fig. 6c for the plume for illustration. The fact that $\overline{\omega'_r u'_\phi}$ is zero in the outer region of the jet/plume means that the azimuthal velocity fluctuations (u'_ϕ) are orthogonal on average to radial vorticity fluctuations (ω'_r). We may suspect, therefore, that u'_ϕ and ω'_r are organized in a similar fashion, although other possibilities may exist. In the core region of the jet/plume, the situation is more complicated and the Q-vortices

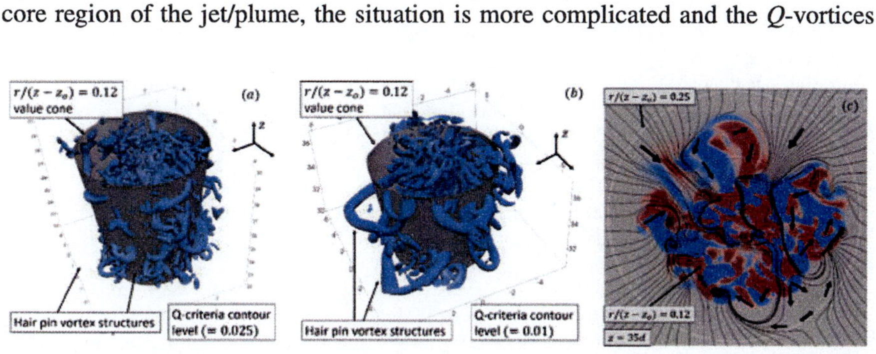

Fig. 6 Iso-surface of the Q-criterion vorticity for **a** jet and **b** plume **c** diametral section of the plume showing contours of ω_r superposed with projected streamlines. The red and blue contours, respectively, represent the positive and negative signed vorticity

show a complex pattern without a clear organization. A closer look at this region is necessary to understand the types of turbulent motions that would result in nonzero values for the two fluxes and the possible origin of such motions, which we plan to do in future.

4 Conclusion

A well-resolved DNS of an axisymmetric turbulent jet and plume has been carried out and is shown to satisfy, to a reasonably good degree, the vector identity relating the Reynolds stress gradient and the vorticity fluxes. The axial derivative term has a negligibly small contribution to the identity and this is particularly the case when the jet/plume is in a self-preserving state. The radial profiles of the two vorticity fluxes for the jet and plume are nearly identical, suggesting a similarity of the underlying structure. We find that $\overline{\omega_r' u_\phi'}$ is zero in the outer region of the jet/plume but reaches large positive values in the core region, which is a significant result. Furthermore, $\overline{\omega_r' u_\phi'}$, in the core of the flow, is equal in magnitude (but opposite in sign) to $\overline{\omega_\phi' u_r'}$. Coherent vorticity in the outer region of the jet/plume is observed in the form of hairpin vortices whereas in the core of these flows it shows a more complex shape. The role of these vortices in determining the non-zero vorticity fluxes is not yet clear.

Acknowledgements The simulations were carried out at the Supercomputer Education and Research Centre at IISc, Bangalore.

References

1. Hussein JH, Capp SP, George WK (1994) Velocity measurements in a high-Reynolds-number, momentum conserving, axisymmetric jet. J Fluid Mech 258:31
2. Wang H, Law AW-K (2002) Second-order integral model for a round turbulent buoyant jet. J Fluid Mech 459:397
3. Plourde F, Pham MV, Kim SD, Balachandar S (2008) Direct numerical simulations of a rapidly expanding thermal plume: structure and entrainment interaction. J Fluid Mech 604:99
4. Reeuwijk MV, Salizzoni P, Hunt G, Craske J (2016) Turbulent transport and entrainment in jets and plumes: a DNS study. Phys Rev Fluids 1:074301
5. Brown GL, Roshko A (2012) Turbulent shear layers and wakes. J Turbul 13:N51
6. Taylor GI (1915) Eddy motion in the atmosphere. Philos Trans A 215:1
7. Taylor GI (1932) The transport of vorticity and heat through fluids in turbulent motion. Proc Roy Soc A 135:685–705
8. Brown GL, Lee M, Moser RD (2015) Vorticity transport: the transfer of viscous stress to Reynolds stress in turbulent channel flow. In: TSFP digital library online. Benel House Inc.
9. Brown GL, Patwardhan SS, Ramesh O (2017) Re-laminarization and re-transition of a turbulent boundary layer from a vorticity point of view. In: Proceedings of the tenth international symposium on turbulence and shear flow phenomena (TSFP10)
10. Craske J, Reeuwijk MV (2013) Robust and accurate open boundary conditions for incompressible turbulent jets and plumes. Comput Fluids 86:284

11. Carazzo G, Kaminski E, Tait S (2006) The route to self-similarity in turbulent jets and plumes. J Fluid Mech 547:137–148
12. Narasimha R, Prabhu A (1972) Equilibrium and relaxation in turbulent wakes. J Fluid Mech 54:1–17
13. Narasimha R, Private communication
14. Chakraborty P, Balachander S, Adrian RJ (2005) On the relationships between local vortex identification schemes. J Fluid Mech 535:189–214

Effect of Strain Rate on Diffusion Flame Structure and Relationship in Scalar Fields

Guguloth Mahesh Nayak, K Nikhil, Pankaj S. Kolhe, and Saravanan Balusamy

Abstract The flow structure of a flickering methane jet diffusion flame is investigated using large eddy simulation employing steady flamelet model to accommodate the strain rate effect. The computed results are processed to obtain a synthetic schlieren image for direct visual comparison with the corresponding experimental schlieren image. Both the experiments and numerical simulations are carried out for jet diameter of 4 mm and Reynolds number of 500. For the accurate scalar field measurements in flames, using optical diagnostic techniques needs a careful investigation of state relationships in refractive index difference and temperature as both are dependent on the local composition. Chemical kinetics and differential diffusion affect the local composition in the flame and hence the scalar field measurements accuracy. It is observed that error in scalar field measurements close to stoichiometric mixture combustion is largest; therefore, modeling the chemical kinetics and species transport of flame is more critical for the scalar field measurements. Therefore, synthetic schlieren image comparison-based validation is employed as the numerical data accounts for strain rate effect on temperature, species concentrations and hence refractive index difference. Additionally, the flame structure is described statistically by mean, root-mean-square (RMS) and probability density function profiles of temperature.

Keywords Flickering flame · Schlieren deflectometry · Strain rate · Large eddy simulations · Steady flamelet model

1 Introduction

It is important to understand the flame dynamics to resolve the complexities involved in it. As practical combustion works on unsteady flame, the nonlinear behavior of flame arouses substantial interest to study. Laminar unsteady flames are known

G. M. Nayak (✉) · K. Nikhil · P. S. Kolhe · S. Balusamy
Department of Mechanical and Aerospace Engineering, Indian Institute of Technology Hyderabad, Kandi, Sangareddy, Telangana 502285, India
e-mail: me15m17p000001@iith.ac.in

L. Venkatakrishnan et al. (eds.), *Proceedings of 16th Asian Congress of Fluid Mechanics*, Lecture Notes in Mechanical Engineering,
https://doi.org/10.1007/978-981-15-5183-3_44

to oscillate because of buoyant acceleration of combustion products on the flame surface. Such a phenomenon is known as the flickering of the flame, which is typically exhibited by the laminar premixed and diffusion flame. The global oscillation frequency often observed around 10–20 Hz [1, 2]. Flame frequency is insensitive to the fuel type, flow rate and diameter, but strongly dependent on the Froude number [2].

In a flickering flame, the flames are stretched due to the interaction between non-uniform flows along with the flame leading to flame front curvature. Variations in the stretch rate induce local variations in the flame temperature and mass burning rate and hence species concentration. As far as species concentration is considered in the reacting flow, the effect of chemical kinetics and differential diffusion rate is important. Combustion occurs due to complex interactions between chemical reactions, mass, heat and momentum transfer. The local flow temperature is a key consequence of these interactions and must, therefore, be accurately measured. Non-intrusive temperature measurements offer important advantages in reacting flows since they do not "insert" a disturbance that can alter the flow characteristics and, consequently, change the temperature field that is to be measured. Optical methods (viz. holographic interferometry, speckle photography, speckle shearing interferometry, rainbow schlieren deflectometry, etc.) have the potential to accurately measure the entire temperature field associated with multidimensional flame, which may be difficult to do using other techniques [3]. These interferometric and deflectometric techniques first determine the refractive index in flames, which is linked to the temperature distribution. In order to measure the flame temperature using the optical diagnostic technique like rainbow schlieren deflectometry (RSD), a prior refractive index difference and temperature relationship are required. The chemical equilibrium assumption for varying mixture fraction or opposed-flow flame reactor model is generally used to obtain state relationships and thus, to relate refractive index difference to the temperature and/or species concentrations using Eq. 1. So, to accurately predict the refractive index and temperature relationship in a diffusion flame, the study of chemical kinetic effect and differential diffusion effect is important. Note that numerical simulation carried out uses the steady flamelet model to account for strain rate effect.

Assuming the ideal gas law, the refractive index, temperature and species concentration of a mixture are related by [4]

$$\delta = \frac{P}{\bar{R}T} \sum_i \kappa_i M_i X_i \tag{1}$$

where κ_i is species Gladstone–Dale constant [5], P is the ambient pressure, $\bar{R}$ is the universal gas constant, X is the mole fraction of the species. The summation is taken over all species present in the mixture. However, a potential source of error arises since the local composition of the flame being studied is usually unknown.

2 Experiment and Numerical Simulation

Figure 1 shows the schematic diagram of the experimental setup that constitutes the schlieren deflectometry system employed for methane diffusion flame visualization. Schlieren deflectometry is a line-of-sight technique that utilizes parallel light rays to detect inhomogeneities in the test medium. Three subparts of schlieren deflectometry apparatus are: (i) light source with aperture placed at focal point of collimating mirror (M1) to obtain parallel light rays, (ii) decollimating mirror (M2) focusing the light rays coming out of test section on to a knife edge at its focal point, and (iii) data acquisition with a high-speed camera with computer interface. Both collimating and decollimating mirrors used are a parabolic mirror of 15.24 cm diameter and 152.4 cm focal length. 100 μm aperture with a LED light source placed at the focal point of the collimating mirror and knife edge placed at the focal point of the decollimating mirror. A high-speed imaging system (Phantom VEO 710L) used to acquire images at 22 kHz sampling rate.

In order to compare the experimental result and capture the flow characteristics in detail, a numerical simulation is performed using a general-purpose CFD code, FLUENT [6]. In large eddy simulation (LES), the large eddies are resolved numerically while a sub-grid scale model computes the fine scale eddies. A reacting flow is simulated within the domain by solving the filtered mass, momentum, energy and species (mixture fraction) equations with appropriate numerical schemes. The boundary conditions are given to match the operating condition of the experiment. The steady-state solution of the $\kappa - \epsilon$ turbulence model is given as the initial guess for the unsteady case. For a grid dependency check, three different grid size of 100, 50 and 25 μm assigned in the fuel jet. The global oscillation frequency does not change significantly with the grid size of 50 and 25 μm. However, when the grid size is 25 μm, there is a marginal change only in oscillation frequency but the computational time of 25 μm is considerably higher than 50 μm grid size; hence, 50 μm grid size is used for the simulation in this study.

To predict the structure of flames from the experimental data, the detailed chemistry of GRI-Mech 3.0 is considered [7]. However, the main motivation for the simulation conducted here is to investigate the scalar measurement accuracy and to support the experiment qualitatively. Therefore, a synthetic schlieren image is constructed to

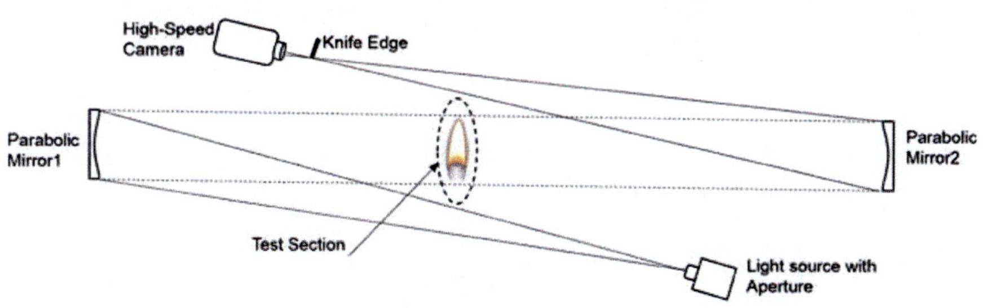

Fig. 1 Schematic diagram of the schlieren deflectometry experimental setup

enable comparison with the experimental results as most of the flickering flame visualization is obtained with the rainbow schlieren deflectometry technique. Local field data of numerical simulation is considered to generate synthetic schlieren images. The procedure to generate synthetic schlieren image from local field data is given in the following passages.

$$\delta = \kappa\rho = \frac{\eta - \eta_o}{\eta_o} \tag{2}$$

$$\Theta(x) = 2\int_{0}^{\sqrt{R^2-X^2}} \frac{\partial\delta}{\partial x}\mathrm{d}y = 2X \cdot \int_{X}^{R} \frac{\partial\delta}{\partial r}\frac{\mathrm{d}r}{\sqrt{r^2 - X^2}} \tag{3}$$

Note that, the schlieren data is path integrated refractive index difference gradient along the line of sight. Gladstone–Dale relationship links the density (ρ) with refractive index difference (δ) in Eq. 2 [4]. Figure 2 shows the flow system schematics with co-ordinate axes for identifying the line of sight. Here, x-direction is the transverse direction and y-direction is considered as line-of-sight direction. Then transverse deflection angle data can be computed using Eq. 3 [8]. The information flow diagram of in-house written MATLAB program for generating synthetic schlieren data consists of various steps listed in the below flow diagram.

Flow Diagram:

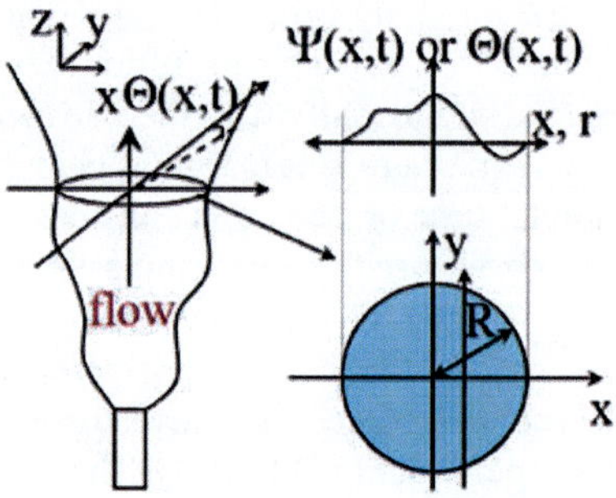

Fig. 2 Schematic for flow, axes and line-of-sight direction

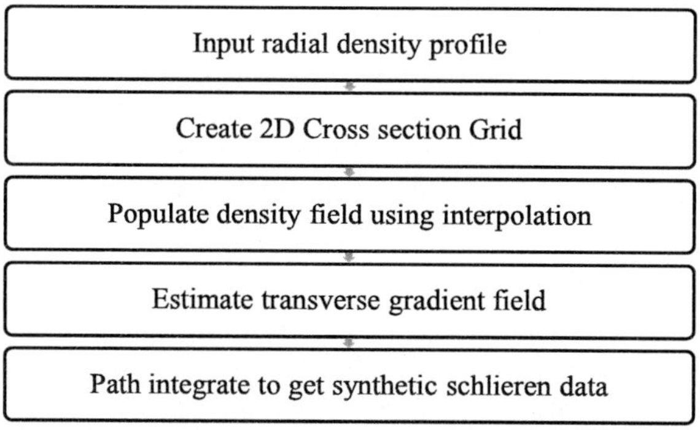

Note that the path integral in Cartesian coordinates is solved using trapezoidal numerical integration method. Benefit of using the synthetic schlieren image for comparison with experimental schlieren image is that it does not involve the tomographic reconstruction errors, uncertainty propagation errors and any limitations of the algorithm employed for reconstruction of local field using path integrated data.

3 Results

Periodic formation and convection of buoyancy-induced outer vortical structure are known to cause flicker in laminar flames. Figure 3 shows the qualitative comparison between synthetic schlieren image created from the numerical result and experimental schlieren image for methane jet diffusion flame with tube diameter 4 mm and

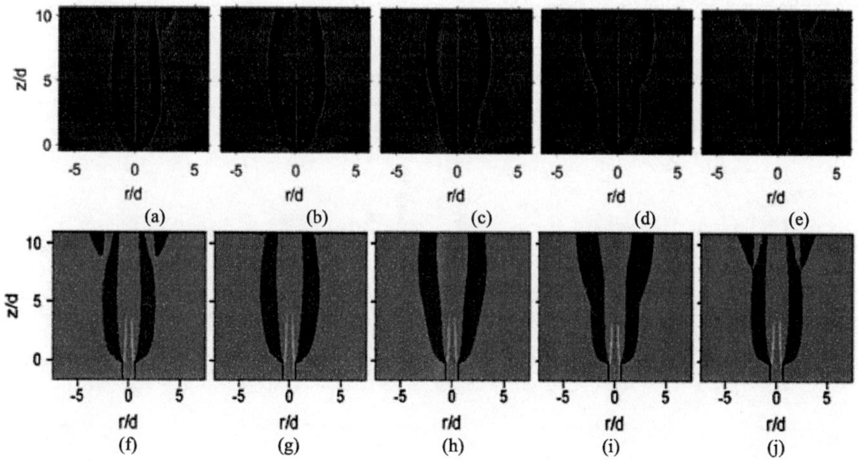

Fig. 3 Comparing instantaneous experimental (**a**–**e**) and synthetic (**f**–**j**) schlieren images for one complete flicker cycle

the Reynolds number of 500. The cold core in the axial region of the flame appears to be narrower in the experiment compared to numerical simulation; the marginal difference can be attributed to numerical diffusion for the relatively thicker cold jet core. At the near field, the instabilities form and it convects downstream to form a large outer vortical structure. It depicts buoyancy-induced outer vortical structure repeatedly squeezing and expanding the flame surface. Figure 3a–e shows the flicker cycle observed in the experimental data and Fig. 3f–j shows the similar cycle from the numerical simulation with time elapsed between two successive images being 20.83 and 19.88 ms, respectively. These periods of the flicker cycle indicate flicker frequency about 12 and 12.57 Hz in the experimental and numerical data, respectively. It is well corroborated with the previous experiments [1, 2]. Flame puff formation occurs at downstream location $z/d > 10$, where the vortical structure quenches the flame reaction by high aerodynamic strain and strong entrainment of relatively much colder ambient air. Note that, the radial and axial extent of the experimental flame structure is almost similar to that of numerical simulation.

3.1 Statistical Description of the Temperature Field

Considering that the flickering flame as a test case, the mean quantities are obtained after a statistically steady state has been achieved. The temperature profiles of the numerical simulation are obtained at a strain rate 5 (1/s). Figure 4 shows the mean and RMS temperature at the axial plane $z/d = 2$, 8, 14, and 20. The mean temperature profile depicts a shape of typical steady diffusion flame. The small RMS temperature at this location indicates a stable flame surface. The decrease in the peak flame temperature from 2170 K at $z/d = 2$–2030 K at $z/d = 20$ is the result of increased fluctuation at a downstream location. The peak RMS temperature has increased on the fuel and oxidizer of the flame. The mean temperature decreases because of periodic

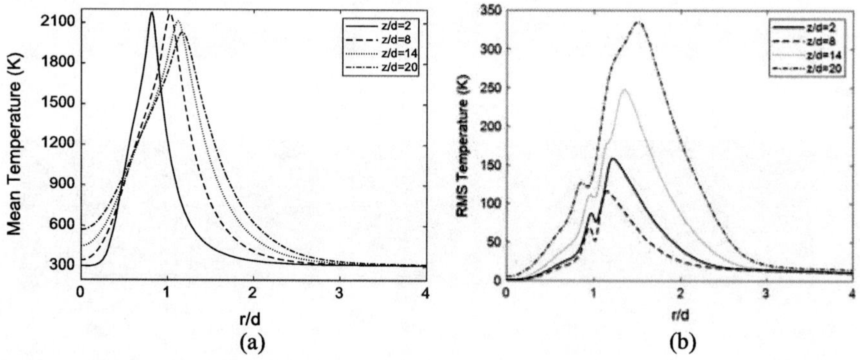

Fig. 4 Mean and RMS temperature profiles at various axial locations

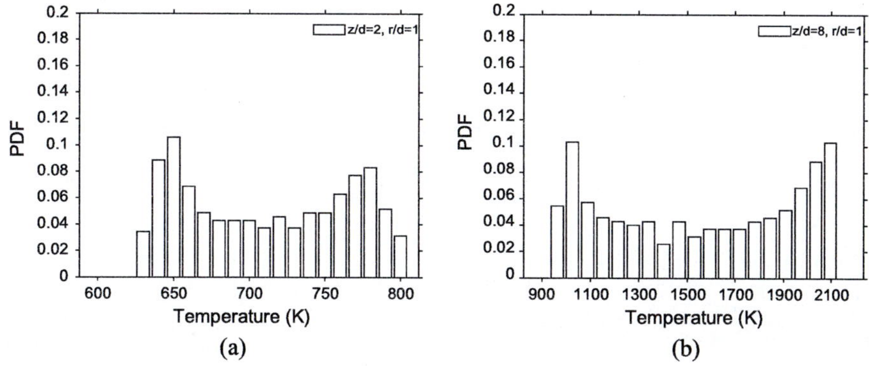

Fig. 5 PDF profile of temperature distribution. **a** $z/d = 2$, **b** $z/d = 8$

quenching of the flame as discussed previously and also reported by Albers and Agrawal [9].

The PDF profiles are obtained using 20 equal intervals between the minimum and maximum temperatures and then normalized by the bin interval. Figure 5 shows the bimodal distribution of the temperature PDF (represented by histogram) at a given location (i.e., $z/d = 2$ and 8 at $r/d = 1$). The PDF temperature is bimodal in shape with peaks at the local minimum and maximum. The periodic sinusoidal oscillation in the outer flow causing the fluid to alternate between hot and cold modes. Note that, this observation of PDF shapes agrees with those of Albers and Agrawal [9].

3.2 Refractive Index Difference and Temperature Relationship

Figure 6 shows the variation of temperature with the refractive index difference distribution in the diffusion flame for equilibrium case and strain rate at 5(1/s). The chemical equilibrium assumption is used to obtain state relationships and thus, to relate refractive index difference to the temperature and/or species concentration [4]. Different curves in fuel and oxidizer sides of the flame imply that the refractive index is affected significantly by combustion species. Refractive difference is minimum (i.e., the temperature is maximum) at the stoichiometric air–fuel ratio. The validity of this approach in the hydrogen diffusion flame is discussed by Shenoy et al. [10], although the non-unity Lewis number and chemical kinetic effects are important in the flame. It can be seen from Fig. 6 that strain rate is deviating the relation from the equilibrium relation. This deviation is significant at the high-temperature stoichiometric regime. The temperature uncertainty in the air side (or fuel lean) is 2 to 3 times higher than that in the fuel side because of the high sensitivity of temperature

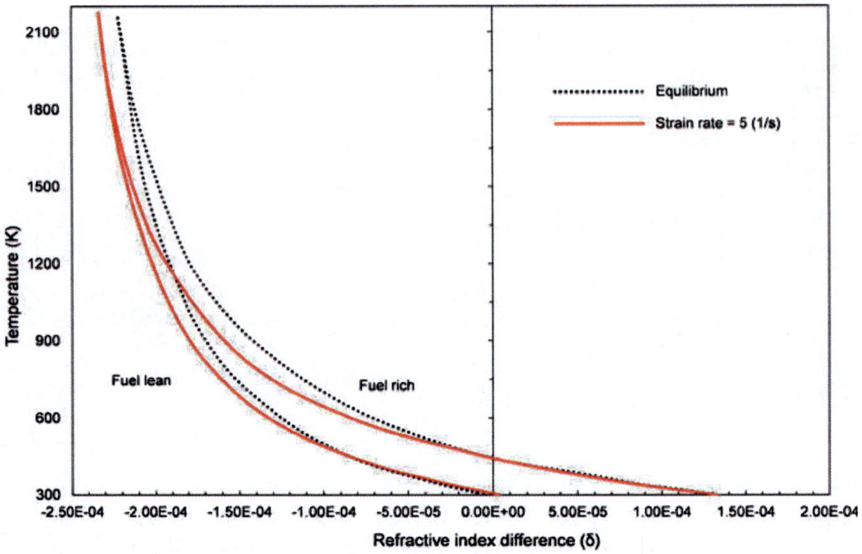

Fig. 6 Temperature versus refractive index difference in methane-air combustion

to the refractive index difference (see Fig. 6). The effect of higher strain rate on the flame needs to be further investigated to predict the scalar quantities accurately.

4 Conclusions

The numerical simulation investigated a flickering methane gas-jet diffusion flame. The instantaneous schlieren images of experimental data are compared qualitatively with the synthetic schlieren images during the flicker cycle. It is observed that the experimental global oscillation is almost close to the numerical simulation. The peak temperature changed as the flame surface stretched/compressed during the flicker cycle. The mean temperature profile had a typical shape of that in steady diffusion flame. The RMS temperature profile shows peak in the inner shear layer and the oxidizer side of the flame surface. The temperature PDF profiles show the bimodal distribution in the flow field, which relates oscillation in the flow field. It is shown that error in scalar field measurements (viz. temperature) close to stoichiometric mixture combustion would be the largest, and hence, the analysis of diffusion flame is more critical, and modeling of kinetic and differential diffusion effect is imperative from the viewpoint of finding the correct maximum temperature. Authors propose the use of synthetics schlieren images for comparison with experimental results as it does not involve the reconstruction errors and limitations of such tomographic algorithms.

Acknowledgements Authors would like to acknowledge the Department of Science and Technology Grant No. ECR/2015/000365 & ECR/2015/000343. Authors would also like to thank MHRD for providing research fellowship for Ph.D. Scholars.

References

1. Chen LD, Seaba JP, Roquemore WM, Goss LP (1989) Buoyant diffusion flames. Symposium (international) on combustion 22(1):677–684
2. Hamins A, Yang, JC, Kashiwagi T (1992) An experimental investigation of the pulsation frequency of flames. Symposium (international) on combustion 24(1):1695–1702
3. Mayinger F (ed) (2013) Optical measurements: techniques and applications. Springer Science and Business Media
4. Agrawal A, Alammar K, Gollahalli S (2002) Application of rainbow schlieren deflectometry to measure temperature and oxygen concentration in a laminar gas-jet diffusion flame. Exp Fluids 32(6):689–691
5. Yates LA (1993) Constructed interferograms, schlieren, and shadowgraphs: a user's manual. NASA CR-194530
6. ANSYS Inc (2017) ANSYS fluent theory guide
7. Smith GP, Golden DM, Frenklach M, Moriarty NW, Eiteneer B, Goldenberg M,… & Lissianski VV (2000) GRImech 3.0 reaction mechanism. Sandia National Laboratory
8. Kolhe PS, Agrawal AK (2009) Abel inversion of deflectometric data: comparison of accuracy and noise propagation of existing techniques. Appl Opt 48(20):3894–3902
9. Albers BW, Agrawal AK (1999) Schlieren analysis of an oscillating gas-jet diffusion flame. Combust Flame 119(1–2):84–94
10. Shenoy AK, Agrawal AK, Gollahalli SR (1998) Quantitative evaluation of flow computations by rainbow schlieren deflectometry. AIAA J 36(11):1953–1960

Excitation of Boundary Layer on the Suction Surface of C-D Compressor Blade

S. Katiyar and S. Sarkar

Abstract Excitation of a separated boundary layer on the suction surface of a controlled-diffusion (C-D) compressor stator blade is studied using large eddy simulation (LES) at a Reynolds number (Re) of 2.1×10^5 following experiment of Hobson et al. (J Propul Power 17(1):154–162, 2001) [1]. At this Re, the boundary layer on the blade remains laminar till the mid-chord, where it separates, undergoes a rapid transition due to high receptivity to free-stream disturbances and then reattaches as a turbulent boundary layer. The features of the flow are the formation of Kelvin–Helmholtz (K–H) rolls and their breakdown, revealing dynamics of coherent vortices and their nonlinear interactions.

Keywords LES · Compressor blade · Separation bubble · Transition and turbulent flow

1 Introduction

The demands of higher efficiency and reduction of cost have led to numbers of studies to understand the flow physics through a compressor blade passage. The lack of certainty in designing an axial compressor blade lies in the nature of its flow. In a compressor cascade, the boundary layer past the rotor or stator blade is subjected to an adverse pressure gradient. If the adverse pressure gradient is strong enough, the boundary layer often separates from the surface, leading to an unsteady flow associated with considerable losses. Further at high altitudes and low flight speeds, Re becomes low deteriorating the performance of a compressor stage. Under these conditions, flow remains laminar over a large percentage of the blade chord until it separates from the surface. The separated boundary layer is unconditionally unstable, initiating a rapid transition and then it reattaches as a turbulent boundary layer.

S. Katiyar · S. Sarkar (✉)
Department of Mechanical Engineering, Indian Institute of Technology Kanpur, Kanpur 208016, India
e-mail: subra@iitk.ac.in

L. Venkatakrishnan et al. (eds.), *Proceedings of 16th Asian Congress of Fluid Mechanics*, Lecture Notes in Mechanical Engineering,
https://doi.org/10.1007/978-981-15-5183-3_45

Several shapes of blade have been studied over the years in an effort to improve the compressor performance and efficiency [2, 3]. Airfoil shapes, that control the diffusion of velocity over the surface, can enhance the amount of laminar flow in relation to turbulent flow. This may delay or avoid flow separation even before the trailing edge is reached. Thus, the aerodynamic loading becomes higher for a C-D blade as compared to the conventional one. The C-D compressor stator blade considered in this study was developed by Gelder et al. [3] and designated as NACA 67B. The design point Re based on the chord of the blade was almost 10^6 at the sea level, which was estimated to decrease to 2×10^5 at 1.5×10^4 m. A detailed experimental study depicting flow on the 67B blade was conducted by Hobson et al. [1] using laser Doppler velocimetry, where, Re was in the range of 2.1–6.4 $\times 10^5$. The boundary layer on the suction surface remained laminar up to the mid-chord, where it separated and reattached within a short distance as a turbulent layer. The features of boundary layer and the bubble size changes significantly with change of Re.

There are numbers of study [4–7] illustrating the transition of a separation bubble. Most of the studies elucidate that transition of a separated boundary layer occurs in a similar fashion of free shear layer with high receptivity to perturbations. Amplification of certain selective frequencies of background disturbances is observed leading to the formation of *K–H* rolls and shedding of large-scale vortices in the vicinity of reattachment. In reality, transition mechanism includes the following features, i.e. receptivity, linear growth of disturbance, and then nonlinear effects leading to breakdown. In this paper, LES with dynamic subgrid-scale (SGS) model is used to study the flow features on the suction surface of NACA 67B blade at Re of 2.1×10^5 (based on the chord and the inlet velocity). The objectives are to resolve the flow separation on the suction surface and excitation of the shear layer developing of three-dimensional motions. Certainly, RANS calculations will not provide the answer. LES can resolve the dynamics of vorticities, their nonlinear interactions and the flow instability. Thus, it is expected that the LES results may help in understanding the flow transition and turbulence augmentation over the suction surface of a highly loaded compressor blade.

2 Computational Details

The filtered, incompressible mass and momentum equations in coordinate independent or covariant form can be expressed as

$$\partial_i(J\overline{u^i}) = 0, \tag{1}$$

$$\partial_t\overline{u^i} = -\partial_j g^{ij}\overline{p}/\rho + (-\overline{u^i u^j} + 2\nu\overline{s^{ij}})_{,j}, \tag{2}$$

where the strain rate is given by

$$\overline{s^{ij}} = \frac{1}{2}\left(g^{ik}\overline{u^j_{,k}} + g^{jk}\overline{u^i_{,k}}\right), \tag{3}$$

g^{ij} is the metric tensor, J is the Jacobian of the transformation, and $\overline{u^i}$ is the filtered contravariant component of velocity. The comma notation denotes covariant differentiation. The term $(\overline{u^i u^j})$ cannot be computed, and the difference between it and the part that can be computed $(\overline{u^i}\,\overline{u^j})$ is the subgrid Reynolds stress as

$$\tau^{ij} = \overline{u^i u^j} - \overline{u^i}\,\overline{u^j} \tag{4}$$

The subgrid Reynolds stress which will be modelled to close the system of equations represents the effect of the subgrid motions on the resolved field of the LES. In the present case, the Germano-Lilly SGS model [8, 9] is used where the model coefficient (C_d) is dynamically calculated instead of input a priori. This is done considering a test filter, which is usually chosen to be twice the size of the mesh filter. Then, the velocity field is filtered on the coarser mesh that effectively results in the subtest-scale stress as

$$T^{ij} = \widehat{\overline{\upsilon^i \upsilon^j}} - \widehat{\overline{\upsilon^i}}\,\widehat{\overline{\upsilon^j}} \tag{5}$$

We now apply the base (Smagorinsky) model at both filter scales,

$$\tau^{ij} - \frac{1}{3}g^{ij}g_{kl}\tau^{kl} = -2C_\mathrm{d}\Delta^2|\overline{s}|\overline{s^{ij}} \tag{6}$$

$$T^{ij} - \frac{1}{3}g^{ij}g_{kl}T^{kl} = -2C_\mathrm{d}\widehat{\Delta}^2|\hat{\overline{s}}|\widehat{\overline{s^{ij}}} \tag{7}$$

Using Germano's identity [9], the Leonard stresses ℓ^{ij} can be expressed in general coordinates as,

$$\ell^{ij} = \widehat{\overline{\upsilon^i}\,\overline{\upsilon^j}} - \widehat{\overline{\upsilon^i}}\,\widehat{\overline{\upsilon^j}} = T^{ij} - \widehat{\tau^{ij}} \tag{8}$$

$$\begin{aligned} L^{ij} &= \ell^{ij} - \frac{1}{3}g^{ij}g_{kl}\ell^{kl} \\ &= \left(T^{ij} - \frac{1}{3}g^{ij}g_{kl}T^{kl}\right) - \left(\widehat{\tau^{ij}} - \frac{1}{3}g^{ij}g_{kl}\,\widehat{\tau^{kl}}\right) = -2C_\mathrm{d}M^{ij} \end{aligned} \tag{9}$$

The models for τ^{ij} and T^{ij} are substituted to define the tensor M^{ij},

$$M^{ij} = \widehat{\Delta}^2|\hat{\overline{s}}|\hat{\overline{s}}^{ij} - \Delta^2\widehat{|\overline{s}|\overline{s^{ij}}} \tag{10}$$

thus,

$$C_{\mathrm{d}} = -\frac{L^{ij}}{2M^{ij}} \tag{11}$$

The least-square estimate of the optimal solution for C_{d}, following Lilly [9], is

$$C_{\mathrm{d}} = -\frac{\langle L^{ij} g_{ik} g_{jl} M^{kl} \rangle}{2\langle M^{ij} g_{ik} g_{jl} M^{kl} \rangle} \tag{12}$$

The angle brackets represent an average over the homogeneous spanwise direction z in which C_{d} will not change. The resulting C_{d} is a function of time and the inhomogeneous coordinates x and y. It should be noted that the computation of C_{d} from Eq. (12) is considerably more complex than in the Cartesian case, owing to the presence of the metric tensor coupling the various components.

The numerical methods employed for the present LES are based on the finite-volume technique using the filtered Navier-Stokes equations in generalized coordinates. The momentum equations are discretized on a staggered mesh and are advanced in time explicitly by the second-order Adams-Bashforth scheme, except for the pressure term, which is solved by a standard projection method. The pressure equation is discrete Fourier transformed in the spanwise direction to obtain a set of decoupled equations that are solved by a multigrid technique. The spatial discretization is second-order central differencing, with the geometric quantities computed to higher order in space in advance of the simulation and stored, as necessary, at several distinct points of a staggered mesh arrangement. The solver used in the present simulation has been extensively validated for variety of transitional and turbulent flows [10–13].

The periodic boundary conditions, for instantaneous and statistical variables, are enforced at upstream and downstream of blades to simulate an infinite row of blades in cascade. On the blade surface, no-slip boundary conditions are imposed. At exit of the computational domain, convective boundary conditions are used, that allow passing of trailing vortices through the outflow boundary without any significant mass redistribution. At the inflow, the velocity components are specified. In the present study, a homogeneous Neumann condition for pressure is applied on the inflow and outflow, as well as on the compressor blade wall while a cyclic matching is applied to the pressure and the associated velocity at upstream and downstream of blades. The blade configuration, the computational domain, and boundary conditions imposed are illustrated in Fig. 1. The domain in the spanwise direction is considered to be 12% of the axial chord. A mesh of $512 \times 256 \times 64$ points is used with $0.5 < \Delta\eta^{+} < 1.5$. An H-grid within the blade passage is developed by applying the power law to the control functions appearing in the elliptic grid-generation scheme [14], which confirms the near-wall orthogonality along the boundaries.

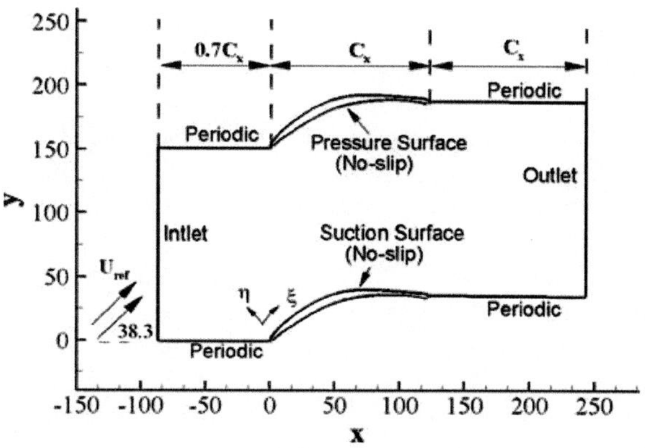

Fig. 1 Computational domain along with boundary conditions, where ξ, η and ζ refer to surface tangent, normal and spanwise coordinates, x, y, z the Cartesian coordinates, C_x the axial chord and U_{ref} inlet flow velocity

3 Results and Discussion

LES results of time-averaged surface pressure (C_{p}) of compressor blade are presented in Fig. 2 and are also compared with the experiment [1]. The experiment indicated a plateau in the distribution of C_{p} implying a separation bubble in the mid-chord region ($\lambda^* = 0.45$–0.7) of the suction surface, where λ^* is λ/C. The pressure predicted by LES illustrates a plateau, followed by a droop and then it rises sharply. The droop in C_{p} distribution, which was absent in experiment, is attributed to the appearance of reverse flow vortex in second-half of the bubble. A substantial difference in C_{p} between the LES and the experiment is evident in the mid-chord region. This is due to the reversal region, where the onset of separation is marginally shifted downstream

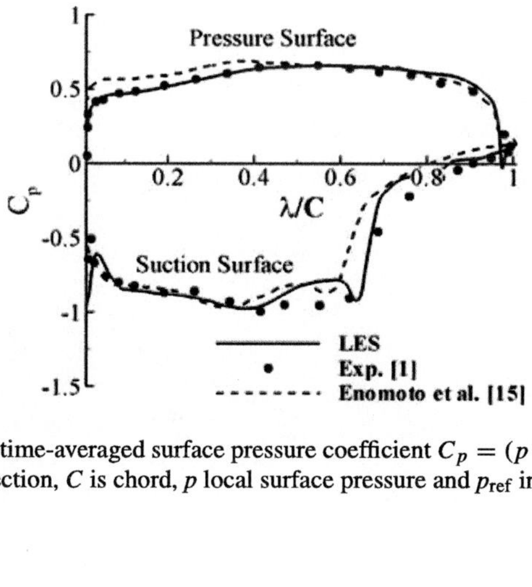

Fig. 2 Distribution of time-averaged surface pressure coefficient $C_p = (p - p_{\text{ref}})/\frac{1}{2}\rho U_{\text{ref}}^2$, where λ refers chordwise direction, C is chord, p local surface pressure and p_{ref} inlet pressure

at $\lambda^* = 0.50$ in the simulation. The pressure distribution predicted by Enomoto et al. [15] is superimposed, depicting the similar trend. However, the evaluation of C_p is better predicted by the present LES, particularly on the pressure surface. A slow diffusion occurs on the pressure surface till 60% of the chord, followed by a favourable pressure gradient.

Figure 3 shows the variation of exit turbulence level along the blade pitch, illustrating a good agreement with the corresponding experiment. The exit plane is being taken at a downstream distance of 20% chord length from the trailing edge of the blade. In the wake region, the turbulence level attains its maximum value 9.38% that reduces to 2% in the core flow region. Further, the time-averaged streamlines superimposed with contours of normalized Reynolds shear stress over the separated region are presented in Fig. 4. The bubble structure can be well appreciated from the streamlines with appearance of a considerable large flow reversal. The onset of separation appears at $x = 55$ ($\lambda^* = 0.50$) and the reattachment at $x = 80$ ($\lambda^* = 0.70$). The concentration of shear stress occurs along the shear layer and becomes maximum near the mean reattachment point, which is attributed to the formation of large-scale structures and their breakdown. The onset of transition appears here at $x = 65$, which is almost at 30% of bubble length following the work of Ol et al. [16]. They reported that the onset of transition can be identified when the value of normalized shear stress is above a critical value of 0.001.

The time-averaged velocity profiles and turbulence level (Tu) at different locations along the suction surface are presented in Fig. 5. LES resolves the boundary layer growth fairly well on the suction surface, although discrepancies exist. Further,

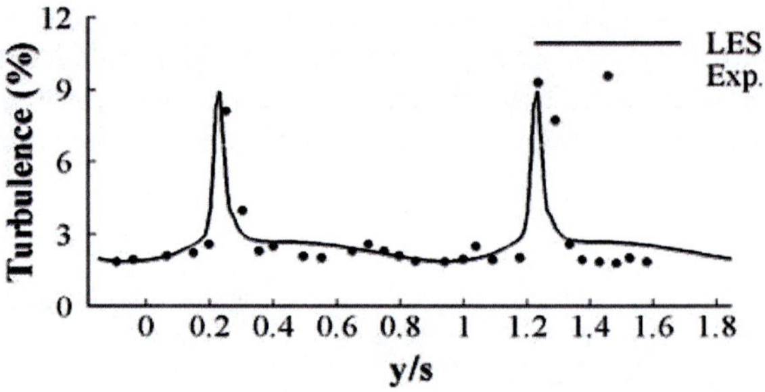

Fig. 3 Exit turbulence level: present LES compared with the experiment [1]

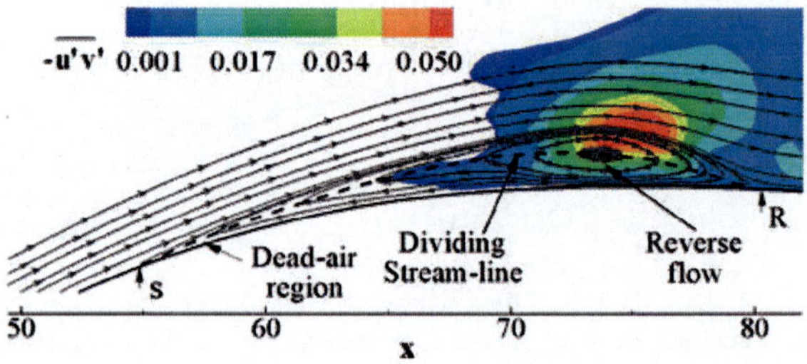

Fig. 4 Mean flow streamlines along with contours of normalized Reynolds shear stress over the separated region

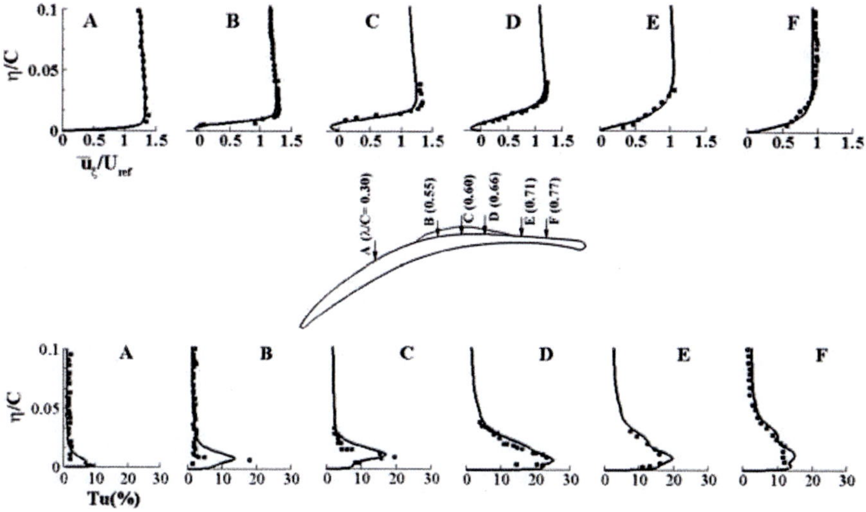

Fig. 5 Velocity profile (top) and turbulence level (bottom) at different stations on suction surface: solid line represents the LES and symbols for the experiment [1]

velocity profiles show a flow reversal at the locations $\lambda^* = 0.55, 0.60$ and 0.66. The estimated turbulence level agrees reasonably well with the experiment in terms of both peak values and their locations, illustrating augmentation of turbulence in the separation region. The double peak feature in Tu profile is very apparent near the middle of the bubble, where, the outer peak is due to the free shear layer and the inner peak is because of wall generated turbulence. Maximum level of Tu is observed near the attachment, that is 0.24 at $\lambda^* = 0.66$.

The spectra of streamwise velocity fluctuations ($u'_\xi = U_\xi - \overline{u}_\xi$) along the maximum u_{rms} line at different streamwise locations within the separation bubble are shown in Fig. 6. In the present study, the shedding frequencies of the shear layer are expressed in term of Strouhal number ($\text{St}_{\theta_s} = f\theta_s/U_e$, where U_e and θ_s are the boundary layer edge velocity and momentum thickness at the separation point). The power spectral density (PSD) has been shifted by an arbitrary amount in the vertical axis. At onset of separation, velocity spectra depict a narrow band of frequencies with low-energy level illustrating a laminar flow, whereas a wide band of energetic frequencies near reattachment point indicates a turbulent flow with abundant small-scale eddies. An amplification of selected frequency at $\text{St}_{\theta_s} = 0.011$ is prominent in between 20 and 40% of the bubble length. The earlier studies [6, 7] indicated that the transition of a separation bubble occurred via *K–H* instability, where the normalized shedding frequency was in the range of $\text{St}_{\theta_s} = 0.01$--$0.017$. Thus, the separated boundary layer on suction surface most probably becomes unstable via inviscid mechanism in the first half of the bubble.

Contours of streamwise velocity fluctuations illustrating the internal growth mechanism of the shear layer are presented in Fig. 7 at different locations: (a) the top view

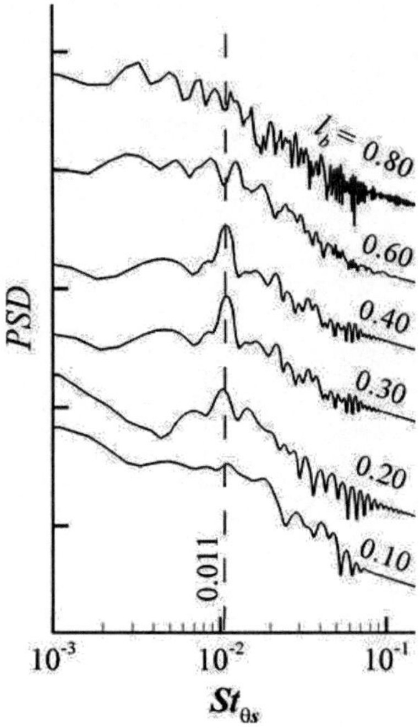

Fig. 6 Spectra of streamwise velocity fluctuations along the maximum u_{rms} line within the separated region, where l_b refers to the bubble length

(ξ--ζ plane) at $\eta^+ = 6$ and (b) the cross-sectional views (η--ζ plane). In the top view, the spanwise *K–H* rolls (blue colour in region A) are observed in first half of the bubble. Further, these *K–H* rolls undergo breakdown, leading to the turbulent reattachment at $\lambda^* = 0.70$ (blue colour in region B). Streamwise streaks, which are the characteristics of the turbulent boundary layer, are also apparent downstream of reattachment. The cross-sectional views depict that the flow ceases to be laminar up to $\lambda^* = 0.5$. The secondary instability appears near $\lambda^* = 0.55$, where the flow becomes three-dimensional with apparent fluctuations in the spanwise direction. The near-wall perturbations are prominent from $\lambda^* = 0.60$. Near to the reattachment, flow becomes highly three-dimensional with large and abundant small-scale eddies. Downstream of reattachment, well-organised mushroom-shaped eruptions illustrating the streaks within the boundary layer are well resolved and the spacing between the two consecutive structures is around 0.018 C. Thus, the selective amplifications of free-stream perturbations (fst), that lead to rollup of shear layer and appearance three-dimensional motions, attributing to reattachment are well resolved by the present LES.

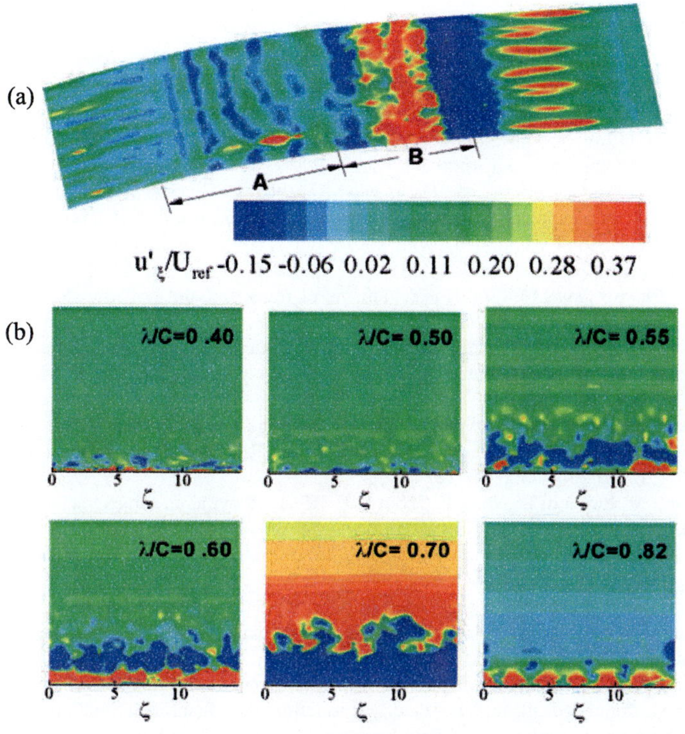

Fig. 7 Streamwise velocity fluctuations over the suction surface: **a** top view and **b** cross-sectional views

4 Conclusions

The present LES resolves the flow features within the blade passage of compressor with a reasonable accuracy, although there exist some discrepancies with the experiment. A separation bubble is depicted in the mid-chord region of the suction surface with a considerable large reverse flow vortex, attributing to a droop in the C_p. The estimated onset of separation and the reattachment point agree well with the experiment. Excitation of separated boundary layer near the mid-chord is mainly governed by the high receptivity of fst, depicting the amplification of selected frequencies. This triggers rollups of shear layer illustrating that transition occurs because of K–H instability in the first half of the bubble. The secondary instability occurs in the second half with development of three-dimensional motions and breakdown attributing to the reattachment, where viscous effect is also significant.

References

1. Hobson GV, Hansen DJ, Schnorenberg DG, Grove DV (2001) Effect of Reynolds number on separation bubbles on compressor blades in cascade. J Propul Power 17(1):154–162
2. Behlke RF (1986) The development of a second generation of controlled-diffusion airfoils for multistage compressors. J Turbomach 108(1):32–41
3. Gelder TF, Schmidt JF, Suder KL, Hathaway MD (1987) Design and performance of controlled-diffusion stator compared with original double-circular-arc stator. SAE Trans 96:1000–1012
4. Watmuff JH (1999) Evolution of a wave packet into vortex loops in a laminar separation bubble. J Fluid Mech 397:119–169
5. Spalart PR, Strelets MK (2000) Mechanisms of transition and heat transfer in a separation bubble. J Fluid Mech 403:329–349
6. Anand K, Sarkar S (2014) Features of a laminar separated boundary layer near the leading-edge of a model airfoil for different angles of attack: an experimental study. ASME J Fluids Eng 139
7. Samson A, Sarkar S (2016) An experimental investigation of a laminar separation bubble on the leading-edge of a modelled aerofoil for different Reynolds numbers. Proc IMechE Part C J Mech Eng Sci 230(13):2208–2224
8. Germano M, Piomelli U, Moin P, Cabot WH (1991) A dynamic subgrid-scale eddy viscosity model. Phys Fluids A 3(7):1760–1765
9. Lilly DK (1992) A proposed modification of the Germano subgrid-scale closure method. Phys Fluids A 4(3):633–635
10. Sarkar S, Voke PR (2006) Large eddy simulation of unsteady surface pressure over a low-pressure turbine blade due to interactions of passing wakes and inflectional boundary layer. J Turbomach 128:221–231
11. Sarkar S (2007) The effects of passing wakes on a separating boundary layer along a low-pressure turbine blade through large-eddy simulation. Proc Inst Mech Eng Part A 221:551–564
12. Sarkar S (2008) Identification of flow structures on a LP turbine blade due to periodic passing wakes. ASME J Fluids Eng 130:061103
13. Sarkar S (2009) Influence of wake structure on unsteady flow in an LP turbine blade passage. ASME J Turbomach 131:041016
14. Hsu K, Lee SL (1991) A numerical technique for two-dimensional grid generation with grid control at all of the boundaries. J Comput Phys 96:451–469
15. Enomoto S, Hah C, Hobsob GB (2000) Numerical and experimental investigation of low Reynolds number effects on laminar flow separation and transition in a Cascade of compressor blades. In: Proceedings of ASME Turbo Expo, 8–11 May 2000, Munich, Germany
16. Ol MV, Hanff E, McAuliffe B, Scholz U, Kahler C (2005) Comparison of laminar separation bubble measurements on a low Reynolds number airfoil in three facilities. In: 35th AIAA fluid dynamics conference and exhibit, 06–09 June 2005, Toronto, Ontario, Canada

Flow Characteristics During the Passing Process of Circulating Tumor Cells Through a Microfluidic Channel

Dipta Saha, Md. S. Rahman, A. K. M. Sadrul Islam, and A. B. M. Toufique Hasan

Abstract Circulating tumor cell (CTC) unfolds an enormous opportunity for precise information about the cancer type and growth. The key challenge is to separate CTCs from blood because of its extreme rarity. Among the existing processes, size and deformation-based separation process offers the best way to isolate cells in a label-free way. The numerical computation gives the opportunity to understand the deformation behavior of CTC and subsequent pressure signature during the flow process over the experimental techniques. In the present study, pressure characteristics are studied for three types of cancer cell considering the surface tension and the viscous property of the cell. Cervical cancer cell shows more resistance during the cell deformation process over the ENB and the brain cancer cell, respectively. Critical and viscous pressure drop pattern with the increase in flow rates is also investigated. A comparison of pressure signature and cell deformation characteristics is made between high and low viscous cervical cancer cells. It is found that critical pressure of highly viscous cell rises significantly due to high shear stress compared to the less viscous cell during the cell passing process. This study provides some important insights into the cell passing process inside the microchannel to design the next-generation CTC microfluidic-based chip.

Keywords CTC · Deformation · Critical pressure · Microfluidic devices · Surface tension

1 Introduction

Cancer is the leading cause of morbidity and mortality worldwide. Cancer is a group of diseases involving abnormal cell growth with the potential to spread to other parts of the body. Circulating tumor cells (CTCs) are shed from cancerous tumors, enter the

---

D. Saha · Md. S. Rahman · A. K. M. Sadrul Islam · A. B. M. Toufique Hasan (✉)
Department of Mechanical Engineering, Bangladesh University of Engineering and Technology (BUET), Dhaka 1000, Bangladesh
e-mail: toufiquehasan@me.buet.ac.bd

L. Venkatakrishnan et al. (eds.), *Proceedings of 16th Asian Congress of Fluid Mechanics*, Lecture Notes in Mechanical Engineering,
https://doi.org/10.1007/978-981-15-5183-3_46

circulatory system, and migrate to distant organs to form metastases that ultimately lead to the death of most patients with cancer.

The clinical value of CTCs as a biomarker for early cancer detection, diagnosis, prognosis, stratification, and pharmacodynamics has been widely explored in recent years. Various microfluidic approaches have already been developed for sorting rare cells including CTC, circulating fetal cells, and stem cells. Multiple cellular properties such as size, deformability, acoustic properties, and di-electrophoretic features can be employed for CTC detection in a microfluidic device [1]. Among the various methods for rare cell capturing; deformation-based CTC filtration offers the advantage of structural simplicity, stable performance, and low cost [2]. To achieve high throughput, high system efficiency, and high isolation purity, the flow regime inside the microchannel and the cell to channel interaction in-depth study is necessary. The numerical approach gives a better understanding of the cell deformation inside the microchannel. Liquid droplet model is implemented for large deformation case. Volume of fluid (VOF) algorithm is used to simulate the deformation of the cell. The surface force of the cell is modeled using continuum surface force model [3]. The effect of cell entry in the microfilter is studied in this research. Other numerical studies showed the effect of 3D channel geometry on the cell deformation and observe that the circular cross-section provides the highest critical pressure and thus is most suitable for high-efficiency CTC separation [4].

The pressure signature and cell deformation behavior inside the microchannel for three different types of cancer cell having different surface tension is introduced in this paper with varying flow rate. The deviation of critical pressure drop and viscous pressure drop from the theoretical value is shown in this study. The effect of an increase in cell to blood viscosity ratio on pressure signature is also captured.

2 Methodology

2.1 Model Description

The microchannel (microfilter) is divided into an entrance chamber, a constriction channel, and an exit chamber as shown in Fig. 1. The entrance and exit chamber radius is 15 μm. For constriction channel, the radius is reported to be within 2.5–6 μm for effective filtering of CTCs according to different cell sorting methods [2].

In this study, channel radius of 2.5 μm and cell radius of 8 μm are used [3]. The viscosity of blood plasma is set as a constant at 0.001 Pa.s. Initially, the viscosity of cancer cells is assumed as same as blood plasma to study the effect of surface tension. For further study of high viscous cervical cancer cell(HeLa Cell), viscosity is considered as 0.044 Pa.s [5].

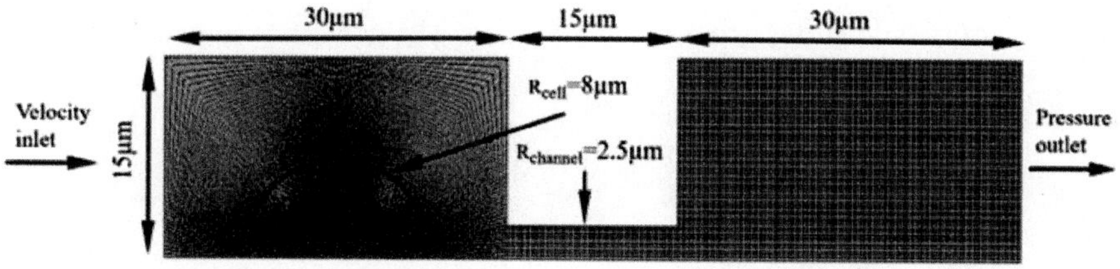

Fig. 1 Axisymmetric 2D mesh structure of the computation domain where the cell is 15 μm from the inlet

2.2 *Theoretical Background*

The total squeezing pressure P_{total} is mainly determined by surface tension and viscous force. Thus, the total inlet pressure can be expressed as,

$$P_{\text{total}} = P_{\text{viscous}} + P_{\text{surface}} \tag{1}$$

in which P_{viscous} represents the viscous pressure drop and P_{surface} represents the surface pressure drop. Viscous pressure drop, P_{viscous} is expressed as,

$$P_{\text{viscous}} = \frac{8\mu L}{R_{\text{channel}}^2}\bar{V} + (K_E + K_C)\frac{\rho\overline{V^2}}{2} \tag{2}$$

where μ is the dynamic viscosity of the carrier fluid, R_{channel} is the radius of the constricted channel, L is the length of the constricted channel, $\bar{V}$ is the average carrier fluid viscosity in the constricted channel, and $K_E + K_C = 1.5$, for a sudden constriction from large to small area.

The pressure drop due to surface tension P_{surface} can be calculated based on Young–Laplace equation assuming the aspirated and non-aspirated part of the cell are spherical which is expressed as,

$$P_{\text{surface}} = 2\sigma\left(\frac{1}{R_{\text{channel}}} - \frac{1}{R_{\text{cell}}}\right) \tag{3}$$

where σ is the surface tension coefficient of droplet carrier fluid and R_{cell} is the radius of the undeformed cell.

2.3 *Numerical Method and Validation*

Among various methods for interface tracking, volume of fluid (VOF) offers the best mass conservation property. Geometric reconstruction scheme is used for interpolation near the interface. The surface tension property of the cell membrane is adopted by using continuum surface force (CSF) method. The governing equations of fluid flow process are given as below:

$$\frac{\partial \alpha}{\partial t} + \nabla.(\alpha \bar{V}) = 0 \quad (4)$$

$$\frac{\partial (\rho \bar{V})}{\partial t} + \nabla.(\rho \bar{V} \bar{V}) = -\nabla p + \nabla.[\mu(\nabla \bar{V} + \nabla \bar{V}^T)] + \bar{F} \quad (5)$$

where $\bar{V}$ is the velocity vector, α is the volume fraction function, ρ is the volume fraction averaged density, μ is the volume fraction averaged viscosity, and $\bar{F}$ is the tension force localized at the surface.

In the present study, transient computation is performed using commercial software ANSYS Fluent. To save computational cost in solving the governing equations, a 2D axisymmetric analysis is employed. The mesh of the computational domain consists of 26,000 quadratic Cartesian grid cells. The discretization of the governing equation is done by the second order implicit differencing scheme.

Boundary conditions are set as velocity inlet at the entrance, pressure outlet at the exit, no-slip boundary wall on the chamber and channel wall, and axisymmetric on the centerline of the microfilter. The contact angle is taken 180° to make non-wetting wall. In this paper, the flow rate for the different case is varied from 7 to 140 nL/s.

A validation of pressure signature and streamline for CTC passing through the microchannel is made between the present results and reference [4]. Figure 2a shows the total inlet pressure at different stage which represents nearly identical results with the reference. Streamlines and cell deformation at different stages of CTC passing through the channel are shown in Fig. 2b. The present flow field also shows an excellent agreement with the reference one [4]. Both of the simulations are done for flow rate 7 nL/s, surface tension 50 mN/m and blood to cell viscosity ratio as 1:1.

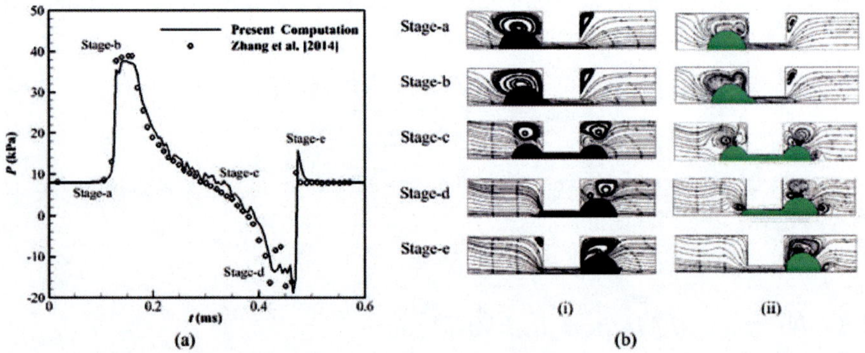

Fig. 2 Flow process during CTC passing through a microfluidic channel; **a** pressure signature, **b** flow field at different stage of the cell (i) present computation and (ii) reference [4]

3 Results and Discussion

Pressure signature is very important for the determination of the microfluidic channel operational condition. Tumor cells are more rigid than normal blood cells. Even cell property variation occurs depending on the type of cancer. In this study, three different type of cancer cells are considered; namely—thebrain cancer cell (surface tension, σ = 10mN/m) [6], the esthesioneuroblastoma (ENB) cell (surface tension, σ = 20 mN/m) [7], and the cervical cancer cell (surface tension, σ = 50 mN/m) [8].

The pressure profiles of these cancer cells at different flow rates are investigated in this study. All these cell passing processes through the microchannel at different flow rate show the same pattern characterized by similar stages like cell entrance, cell occupation in the filtration channel, and cell squeezing out.

But from Fig. 2a, it is found that the pressure signature varies significantly with the flow time for a constant flow rate 7 nL/s. At stage-a, when the cell progress toward the filtration channel, pressure drop remains constant because only the viscous pressure is available in this region, no effect of surface tension in this region. At stage-b, during the entrance in the microchannel, cell goes through a significant deformation and pressure drop suddenly rises significantly because of the resistance created by the surface membrane of the cell. Interfacial surface tension property of the cell membrane resists the cell not to undergo throughout the deformation process. The peak of the pressure signature graph is denoted by the critical pressure (P_{critical}). A cell must have to overcome this pressure for the successful passing through the filtration channel. At stage-c, cell fully occupies both end of the channel, and net pressure due to surface tension becomes zero. As a result, the total pressure at inlet reaches the same level of the viscous pressure. The cell tries to pull out itself from the microchannel in stage-d. So, the surface force changes the direction results the reduction in total pressure. At stage-e, total pressure suddenly rises due to the bouncing of the rear part of the cell.

As shown in case of Fig. 3a, it is observed that cell with surface tension 50, 20, and 10 mN/m makes the peak pressure 37.6, 20.9, and 14.9 kPa, respectively, for flow rate 7 nL/s. The critical pressure reduced significantly due to the softer cell membrane with reduced surface tension. From the observation of cell deformation, it reveals that cell with less surface tension provides a smaller driving force to regain own shape.

From Fig. 4, it is observed that at flow rate 49 nL/s during the entrance period, the frontal shape of the cell with surface tension 50 mN/m remains spherical but the frontal shape of the cell with surface tension 20 and 10 mN/m trends to a cylindrical shape. At this flow rate, it is also found that less stiff cells with surface tension 20 mN/m and 10 mN/m detach from the filtration channel and allow the carrier fluid to co-flow through the filtration channel due to the dominance of inertia force over the surface tension force.

For comparison, we have studied highly viscous cervical cancer cell with the previously assumed less viscous cervical cancer cell. HeLa cell with viscosity 0.044 Pa.s [5] is adopted as a highly viscous cervical cancer cell in this study. Figure 5 shows

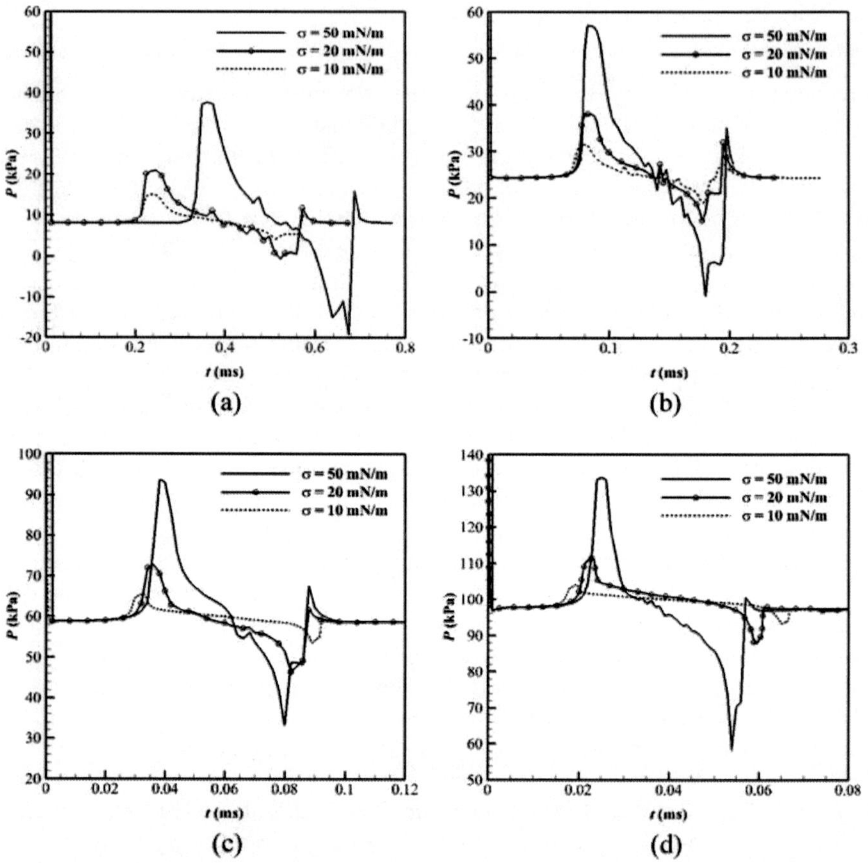

Fig. 3 Pressure signature during the deformation process of CTC inside the microchannel for flow rate **a** 7 nL/s, **b** 21 nL/s, **c** 49 nL/s, **d** 77 nL/s for viscous ratio μ_{cell}: $\mu_{\text{blood}} = 1{:}1$

the effect of the high viscous cell on the pressure profile and the cell deformation. From this Fig. 5, it is observed that the viscous pressure is same for both types of cell when the cell is at inlet and exit chamber because there is no effect of shear stress on the cell during this state. High viscous cell maintains a high level of total pressure when the cell fully occupies the channel due to more adhesion with the channel wall (H1, H3). This pressure remains nearly constant during whole-cell passing process through the microfilter. In case of less viscous cell, the critical pressure occurs when the cell just enters into the channel (L1, L3). If we observe the exit stage of the cell, it is found that at flow rate 7 nL/s, the shape of rare part of the cell is spherical (L2) in case of a less viscous cell but the rare part of the high viscous cell is concave (H2). At flow rate 35 nL/s, it is observed that the rare shape of less viscous cell still remains spherical (L4) but cell splitting occurs in case of the high viscous cell due to the development of high shear stress (H4).

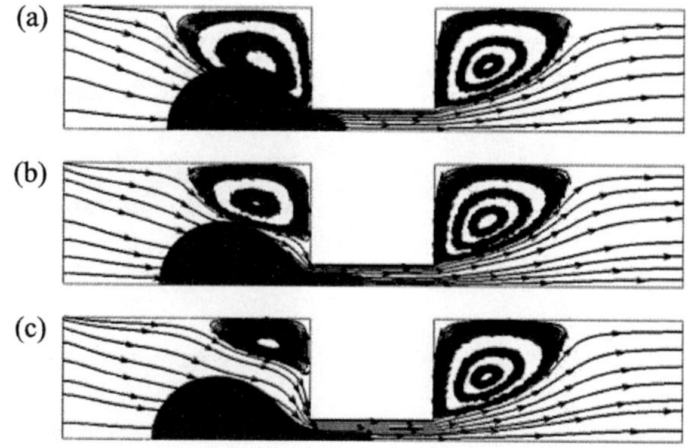

Fig. 4 Streamline and cell deformation behavior at constant flow rate 49 nL/s for surface tension **a** 50 mN/m, **b** 20 mN/m, **c** 10 mN/m, respectively keeping the viscous ratio $\mu_{\text{cell}}: \mu_{\text{blood}} = 1{:}1$

From Fig. 6, it is observed that the total critical pressure and the viscous pressure at inlet increase with the increase in the flow rate. We also find that at low flow rate, the deviation of simulated result from the theoretical result is negligible but at high flow rate, derivation becomes significant. This deviation indicates that at higher flow rate cases, cell tends to detach from the channel and co-flow occurs.

4 Conclusions

In this study, the cell passing process of circulating tumor cell (CTC) through a microfluidic channel is investigated for three different types of tumor cells. The results obtained from the present study are:

- It is observed that cervical cancer cell with higher surface tension gives the maximum critical pressure compared to the cell with lower surface tension like ENB and brain cancer cell, respectively.
- With the increase in flow rate, it is observed that co-flow occurs earlier in case of less stiff tumor cells compared to the higher stiff tumor cells.
- By studying the viscous effect of the cervical cancer cell, it is observed that critical pressure rise significantly in case of the highly viscous cell than the lower one because of the adhesion with the channel. Highly viscous cell shows cell splitting phenomena while leaving the channel at a higher flow rate due to high shear stress.
- Critical pressure and viscous pressure slightly deviate from theoretical values at high flow rate cases due to non-spherical cell deformation and co-flow.

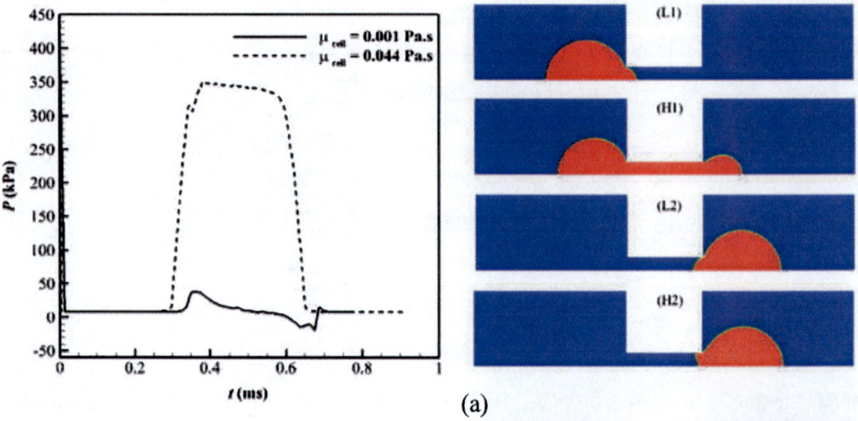

(a)

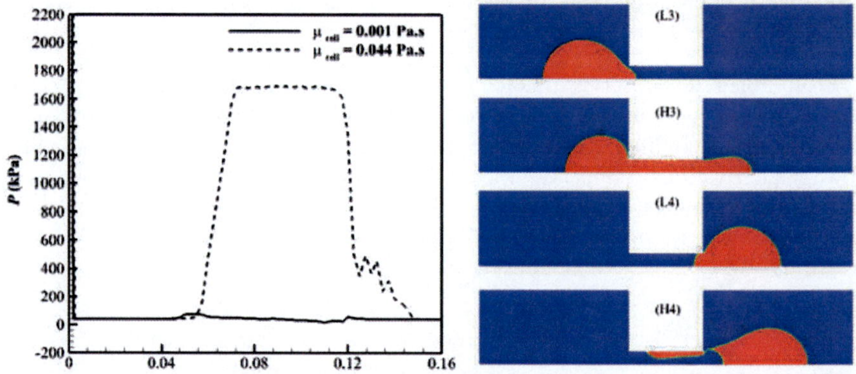

(b)

Fig. 5 Pressure signature of high viscous and low viscous cell for flow rate **a** 7 nL/s and **b** 35 nL/s with surface tension 50 mN/m; **a** L1, H1 indicates the stage of maximum critical pressure and L2, H2 represents the stage of cell exit at flow rate 7 nL/s and **b** L3, H3 indicates the stage of maximum critical pressure, and L4, H4 represents the stage of cell exit at flow rate 35 nL/s

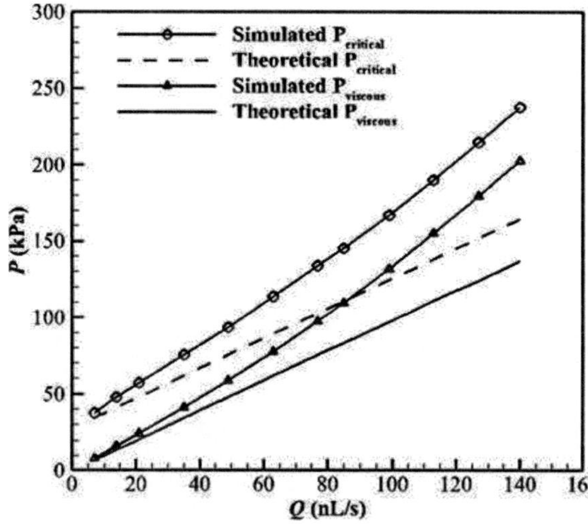

Fig. 6 The effect of flow rates on pressure components for surface tension, $\sigma = 50$ mN/m and viscous ratio, $\mu_{\text{cell}}: \mu_{\text{blood}} = 1:1$

References

1. Gossett DR, Weaver WM, Mach AJ, Hur SC, Tse HTK, Lee W, Amini H, Di Carlo D (2010) Label-free cell separation and sorting in microfluidic systems. Anal Bioanal Chem 397(8):3249–3267
2. Mohamed H, Murray M, Turner JN, Caggana M (2009) Isolation of tumor cells using size and deformation. J Chromatogr A 1216(47):8289–8295
3. Zhang Z, Chen X, Xu J (2015) Entry effects of droplet in a micro confinement: Implications for deformation-based circulating tumor cell microfiltration. Biomicrofluidics 9(2):024108
4. Zhang Z, Xu J, Hong B, Chen X (2014) The effects of 3D channel geometry on CTC passing pressure–towards deformability-based cancer cell separation. Lab Chip 14(14):2576–2584
5. Kalwarczyk T, Ziebacz N, Bielejewska A, Zaboklicka E, Koynov K, Szymanski J, Wilk A, Patkowski A, Gapinski J, Butt HJ, Hołyst R (2011) Comparative analysis of viscosity of complex liquids and cytoplasm of mammalian cells at the nanoscale. Nano Lett 11(5):2157–2163
6. Winters BS, Shepard SR, Foty RA (2005) Biophysical measurement of brain tumor cohesion. Int J Cancer 114(3):371–379
7. Gonzalez-Rodriguez D, Guevorkian K, Douezan S, Brochard-Wyart F (2012) Soft matter models of developing tissues and tumors. Science 338(6109):910–917
8. Preetha A, Huilgol N, Banerjee R (2005) Interfacial properties as biophysical markers of cervical cancer. Biomed Pharmacother 59(9):491–497

A Computational Study of the Effect of *g*-Forces on Busemann Biplane Aerodynamics

Javers Argho Sarder, Ahmad Asif Sami, and A. B. M. Toufique Hasan

Abstract The main reason that discontinued the commercial supersonic transports for civilians is the generation of strong sonic booms. Busemann biplane is a two-dimensional (2D) supersonic biplane configuration which is very effective in reducing the effect of shock wave and wave drag significantly at its design Mach number. Although, in the off-design conditions, there are problems such as choked flow phenomena and flow hysteresis at zero angle of attack (AOA). In this study, the effects of *g*-forces on the flow hysteresis of Busemann biplane are analyzed using computational fluid dynamics (CFD) in inviscid flow (Euler) mode. For this purpose, the drag characteristics of the biplane structure are quantified under different acceleration (and deceleration) parameter of 5, 10, 40, 80 and 160*g*. It is expected that these data will contribute to achieve an effective design of the supersonic biplane in the near future.

Keywords Supersonic biplane · Busemann biplane · *g*-Forces · Flow hysteresis

1 Introduction

From the Wright Flyer (1903) till today, airplanes are playing an important role for achieving economic growth and development. Concorde (1969–2003) was the first and last commercial supersonic transport (SST) in the history of mankind which was retired before building the next generation of SST. One of the fundamental problems that is preventing supersonic commercial transport is the generation of strong sonic boom. It generates enormous amounts of sound energy, sounding similar to an explosion or a thunderclap to the human ear. Busemann Biplane (Fig. 1) is a 2D biplane design invented by Adolf Busemann (1901–1986) which avoids the formation of N-type shock waves and thus does not create a sonic boom.

J. A. Sarder · A. A. Sami · A. B. M. Toufique Hasan (✉)
Department of Mechanical Engineering, Bangladesh University of Engineering and Technology (BUET), Dhaka 1000, Bangladesh
e-mail: toufiquehasan@me.buet.ac.bd

L. Venkatakrishnan et al. (eds.), *Proceedings of 16th Asian Congress of Fluid Mechanics*, Lecture Notes in Mechanical Engineering,
https://doi.org/10.1007/978-981-15-5183-3_47

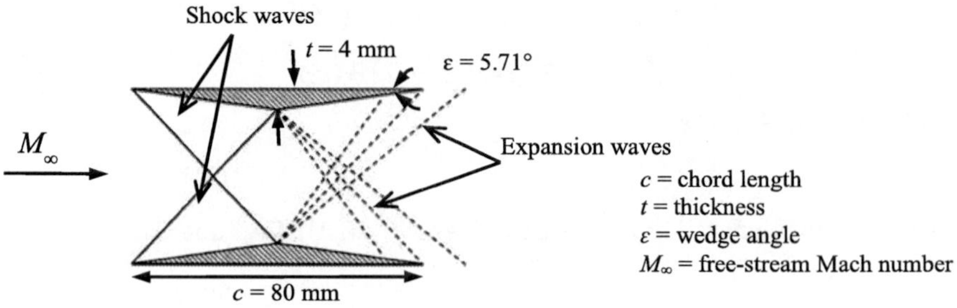

Fig. 1 Supersonic biplane concept (busemann biplane) [1]

In a real flight, it is not easy to design an ideal boomless biplane that can effectively fly from subsonic to supersonic speeds. For a Busemann biplane placed in supersonic flow, oblique shock waves are generated at the leading edge (Fig. 1). These shock waves are cancelled out by the expansion waves generated at the apex (throat) of the biplane elements which reduces the drag. This wave cancellation of Busemann biplane occurs only at its designed Mach number. However, in the off-design conditions, choked flow phenomena and flow hysteresis can be observed which significantly increases the drag of the biplane body. In the other studies, acceleration and deceleration of Busemann Biplane was observed in a quasi-unsteady mode by solving Euler equation. None of them describes what happens when the biplane is accelerated or decelerated in unsteady conditions. To fill the research gap, the vision of this study is to see the effect of '*g*-forces' on the flow hysteresis of the Busemann biplane in unsteady conditions for both accelerating and decelerating cases. '*g*-Force' is a measurement of the type of force per unit mass, typically 'acceleration' that causes a perception of weight, equal to the conventional value of gravitational acceleration (g) on Earth. Human body can sustain any velocity as long as it is constant. It is also flexible enough to survive a huge amount of *g*-force for a small amount of time. Horizontal *g*-Forces are better sustained by human body than the vertical *g*-Forces. Here, higher values of *g*-Forces were chosen to enhance the differences between them to better understand the biplane characteristics under different circumstances as done in Roohani and Skews [2]. Hopefully, the data of this study will contribute to the development of the next generation of supersonic transport.

2 Computational Methods and Validation

2.1 Computational Methods

For the computational analysis, 2D Euler (inviscid) equations were used. For the acceleration and deceleration of the Busemann biplane under several *g*-forces, user defined functions (UDF) were used along with momentum and energy source terms

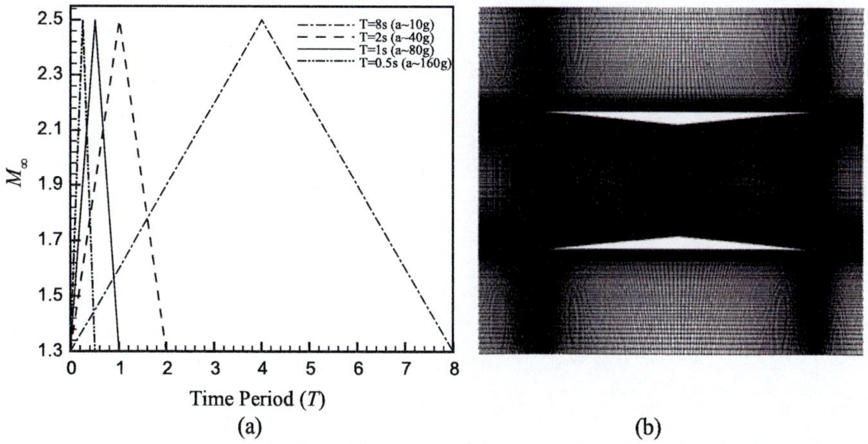

Fig. 2 **a** Variation of Mach number with time and **b** close view of computational domain with grids

in ANSYS Fluent 14.5. The biplane was accelerated and decelerated linearly from a Mach number range of $1.3 < M_\infty < 2.5$ for different time periods ($T = 16$, 8, 2, 1 and 0.5 s) as shown in Fig. 2a. Figure 2b shows the computational domain of the Busemann biplane along with nomenclature.

2.2 *Computational Validation*

In Fig. 3a, the drag characteristics of a Busemann biplane are shown for a Mach number range of $0.3 < M_\infty < 3.0$. At the Mach number of $M_\infty = 0.3$ the value of drag coefficient, C_d is lower which suddenly increases at a Mach number of $M_\infty =$ 0.5. Then it eventually starts to decrease with the increase of Mach number. At a Mach number of $M_\infty = 1.63$, C_d suddenly decreases to a very much lower value and goes to its minimum at a Mach number of $M_\infty = 1.7$ (design Mach number). Around a range of $1.64 < M_\infty < 2.0$, the drag coefficient almost remains to its minimum value. Thus, the biplane is intended to fly in this range. We validated these results with that of Kusunose et al. [3]. Further, in Fig. 3b, drag characteristics are shown for the accelerating and decelerating conditions (quasi-unsteady) for a Mach number range of $1.3 < M_\infty < 2.5$. While accelerating, C_d has a higher value at a Mach number of $M_\infty = 1.3$ which gradually reduces with the increase of Mach number. At a Mach number of $M_\infty = 2.18$, there is a sudden decrease in the drag coefficient. While decelerating from $M_\infty = 2.5$ though, the value of drag coefficient decreases with the decrease of Mach number. At a Mach number of $M_\infty = 1.63$, C_d suddenly increases to a much higher value and matches with that of accelerating condition. Thus, a flow hysteresis is created. We validated these results with that of Yamashita et al. [4].

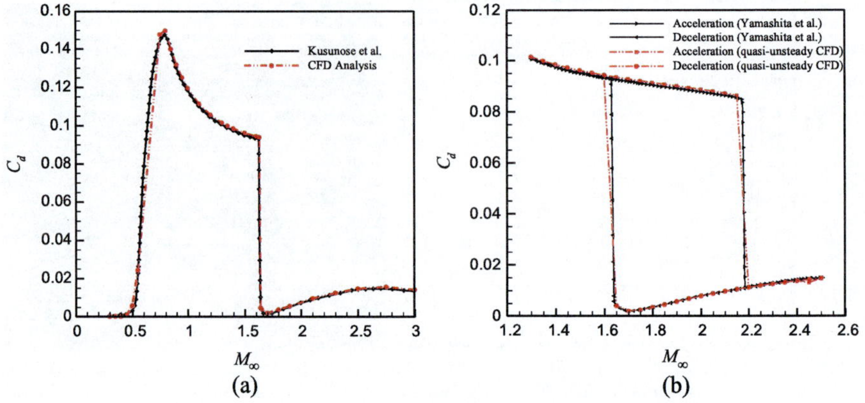

Fig. 3 **a** Comparison of C_d vs M_∞ at zero lift condition between present CFD analysis and Kusunose et al. [3], and **b** Comparison of C_d vs M_∞ in accelerating and decelerating condition between CFD analysis and Yamashita et al. [4]

The flow structures are also compared with available experimental results. In Fig. 4a, b, Schlieren flow visualizations of the wave cancellation effect are shown respectively for $M_\infty = 1.7$ and $M_\infty = 1.9$. Oblique shock waves are seen at the leading edge of the biplane which are cancelled out by the expansion waves generated at the apex (throat) of the inner side of the biplane body. This wave cancellation effect reduces the drag at these Mach numbers. We validated our CFD results of Schlieren images (x-density gradient) with the experimental Schlieren images of Kuratani et al. [5].

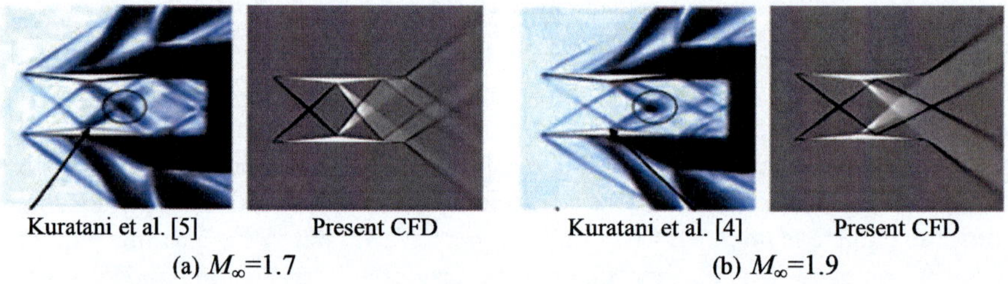

Fig. 4 Comparison of schlieren flow fields between our CFD analysis and experimental data from Kuratani et al. [5]

3 Results and Discussions

3.1 Drag Characteristics of Busemann Biplane for Different g-Forces and Flow Times

In the present study, the Busemann biplane was accelerated and decelerated in a Mach number range of $1.3 < M_\infty < 2.5$ and the drag characteristics (C_d) along with the Schlieren images (*x*-density gradient) were observed.

In Fig. 5, Busemann biplane is accelerated under an acceleration, *a* of 10*g* form a Mach number of $M_\infty = 1.3$ to a Mach number of $M_\infty = 2.5$ in 4 s and then decelerated back to $M_\infty = 1.3$ (again in 4 s). The total time necessary to complete the acceleration and deceleration cycle is thus 8 s ($T = 8$ s). From the Schlieren images shown in Fig. 5, it can be observed that while accelerating, at the Mach number of $M_\infty = 1.6$ (point—1), there is a strong bow shock wave in the upstream of the biplane body which is responsible for a higher value of drag coefficient at that Mach number. With the increase of Mach number, this bow shock wave moves forward and starts to disappear (point—2) gradually between the biplane elements causing a gradual decrease of drag until a Mach number of $M_\infty = 2.21$ or higher. At that Mach number, the bow shock wave completely disappears (similar to point—3) which suddenly and significantly reduces the drag of the biplane body. In case of decelerating, as the Mach number approaches the design Mach number $M_\infty = 1.7$, the drag coefficient decreases due to the wave cancellation effect. However, when the Mach number is decreased further, the drag starts to increase due to preparation of the reappearance of the bow shock wave (point—4) until a Mach number of $M_\infty = 1.61$ or lower, where the bow shock wave appears again in the upstream of the biplane (point—5). This causes a sudden and drastic increase in C_d. After increasing

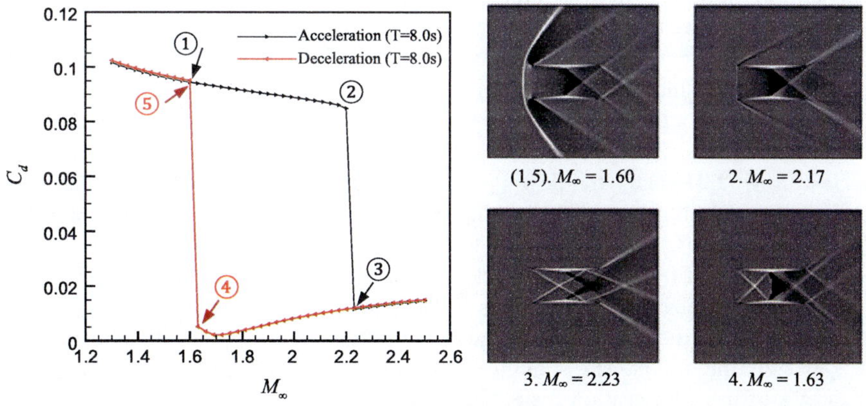

Fig. 5 C_d versus M_∞ for T = 8 s (a ~ 10*g*) along with Schlieren images at different significant conditions (right side figures)

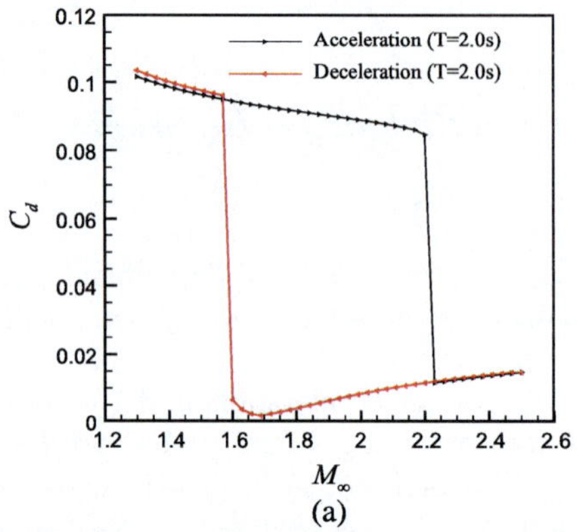

Fig. 6 C_d versus M_∞ for **a** T = 2 s (a ~ 40g) and **b** T = 1 s (a ~ 80g)

the values of C_{d},, it matches with that of the accelerating condition at those Mach numbers. Accordingly, a flow hysteresis is created at the Mach number range of 1.61 < M_∞ < 2.21. For acceleration of a ~ 5g and total flow time of T = 16 s, almost same result in both accelerating and decelerating conditions was observed which is not separately shown; however, this is reflected in later figure of comparison (Fig. 8).

Figure 6a shows the drag characteristics of Busemann biplane at an acceleration of a ~ 40g and the corresponding total flow time of T = 2 s for the same Mach number range as discussed above. The acceleration cycle is same as previously described. Nevertheless, in the deceleration cycle, the sudden and drastic increase in C_d occurs at a lower Mach number of M_∞ = 1.59 because of the delayed appearance of bow shock in the upstream of biplane body. Thus, the new flow hysteresis range becomes 1.59 < M_∞ < 2.21. In Fig. 6b, acceleration is a ~ 80g and total flow time is T = 1 s and a different flow hysteresis range is found as 1.56 < M_∞ < 2.21. This is because the drastic C_d increase occurs at a Mach number of M_∞ = 1.56 while decelerating, whereas the acceleration cycle is completely similar to that of 10g. So in both cases, flow hysteresis is increased.

Similar observations have been found as shown in Fig. 7 where the acceleration is ~160g and total flow time is T = 0.5 s. Now, the new flow hysteresis range is 1.50 < M_∞ < 2.21. Thus, the flow hysteresis is further increased. However, taking a closer observation of the graph and the Schlieren images, some different things can be identified. The C_d values between the acceleration and deceleration curve before and after the flow hysteresis region do not match with one another. In both cases, the C_d of the decelerating condition is higher than that of the accelerating condition. This can be explained using those Schlieren images. At the higher Mach number of M_∞ = 2.26, the oblique shock angle of the acceleration cycle (Point—1) is greater than that of the deceleration cycle (Point—2). So the expansion waves generated at the apex (throat) inside biplane cannot cancel out the oblique shock wave in the deceleration cycle as much as it can in the acceleration cycle. Therefore, the drag coefficient of the deceleration cycle is a bit higher there. At the lower Mach number

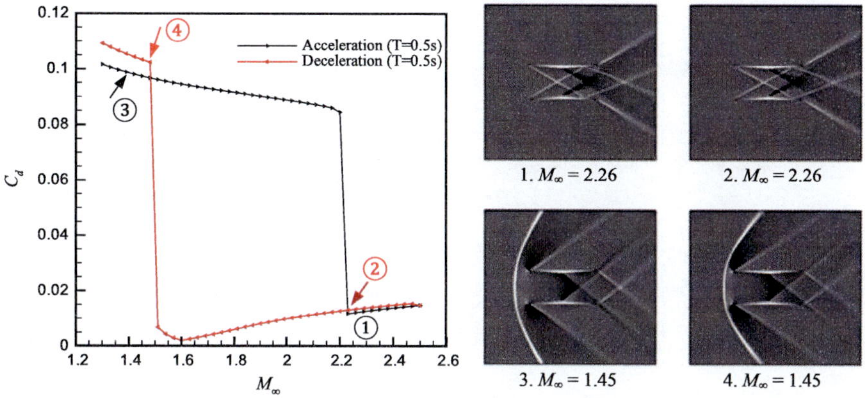

Fig. 7 C_d versus M_∞ for T = 0.5 s (a ~ 160g) along with Schlieren images at different significant conditions (right side figures)

of $M_\infty = 1.45$, if we observe Schlieren images of the accelerating (Point—3) and decelerating (Point—4) condition, we can see that the bow shock wave is closer to the biplane body in the decelerating condition as it just appeared (at $M_\infty = 1.5$). In the accelerating condition, the bow shock wave is far from the body and weak as it is disappearing with the increase of Mach number. So, the drag coefficient is also higher in the decelerating condition than that of accelerating one. This phenomena was also observed in the previous two cases (a ~ 40g and a ~ 80g) in a smaller scale because of the same reason. But in the case of a ~ 10g and a ~ 5g, this phenomena is negligible. So, it can be said that, with the increase of g-force (acceleration), the difference of drag coefficient between the acceleration and deceleration curve before and after the flow hysteresis region increases.

3.2 *Comparison Between Different Cases*

In Fig. 8, all the things that were previously described are summarized and drawn in a single plot. From this graph, it can be said that in all the unsteady cases, the values of C_d completely matches with one another in the accelerating condition. But in the decelerating conditions, with the increase of g-force, the sudden and drastic increase of the value of C_d shifts to a lower Mach number because of the later appearance of the strong bow shock wave in the upstream of the biplane. This increases the flow hysteresis with the increase of g-force. Another thing to observe is, with the increase of g-force, the difference between C_d values between the accelerating and decelerating conditions before and after the flow hysteresis region is increased and has a higher value in the decelerating condition.

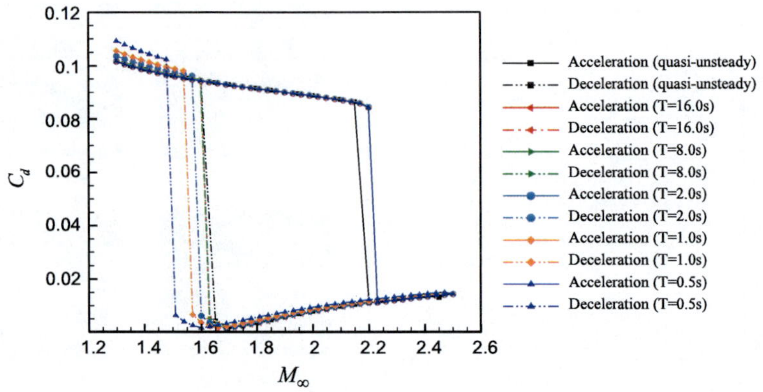

Fig. 8 C_d versus M_∞ in both accelerating and decelerating conditions for different flow times (& under different *g*-forces) and at quasi-unsteady condition

4 Conclusions

The following conclusions can be summarized based on the present computational analysis of acceleration and deceleration effects on the aerodynamics of Busemann biplane:

i. With the increase of *g*-force, there is no significant change in the drag characteristics of the Busemann biplane in the accelerating condition. But while decelerating, with the increase of *g*-force, the drastic drag increase shifts to a lower Mach number.
ii. With the increase of *g*-force acting on the biplane body, the flow hysteresis increases.
iii. With the increase of *g*-force, the difference between C_d values between accelerating and decelerating condition before and after the flow hysteresis region increases having a higher value while decelerating.

References

1. Yonezawa M, Obayashi S (2009) Reducing drag penalty in the three-dimensional supersonic biplane. Proc Inst Mech Eng Part G J Aerosp Eng 223(7):891–899
2. Roohani H, Skews BW (2008) Unsteady aerodynamic effects experienced by aerofoils during acceleration and retardation. Proc Inst Mech Eng Part G J Aerosp Eng 222(5):631–636
3. Kusunose K, Matsushima K, Maruyama D (2011) Supersonic biplane—a review. Prog Aerosp Sci 47(1):53–87
4. Yamashita H, Yonezawa M, Obayashi S, Kusunose K A study of busemann-type biplane for avoiding choked flow. In 45th AIAA aerospace sciences meeting and exhibit, American Institute of Aeronautics and Astronautics

5. Kuratani N, Ogawa T, Yamashita H, Yonezawa M, Obayashi S Experimental and computational fluid dynamics around supersonic biplane for sonic-boom reduction. In 13th AIAA/CEAS aeroacoustics conference (28th AIAA aeroacoustics conference), American Institute of Aeronautics and Astronautics

Interfacial Waves in Dual Stream Granular Flows

Aqib Khan, Praneeta B. Sachan, Akhil K. Mathews, Shivam Verma, Priyanka Hankare, Rakesh Kumar, and Sanjay Kumar

Abstract Granular matter exhibits interesting patterns and instabilities that in many practical situations become critically important and require a proper understanding. Many of these phenomena arise because of small perturbations and instability waves that are present within the flow system. In the present work, we study the characteristics of interfacial waves that are generated when two streams of granular material move together side-by-side on an inclined chute. The streams are gravity driven and are composed of monodisperse grains of same material but different grain size. As the grains accelerate down, the chute instabilities appear on the interface, which grow into finite, quasi-periodic waves. Under certain experimental conditions, the waves are shown to locally induce the cross-stream mixing of grains. The mechanism of formation of interfacial waves is found to be sensitive to grain size and velocities of granular streams, which are varied by changing the chute inclination.

Keywords Granular flows · Shear instabilities and mixing

1 Introduction

Instabilities, chaos, transition, and turbulence are some of the fundamental topics of fluid dynamics, which despite having consumed overwhelming efforts are far from being understood very well. If the flowing medium is, in its very first place, complex, such as blood, polymer melts, slurries, plasmas, paints, and emulsions, which are governed by the non-Newtonian mechanics, then understanding of the underlying physics behind the complex flow phenomena is even more challenging. One important fluid flow phenomenon is the shear layer which is formed when different fluid streams interact. The presence of these shear layers has been widely studied in fluid flows and is shown to be a precursor to instabilities leading to turbulent flows [1, 2].

A. Khan (✉) · P. B. Sachan · A. K. Mathews · S. Verma · P. Hankare · R. Kumar · S. Kumar
Indian Institute of Technology Kanpur, Kanpur, Uttar Pradesh 208016, India
e-mail: khanaqib@iitk.ac.in

L. Venkatakrishnan et al. (eds.), *Proceedings of 16th Asian Congress of Fluid Mechanics*, Lecture Notes in Mechanical Engineering,
https://doi.org/10.1007/978-981-15-5183-3_48

Granular materials exhibiting fluid-like behavior share many similarities with the conventional fluid flows. Shear layers and instabilities that are commonly studied for conventional fluid flows have, however, received little attention for granular flows. Since granular materials are commonly encountered in natural flows like landslides, avalanches, rivers, and industrial processes, understanding the behavior of shear layers and instabilities is of much importance. For example, the presence of interfacial waves and disturbances in granular streams can destabilize the flow resulting in enhanced mixing. The presence of shear and instabilities can influence the momentum of the bulk of landslides and avalanches and thereby influence their impact on the downstream structures [3, 4].

In an experimental study performed by Goldfarb et al. [5] on two parallel granular streams flowing adjacent to each other, it was observed that instability waves were generated at the interface of two granular layers. The amplitude of the waves was found to be sensitive to physical parameters, such as the surface finish of the channel and whether grains are monodisperse or polydisperse. Similar kind of instability waves were observed in fluid flows that are referred to as Kelvin–Helmholtz (K-H) instability [6]. The K-H instability in fluid flows has helped in explaining many physical phenomena such as the formation of waves on the water surface in rivers, swirling of clouds, and formation of vertical structures in jets. In a separate study, Ciamarra et al. [7] using computer simulations of binary granular mixtures in two different arrangements, subjected to horizontal vibrations, demonstrated the existence of a unified interfacial dynamic instability. They argued that this type of surface shear instability is the basic common mechanism that is responsible for different phenomena such as segregation [8, 9] and wave's formation in granular layers [5].

In the present work, we aim to study some of the fundamental aspects of the shear layer that is formed when two gravity-driven granular streams are moved together on an inclined plane. Shear waves that generate due to the shear on the stream interface results in the formation of waves that evolve with the flow. These waves are characterized for different experimental conditions.

2 Experimental Setup

Figure 1 shows the schematic of the experimental setup. It consists of a chute that is inclined at an angle, Φ, to the horizontal. A rectangular hopper that is integrated to the chute is situated at the top of the chute. The channel is 300 mm wide and 900 mm long. The hopper is divided into two chambers by using a splitter plate. Granular material is composed of spherical glass beads manufactured from high-grade lime silica with refractive index 1.525. The combination of granular material used for the two streams is as follows: (i) stream one with fine grains having nominal diameter of 125 μm and white color, and stream two with coarse grains of nominal diameter 425 μm and black in color; (ii) stream one with fine grains of nominal diameter 125 μm and white color, and stream two with coarse grains of nominal diameter

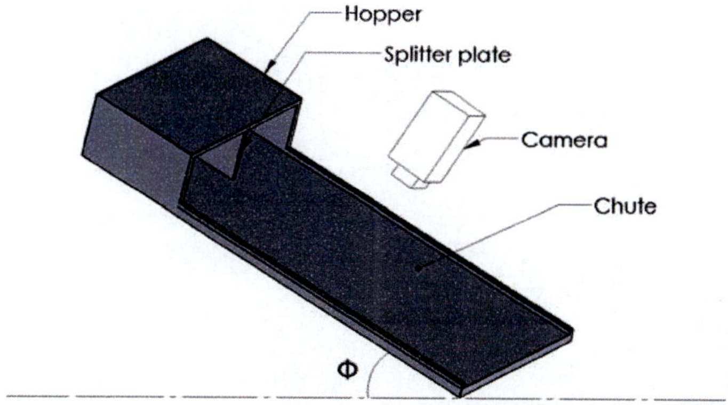

Fig. 1 Schematic of the experimental setup. Φ is the inclination of the channel with the horizontal

825 μm and black in color. The two chambers are loaded with different grains before each experiment. Stationary grains in the hopper are released on the inclined surface using a gate mechanism. Velocity of the stream is changed by changing the inclination (Φ) of the chute, as shown in Fig. 1. Flow visualization is done by using a DSLR camera with high resolution and a frame rate of 60 frames per second. Several movies are recorded at each setting for post processing.

3 Results

Figure 3 shows the flow field on the inclined chute for $\Phi = 20°$ with stream of fine grains in white and that of coarse grains in black color, as mentioned above in Sect. 2. As the gate is removed, grains freely accelerate on the inclined chute. The interface is filmed continuously, so that a generalized behavior can be inferred under the specified experimental conditions. As soon as the two streams, which initially remain separated for small distance downstream of the hopper come in contact, disturbances in the form of weak waves start appearing on the interface. The waviness observed in the experiments is not continuous, rather intermittent waves seemingly appear randomly at locations slightly downstream of the separator. Figure 2a shows the instantaneous flow field of the shear layer at three time instants. One set of such waves are marked with white box and magnified in the inset in Fig. 2. The flow is from right to left as marked in the figure. The instability waves are tracked in successive frames of the video footage and are marked in the main figure and highlighted in the insets at three different time instants. As the two granular streams merge downstream of the splitter, weak disturbances appear on the interface as shown in Fig. 2a, at time instant referenced as $t = 0$ s. As the flow continues down the chute, the disturbances, while growing in amplitude, propagate along the interface, as shown in Fig. 2b at time instant $t = 0.17$ s. As the streams move further downstream, the disturbances

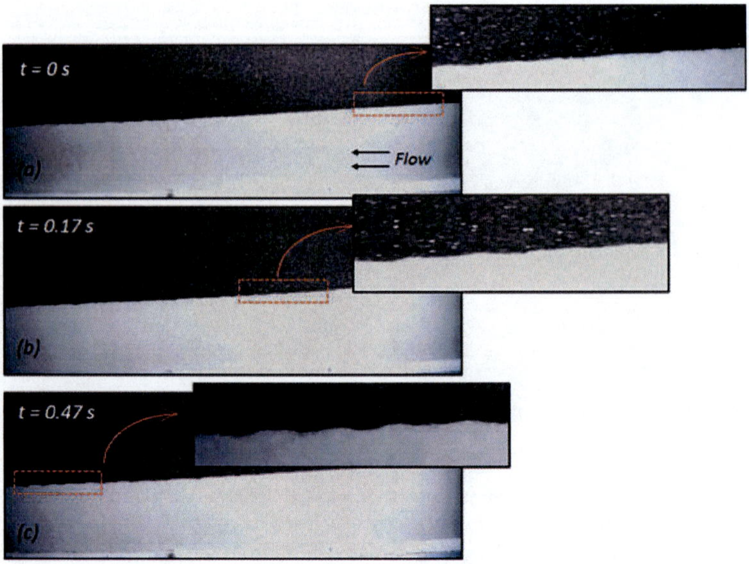

Fig. 2 Evolution of interfacial waves on the interface of two granular streams for chute angle and higher mass flow rate

finally evolve into quasi-periodic waves that are easily observable in Fig. 2c at time instant $t = 0.47$ s.

Experiments are then repeated with the same set of grains and same inclination but with lower mass flow rate. For lower mass flow rate, qualitatively same behavior is observed with subtle differences. With decrease in the mass flow rate, the depth of the granular layer decreases resulting in lower velocity of the granular layer near the free surface. The flow accelerates rather slowly down the channel, and the disturbances grow into finite waves in relatively small distance on the chute surface as can be seen from Fig. 3 at different time instants. Another important observation for smaller mass flow rate is that the waves near the further downstream location which is marked by a box and magnified in the inset in Fig. 3c get smeared and show the early sign of concentration diffusion across the interface of the two granular streams.

In order to study the effect of channel inclination, we have performed experiments with changing the chute inclination (Φ) while keeping the mass flow rate constant. The effect of inclination on the interface is shown in Fig. 4. At low inclinations (Fig. 4a, b), with lower flow velocity, waves are mild with smaller amplitude and generated at smaller frequency. If the channel inclination is high ($\Phi = 35°$), as shown in Fig. 4c, waves are stronger, and the interface appears to be more distorted as compared to the case where inclination is small.

In the present experiments, channel inclination is found to be the main governing parameter other than the mass flow rate. Even though the inclination is same for both the streams, the velocity of the individual stream is different. The difference in

Fig. 3 Evolution of interfacial waves on the interface of two granular streams for chute angle and small mass flow rate

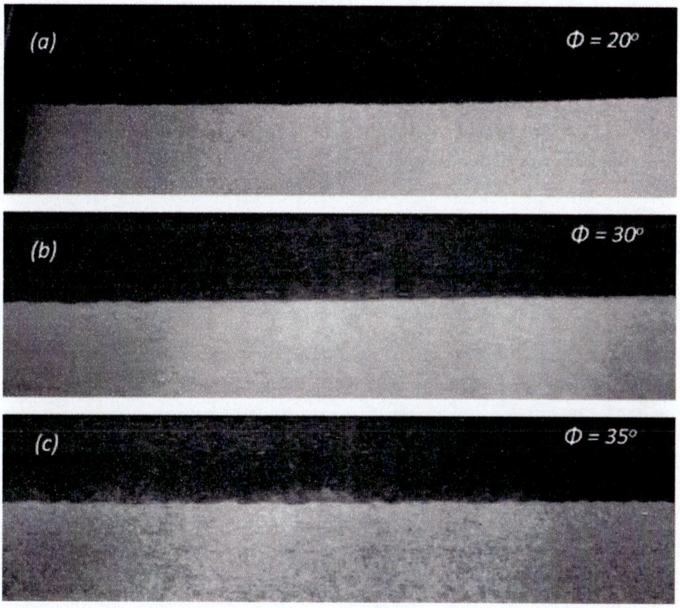

Fig. 4 Effect of channel inclination, Φ, on the shear layer

velocity that arises due to the difference in the grain size results in unequal dilatation of the flow. This unequal dilatation at the interface introduces shear in the form of extensional strain that causes the interface to turn locally, which ultimately appears as disturbances/waves to the observer. As the flow dilates further while accelerating due to external body force, the disturbances grow into finite quasi-periodic waves. These shear-induced surface instability waves, as earlier discussed by Ciamarra et al. [7], play an important role in pattern formation and the granular segregation. It seems plausible that these instabilities will play their role in modifying the mixing characteristics of granular material. Mixing in granular materials has been a challenging task, especially when the constituents of the mixture differ in size and/or density. When a granular system with different constituents is agitated at certain frequencies, it has a tendency to segregate which is counter-intuitive to the observations in conventional molecular fluids. Thus, it would be interesting to explore other possibilities that can be used to enhance the mixing capabilities in granular mixtures.

4 Conclusions

We have performed experimental study of the shear formed at the interface of two adjacent flowing granular streams with different grain size on an inclined plane. We have examined the growth of fine disturbances into observable quasi-periodic waves as the shear layer evolves down the inclined chute. Two experimental parameters namely the chute inclination and mass flow rate from the hopper dominate the entire flow field and in particular the shear instability phenomenon.

References

1. Hussain F (1986) Coherent structures and turbulence. J Fluid Mech 173:303–356
2. Yoda M, Hesselink L, Mungal MG (1992) The evolution and nature of large-scale structures in the turbulent jet. Phys Fluids A 4:803–811
3. Faug T, Childs P, Wyburn E, Einav I (2015) Standing jumps in shallow granular flows down smooth inclines. Phys Fluids 27:073304
4. Gray JMNT, Cui X (2007) Weak, strong and detached oblique shocks in gravity-driven granular free-surface flows. J Fluid Mech 579:113–136
5. Goldfarb DJ, Glasser BJ, Shinbrot T (2002) Shear instabilities in granular flows. Nature 415:302
6. Andereck CD, Liu SS, Swinney HL (1986) Flow regimes in a circular Couette system with independently rotating cylinders. J Fluid Mech 164:155–183
7. Ciamarra MP, Coniglio A, Nicodemi M (2005) Shear instabilities in granular mixtures. Phys Rev Lett 94:188001
8. Makse HA, Shlomo H, Peter RK, Stanley HE (1997) Spontaneous stratification in granular mixtures. Nature 386:379
9. Ottino JM, Khakhar DV (2000) Mixing and segregation of granular materials. Ann Rev Fluid Mech 32:55–91

Experimental Investigation of Vortex Breakdown Structure Over a Nonslender Flying Wing Configuration

Vivek Kumar, Alakesh Chandra Mandal, and Kamal Poddar

Abstract An experimental study has been carried out to investigate the nature of vortex breakdown structure of the complex flow field over a flying wing configuration with sweep angle, $\Lambda = 53^{\circ}$ at the Reynolds number (Re) range, 2.5×10^5–7.5×10^5. Surface oil flow visualization, surface pressure measurements, and the particle image velocimetry (PIV) measurements have been carried out to characterize the flow field over the flying wing. This study reveals the separated vortical flow with vortex breakdown and a wave-like structure in the flow visualization images. This wave-like structure is found to be related to the onset of vortex breakdown over the flying wing configuration.

Keywords Vortex breakdown · Nonslender flying wing · Vortical flows · Separated vortical flow with vortex breakdown

1 Introduction

Earlier investigations on the leading-edge vortex (LEV) breakdown have mainly been dedicated on the slender delta wings. However, recent developments in micro-air vehicles (MAVs) and unmanned air vehicles (UAVs) have encouraged the investigations on nonslender delta wings. The nonslender delta wings, distinguished by a sweep angle less than 55° [1], show quite different flow structures, compared to the slender ones. One characteristic of the nonslender delta wing is vortex breakdown. Vortex breakdown is a sudden expansion of vortices, which were first observed over delta wings. This phenomenon affects both steady and unsteady aerodynamics of delta wings.

The theoretical framework and the experimental investigations of vortex breakdown have been addressed in the reviews and investigations of Harvey [2], Sarpkaya [3], Hall [4], Escudier [5], Brown and Lopez [6], Lopez and Perry [7], Delery [8]. The work of Lambourne and Bryer reveals two typical types of breakdown over a delta

V. Kumar (✉) · A. C. Mandal · K. Poddar
Department of Aerospace Engineering, Indian Institute of Technology, Kanpur 208016, India
e-mail: vivekk@iitk.ac.in

L. Venkatakrishnan et al. (eds.), *Proceedings of 16th Asian Congress of Fluid Mechanics*, Lecture Notes in Mechanical Engineering,
https://doi.org/10.1007/978-981-15-5183-3_49

wing: bubble type and spiral type. The former appears to be axisymmetric, while the latter appears to be strongly non-axisymmetric with a spiral or corkscrew shape. Subsequent experimental investigations have been focused on the characteristics of the breakdown phenomena and its onset position above the wing. These experiments reveal that while the two types of vortex breakdown can switch to each other from time to time, the spiral type occurs more often than the bubble type [9].

Although the vortex breakdown phenomena were extensively studied in the past for a slender delta wing configuration, a nonslender delta wing, in this regard, received comparatively less attention [1]. Nonetheless, there have been some studies on nonslender delta wings (see [1], for details). Similar such studies on a nonslender flying wing configuration are important for its detailed aerodynamic characterizations, as this configuration can potentially be important for designing the next-generation UAVs and MAVs.

Following our previous work on this configuration [10], we focus here on a wave-like structure seen in the surface oil flow visualization over the flying wing, and its relationship with the vortex breakdown phenomenon and the associated flow physics. The vortex breakdown phenomenon is directly involved with the aerodynamic performances, such as loss of lift, and therefore, it is necessary to understand the vortex breakdown phenomenon in detail for a flying wing configuration and its association with the wave-like flow visualization patterns over the wing. Detailed experimental investigations using the surface pressure and the flow field measurements along with the surface flow visualization are performed to gain knowledge about the vortex breakdown phenomena over a nonslender, complex geometric profile of a flying wing.

2 Experimental Setup

All the experiments presented in this paper were performed in a closed-circuit wind tunnel available at the Low-Speed Aerodynamics Laboratory, IIT Kanpur. Test section size of the tunnel is 1.68 m $\times$ 0.9 m $\times$ 0.6 m (length $\times$ width $\times$ height). It has a contraction ratio of 6.5. The turbulence level in the test section is measured to be 0.3% of the freestream velocity. Siemens-made speed controller is used to control the velocity in the test section. The maximum velocity that can be achieved in the test section is about 35 m/s. The flying wing model with a rounded leading edge and complex profile was mounted on a pitching mechanism in the test section, as shown in Fig. 1. The model was manufactured by the selective laser sintering (SLS) process with two parts, i.e., an upper part and a base part; micro-sized polymer powder (PA 2200) was used for smooth surface finish. The leading-edge sweep angle, Λ, of the model was 53°, and it is categorized as a nonslender flying wing configuration. The

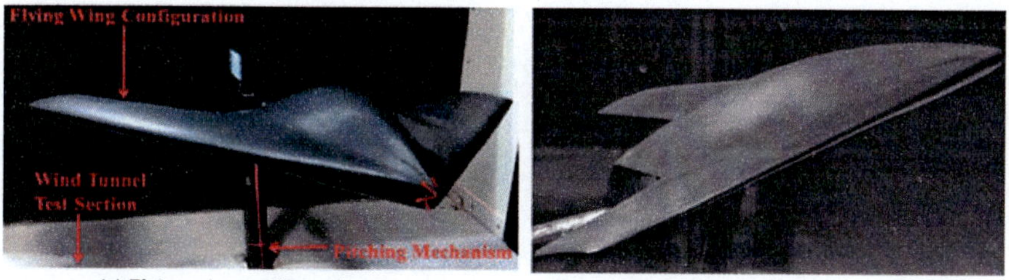

(a) Flying wing configuration (b) Starboard side of flying wing configuration

Fig. 1 Experimental model mounted in the wind tunnel test section

root chord (C) and the maximum span of the model are 390 mm and 551 mm, respectively. The planform area and the aspect ratio are found to be 0.1074 m^2 and 2.84, respectively. The blockage ratio is 9.65% at $\alpha = 30^o$.

Particle Image Velocimetry (PIV) Technique

The 2D-PIV measurements were performed in a plane passing through the vortex centerline and perpendicular to the surface of the model, as schematically shown in Fig. 2. Smoke particles in the flow field were illuminated by a 200 mJ Nd-YAG laser at a wavelength of 532 nm. The timing control for the laser pulses was done by a synchronizer (IDTvision) with an average delay time of 8 μs. The particle images were acquired using the MicroVec software (Vision Asia, Singapore) with a CCD camera (8MP, Imprex). A lens of 85 mm focal length was used in the camera while acquiring the images. The seeding in the tunnel was done using a commercial fog generator (Antari Z1500II-R) with Antari FLG fog liquid. The fog generator which

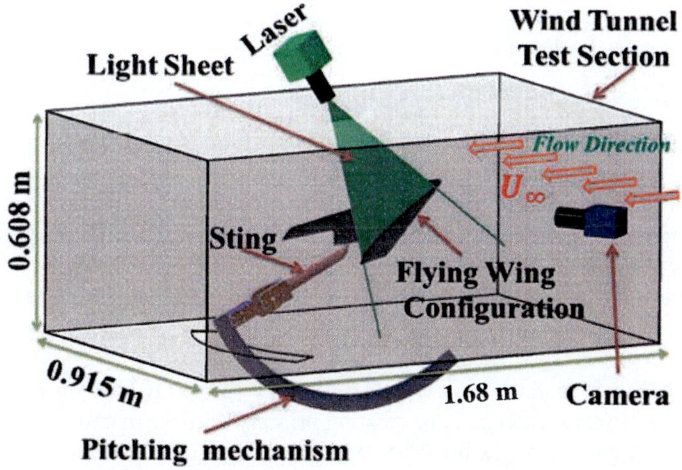

Fig. 2 Schematic of experimental setup for the particle image velocimetry measurements in a plane passing through vortex centerline

was placed near the diffuser section of the closed-circuit wind tunnel can produce particles ~1 micron in size. The acquired PIV images were processed using the proVISION XS package (IDTvision, USA) software, and further post-processing of the PIV data was carried out using MATLAB.

Pressure Measurement

For surface pressure distribution along the spanwise direction (i.e., Y direction, as described below), various static pressure ports were made at various chordwise locations (X/C locations); note that here X denotes the chordwise distance from the apex of the flying wing model, i.e., the streamwise distance. A total of 109 pressure ports were made on the suction side of the model at various chordwise locations, that is, X/C = 0.2, 0.3, 0.4, 0.5, 0.6, 0.7, 0.8.

The pressure measurement system consists of the followings: four electronically scanned pressure (ESP) scanners, a multiplexer unit, a high-speed 14-bit DAQ module (PXI-6133) with PXI system, and a workstation with an interface board (PCI-8336). For surface pressure measurements in the present study, we have used total four ESP scanners (three with 20-inch water column range and one with 10-inch water column range), each having 32 pressure ports. We may also mention that the maximum error of each scanner is within 0.05% of the full-scale pressure range. Also, output of each ESP is selectively routed to the DAQ module (PXI-6133) by applying its unique binary address to the multiplexers which can switch between ports at maximum rates up to 20,000 ports/s. In the present experiment, pressure ports were scanned at the rate of 19,520 ports/s, leading to an effective sampling rate of 610 samples/sec/port. After converting raw data from analog to digital form, the output raw pressure data (in volts) was normalized with the excitation voltage. LabVIEW-based virtual instrumentation program was used to convert the normalized data to the pressure data by utilizing the calibration coefficients. The raw data (V/V_{ex}) and the mean pressure and the root-mean-squared pressure data were stored in a computer for further analyses.

Oil Flow Visualization

In the present work, the surface oil flow visualization was accomplished using a mixture of titanium dioxide, oleic acid, and transformer oil. This mixture was then applied uniformly to the suction side of the model, which was also illuminated using a halogen lamp for photography purpose. A DSLR camera (Nikon, D750), equipped with a lens (AF-S Nikkor 24–120 mm f/4G ED VR), was used to capture the images of the surface oil pattern at different Reynolds numbers and angle of attacks.

Results

The surface flow visualization at $\alpha = 20^{\circ}$ for $Re = 3.7 \times 10^5$, as shown in Fig. 3, reveals the signature of two counter-rotating primary vortices that are formed on the flying wing before the vortex breakdown, along with axially attached flow over the fuselage. One may also notice in Fig. 3 that there exists a wave-like structure near the leading edge of the flying wing configuration in the range of $0.4 \le X/C \le 0.6$, as better seen in a zoomed view, added as an inset of this figure. In the range of chordwise

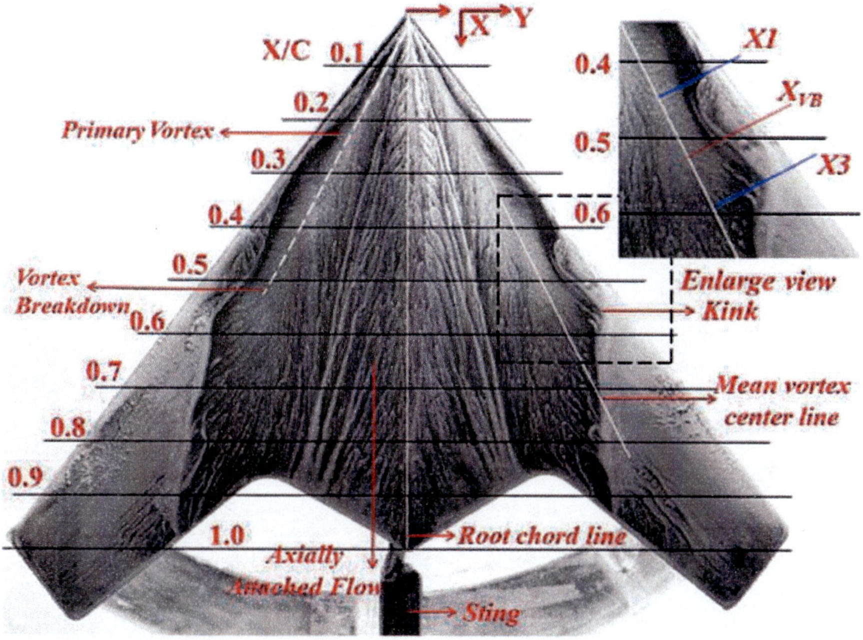

Fig. 3 Surface oil flow visualization on the upper surface of the flying wing at $\alpha = 20°$ for Re = 3.7×10^5

distance, $0.5 \le X/C \le 0.6$, sudden expansions of the skin friction lines on the flying wing most likely indicate vortex breakdown. The white line over the starboard side of the flying wing in Fig. 3 indicates the mean vortex centerline. Further, along the white line, X1, X_{VB}, and X3 in inset figure denote the starting of the wave-like pattern, vortex breakdown location, and the ending of the wave-like pattern, respectively.

The coefficient of pressure distributions over the starboard side of the flying wing configuration at $\alpha = 20°$ for Re $= 3.7 \times 10^5$ at various chordwise locations, as displayed in Fig. 4, reveals that the suction goes to be stronger in the range of X/C = 0.2 to 0.4. This in turn indicates that the primary vortex is becoming stronger. Beyond X/C = 0.4, a gradual decrease in suction can be noticed in Fig. 4. One may also notice flat pressure distribution beyond X/C = 0.6 in Fig. 4. This indicates that the vortex breakdown phenomena occur in the range of $0.4 \le X/C \le 0.6$.

Thus far, the flow visualization (Fig. 3) and the pressure distribution (Fig. 4), as discussed above at $\alpha = 20°$ for Re $= 3.7 \times 10^5$, suggest the presence of the vortex breakdown of the leading vortices for the flying wing configuration in the range of $0.4 \le X/C \le 0.6$. To confirm the vortex breakdown and to investigate the characteristics of the breakdown in detail, the PIV measurements in a plane normal to the surface and along the white line, shown in Fig. 3, have been carried out at $\alpha = 20°$ for Re = 3.7×10^5. This plane of measurements along the mean vortex centerline has been chosen based on the surface oil flow visualization. The velocity contours in Fig. 5(a)

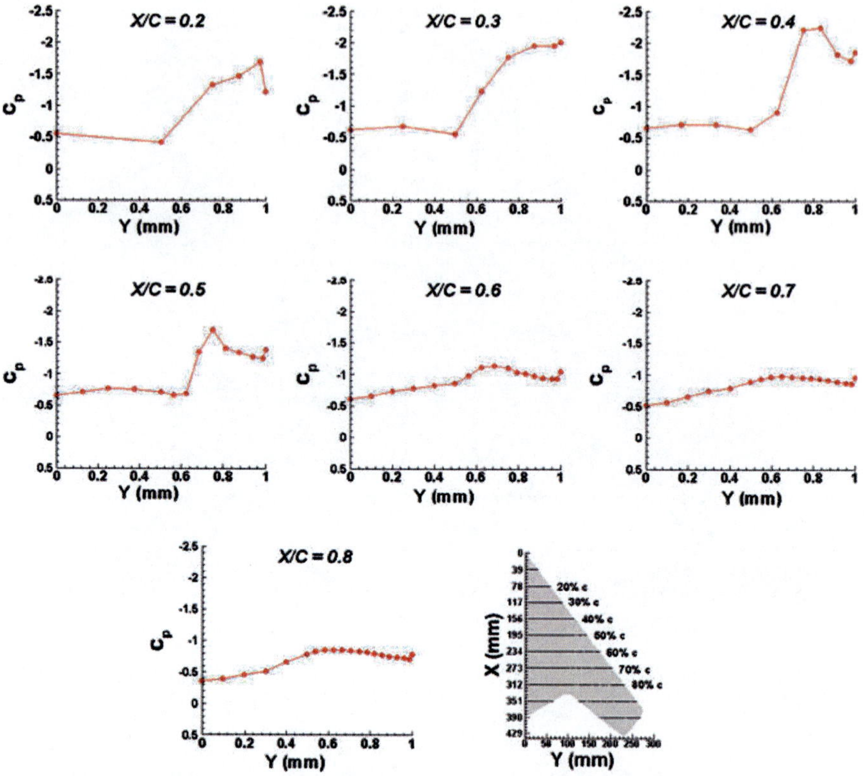

Fig. 4 Surface pressure distribution on the upper surface of the flying wing (starboard side) at $\alpha = 20^{o}$ for Re = 3.7×10^5 at various chord-wise locations

and the streamlines in Fig. 5(b) show that there exists a stagnation point, which is better seen in the zoomed view in Fig. 5(c). Axial drop of the mean axial velocity after vortex breakdown and existence of a stagnation point followed by reversed flow indicate a phenomenon of the helical swirl at the downstream of breakdown [1, 4]. A sudden expansion of LEV typically characterizes the vortex breakdown (see [5], for details). A stagnation zone occurs downstream of the breakdown position, with a small recirculating region ($U_{axial}/U \leq 0$) [11]. Two opposite sign mean axial vorticity zones, as seen in Fig. 5(d), appear at the vortex breakdown location along the vortex centerline, in comparison with the sign change of the axial vorticity along the vortex centerline and the occurrence of zero axial velocity along the centerline of the vortex [6]. Hence, the PIV measurements in Fig. 5 along vortex centerline clearly show the starting point of the wavelike structure (X1), the vortex breakdown location (X_{VB}), and the end point of the wavelike structure. It is also clear from these observations that the vortex breakdown phenomena occur in the range of chordwise distance, $0.4 \leq X/C \leq 0.6$.

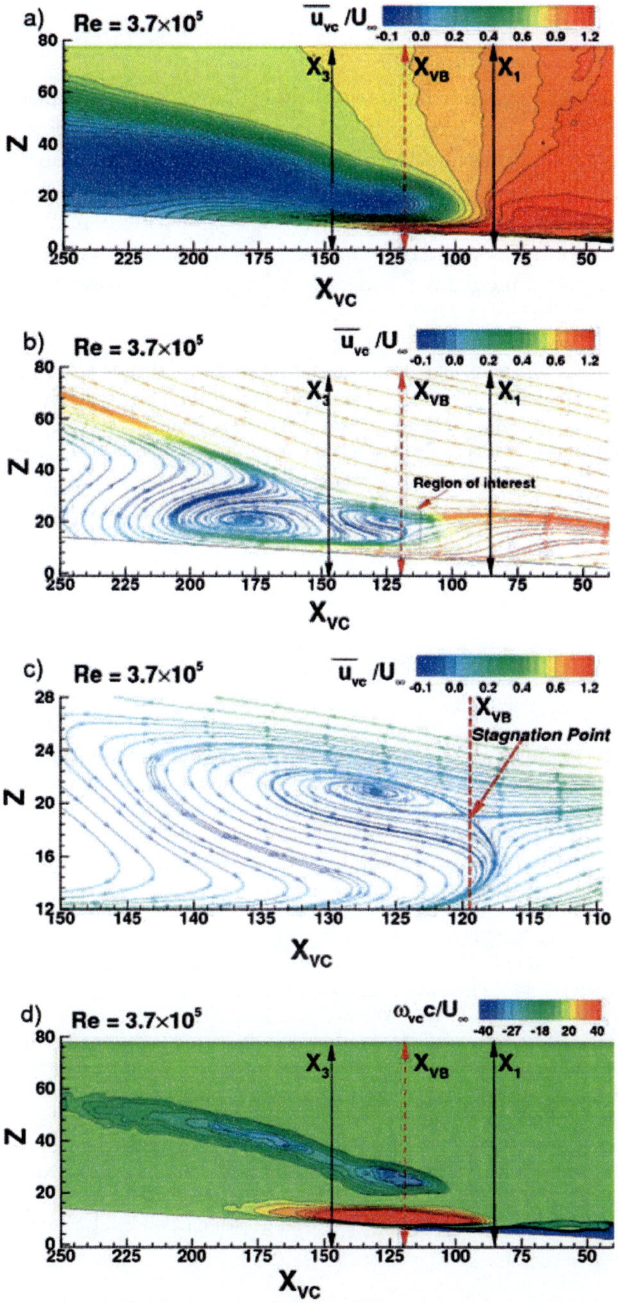

Fig. 5 (**a**) Contours of mean axial velocity, $\overline{u_{vc}}/U_{\infty}$, at α = 20° for Re = 3.7 × 10^5. (**b**) Sectional streamlines on plane through mean vortex centerline. (**c**) Enlarge view of region of interest; streamlines are multi-colored by mean axial velocity ($\overline{u_{vc}}/U_{\infty}$) contour. (**d**) Contours of azimuthal vorticity, ($\omega_{vc}C/U_{\infty}$)

3 Conclusions

An experimental investigation has been carried out to understand the flow field over a flying wing configuration. To accomplish this task, we have performed surface flow visualization, surface pressure measurements, and the PIV measurements on the flying wing configuration. From the results presented above, we find that the surface flow pattern or the wave-like structure over the flying wing is related to the vortex breakdown phenomena, occurring in the range of chordwise distance, $0.4 \leq X/C \leq 0.6$; that is, the vortex breakdown and surface flow patterns (wave-like structure) are closely linked together.

Acknowledgements We are thankful to the technical staff for valuable help in the experimental activity. We are also grateful to the other graduate students of the Low-Speed Aerodynamics Laboratory of Aerospace Engineering Department, especially Mr. Balamurgun and Mr. Anifa, for valuable help while conducting PIV experiments.

References

1. Gursul I, Gordnier R, Visbal M (2005) Unsteady aerodynamics of nonslender delta wings. Prog Aerosp Sci 41(7):515–557
2. Harvey JK (1962) Some observations of the vortex breakdown phenomenon. J Fluid Mech 14(4):585–592
3. Sarpkaya T (1970) An experimental investigation of the vortex-breakdown phenomenon, Monterey, California. Naval Postgraduate School. https://calhoun.nps.edu/handle/10945/29294
4. Hall MG (1972) Vortex breakdown. Ann Rev Fluid Mech 4(1):195–218
5. Escudier M (1988) Vortex breakdown: observations and explanations. Prog Aerosp Sci 25(2):189–229
6. Brown GL, Lopez JM (1990) Axisymmetric vortex breakdown Part 2. Physical mechanisms. J Fluid Mech 221:553–576
7. Lopez JM, Perry AD (1992) Axisymmetric vortex breakdown. Part 3 onset of periodic flow and chaotic advection. J Fluid Mech 234:449–471
8. Delery JM (1994) Aspects of vortex breakdown. Prog Aerosp Sci 30(1):1–59
9. Gursul I (2005) Review of unsteady vortex flows over slender delta wings. J Aircraft 42(2):299–319
10. Kumar V, Mandal AC, and Poddar K (2020) Experimental study of the complex vertical flow structures over a flying wing configuration. J Flow Vis Image Process 27(1)
11. Wang C, Gao Q, Wei R, Li T, Wang J (2016) 3D flow visualization and tomographic particle image velocimetry for vortex breakdown over a non-slender delta wing. Exp Fluids 57(6):98
12. Lambourne NC, Bryer DW (1961) The bursting of leading-edge vortices-some observations and discussion of the phenomenon. Aeronautical Research Council, Reports and Memoranda No. 3282

Experimental Investigation of Three-Dimensional Modes in the Wake of a Rotationally Oscillating Cylinder

Soumarup Bhattacharyya, Shivam Verma, and Sanjay Kumar

Abstract Three dimensionalities in the wake of the flow past a circular cylinder executing sinusoidal rotary oscillations about its own axis are studied experimentally. The results of water tunnel experiments on a rotationally oscillating cylinder in cross-flow for Re = 250, amplitude $\pi/2$, and non-dimensional forcing frequency $0.5 \leq FR \leq 5$ are discussed. Observation made for Re = 250 with amplitude $\pi/2$ and a forcing frequency of FR = 0.5, confirmed a mode having a spanwise non-dimensional wavelength $\lambda/d \simeq 2$ which is also observed for a stationary cylinder at lower Re number. On increasing forcing frequency, this mode disappears and a new unreported mode with a much smaller wavelength of $\lambda/d = 0.5$ appears at around FR = 1. This mode almost remains stable till FR = 1.5. At higher forcing frequency FR = 2, a mode with mushroom-shaped vortices and a wavelength of $\lambda/d = 1.2$ is identified. At higher forcing frequencies, the far wake becomes very stable and mixing is highly reduced which may be useful for a lot of practical applications. Effects on spanwise wavelength by varying the amplitude at a forcing FR = 1 and a high FR = 5 are also discussed.

Keywords Wakes · Vortex shedding · Three-dimensional modes

1 Introduction

Flow past a circular cylinder can be considered as a mainstream study of incompressible bluff body flows. Past research has been focused particularly on the cylinder wake as it exhibits diverse phenomenon at various Reynolds number (Re). The wake study is important as it develops into a series of distinct regimes, which further affects mixing of the flow and changes overall pressure drag. Study of stationary cylinder by Williamson [1] showed the presence of three-dimensional modes beginning with Mode A at a Re = 180. Interestingly, after Re = 220, Mode B coexists with mode

S. Bhattacharyya (✉) · S. Verma · S. Kumar
Department of Aerospace Engineering, Indian Institute of Technology Kanpur, Kanpur, Uttar Pradesh 208016, India
e-mail: soumarup@iitk.ac.in

L. Venkatakrishnan et al. (eds.), *Proceedings of 16th Asian Congress of Fluid Mechanics*, Lecture Notes in Mechanical Engineering,
https://doi.org/10.1007/978-981-15-5183-3_50

A which becomes dominant after Re = 250. Norberg et al. [2] studied the effect of aspect ratio of the cylinder on the wake. Further, computational studies by Mittal and Kumar [3] backed with experimental studies of Kumar et al. [4] were made for rotating cylinder with a non-dimensional rotation rate of $\alpha = \Omega d/2U$ varying from 0 to 5, where Ω is the angular velocity of the cylinder rotation, d is the cylinder diameter, and U is the free stream velocity. Radi et al. [5] confirmed experimentally that at higher rotation rates three dimensionionality becomes increasingly complex. Mode C with a wavelength of $\lambda/d = 1.2$ and Mode D with a wavelength of $\lambda/d = 2$ arise from the unsteady wake, whereas Mode E with a wavelength $\lambda/d = 2$ and Mode F with a wavelength of $\lambda/d = 0.45$ emerges as a result of steady wake where vortex shedding is suppressed. These modes are also used as a validation of the present experimental setup. Rotational oscillations about the cylinder axis by Thiria et al. [6] provide a very enriching way to control the wake and have gained recent attention. Experimental studies on the wake of an oscillating cylinder by Kumar et al. [7] predict three-dimensional behavior of the forced wake. In this study, we investigate the three-dimensional modes in the wake of the rotationally oscillating cylinder at various amplitude and forcing frequencies. This investigation would provide an insight on fluid mixing or unmixing and wake modification due to cylinder oscillation about its own axis. The cylinder was forced to oscillate according to

$$\theta = \theta_0 \text{ Sin}(2\pi \text{ft})$$

where θ_0 is the amplitude of oscillation, θ is the angular position of the cylinder, f is the forcing frequency, and t is the time. The vortex shedding frequency for stationary cylinder is found using CTA hot wire analysis and is denoted by f_0. The frequencies in this study have been non-dimensionalized as FR $= f/f_0$.

2 Experimental Setup and Validation

The experiments were conducted in a water tunnel (Rolling Hills Inc:model 0710) with a test section depth of 0.25 m, length 0.46, and width 0.18 m. The cylinder had a diameter of 8 mm. The flow uniformity was checked both by PIV and DANTEC CTA hot wire anemometry. The tunnel floor had a black anodized plate with a bearing to facilitate rotation and oscillation. Moons Servomotor was used to rotate and oscillate the cylinder, and the signal for rotation and oscillation was provided by M Servo Suite software. Flow visualization was done using hydrogen bubble flow visualization technique. A 50-μm-platinum wire of 150 mm length was positioned slight downstream depending on the flow for hydrogen bubble flow visualization. A metal anode was placed 320 mm downstream of the cylinder, and a voltage of 10 volts was applied cross the platinum wire and anode of produce the stream of bubbles.

Figure 1 shows the end view, side view, and top view the for flow visualization setup. Variation of Strouhal number with Reynolds number in the range $175 \leq \text{Re} \leq 275$ for a stationary cylinder as measured by Williamson [1] was used

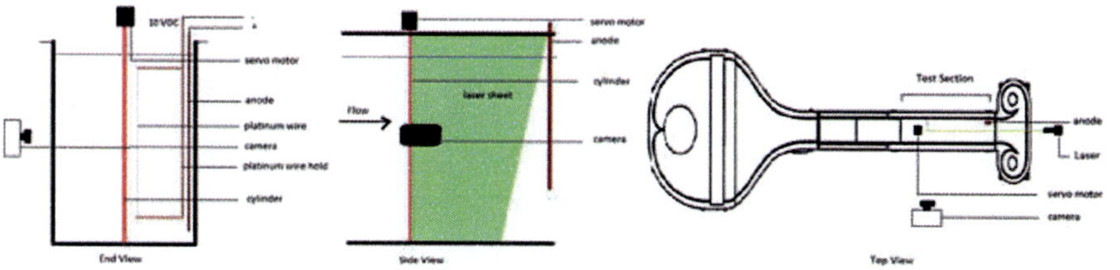

Fig. 1 Schematic views of the experimental setup in the water tunnel used for flow visualization

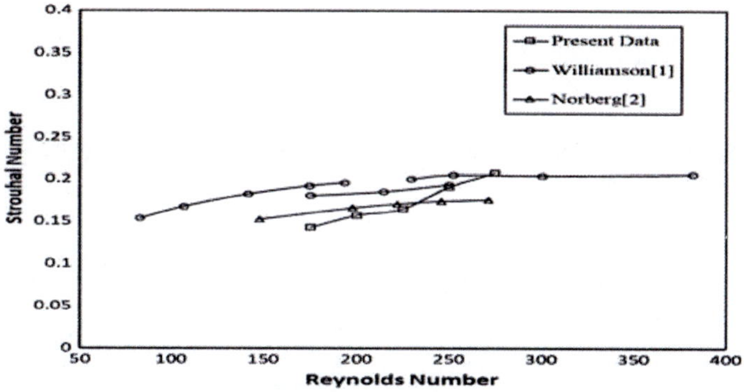

Fig. 2 Variation of Strouhal number with Reynolds number

as a validation. CTA hot wire measurements in the current setup showed slightly lesser values of shedding frequencies as compared to Williamson [1] which can be attributed to the observation made by Norberg et al. [2] for cylinders with low aspect ratios (l/d). Figure 2 shows the comparison of Strouhal number versus the Reynolds number, and Fig. 3 shows the shedding frequency $f_0 = 0.74$ Hz at Re 250, the value of which is further used in the experiments for finding the non-dimensional FR.

Flow visualization results for stationary cylinder were validated using the images from Radi et al. [5]. All the flow visualization images are shown such that the flow directed from left to right. The hydrogen bubble creates a bright streakline when it crosses the laser sheet. At Re = 175 oblique shedding and emergence of Mode A instabilities patterns are observed with a wavelength of $\lambda/d = 3$ and the wake is highly three-dimensional in nature. Increasing Re results to decay of this mode and at Re = 250 Mode B with a wavelength of $\lambda/d = 1$ becomes more dominant. Mode A and Mode B coexisted at a range of $220 \leq \text{Re} \leq 250$. Validation image of flow visualization of Mode B at Re = 275 is presented in Fig. 4 with image of Radi et al. [5] on the left (Fig. 4a) and image of the present experiment on the right (Fig. 4b).

Validation with Radi et al. [5] for three-dimensional modes found in wake of a rotating cylinder was carried out. A subharmonic Mode C and a spanwise periodicity and a wavelength of $\lambda/d = 1.1$ were confirmed at Re = 275 and $\alpha = 1.2$. The spanwise wavelength was inversely proportional to the rotation rate in this mode for

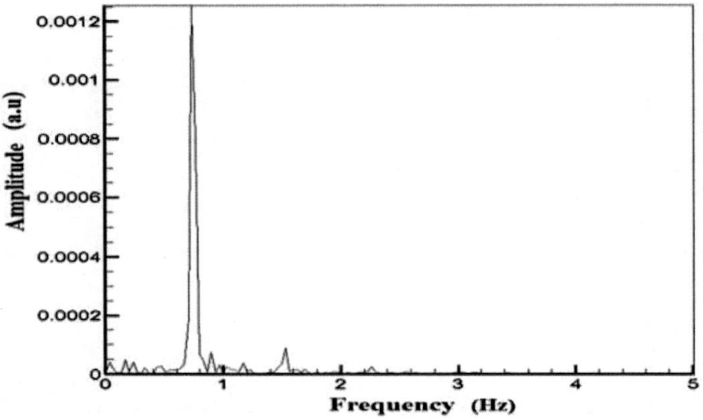

Fig. 3 Amplitude spectrum in the wake of a stationary cylinder at Re = 250 giving $f_0 = 0.74$ Hz as the shedding frequency

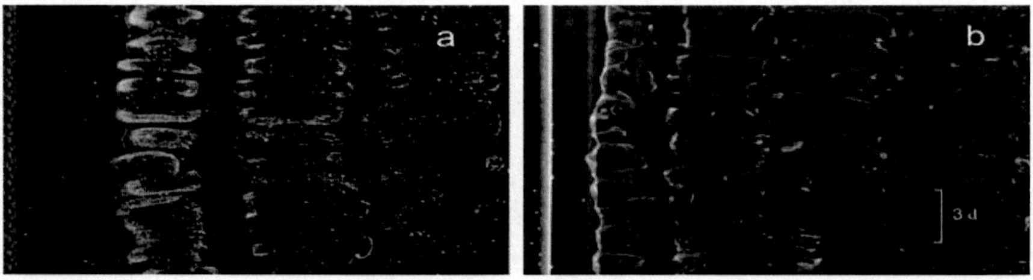

Fig. 4 Validation of present flow visualization technique and experimental set-up. (**a**) Visualization of Mode B by Radi et al. at Re = 275 for a stationary cylinder (**b**) Present data in same conditions. The vertical extent of the image in (**b**) is 15 cm. Platinum wire cannot be seen as it is placed upstream. Scaling is same for both the images

$0.7 \leq \alpha \leq 1.7$. For rotation rate below $\alpha = 1$, the flow becomes highly irregular and for rotation rates above $\alpha = 1.85$ Mode D is observed. Vortex shedding is suppressed for $\alpha \geq 1.9$; hence, a new Mode E is described for vortex suppressed wake at $\alpha \geq 1.9$ with a similar wavelength of Mode D.

As this mode originates from a different base flow, i.e., vortex shedding ceases and the wake becomes stable, in spite of having a similar wavelength the structure of the mode is different. The visualization of this mode is presented in Fig. 5 with validation image of Radi et al. [5] on the left (Fig. 5a) and present experiment (Fig. 5b) on the right. For $\alpha \geq 2.25$, the streamwise vortices are no more stationary which can be attributed to centrifugal instability observed in three-dimensional computation of Mittal et al. [3]. Second shedding may be observed if the platinum wire is kept up stream as the shedding occurs only at the near vicinity of the cylinder.

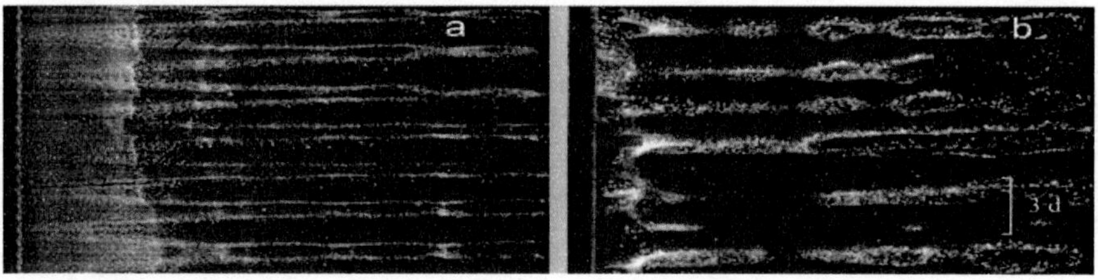

Fig. 5 Validation of Mode E at Re = 250 and α = 2.1 for a rotating cylinder. (**a**) Radi et al. [1] (**b**) Present work. Platinum wire is not visible as it is placed 3D downstream. The vertical extent of the image in (**b**) is 15 cm. Scaling is same for both the images

3 Results and Discussion

The present section discusses the effect of forcing frequency and forcing amplitude on the three-dimensional (spanwise) structure of the wake of a rotationally oscillating cylinder at Re = 250. Spanwise oscillations (three dimensionalities) in the wake are visualized, and their wavelength is quantified using image processing. The effect of forcing frequency on the spanwise wake at a fixed oscillation amplitude of π/2 is discussed followed by the effect of oscillation amplitude at a fixed forcing frequency of FR = 1 and FR = 5.

Effect of Oscillation Frequency:

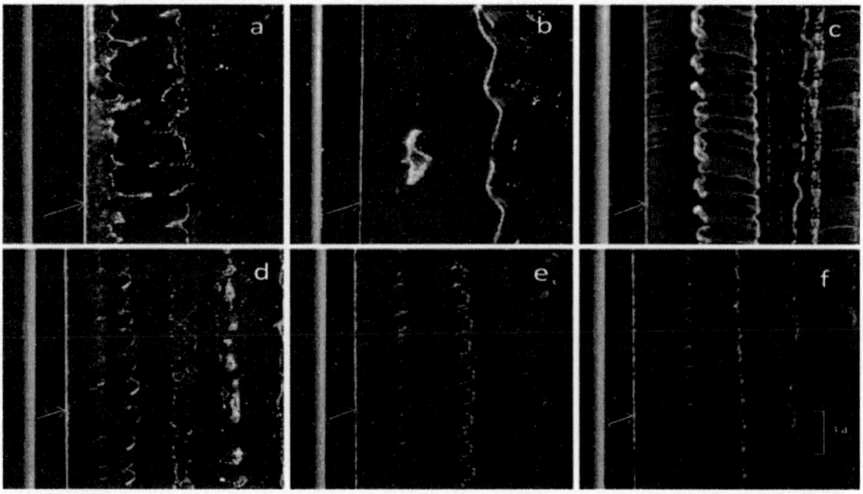

Fig. 6 Effect of cylinder oscillation frequency on the spanwise structure of the wake at oscillation amplitude of π/2. (**a**) FR = 0; (**b**) FR = 0.5; (**c**) FR = 1; (**d**) FR = 1.5; (**e**) FR = 2; (**f**) FR = 2.5. Scaling is same for all the images. The position of the platinum wire is denoted by an arrow in each image

Figure 6 shows the spanwise structure of the wake at six values of forcing frequencies in the range $0 \leq \text{FR} \leq 2.5$ at $\theta = \pi/2$. Figure 6(a) shows the spanwise structure of a stationary cylinder (FR = 0). Mode B instability is clearly visible with a characteristic wavelength of $\lambda/d \sim 1.18$. Increasing the forcing frequency to FR = 0.5 results in a flow which lacks spanwise coherence and intermittently changes structure. Figure 6(b) shows an instant where there is a large wavelength structure along the span similar to Mode A instability of a stationary cylinder. The wavelength of this mode is $\lambda/d \sim 1.81$ as determined using image processing. For FR = 0.5, similar nonlinear interactions along the span may be at work as for the case of Mode A instability of stationary cylinder although Mode B is the dominant mode at this Re. As the forcing frequency is increased to FR = 1.0 (Fig. 6(c)), it appears that nonlinear interactions along the span are reduced as the resulting three-dimensional wake structure consists of straight vortex column with well-defined wavelength of $\lambda/d \sim 0.6D$ riding on it. The spatial coherence is good, and the structure is very repeatable. The bubble sheet enters the near wake and crosses the laser sheet where it gets illuminated. At this location, the sheet splits up with one section moving downstream and the other slightly upstream for a small distance. The upstream moving part has oval cross-sectional vortices with wavelength of $\lambda/d \sim 0.6$. This new mode can be named as Mode Z and is visualized in Fig. 6(c). Cross-stream movement of the platinum wire made this mode visible, and hence, platinum wire is placed in a location where the three-dimensional patterns are clearly visible.

Further increase in FR gives rises to mushroom-shaped three-dimensional columns like we notice in the wake of a rotating cylinder at a rotating rate of $\alpha = 1.2$. Hence, we can confirm the presence of mode C in an oscillating cylinder just like we observe for a rotating cylinder. At FR = 2, the flow visualization image in Fig. 6e shows strict spanwise periodicity with a wavelength of $\lambda/d = 1.14$.

Increasing the FR to 2.5 does not have any significant change in the wave length which has a value of $\lambda/d = 1.11$, and the flow visualization image is shown in Fig. 6f. From the literature study, Kumar et al. [6], we get to predict that there is vortex roll up and shear layer shedding which may result to these kinds of spanwise periodicity and unmixing characteristics. The procedure of estimation of spanwise wavelength is shown in Fig. 7 for FR = 1 and amplitude of $\theta = \pi/2$. For FR = 1, at an amplitude of $\theta = \pi/2$, Fig. 7b shows the pixel intensity variation along the length and the FFT of the amplitude spectrum gave the wavelength of the spanwise periodicity in Fig. 7c. The yellow line corresponded to the specific downstream location where the pixel intensity was measured and the position of the peak in power spectrum gives us the wavelength.

Effect of Forcing Amplitude:

At constant forcing frequency variation of amplitude θ shows interesting wake response. Figure 8 shows the change in wave length and three-dimensional wake patterns when the oscillation amplitude is changed keeping FR = 1. Figure 9 shows changes in wavelength λ/d at FR = 1 by varying the θ at $\pi/4$, $\pi/2$. We notice that the spanwise wavelength reduces with increase in amplitude. Figure 10 shows us the changes in wavelength at $\theta = \pi/2$ by varying the FR. In addition, it is observed that

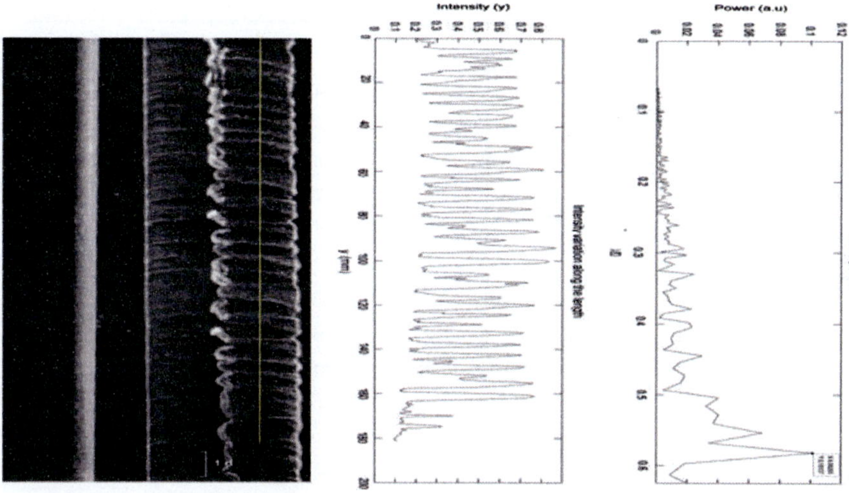

Fig. 7 Estimation of spanwise wavelength at $\theta = \pi/2$ an FR = 1. (**a**) flow visualization digital image; (**b**) pixel intensity variation along the yellow line in (**a**); (**c**) power spectrum obtained by FFT of (**b**) which gives $\lambda/d = 0.60$

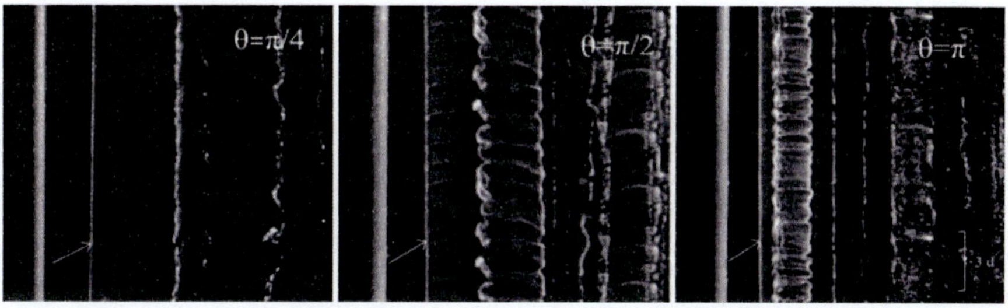

Fig. 8 Effect of oscillation amplitude on the three-dimensional wake structure at FR = 1 at Re = 250. Scaling is same for all the images. The position of the platinum wire is denoted by an arrow in each image

the vortex columns become straighter and more compact as oscillation amplitude is increased as can be seen in Fig. 8. This could have implications in mixing of flow.

It is observed from flow visualization that the far wake (20 D) becomes more two dimensional for certain amplitude and forcing frequency. Figure 11 shows the spanwise wake structure at a fixed forcing frequency of FR = 5 and three forcing amplitudes of $\pi/4$, $\pi/2$, and π at FR = 5, at amplitude $\theta = \pi/4$ and $\theta = \pi/2$; the wake has minimum distortion and shedding is strictly two dimensional as evidenced from nearly straight or gently curved vortex columns. The wake is steady, and mixing is limited.

On the contrary at amplitude, $\theta = \pi$, the wake has distortions and three-dimensional natures which can be attributed to wake evolution resulting from merging

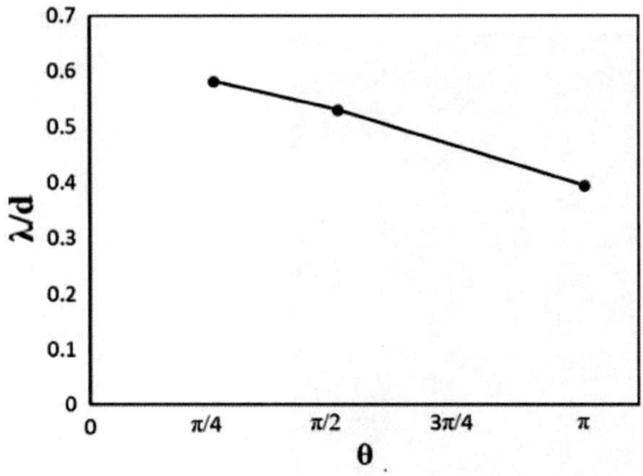

Fig. 9 Variation of wavelength with amplitude

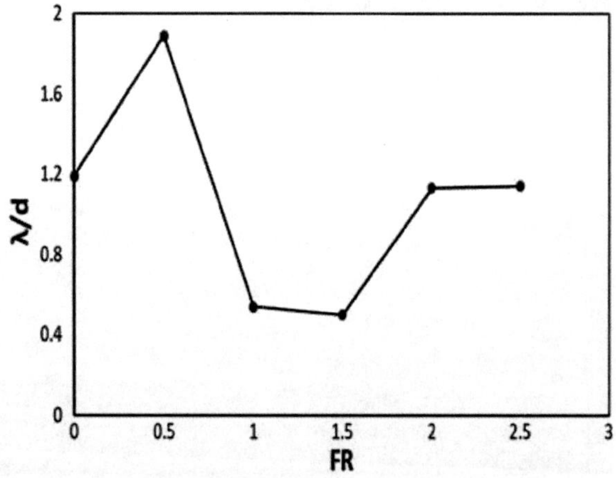

Fig. 10 Variation of wavelength with forcing frequency

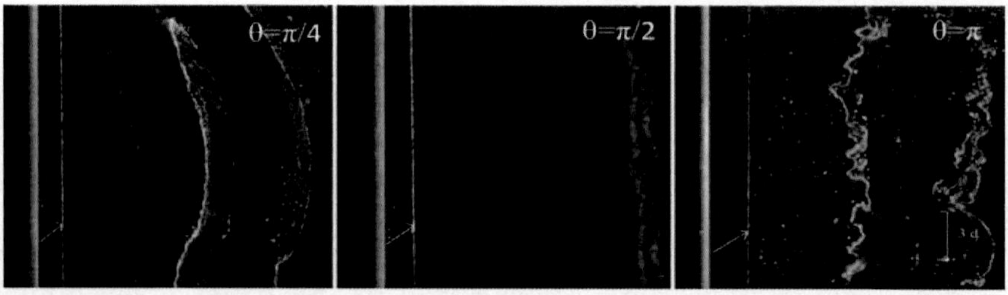

Fig. 11 Effect of oscillation amplitude on the three-dimensional wake structure at FR = 5 at Re = 250. Scaling is same for all the images. The position of the platinum wire is denoted by an arrow

of the near wake consisting of shed shear layers with small-scale vortices as observed by Kumar et al. [6]. Functional relationship between the wavelength and forcing frequency that facilitates far wake studies can be a scope for future investigation.

4 Conclusions

The results of water tunnel experiments on a rotationally oscillating cylinder in cross-flow, for Reynolds numbers 250, amplitude $\pi/2$, and non-dimensional forcing frequency $0.5 \le \mathrm{FR} \le 5$ are closely observed. Investigation made for Re = 250 with amplitude $\theta = \pi/2$ and a forcing frequency of FR = 0.5, confirmed a mode having a spanwise non-dimensional wavelength $\lambda/d = 2$ and characteristics of Mode A described for a stationary cylinder. On increasing forcing frequency, this mode disappears and a new Mode Z having a wavelength of $\lambda/d = 0.6$ and an oval structure appears at around FR = 1. This mode almost remains stable till FR = 1.5. At higher forcing frequency FR = 2, a mode similar to Mode C with mushroom-shaped vortices and a wavelength of $\lambda/d = 1.2$ is observed. At higher forcing frequencies, the far wake becomes very stable and mixing is highly reduced. Effects on wavelength at different amplitude θ at forcing frequency of FR = 1 and FR = 5 have been visualized.

Acknowledgements The authors would like to acknowledge the support provided by the grant ARDB/AE/2018189. In addition, the help from our laboratory colleagues in setting up the experiments is gratefully acknowledged. Special thanks go to Prof. Kamal Poddar for providing continuous laser from his laboratory for flow visualization purposes.

References

1. Williamson CHK (1996) Three-dimensional wake transition. J Fluid Mech 328:345–407. https://doi.org/10.1017/s0022112096008750
2. Norberg C (1994) An experimental investigation of the flow around a circular cylinder: influence of aspect ratio. J Fluid Mech 258:287–316. https://doi.org/10.1017/s0022112094003332
3. Sanjay M, Kumar B (2003) Flow past a rotating cylinder. J Fluid Mech 476:303–334. https://doi.org/10.1017/s0022112002002938
4. Kumar SS, Cantu CC, Gonzalez BB (2011) Flow past a rotating cylinder at low and high rotation rates. ASME. J Fluids Eng 133(4):041201–041201-9. https://doi.org/10.1115/1.4003984
5. Radi A, Thompson MC, Rao A, Hourigan K, Sheridan J (2013) Experimental evidence of new three dimensional modes in the wake of a rotating cylinder. J Fluid Mech 734:567–594. https://doi.org/10.1017/jfm.2013.486
6. Thiria B, Goujon-Durand S, Wesfried J (2006) The wake of a cylinder performing rotary oscillations. J Fluid Mech 560:123–147. https://doi.org/10.1017/s0022112006000656
7. Kumar S, Lopez C, Probst O, Francisco G, Askari D, Yang Y (2013) Flow past a rotationally oscillating cylinder. J Fluid Mech 735:307–346. https://doi.org/10.1017/jfm.2013.469

Numerical Investigation of Pulsating Flow Characteristics of Fluid in a Rough Circular Microchannel

Abdul Aziz Shuvo and A. K. M. Monjur Morshed

Abstract In this study, pulsating flow of fluid in rough and smooth circular microchannel is investigated numerically. Pulsating flow amplitude (A), frequency (ω), and roughness of microchannel effect on heat transfer and pressure drop are investigated compared with smooth microchannel. Reynolds number ($\mathrm{Re} = 200$) is constant throughout the simulation. Heat transfer performance of circular microchannel and pressure drop is found increasing with increment of pulsating amplitude. However, pulsating flow frequency impact in heat transfer of circular microchannel is found inconsequential. The whole simulation is repeated for three different pulsating flow amplitudes (A) and roughness spacing frequency (s).

Keywords Pulsating flow · Circular microchannel · Roughness

1 Introduction

Heat dissipation is a matter of great concern in the last decade. With the advanced use of computer, large heat flux removal from CPU, graphics and other small area, etc., is a must; otherwise, the device will be damaged. Different researchers have already developed many methods to increase heat transfer rate from this small area. Microchannel is one of the leading techniques to dissipate large amount of heat from a limited space and have great contribution in enhancement of heat transfer.

Pulsating flow also has great impact in heat transfer enhancement provided that pulsating flow is under a suitable condition. When pulsating flow is applied in a microchannel, it alters both thermal and hydrodynamic boundary layer thickness causing alteration in heat transfer performance of the channel [1]. Both increment and decrement of Nu have been found for pulsating flow. Zahir et al. [2] noticed

A. A. Shuvo · A. K. M. Monjur Morshed (✉)
Department of Mechanical Engineering, Bangladesh University of Engineering and Technology (BUET), Dhaka 1000, Bangladesh
e-mail: monjur_morshed@me.buet.ac.bd

A. A. Shuvo
e-mail: abdulazizshuvo@gmail.com

L. Venkatakrishnan et al. (eds.), *Proceedings of 16th Asian Congress of Fluid Mechanics*, Lecture Notes in Mechanical Engineering,
https://doi.org/10.1007/978-981-15-5183-3_51

enhancement in heat transfer for pulsating flow in case of laminar and turbulent flow regimes. Kearney et al. [3] performed the time resolved structure of a thermal boundary layer for pulsating flow. They reported that flow reversal is not an only reason for enhancement of heat transfer. Hemida et al. [4] investigated heat transfer performance in a duct for laminar pulsating flow. They reported enhancement of heat transfer with amplitude but decrement with frequency and Prandlt number. Heat transfer for pulsating turbulent flow was studied experimentally and Martinelli et al. [5] reported pulsating frequency had no direct effect on Nu.

Basically, wall roughness is neglected for thermally developed Nu and pressure drop. In case of mini and microchannels ($10\ \mu\text{m} \leq d \leq 3$ mm) [6], the ratio of roughness and the diameter is significantly different from those in conventional channels. So, roughness can have an effect on heat transfer characteristics of microchannels. Wu and Cheng [7] found increment in heat transfer and pressure drop varying surface conditions in microchannel. Kandlikar et al. [8] proposed a design of microchannel for fluid flow which is called microchannels with saw-tooth profile roughness with structures.

Whether heat transfer enhances or not due to pulsating flow, it can be clearly declared from the above literature. Roughness in microchannel is also neglected due to complexity of pulsating flow. So, studies are needed to understand the heat transfer characteristics of microchannel under pulsating flow as well as pulsating flow effect in microchannel in the presence of roughness.

2 Problem Statement

Present study investigates heat transfer and pressure drop effect on circular microchannel due to the roughness spacing frequency (s) and pulsating amplitude (A). Pulsating flow frequency (ω) is also investigated for pulsating amplitude. Figure 1 shows the schematic diagram of the saw-tooth circular microchannel where L is the geometry length, R is the radius, l is the spacing between saw-tooth. Pulsating flow is implemented in the inlet of the microchannel using UDF. Constant heat flux and no-slip condition are considered in microchannel wall pressured outlet is considered in outlet. Water is the working fluid in the study and initial temperature was assumed 300 K. Viscous dissipation was not considered in the study. Viscous dissipation becomes prominent in case of smaller size micro-conduit when $D < 50\ \mu\text{m}$ [9].

2.1 Mathematical Formulation

The governing equations mass continuity, momentum equation, and energy equation are solved assuming constant fluid property with temperature, no-slip condition in wall, and laminar flow.

Continuity equation:

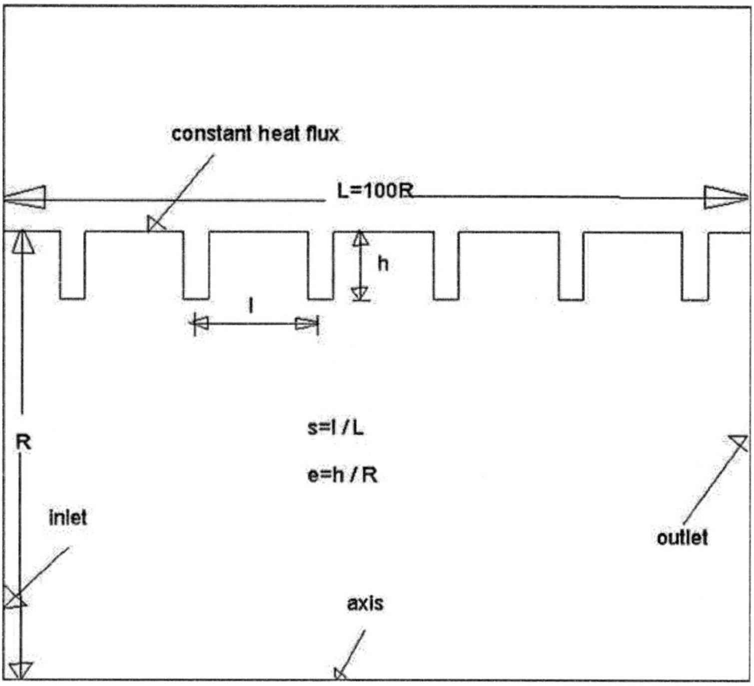

Fig. 1 Schematic diagram of the present problem

$$\frac{1}{r}\frac{\partial(rv)}{\partial r}+\frac{\partial u}{\partial x}=0 \tag{1}$$

Momentum equation:

$$\frac{\partial u}{\partial t}+u\frac{\partial u}{\partial x}+v\frac{\partial u}{\partial r}=-\frac{1}{\rho}\frac{\partial P}{\partial x}+\vartheta\left(\frac{1}{r}\frac{\partial}{\partial r}\left(r\frac{\partial u}{\partial r}\right)+\frac{\partial^2 u}{\partial r^2}\right) \tag{2}$$

$$\frac{\partial v}{\partial t}+u\frac{\partial v}{\partial x}+v\frac{\partial v}{\partial r}=-\frac{1}{\rho}\frac{\partial P}{\partial r}+\vartheta\left(\frac{1}{r}\frac{\partial}{\partial r}\left(r\frac{\partial v}{\partial r}\right)+\frac{\partial^2 v}{\partial r^2}-\frac{v}{r^2}\right) \tag{3}$$

Energy equation:

$$\frac{\partial T}{\partial t}+u\frac{\partial T}{\partial x}+v\frac{\partial T}{\partial r}=\frac{\alpha}{r}\frac{\partial}{\partial r}\left(r\frac{\partial T}{\partial r}\right) \tag{4}$$

Pulsating flow is implemented in the inlet by the user-defined function commercial software FLUENT. Pulsating flow can be expressed:

$$u=U_{\mathrm{m}}(1+A\sin(\omega t)) \tag{5}$$

where u is the pulsating flow, U_m is the mean velocity, A is the pulsating amplitude.

Reynolds number:

$$\mathrm{Re} = \frac{\rho U_m (2R)}{\mu} \quad (6)$$

Local Nusselt number is function of length and time due to pulsating flow. Non-dimensional time $t^* = \frac{t}{T}$, t is flow time and T is the period of the pulsating flow.

Cycle average Nusselt number defines:

$$\mathrm{Nu} = \frac{1}{t^* L} \int_0^{t^*} \int_0^{L} \mathrm{Nu}(x, t^*) \mathrm{d}x \mathrm{d}t^* \quad (7)$$

$$\mathrm{Nu} = \frac{q(2R)}{K(T_w - T_m)} \quad (8))$$

Where q is the heat flux, K is the thermal conductivity of the fluid, T_w is the cycle average wall temperature of the circular microchannel, and T_m is defined as:

$$T_m = \frac{\int_0^R \rho u(x, t) T(x, t) y \mathrm{d}y}{\int_0^R \rho u(x, t) y \mathrm{d}y} \quad (9)$$

The following assumptions are considered to the mathematical model:

1. Laminar, inviscid, and incompressible flow and constant fluid properties with temperature;
2. Rectangular saw-tooth circular microchannel profile roughness and constant roughness height;
3. Negligible gravity.

2.2 *Grid Independent Test and Validation*

Three different grid resolutions 10 × 800, 10 × 1000, 15 × 1500 were used to ensure the grid independent solution. Smooth circular microchannel is used for validation and grid independent test due to absence of the literature for the present studied flow. The local Nusselt number for smooth circular microchannel is shown in Fig. 2b. The thermally developed Nu was found 4.72 which is found to be good agreement with literature [10].

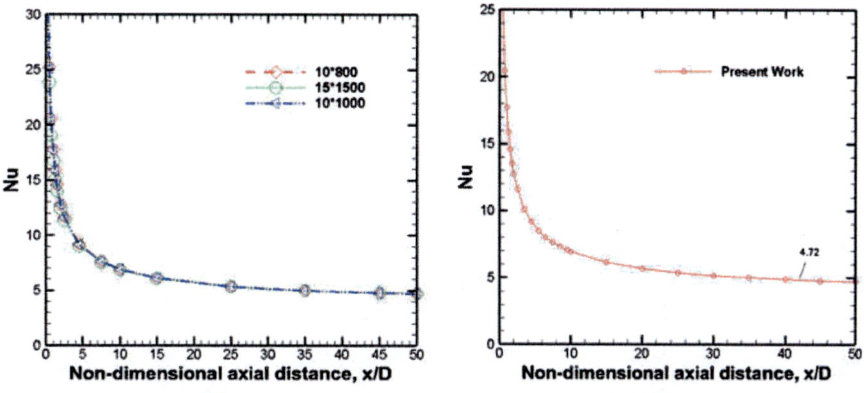

(a) (b)

Fig. 2 **a** Grid independent test for three different grid resolutions. **b** Variation of the Nu with the non-dimensional axial distance, *x*/*D*

3 Results and Discussion

Circular microchannel with roughness and smooth surface has been investigated under pulsating flow condition. In case of microchannel ratio of roughness and diameter is significant than those in conventional channels. Heat transfer performance of rough microchannel under pulsating is important to investigate. To investigate the pulsating flow and heat transfer effect in rough microchannel, three different roughness spacing frequencies are chosen s =0 .025, s = 0.035, s = 0.05. Pulsating flow amplitude (A) is claimed to have effect on heat transfer. So, three different amplitude $A = 1, A = 1.5, A = 2$. Firstly, simulation is performed with rough circular microchannel and smooth circular microchannel. In case of smooth microchannel, Nu for thermally and hydro dynamically developed region is smooth but fluctuation is visible in rough microchannel shown in Fig. 3. Due to the presence of the roughness in microchannel, it alters the fluid flow in the channel; so, the wall flow recirculation is induced which results in fluctuation. Both average pressure drop and heat transfer are enhanced in the presence of roughness.

Sinusoidal pulsating flow is applied in the inlet of the circular microchannel. Pulsating flow of two different frequencies, $\omega = \pi$, 2π are applied for different pulsating amplitudes in constant roughness spacing frequency s =0.025 shown in Fig. 5a–d. It is noticed that changing frequency (ω) of the pulsating flow have no effect in circular microchannel for the specific roughness spacing frequency when the flow duration is considered for the respective period of sinusoidal flow. Thermally developed Nu of circular microchannel is also changing with non-dimensional time, t^*. In spite of sinusoidal pulsating flow, Nusselt number is not sinusoidal but pressure in inlet of the microchannel follows the pattern of the inlet sinusoidal pulsating flow shown in Fig. 4a–c. Nu reaches maximum value around $t^* = 0.3$ and $t^* = 0.74$ as the pulsating velocity reaches peak value at that moment. Nu decreases when pulsating

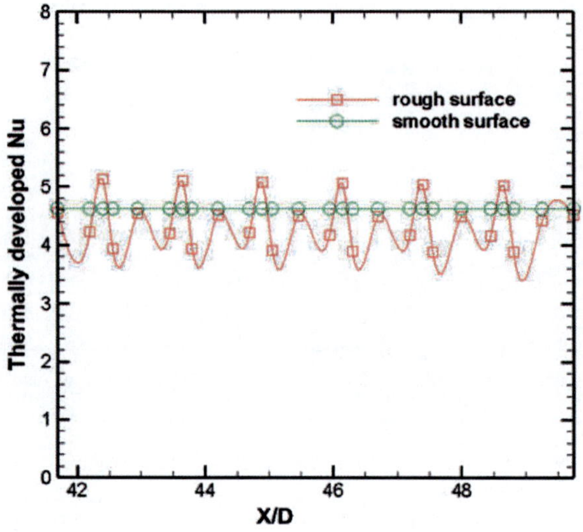

Fig. 3 Variation of Nu with the non-dimensional axial distance in thermally developed region of the circular microchannel

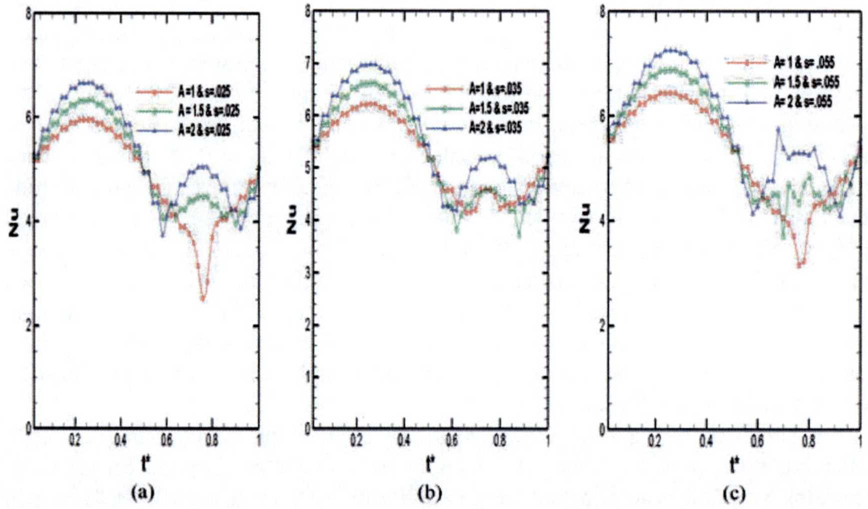

Fig. 4 Variation of Nu with non-dimensional time t^* for **a** $s = 0.025$; **b** $s = 0.035$; **c** $s = 0.055$

flow velocity becomes negative or lower value. However, overall more heat transfer is found higher in case of pulsating flow in expense of slightly higher pumping power. While pulsating flow is applied in the inlet of the circular microchannel, flow reversal effect declines the boundary layer thickness. Hence, heat transfer enhances in the circular microchannel. With the increment of pulsating flow amplitude, heat transfer

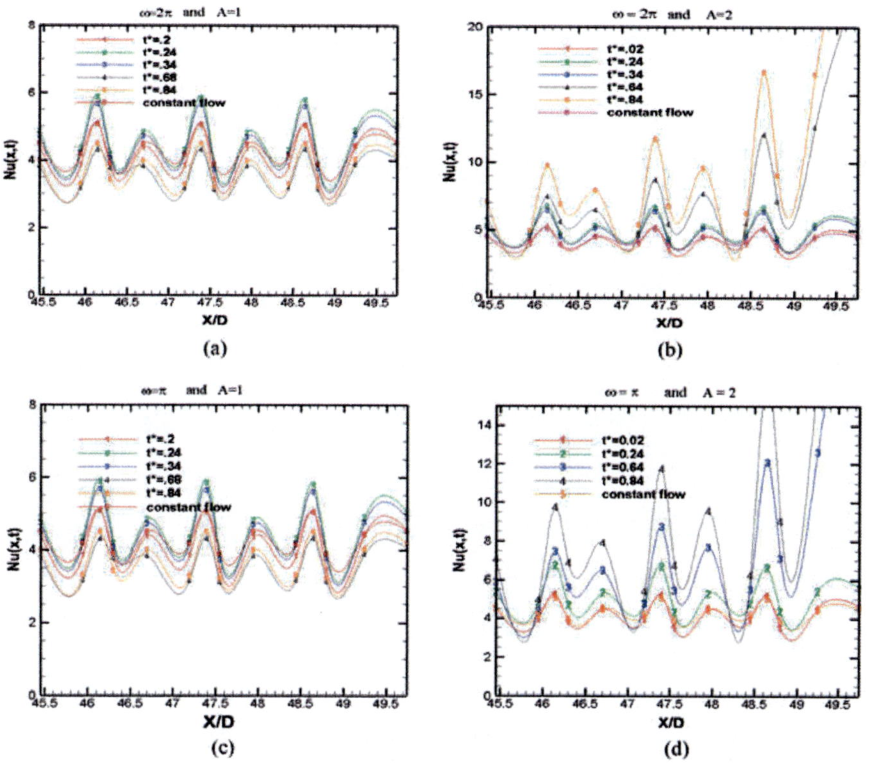

Fig. 5 Variation of Nu with *x*/*D* in thermally developed region of the circular microchannel. **a** $\omega = 2\pi$, $A = 1$; **b** $\omega = 2\pi$, $A = 2$; **c** $\omega = \pi$, $A = 1$; **d** $\omega = 2\pi$, $A = 2$

as well as pressure drop increases (Fig. 6). Heat transfer effectiveness, α defines as the ratio of pulsating cycle average Nu in rough microchannel and constant flow Nu in smooth circular microchannel. Relative pressure drop, β defines as pulsating average pressure drop in rough microchannel and constant flow pressure drop in smooth circular microchannel. $\alpha > 1$ represents enhancement of heat transfer. It is noticed that pulsating flow enhances heat transfer in expense of pressure drop with increment of pulsating flow amplitude shown in Fig. 7a, b. Heat transfer enhances more in smooth surface than rough surface under pulsating flow condition. As the roughness spacing frequency increases, heat transfer effectiveness α increases, and pressure drop ratio β decreases.

The fluid flow streamlines shown in Fig. 6 show that when the roughness spacing enlarges, it lets the fluid to get closer and more contact with the microchannel heated wall. As a result, more heat transfer is found in the microchannel with higher roughness spacing. This is why the more wall temperature and less fluid mass weighted temperature were found in rough surface than smooth surface. Thus, Nu of smooth surface is found higher than the rough circular microchannel.

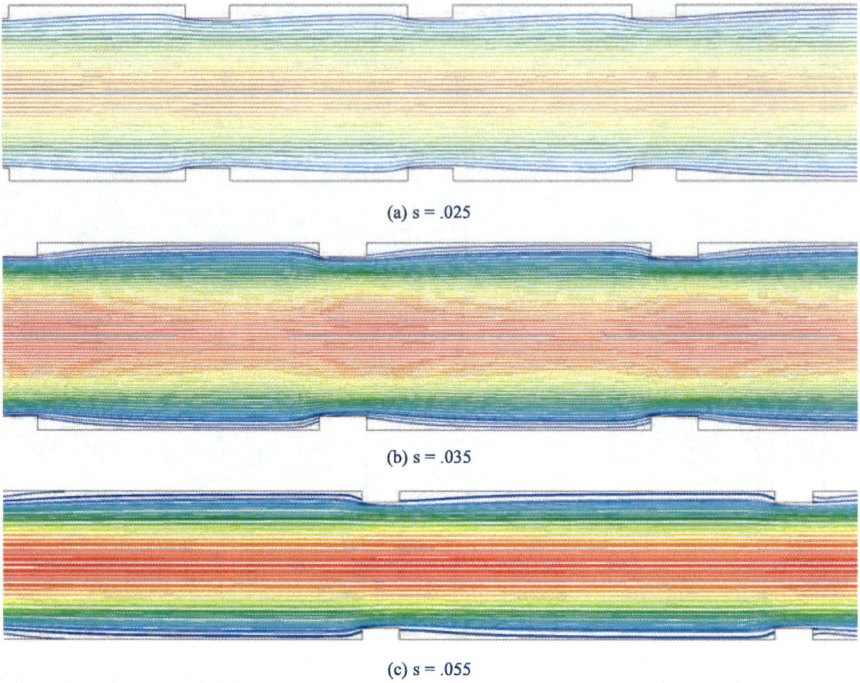

Fig. 6 Streamlines of the fluid flow in the microchannel for **a** roughness spacing frequency $s = 0.025$; **b** roughness spacing frequency $s = 0.035$; **c** roughness spacing frequency $s = 0.055$

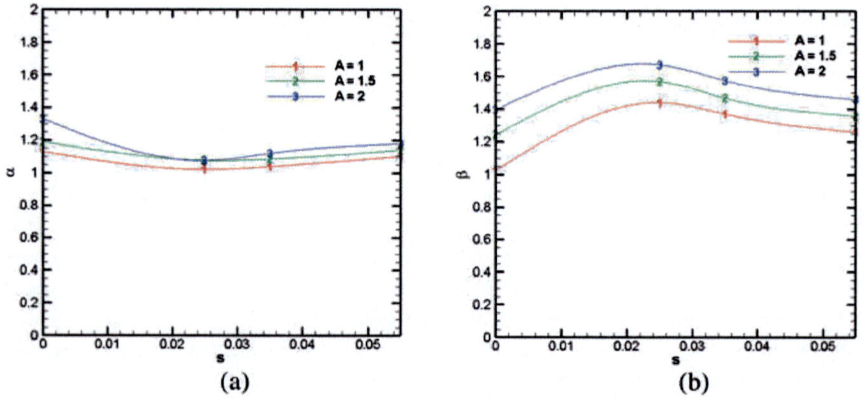

Fig. 7 **a** Variation of heat transfer effectiveness (α) with roughness spacing frequency (s); **b** Variation of pressure ratio (β) with roughness spacing frequency (s)

Table 1 Thermal performance factor

Surface condition	$A = 1$	$A = 1.5$	$A = 2$
Smooth	1.13	1.11	1.19
$S = 0.055$	1.00	1.01	1.04
$S = 0.035$	0.97	0.96	0.98
$S = 0.025$	0.91	0.90	0.93

To evaluate thermal performance of microchannel. There should be a correlation between heat transfer and the pressure drop. Guo et al. [11] used the thermal performance factor, pf to obtain the comprehensive effect on heat transfer and the flow resistance.

$$\mathrm{pf} = \left(\frac{\mathrm{Nu}_d}{\mathrm{Nu}_s}\right) \div \sqrt[3]{\left(\frac{\Delta P_d}{\Delta P_s}\right)} \quad (10)$$

Here, Nu_s and ΔP_s are the Nusselt number and pressure drop of the channel with smooth surface and steady flow. Nu_d and ΔP_d are Nusselt number and pressure drop of the channel with pulsating flow, respectively. Thermal performance factor is listed for both rough and smooth surface for different amplitudes in Table 1.

4 Conclusions

Present study considered rough surface of circular microchannel under pulsating flow condition and this study is investigated using commercial FLUENT software. In the study, roughness height and profile were considered constant. The effect of roughness spacing frequency, pulsating amplitude on the heat transfer performance, and pressure drop of circular microchannel are investigated. The important investigations is listed below:

- Constant fluctuation of thermally developed Nu is found in the rough circular microchannel for both constant flow and pulsating flow. Nu is found slightly higher in rough surface than smooth surface both in pulsating and constant flow.
- Average Nu(t) does not follow the input pattern of the pulsating flow. Nu(t) is found higher around $t^*= 0.3$ and $t^* = 0.74$.
- Pulsating frequency effect is found negligible in circular microchannel for constant roughness spacing frequency (s).
- Heat transfer performance and pressure drop increase with increasing pulsating amplitude.
- Maximum heat transfer effectiveness α is found for smooth surface with minimum relative pressure drop β under pulsating flow condition.
- Thermal performance factor, pf is higher for smooth circular microchannel than rough microchannel under pulsating flow condition.

References

1. Habib M et al (2002) Convective heat transfer characteristics of laminar pulsating pipe air flow. Heat Mass Transf 38(3):221–232
2. Zohir AE et al (2005) An experimental investigation of heat transfer to pulsating pipe air flow with different amplitudes. Heat Mass Transf 42(7):625
3. Kearney SP, Jacobi AM, Lucht RP (2001) Time-resolved thermal boundary-layer structure in a pulsatile reversing channel flow. J Heat Transfer 123(4):655–664
4. Hemida H et al (2002) Theoretical analysis of heat transfer in laminar pulsating flow. Int J Heat Mass Transf 45(8):1767–1780
5. Martinelli R et al (1943) Heat transfer to a fluid flowing periodically at low frequencies in a vertical tube. Trans. Asme 65(7):789–798
6. Taylor JB, Carrano AL, Kandlikar SG (2006) Characterization of the effect of surface roughness and texture on fluid flow—past, present, and future. Int J Therm Sci 45(10):962–968
7. Wu H, Cheng P (2003) An experimental study of convective heat transfer in silicon microchannels with different surface conditions. Int J Heat Mass Transf 46(14):2547–2556
8. Kandlikar SG et al (2005) Characterization of surface roughness effects on pressure drop in single-phase flow in minichannels. Phys Fluids 17(10):100606
9. Koo J, Kleinstreuer C (2004) Viscous dissipation effects in microtubes and microchannels. Int J Heat Mass Transf 47(14):3159–3169
10. Kays WM (2012) Convective heat and mass transfer. Tata McGraw-Hill Education
11. Guo L, Xu H, Gong L (2015) Influence of wall roughness models on fluid flow and heat transfer in microchannels. Appl Therm Eng 84:399–408

Wake Structure of an Oscillating Cylinder with an Attached Filament

Puja Sunil, Sanjay Kumar, and Kamal Poddar

Abstract Experiments are conducted on a rotationally oscillating cylinder with an attached flexible filament at a Reynolds number of 150. The cylinder is forced to oscillate at an oscillation amplitude of $\theta = \pm 5^\circ$ and a fixed forcing frequency ratio. Two filaments of different stiffness are considered in the study. Wake flow visualization is carried out using LIF technique, and time-averaged velocity and vorticity measurements are carried out using particle image velocimetry. An attached filament modifies the wake structure past a rotationally oscillating cylinder from Kármán vortices to reverse Kármán vortices. Filament stiffness plays a crucial role in determining the wake structure, with the stiffer filament resulting in a wider wake and the flexible filament resulting in a narrow wake.

Keywords Rotationally oscillating cylinder · Flexible filament · Flow visualization · Reverse Kármán vortices · Momentum excess · Particle image velocimetry

1 Introduction

Flow control past bluff bodies is important due to its practical significance in everyday life. Control of flow past a circular cylinder has been widely studied. Vortex shedding past circular cylinders can give rise to unsteady forces acting on the cylinder, possibly leading to structural damage due to vortex induced vibration. Most of the literatures have, therefore, focused on controlling/suppressing the vortex shedding past cylinders, by employing both passive and active methods of flow control. Some examples of passive methods include the placement of control rods in the cylinder wake, surface roughness, or the use of a rigid splitter plate. Active methods of flow control include forcing the cylinder at certain parameters, blowing, use of jet actuators, among others. Strykowski and Sreenivasan [1] have shown that vortex shedding

P. Sunil (✉) · S. Kumar · K. Poddar
Department of Aerospace Engineering, Indian Institute of Technology Kanpur, Kanpur 208016, India
e-mail: pujas@iitk.ac.in

L. Venkatakrishnan et al. (eds.), *Proceedings of 16th Asian Congress of Fluid Mechanics*, Lecture Notes in Mechanical Engineering,
https://doi.org/10.1007/978-981-15-5183-3_52

behind a cylinder can be altered and suppressed by placing a smaller control cylinder in the wake. Forced oscillations of the cylinder result in a variety of wake structures. Controlled rotational oscillations of circular cylinder produce interesting effects on its wake, like the phenomenon of lock-in when the vortex shedding frequency in the wake is the same as the forcing frequency of the bluff body. The forcing is known to reduce drag significantly at certain forcing parameters. Kumar et al. [2] and Thiria et al. [3] observed that rotary oscillations have significant effect on the wake structure depending on the forcing parameters. Further, modifications of the cylinder wake can be achieved by attaching a rigid splitter plate to a cylinder as shown in Gerrard [4], Anderson [5] and Bearman [6]. The vortex shedding frequency was altered due to the presence of the splitter plate. For a cylinder with an attached rigid splitter plate, the interaction between the shear layers behind the cylinder was delayed due to the presence of the splitter plate, eventually resulting in weaker vortices in the wake. It was also noted that three dimensionalities in the flow were reduced by employing a splitter plate.

Lauder et al. [7] presented an overview of animal locomotion noting how fish often propel themselves through a fluid by employing flexible appendages for locomotion. Deriving inspiration from fish and birds, studies have been conducted on the use of flexibility in modifying the flow past streamlined and bluff bodies. Wu and Shu [8] found that the wake of a stationary cylinder is strongly affected by the flapping amplitude and frequency of the flexible plate at a $Re = 100$. Wu et al. [8] showed that a flexible splitter plate attached to a stationary cylinder can suppress vortex shedding past a stationary cylinder. Consequently, the fluctuating lift force and the drag acting on the cylinder are reduced. Dewey et al. [9] and Moored et al. [10] showed that pitching flexible panels attached to streamlined forebodies resulted in improvements in thrust. Sunil et al. [11] showed that a filament attached to an oscillating cylinder modifies the wake structure from a Kármán vortex street to a reverse Kármán vortex street.

The problem of the effect of a flexible filament attached on the downstream stagnation point of a circular cylinder exhibiting rotational oscillations has not been addressed in the existing literature to the best knowledge of the authors. The present study aims to address the following questions: (1) How does filament stiffness affect the flow past a rotationally oscillating cylinder? (2) Is there still a lock-in of the vortices in the wake? In this study, an attempt is made to understand the flow physics past a rotationally oscillating cylinder with attached filaments of different stiffness through laser-induced flow visualizations of the flow structure as well as through velocity measurements using time-averaged PIV at $Re_D = 150$, where Re_D is the Reynolds number based on cylinder diameter, D. Initially, the effect of a flexible filament on the flow past a rotationally oscillating cylinder is examined. This is followed by a discussion on the effect of filament stiffness on the resultant flow structure.

2 Experimental Setup and Diagnostics

The experiments for the present study were carried out in a water tunnel with a test section of size 0.27 m × 0.46 m × 0.18 m. The water velocity needed to achieve Re_D = 150 in the present experiment was 1.42 cm/s. The cylinder model was inserted through a slot in the Plexiglas sheet which covers the top of the test section. It was flush mounted on an anodized plate placed at the bottom wall of the test section. A bearing inserted in the anodized plate ensured smooth motion of the cylinder without any wobbling. The cylinder model was a brass rod of wetted length 250 mm and diameter, $D = 10$ mm. Figure 1(a) and (b,c) shows the schematic of the problem and the experimental setup, respectively. The filaments used were flexible plastic sheets of fixed aspect ratio, $L/D = 5$ (L—streamwise length of the filament) and varying stiffness (achieved by varying the sheet thickness), filament 1 (thickness—110 microns, Areal density = 36.7 mg/cm^2, Young's modulus = 2.46 GPa) and filament 2 (thickness—20 microns, Areal density = 1.14 mg/cm^2, Young's modulus = 1.31GPa)). The sheets spanned the cylinder height. Figure 2 is a superimposed image which shows the cylinder with the attached filaments of varying stiffness.

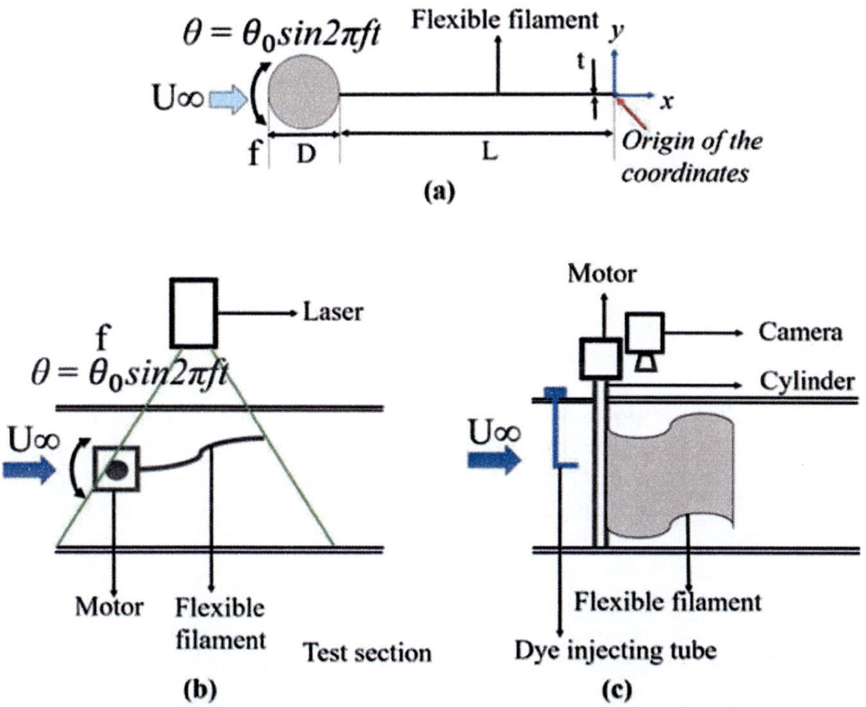

Fig. 1 (**a**) Schematic of the problem, (**b,c**) Top view and side view of the experimental setup (Sunil et al. [11])

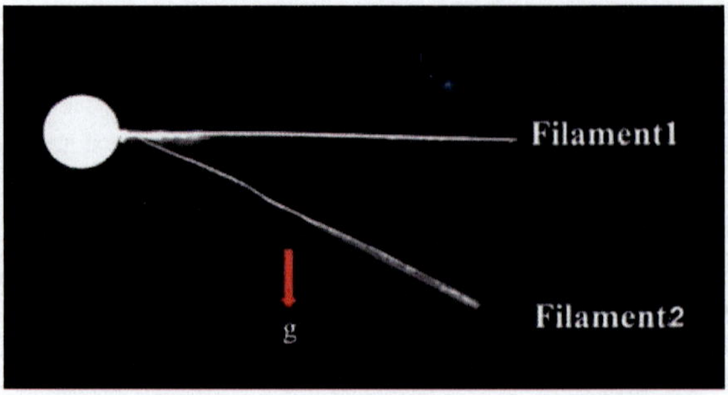

Fig. 2 Filament deflection under self-weight

This picture shows the deflection of the filaments under self-weight and gives an idea about the flexibility of each filament, and the tip deflection of filament 2 is ~18 times that of filament 1.

The cylinder was imparted sinusoidal rotary oscillation at various forcing frequencies using a servo motor. The motor was coupled to a signal generator which provided analog sinusoidal signal of known frequency. The cylinder was forced to oscillate according to $\theta = \theta_0 \sin 2\pi ft$ where θ is the angular position of the cylinder, θ_0 is the oscillation amplitude, f is the forcing frequency, and t is time. The flow visualization was performed using laser-induced fluorescence technique using Rhodamine—B as the dye. The dye was injected through two thin tubes (of 0.9 mm diameter) placed ~10*D* upstream of the cylinder. The illumination was provided using a 3 W continuous laser from the side of the tunnel, and the videos were recorded using a Nikon DSLR camera (at 60 frames per second and capturing a total length of 19*D* along the flow direction) viewing perpendicular to the illumination and flow direction.

The experimental setup for PIV measurements and flow visualization was similar, with the illumination for PIV measurements being provided by a TSI evergreen laser from the side of the tunnel at the midspan of the model. A laser light sheet of thickness less than 1 mm was realized using light sheet optics. The flow was seeded with 10 μm diameter silver-coated hollow glass spheres. An 8MP TSI Powerview plus camera captured images from the top. The images were of size 3312 pixels $\times$ 2488 pixels corresponding to a physical area of 170 mm $\times$ 110 mm viewed through the camera, resulting in a resolution of 0.05 mm/pixel along the streamwise direction and 0.04 mm/pixel along the cross-stream direction. The camera and the laser were synchronized using a TSI 610,036 laser pulse synchronizer. The data was processed using a correlation window size of 32 pixels $\times$ 32 pixels using PIVlab. Mean vorticity and velocity data were obtained by averaging the instantaneous frames acquired through time-averaged PIV. The non-dimensional parameter in the study is the forcing frequency ratio, $\mathrm{FR} = f/f_0$, where f is the frequency at which the cylinder is forced to oscillate about its own axis and f_0 is the natural vortex shedding frequency of the stationary cylinder with no filament (which for $\mathrm{Re}_D = 150$ is 0.25 Hz as measured using the hot wire anemometer). This corresponds to a Strouhal number, $\mathrm{St} = 0.176$ for the present study, which is in accordance with the published work of $\mathrm{St} = 0.178$ (Roshko [12]).

3 Results and Discussion

The following section presents a comparison of the flow structure past a rotationally oscillating cylinder with and without an attached filament. In the subsequent section, the effect of filament stiffness on the flow past a rotationally oscillating cylinder is explored. The modifications in the wake structure are observed through flow visualization. Further, velocity and vorticity data obtained through time-averaged PIV are presented to examine the effect of an attached flexible filament on the flow past a rotationally oscillating cylinder.

Flow Past a Rotationally Oscillating Cylinder with and Without Filament The modification in the wake structure past an oscillating cylinder with and without an attached filament is presented in Fig. 3. For an oscillating cylinder, the CW vortex shed from the cylinder upper surface is aligned above the wake centreline, whereas the CCW vortex shed from the cylinder lower surface is aligned below the wake centreline, resulting in the formation of Kármán vortex street, which is indicative of momentum deficit in the wake, resulting in drag acting on the forebody. However, for an oscillating cylinder with an attached flexible filament, the presence of the filament results in the CW vortex being shed below the wake centreline and the CCW vortex being shed above the wake centreline, thereby resulting in the formation of reverse Kármán vortex street, which is indicative of momentum excess in the wake possibly resulting in drag reduction. It can, therefore, be concluded that, an attached filament considerably modifies the wake structure of an oscillating cylinder.

Effect of Filament Stiffness on the Flow Past a Rotationally Oscillating Cylinder This section discusses the effect of filament stiffness on the flow past a rotationally oscillating cylinder. Filament 1 is stiffer, whereas filament 2 is very flexible. Here, the cylinder is forced to oscillate at a cylinder oscillation amplitude, $\theta = \pm 5°$ and FR = 2.5. The filaments are of length, L = 5*D*.

As observed from Fig. 4, in both the cases, the wake structure is an asymmetric P + S wake (Williamson and Roshko [13]); that is, a single reverse Kármán vortex

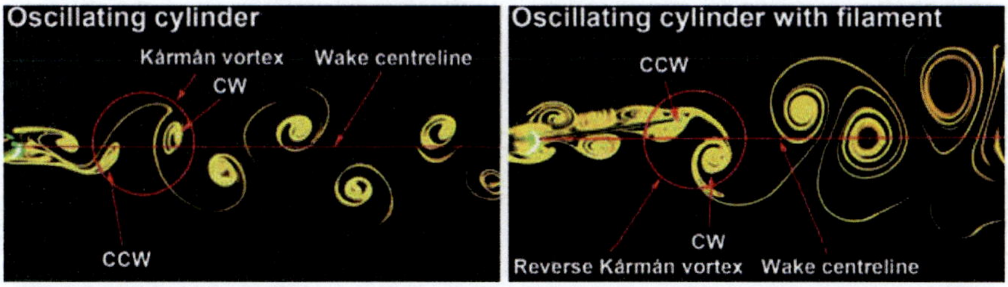

Fig. 3 Comparison of the flow structure past a rotationally oscillating cylinder with and without an attached flexible filament at Re*D* = 150. The cylinder is forced to oscillate at an amplitude, $\theta = \pm 5°$ and forcing frequency ratio, FR = 1.7. A filament of length, L = 5*D*, is used in the study (Sunil et al. [11])

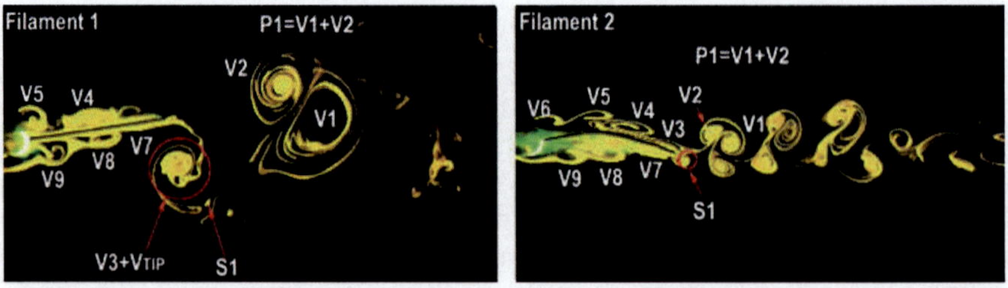

Fig. 4 Wake structure of a rotationally oscillating cylinder with attached filaments of varying stiffness. Here, filament 1 is stiffer, whereas filament 2 is very flexible

pair (P) and a single counter clockwise vortex (S) are shed during an entire cycle of the cylinder oscillation. In both cases, the small-scale vortices (shed from the cylinder and riding along the filament) as well as the wake vortices are locked-into the cylinder forcing frequency. The images are shown at a time instant of t = T/4 of the cylinder oscillation (the cylinder is at its extreme position during its counter clockwise motion). Filament 2, being highly compliant with the flow, results in a narrower wake, with the asymmetric vortices aligned along the wake centreline, indicating that the pair of vortices, P1 (Fig. 4) and the single vortex, S1 may be of similar strength. For the stiffer filament 1, there is a marked asymmetry in the wake structure with the vortex pair, P1 convecting away quickly from the wake centreline, indicating that in these cases, P1 may in fact be stronger than S1. Consequently, the excess thrust indicated by the presence of the reverse vortex pair is aligned along the wake centreline for the cylinder with filament 2, whereas for the case of the cylinder with filament 1, there may be a side force resulting from the asymmetry of the vortices. The phase lag between the cylinder oscillation and the filament tip is nearly π for filament 2, whereas filament 1, being stiffer has the least phase lag.

For filament 2, during the upstroke of the filament tip, a CW vortex is formed at the filament tip which then merges with the CW vortex rolling off from the upper side of the filament into the wake. During the down stroke of the filament tip, a CCW vortex is shed from the filament tip which then merges with a CCW vortex from the lower side of the filament. The CCW and CW vortices then merge to form a vortex pair in the wake. In addition, a small single CCW vortex is shed from the filament tip which then merges with the vortex pair resulting in a P + S wake. In Fig. 4, the vortex pair, P1 = V1 + V2, was shed in the previous cycle of cylinder oscillation. Here, V1 consists of CW vortex formed from the filament tip and that from the upper surface of the filament and V2 is made up of CCW vortex from the filament tip and that from the lower side of the filament. S1 is CCW vortex shed from the filament tip for filament 2. There are 6 vortices (3CW—V4, V5, V6 and 3 CCW—V7, V8, V9) riding along the filament length at a given instant of time for filament 2, whereas for filament 1, there are only 4 vortices (2CW—V4, V5 and 2CCW—V8, V9) riding along the filament length. It takes approximately 4 cycles of the cylinder oscillation for the vortices formed near the cylinder (e.g., V5 and V9 for filament 2) at $t = 0$ to shed into the wake (as a P + S entity), whereas for filament 1, it takes only 3 cycles of the cylinder oscillation. This suggests that the filament flexibility has a strong impact on the wake structure.

From the mean vorticity contours in Fig. 5, it can be concluded that for the oscillating cylinder with filament 1, the vorticity magnitude associated with reverse vortices is stronger than the vorticity magnitude associated with the counterclockwise vortices (present below the mean centreline), as speculated from flow visualization images (for the reverse vortex pair, P1 and the single counterclockwise vortex, S1, Fig. 4). For the oscillating cylinder with filament 2, it can be observed that the vorticity magnitude associated with reverse vortices and that of the counter clockwise vortices (present below the mean centreline) are comparable, thereby leading to a narrower wake as seen from the flow visualization (Fig. 4).

The velocity vectors superimposed on the velocity magnitude contour presented in Fig. 6 confirm that for the oscillating cylinder with filament 1, the momentum excess associated with reverse vortex street is aligned asymmetric to the mean centreline, possibly resulting in a side force acting on the cylinder as conjectured from flow visualization. There is a wide recirculation region (Y ~ 0.4L) associated with the formation of the counterclockwise vortices (below the mean centreline) that extends from the filament tip to X/L ~ 0.4, beyond which there is a wake deficit at all downstream distances. For the oscillating cylinder with filament 2, similar to the case with filament 1, there is a momentum excess associated with the formation of reverse vortex street and a momentum deficit related to the counterclockwise vortices. However, since the reverse vortices and counterclockwise vortices formed are similar in strength, they may tend to cancel each other possibly resulting in a zero net streamwise force acting on the cylinder.

The streamwise velocity profiles at several downstream distances for an oscillating cylinder with filament 1 and 2 are shown in Fig. 7. As expected and observed from the velocity magnitude contour, for both the cases, there is a momentum excess (due

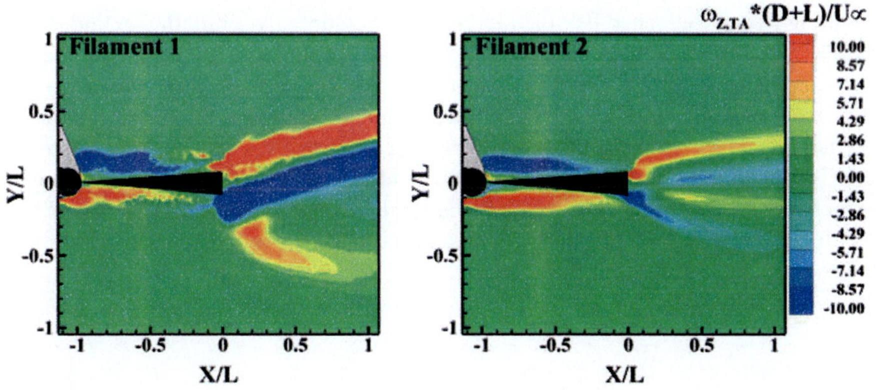

Fig. 5 Mean vorticity contours of a rotationally oscillating cylinder with attached filaments 1 and 2. The cylinder is forced to oscillate at $\theta = \pm 5°$ and FR = 1.7 at ReD = 150. The shaded area is the region where no data is available due to the filament excursion in the flow

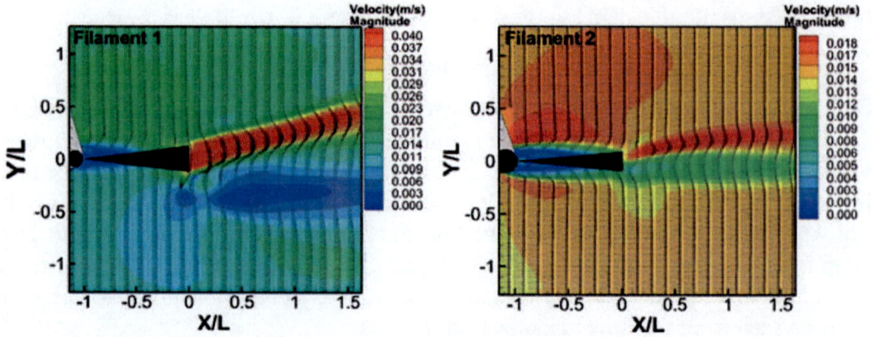

Fig. 6 Velocity magnitude contours of a rotationally oscillating cylinder with attached filaments 1 and 2. The cylinder is forced to oscillate at $\theta = \pm 5°$ and FR = 1.7 at ReD = 150

to the presence of reverse von Kármán street) and a momentum deficit (due to the presence of the counter clockwise vortices) associated with the formation of the P + S wake structure. For the filament 1, at X/L = 0.5, in the deficit region, $U/U\infty$ ~ 0, indicating the presence of a strong recirculation zone. Further, it can be noted that at all downstream distances up to X/L = 2, $U/U\infty$ does not exceed 0.5. The wake vortices for the oscillating cylinder with filament 2 are weaker as compared to the case with filament 1; thereby, the associated gain or deficit in the momentum is smaller in this case.

4 Conclusion

Experiments were conducted to study the wake structure and its modification resulting from an attached flexible filament on a rotationally oscillating cylinder at a $Re_D = 150$. A flexible filament desirably modifies the flow structure from a standard von Kármán vortex street, which is indicative of drag acting on the forebody to reverse vortex street which is associated with thrust generation. It was observed that filament stiffness is crucial in determining the wake structure. A stiffer filament attached to an oscillating cylinder results in a wider wake with stronger vortices convecting away from the mean centreline, hinting at a side force acting on the cylinder, whereas the flexible filament results in a narrow wake, with weaker vortices aligned along the mean centreline, indicating that the resultant force is also aligned along the mean centreline in this case.

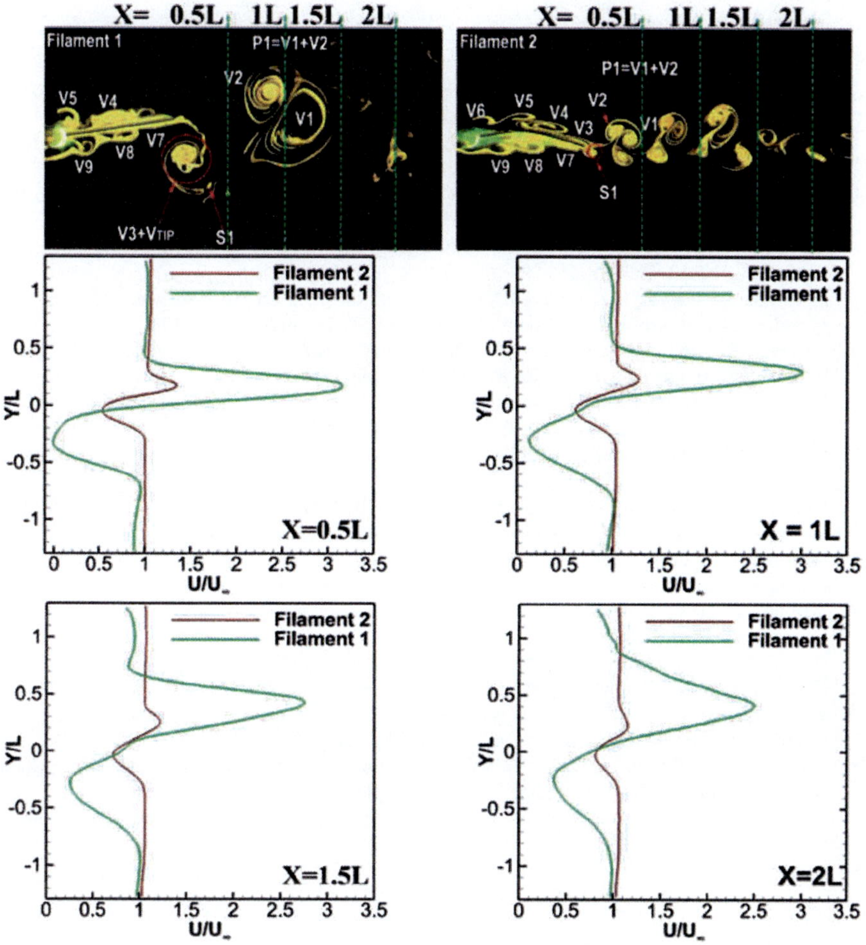

Fig. 7 Velocity magnitude contours of a rotationally oscillating cylinder with attached filaments 1 and 2. The cylinder is forced to oscillate at $\theta = \pm 5°$ and FR = 1.7 at ReD = 150

Acknowledgements The authors would like to acknowledge the help received from the staff and the graduate students from Low Speed Aerodynamics Laboratory and the Fluid Dynamics Laboratory of the department in conducting the experiments.

References

1. Strykowski P, Sreenivasan K (1990) On the formation and suppression of vortex shedding at low Reynolds numbers. J Fluid Mech 218:71–107
2. Kumar S, Lopez C, Probst O, Francisco G, Askari D, Yang Y (2013) Flow past a rotationally oscillating cylinder. J Fluid Mech 735:307–346
3. Thiria B, Goujon-Durand S, Wesfreid J (2006) The wake of a cylinder performing rotary oscillations. J Fluid Mech **560**:123–147
4. Gerrard J (1966) The mechanics of the formation region of vortices behind bluff bodies. J Fluid Mech 25:401–413
5. Anderson E, Szewczyk A (1997) Effects of a splitter plate on the near wake of a circular cylinder in 2 and 3-dimensional flow configurations. Exp Fluids 23:161–174
6. Bearman P (1965) Investigation of the flow behind a two-dimensional model with a blunt trailing edge and fitted with splitter plates. J Fluid Mech 21:241–255
7. Lauder GV, Lim J, Shelton R, Witt C, Anderson E, Tangorra JL (2011) Robotic models for studying undulatory locomotion in fishes. Marine Technol Soc J 45:41–55
8. Wu J, Shu C (2011) Numerical study of flow characteristics behind a stationary circular cylinder with a flapping plate. Phys Fluids 23:073601
9. Dewey PA, Boschitsch BM, Moored KW, Stone HA, Smits AJ (2013) Scaling laws for the thrust production of flexible pitching panels. J Fluid Mech 732:29–46
10. Moored K, Dewey P, Boschitsch B, Smits A, Haj-Hariri H (2014) Linear instability mechanisms leading to optimally efficient locomotion with flexible propulsors. Phys Fluids 26:041905
11. Williamson CHK, Roshko A (1988) Vortex formation in the wake of an oscillating cylinder. J Fluids Struct 2:355–381
12. Roshko A (1954) On the development of turbulent wakes from vortex streets, Technical report, National Advisory Committee for Aeronautics, 1191
13. Wu J, Shu C, Zhao N (2014) Investigation of flow characteristics around a stationary circular cylinder with an undulatory plate. Eur J Mech B/Fluids 48:27–39
14. Sunil P, Kumar S, Poddar K (2019) Wake modification of a forced circular cylinder with an attached filament. J Flow Vis Image Process (accepted for publication)

Thermal Transportation Behavior Prediction of Water Molecules by Different Rigid Water Models: A Molecular Dynamics Study

A. K. M. M. Morshed, Muhammad Rubayat Bin Shahadat, Md. Rakibul Hasan Roni, and Md. Ferdous Alam

Abstract Eight most prominent rigid water models were selected in order to evaluate the temperature dependent thermal conductivity of liquid water. Equilibrium molecular dynamics (EMD) simulation was the key tool to serve the purpose and the temperature was varied in between 283 and 363 K. NVT ensemble has been introduced in all the simulations to evaluate the properties. The results got from the simulation were compared with the experimental results in order to find out the best suitable model for simulating water. From the comparison, it is prominent that every rigid water model over predicts the value of thermal conductivity of water and no single model can predict the increasing trend of thermal conductivity of water with the buildup of temperature. It may also be inferred from the comparison with experimental values that the five-site models (TIP5P-Ew) can presage the thermal conductivity most precisely among the existing models because of its improved geometric parameters.

Keywords Thermal conductivity · Rigid water models · Equilibrium molecular dynamics simulation

---

A. K. M. M. Morshed · Md. R. H. Roni
Department of Mechanical Engineering, Bangladesh University of Engineering and Technology (BUET), Dhaka, Bangladesh

M. R. B. Shahadat (✉)
Department of Mechanical Engineering, Hajee Mohammad Danesh Science and Technology University (HSTU), Dinajpur, Bangladesh
e-mail: rubayat37.me@hstu.ac.bd

Md. F. Alam
Department of Mechanical Engineering, Shahjalal University of Science and Technology (SUST), Sylhet, Bangladesh

L. Venkatakrishnan et al. (eds.), *Proceedings of 16th Asian Congress of Fluid Mechanics*, Lecture Notes in Mechanical Engineering,
https://doi.org/10.1007/978-981-15-5183-3_53

1 Introduction

Molecular dynamics (MD) technique is extensively used to simulate molecular systems [1]. Among other molecular systems, diatomic molecular liquid has also been modeled using MD [2, 3]. Later Monte Carlo (MC) [4], and then MD [5] simulation have been used to model liquid water. Besides, quantum (AIMD) molecular dynamics has been used to investigate water properties as well [6]. A review study has shown that there are 46 water models [7] and these water models were labeled as rigid, flexible, and polarizable models [8]. Later, scientists have aimed for a new general purposed model like TIP4P/2005 [9] shows very better accomplishment in calculating numerous thermal properties. Recently, improvement in predicting water properties has been achieved through another improved and general purposed four-site water model (TIP4P-Ew) [10]. Another model, TIP5P-Ew [11], has been reparametrized on the basement of TIP5P model and its great performance has been proved for producing experiment results for liquid water. It is evident from the previous studies that though there are many water models for predicting the properties of liquid water, none of these models is too effective to presage all the properties of liquid water. It is mentionable that many studies depend on which model has been chosen to carry out the simulations. For example, water model basically fixes the thermal properties as well as structure of clathrate hydrates [12]. In order to investigate different properties of water models, a lot of studies have been carried out and these studies have reported the advantages and disadvantages of the respected models as well. But performance-based study among these models has been reported for very few times. The equilibrium and dynamic properties have been investigated by using SPC/E and TIP4P models [13]. Besides, structure of different three-site models (TIP3P, SPC, and SPC/E) have also been tested at room temperature [14]. Recently, MD simulation has been applied in many studies for investigating the thermal conductivity using various water models [15–17].

Two major methods, equilibrium molecular dynamics (EMD) and nonequilibrium molecular dynamics (NEMD), had been employed in order to study thermal properties using MD. Green–Kubo [18] is a method that uses EMD which works in the conjunction of properties related to transport phenomena and time integral of a correlation function. Green–Kubo usually can give a reasonable accurate and precise result. For the NEMD simulation, to study thermal conductivity, temperature gradient or velocity gradient established. There is another NEMD approach which was formulated by Muller-Plathe is—reverse nonequilibrium molecular dynamics (RNEMD) [19]. It is also effectively used for simulating monoatomic fluids [20] as well as polyatomic fluids [21]. Mao and Zhang [22] in there study calculated thermal conductivity by eight rigid water models (SPC, SPC/E, TIP3P, TIP4P, TIP4P-Ew, TIP4P-2005, TIP5P, and TIP5P-Ew) at two different temperatures (298 K and 318 K) using reverse non-equilibrium MD (RNEMD) simulations and reported relative errors of 2–53% at 298 K and 8–49% at 318 K with respect to the experimental measures [23]. Though NEMD simulation is very common in predicting thermal properties of water, in very few studies, EMD have been employed to compare the thermal conductivities of different water models. Recently, EMD has been used to study thermal conductivity of water with the change of temperature by using the SPC/E model [25].

In the present study, EMD (the Green–Kubo method) has been selected for the calculation of thermal conductivities of eight different water models (SPC, SPC/E, TIP3P, TIP4P, TIP4P-2005, TIP4P-Ew, TIP5P, and TIP5P-Ew) with a temperature range of 283–363 K. This study aims to calculate thermal conductivity of water and then to compare the findings with the experimental measures at various temperatures.

2 Simulation Methodology

2.1 Green Kubo Formulation

Thermal conductivity λ by Green–Kubo method can be written as,

$$\lambda = \frac{V}{3KT^2}\int_0^\infty \mathrm{d}t\langle \dot{q_\alpha}(0).\dot{q_\alpha}(0)\rangle \tag{1}$$

where

$$\dot{q_\alpha}(t) = \frac{1}{V}\left[\sum_i \varepsilon_i(t)\cdot v_{i\alpha}(t) + \frac{1}{2}\sum_i\sum_{j\neq i} r_{ij\alpha}(t)\cdot\left[v_i(t)\cdot f_{ij}(t)\right]\right] \tag{2}$$

with the energy of molecule i,

$$\varepsilon_i(t) = \frac{m_i v_i(t)^2}{2} + \frac{1}{2}\sum_{j\neq i}\varnothing\left[r_{ij}(t)\right] \tag{3}$$

and $\alpha = x$, y and z. $\varphi[r_{ij}(t)]$ can be modeled as potential energy among the particles i and j at time t. Heat flux (Eq. (2)) with the energy of each molecule (Eq. (3)) comprise three terms:

$$\dot{q_\alpha} = \dot{q}_\alpha^{tm} + \dot{q}_{i\alpha}^{pm} + \dot{q}_\alpha^{ti} \tag{4}$$

where

$$\dot{q}_\alpha^{tm} = \frac{1}{V}\sum_i\left[\frac{m_i v_i^2}{2}\right]v_{i\alpha} \tag{5}$$

$$\dot{q}_{i\alpha}^{pm} = \frac{1}{V}\sum_{i}\left[\frac{1}{2}\sum_{j\neq i}\emptyset(r_{ij})\right]v_{i\alpha} \tag{6}$$

$$\dot{q}_{\alpha}^{ti} = \frac{1}{V}\sum_{i}\left[\frac{1}{2}\sum_{j\neq i}r_{ij\alpha}(v_i\cdot f_{ij})\right] \tag{7}$$

q_{α}^{tm} and $\dot{q}_{\alpha}^{pm}$ are the translational and the potential energy because of molecular movement, and $\dot{q}_{\alpha}^{ti}$ is the transfer of translational energy because of molecular-level interaction. Therefore, the thermal conductivity (Eq. (1)) can be approximated by the following terms

$$\lambda \approx \lambda_{\text{tot}} = \lambda_{tm} + \lambda_{pm} + \lambda_{ti} \tag{8}$$

2.2 *Molecular Dynamics Simulation*

The main differences among the water models are differences in potentials, differences in charge values, and differences in geometric potentials. As all the models are rigid water model, shake algorithm has been introduced to make the water rigid. This algorithm is a part of lammps RIGID package. So, in order to use this algorithm in the code, RIGID package has to be installed. NVT ensemble has been introduced in all the simulations to evaluate the properties using Nose-Hoover thermostat. Periodic boundary condition was implemented at all boundaries. The dimension of the simulation cell was 20 Å × 20 Å × 20 Å and 900 water molecules were created in the cubic system as shown in the Fig. 1. Figure 2 shows the two dimensional view of the simulation domain at two different time. The dimension of the box is taken such that it can eliminate the size effect. The density of the was fixed at 1004 kg/m^3 and it is represented in Fig. 3. The timesteps of the simulation was 1 fs and the total simulation was run for 1 ns. The system came to an equilibrium at 0.2 ns. The cut-off distance was 3.54 nm. PPPM kspece solver was used in the simulation as it gives a very little variation from Ewald solver which is shown in Fig. 4. As TIP4P-Ew and TIP5P-Ew are specially designed for Ewald solver, Ewald solver is used only in these cases.

Five temperatures—283 K, 300 K, 318 K, 353 K, and 363 K were taken in the simulation and the thermal equilibrium of the simulation cells are added in Fig. 5. All the simulations were performed in LAMMPS and all the visualizations were done using OVITO. Here, in this literature, eight most prominent rigid water models have been considered. Three-site models (SPC, SPC/E, and TIP3P) basically mean there are three interaction sites. These three interaction sites basically indicate three

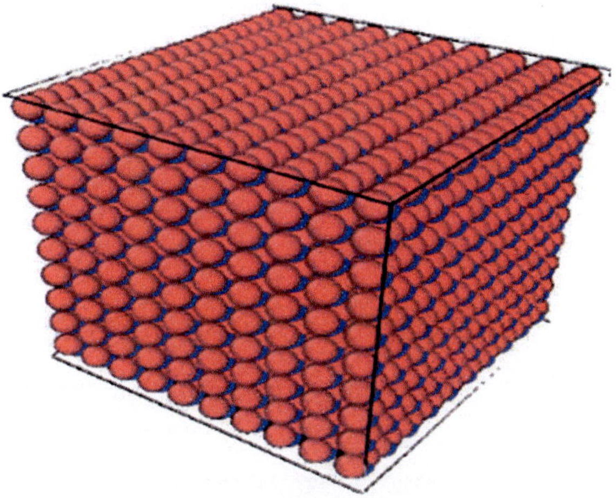

Fig. 1 Water box created for simulation

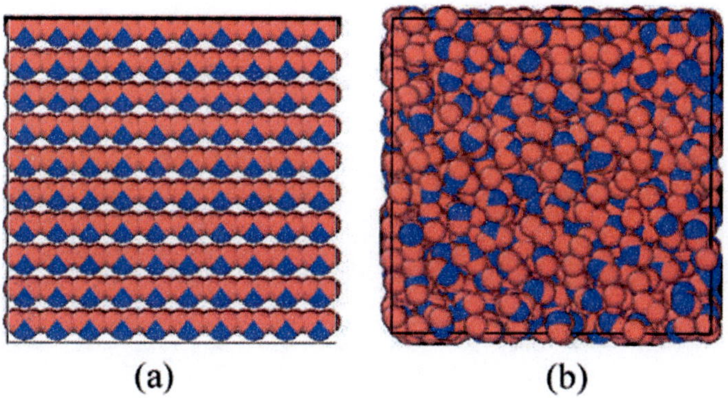

Fig. 2 Water molecules at **a** $t = 0$ ns and **b** $t = 10$ ns

atoms of water molecules. Point charge is considered in each three interaction sites. Because of simple geometry these three site models have very excellent computational efficiency. But this geometry is not the exact representation of practical water molecule due to the absence of any lone pair electron. Another exception is here to use the angle 109.47° instead of 104.5°. Four-site models (TIP4P, TIP4P-2005, TIP4P-Ew) mean there are four interaction points. That means there is an extra pseudo site. This site corresponds a negative-charged atom near oxygen. Because of this negative charge, electrostatic distribution has been improved which make its geometrical consideration better. Five-site models (TIP5P, TIP5P-Ew) actually work with five interaction points. Two negative-charged pseudo sites have been added with oxygen which represent the lone pair electrons of oxygen. As lone pair electrons are added, the electrostatic interactions are more developed here. Hence, it has the most

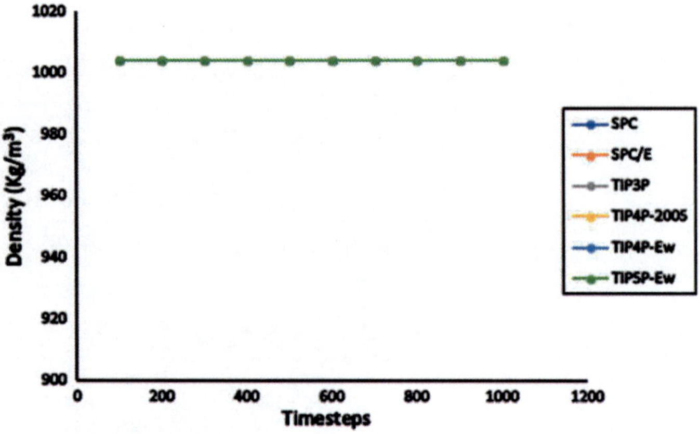

Fig. 3 Constant density of water with timesteps

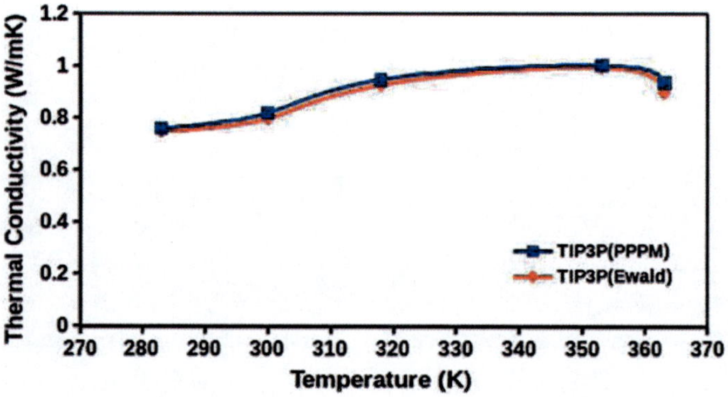

Fig. 4 Variation of conductivity between PPPM and Ewald solver

accurate geometric consideration which makes it more accurate representation of actual water molecules.

The potential functions have the similar form for all the rigid water models. It mainly consists of the alms of three forces. They are electrostatic force, dispersion force, and repulsive force.

$$E_{ab} = \sum_{i}^{\text{on}\,a}\sum_{j}^{\text{on}\,b}\frac{K_c q_{ai} q_{bj}}{r_{aibj}} + \sum_{i}^{\text{on}\,a}\sum_{j}^{\text{on}\,b} 4\varepsilon_{aibj}\left[\left(\frac{\sigma_{aibj}}{r_{aibj}}\right)^{12} - \left(\frac{\sigma_{aibj}}{r_{aibj}}\right)^{6}\right]$$

where a and b denote two dissimilar atoms, i and j stands for atom i in one individual molecule for all three-site rigid water molecules, and kc is the electrostatic constant. For both four-site and five-site, i and j stands for massless but charged site at first

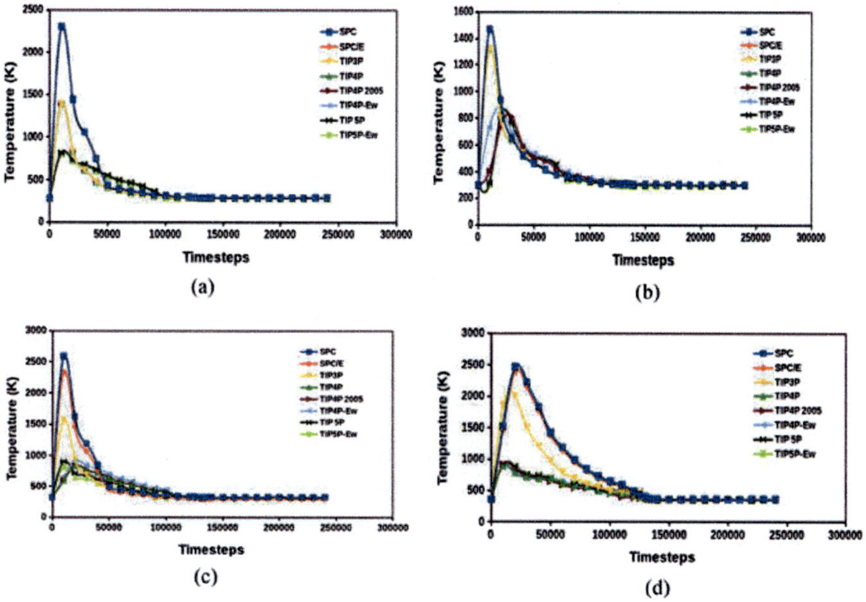

Fig. 5 Equilibrium of temperature with timesteps for **a** 283 K, **b** 300 K, **c** 318 K, **d** 353 K

Table 1 Parameters of different potentials [17]

Model	σ (Å)	ε (kcal/mol)	q_H (e)	q_O (e)	θ (°)	ϕ (°)	L_M/L_L (Å)	L_B (Å)
SPC	3.16600	0.15535	0.41000	−0.82000	109.47	N/A	N/A	1.00000
SPC/E	3.16600	0.15535	0.42380	−0.84760	109.47	N/A	N/A	1.00000
TIP3P	3.15061	0.15210	0.41700	−0.83400	104.52	N/A	N/A	0.97520
TIP4P	3.15365	0.16348	0.52000	−1.04000	104.52	N/A	0.15000	0.97520
TIP4P-Ew	3.16435	0.16275	0.52422	−1.04844	104.52	N/A	0.12500	0.97520
TIP4P-2005	3.15890	0.18521	0.55640	−1.11280	104.52	N/A	0.15460	0.97520
TIP5P	3.12000	0.15999	0.24100	−0.24100	104.52	109.47	0.70000	0.97520
TIP5P-Ew	3.09700	0.17801	0.24100	−0.24100	104.52	109.47	0.70000	0.97520

summation; and denote an atom in the second summation. All parameters for different potentials are summarized in Table 1 [17].

3 Results and Discussion

Table 2 shows the final values of thermal conductivity of water at different temperatures using different rigid water models and Table 3 indicates the percentage of

Table 2 Thermal conductivity of water using different rigid water models at various temperature (W/mK)

Temperature (K)	SPC	SPC/E	TIP3P	TIP4P	TIP4P-2005	TIP4P-Ew	TIP5P	TIP5P-Ew	Experimental
283	0.814 ± 0.018	0.842 ± 0.023	0.76 ± 0.021	0.823 ± 0.017	0.875 ± 0.019	0.773 ± 0.021	0.704 ± 0.015	0.712 ± 0.011	0.581
300	0.857 ± 0.021	0.871 ± 0.021	0.82 ± 0.022	0.85 ± 0.018	0.895 ± 0.017	0.84 ± 0.020	0.713 ± 0.012	0.728 ± 0.008	0.61
318	0.861 ± 0.02	0.884 ± 0.015	0.949 ± 0.013	0.93 ± 0.020	0.9934 ± 0.021	0.998 ± 0.018	0.5291 ± 0.011	0.742 ± 0.009	0.637
353	0.851 ± 0.024	0.891 ± 0.018	1.005 ± 0.021	0.952 ± 0.017	0.9238 ± 0.02	0.96 ± 0.019	0.7521 ± 0.01	0.735 ± 0.012	0.67
363	0.854 ± 0.021	0.864 ± 0.02	0.939 ± 0.024	0.941 ± 0.021	0.935 ± 0.022	0.97 ± 0.022	0.7812 ± 0.009	0.775 ± 0.01	0.675

Table 3 Percentage of errors of predicting thermal conductivity of water using different rigid water models at various temperature

Temperature (K)	SPC	SPC/E	TIP3P	TIP4P	TIP4P-2005	TIP4P-Ew	TIP5P	TIP5P-Ew
283	40.1032702	44.92254733	30.80895009	41.65232358	50.60240964	33.0464716	21.17039587	22.5473321
300	40.4918033	42.78688525	34.42622951	39.3442623	46.72131148	37.70491803	16.8852459	19.3442623
318	35.1648352	38.7755102	48.97959184	45.99686028	55.94976452	54.67189953	16.93877551	16.4835164
353	27.0149254	32.98507463	50.00	42.08955224	37.88059701	43.28358209	12.25373134	9.70149253
363	26.5185185	28.00	39.1111111	39.40740741	38.51851852	43.7037037	15.7362963	14.814814

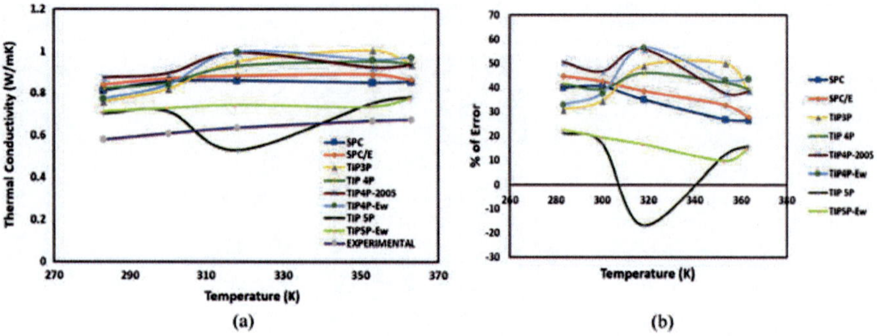

Fig. 6 **a** Thermal conductivity of water with temperature by different rigid water models, **b** percentage of errors from the experimental values

error from the experimental values for using rigid water model. From the results, it is evident that the percentage of relative errors is less than 5% which indicates the reliability of our results. Equilibrium molecular dynamics (EMD) simulation was used to produce all these results. From Tables 2 and 3, it is clear that five-site model predicts the thermal conductivity close to the experimental values. Mao et al. used reverse non-equilibrium molecular dynamics (RNEMD) simulation and found the five-site model as the best model for predicting thermal conductivity of water. Hence, it is clear that whatever the simulation method is, two five-site models have the better performance in predicting thermal conductivity of water. Figure 6a shows the thermal conductivity of water with temperature by different rigid water models and Fig. 6b shows the percentage of errors from the experimental values.

From Table 2, it is evident that very little differences are there between two five-site models at 284 K, 353 K, and 363 K temperatures. So, it can be referred that both the five-site models can predict thermal conductivity best among all the model. Hence, if in any simulation, finding out thermal conductivity of water is important, one should use five-site models. The main reason of five-site model being best that a five-site model places negative charges on atoms which represents lone pair electrons on oxygen atom. That is why it builds a tetrahedral like geometry. Compared with the other previous models, five-site models improve the geometry and hence it produces data very close to experimental values. However, five-site models cannot produce the exact trend of change of thermal conductivity with temperature. From Table 2, it can be inferred that with the buildup of temperature, the experimental value of thermal conductivity of water increases. But both TIP5P-Ew model and TIP5P cannot reproduce the trend of experimental values. The four site models-TIP4P, TIP4P-2005, and TIP4P-Ew also cannot reproduce the increasing trend of thermal conductivity of water with buildup of temperature and the values are so much over predicted. Similar conclusions can be drawn for two-site models and three-site models. Therefore, none of the models can precisely approximate the experimental values of thermal conductivity of water.

Another decision can be made from Table 2. All the rigid water models used here basically overestimate the value of thermal conductivity of water. Among all models, the five-site models overestimate the value of conductivity less than the other models. Again, the uncertainty of the data for all the temperature is within 5%. The percentage of relative error is also comparatively low. All these things make five-site models as the best model to predict thermal conductivity.

4 Conclusions

Equilibrium molecular dynamics (EMD) simulation has been employed with a view to calculate the thermal conductivity of water by using eight most prominent rigid water models and also find out the best model. Five different temperatures were taken in consideration and the systems were equilibrated at those temperatures before calculating thermal conductivity of water. Several decisions can be inferred from the study

- The five-site models (TIP5P and TIP5P-Ew) are the best models to find out the thermal conductivity of liquid water because of its improved geometric parameters. These five site models approximate the values of thermal conductivity close to experimental values.
- No single rigid water model can reproduce the increasing tendency of thermal conductivity with the buildup of temperature.
- All the models used here basically overestimate the values of thermal conductivity from the experimental ones at all the temperatures.

References

1. Allen MP, Tildesley DJ (1987) Computer simulation of liquids. Oxford University Press, Oxford, p 2
2. Harp GD, Berne BJ (1968) Vibrational relaxation of diatomic molecules in gases and liquids. J Chem Phys 49:1249
3. Berne BJ, Harp GD (1970) On the calculation of time correlation functions. Adv Chem Phys 17:63
4. Barker JA, Watts RO (1969) Structure of water; A Monte Carlo calculation. Chem Phys Lett 3:144–145
5. Rahman A, Stillinger FH (1971) Molecular dynamics study of liquid water. J Chem Phys 55:3336
6. Lee HS, Tuckerman ME (2007) Dynamical properties of liquid water from ab initio molecular dynamics performed in the complete basis set limit. J Chem Phys 126:164501
7. Guillot B (2002) A reappraisal of what we have learnt during three decades of computer simulations on water. J Mol Liq 101:219
8. Caleman C (2007) Molecular dynamics simulation study for shear viscosity of water. J Chem Phys 22:709

9. Abascal JLF, Vega C (2005) A general-purpose model for the condensed phases of water: TIP4P/2005. J Chem Phys 123:234505
10. Horn HW, Swope WC, Pitera JW (2004) Development of an improved four-site water model for biomolecular simulations: TIP4P-Ew. J Chem Phys 120(20):9665–9678
11. Rick SW (2004) A Re-Optimization Of The Five-Site Water Potential (TIP5P) for use with Ewald sums. J Chem Phys 120(13):6085–6093
12. Chialvo AA, Houssa M, Cummings PT (2002) Molecular dynamics study of the structure and thermophysical properties of model SI clathrate hydrates. J Phys Chem B 106(2):442–451
13. Bertolini D (1997) Thermal conductivity of water: molecular dynamics and generalized hydrodynamics results. Phys Rev E 56(4):4135–4151
14. Mark P, Nilsson L (2001) Structure and dynamics of the TIP3P, SPC, and SPC/E water models at 298 K. J PhysChem A 105:9954–9960
15. Haile JM (1992) Molecular dynamics simulation: elementary Methods, 1st edn. Wiley, Chichester
16. Muller-Plathe F (1997) A Simple Nonequilibrium Molecular Dynamics Method for Calculating the Thermal Conductivity. J. Chem. Phys. 106:6082–6085
17. Zhang M, Lussetti E, Souza LESD, Muller-Plathe F (2005) Thermal conductivities of molecular liquids by reverse nonequilibrium molecular dynamics. J Phys Chem B 109:15060–15067
18. Bedrov D, Smith GD (2000) Thermal conductivity of molecular fluids from molecular dynamics simulations: application of a new imposed flux method. J Chem Phys 113(18):8080–8084
19. Bertolini D, Tani A (1995) Generalized hydrodynamics and the acoustic modes of water: theory and simulation results. Phys Rev E 51:1091
20. Isachenko VP, Osipova VA, Sukomel AS (1980) Heat transfer. Mir, Moscow
21. Kataoka Y (1989) Thermodynamic data: systematics and estimation. Bull Chem Soc Jpn 62:1421
22. Haynes WM (2014) Liquid properties: handbook of chemistry and physics, 91st edn. Taylor & Francis, New York
23. Romer F, Lervik A, Bresme F (2012) Nonequilibrium molecular dynamics simulations of the thermal conductivity of water: a systematic investigation of the SPC/E and TIP4P/2005 models. J Chem Phys 137:074503
24. Mao Y, Zhang Y (2012) Thermal conductivity, shear viscosity and specific heat of rigid water. Chem Phys Lett 542:37
25. Lee SH (2014) Temperature dependence of the thermal conductivity of water: a molecular dynamics simulation study using the SPC/E model. Mol Phys 112(16):2155–2159

Global Stability Analysis of the Spatially Developing Boundary Layer: Effect of Wall Suction and Injection

Ramesh Bhoraniya and Vinod Narayanan

Abstract The laminar boundary layer's global temporal modes have been computed under the effect of suction and injection at the wall. The base velocity profile of the boundary layer has affected significantly by the mass transpiration in through the wall in the normal to flow direction. The governing stability equations have been derived using standard procedures. A spectral collocation method with the Chebyshev polynomials have been used for the discretization of the stability equations. The two-dimensional eigenvalues problem has been formed and solved using Arnoldi's algorithm. The different rates of suction and injection ($V_w = 0.01, 0.025$, and 0.050) at $Re = 226$ were considered to study the temporal and spatial growth of the small disturbances. It has been observed that an increased rate of suction has a stabilization effect, and injection has destabilization effect.

Keywords Global stability · Suction · Injection · Boundary layer

1 Introduction

The boundary layer stability and the transition is a wide field of research due to its significant impact on the fundamental physics and applications. The suction delays the flow separation extends the laminar regime and thus reduces drag on the solid surfaces of the propelling bodies through a viscous fluid. The suction and injection have a significant effect in industrial applications like heat exchangers, recovery of petroleum resources, catalytic reactors, chemical reactions, and chemical vapor deposition on the solid surfaces. It also has an important role to control the boundary layer flow over aircraft wings and turbine blades. The suction/injection in wall-normal direction can be made possible through discrete spanwise slots or porous strips.

R. Bhoraniya (✉)
Department of Mechanical Engineering, Marwadi University, Rajkot, India
e-mail: rameshkumar.bhoraniya@marwadieducation.edu.in

V. Narayanan
Department of Mechanical Engineering, Indian Institute of Technology Gandhinagar, Palaj, Gandhinagar, India

L. Venkatakrishnan et al. (eds.), *Proceedings of 16th Asian Congress of Fluid Mechanics*, Lecture Notes in Mechanical Engineering,
https://doi.org/10.1007/978-981-15-5183-3_54

Reynolds and Saric [1] experimentally observed that the velocity profile moved toward the wall in a region of high dissipation due to suction [2]. Saric and Reed [3] experimentally investigated the effect of mass injection and suction and found that suction has stabilizing effect while weak blowing reduces the critical Reynolds number for a boundary layer [3]. Watanabe [4] studied the effect of suction and injection of uniform rate for an incompressible boundary layer subjected to the streamwise pressure gradient. The electrically conducting fluid was considered for this theoretical study. The author found that the skin friction increased, and displacement thickness reduced with increased suction/injection parameter. The critical Reynolds number increased with the increased suction/injection parameter, which makes the flow stable [4]. Lingwood [5] studied the onset of the boundary layer formed on a rotating disk with uniform suction and found that suction delay onset of the absolute instability while injection promotes the onset. At the same magnitude of suction/injection, the effect of suction is found more effective in stabilization than that of injection in destabilization. It is also found that suction damps both stationary and traveling modes [1]. Fransson and Alfredsson [6], experimentally and theoretically investigated the effect of suction on the transition of the boundary layer. They found that under the influence of free-stream turbulence, suction prevents boundary layer transition, while without suction, it gave rise to transition [6]. Aydln and Kaya [2] studied the effect of suction and injection on the heat transfer rate on a porous flat-plate using a similarity solution. They found an enhancement in heat transfer with suction and reduction with wall injection. Wang et al. [7] investigated the effect of cooling and suction on the supersonic boundary layer at a Mach number of 4.5. They found that cooling can delay the onset of the transition, and flow control becomes active with a reduction in wall temperature. The suction hampers the transition up to some intensity of it. However, it also increases the friction coefficient [7]. Huang and Wing [8] investigated the effect of local suction, and they found that a weak suction rate of 0.086 m/s has a significant impact on base flow, and T-S waves have been reduced to 90% which has a strong effect on the transition process. The total suction rate is a crucial parameter, and suction location near the lower branch of the neutral curve found more effective [8]. Hinvi et al. [9] studied the effect of small constant suction at lower and small constant injection at the upper wall in a plane Poiseuille flow with the porous wall. They found that the instability of a small disturbance can be studied by the modified Orr–Sommerfeld equation, and small suction or injection is important for control [9].

The main objectives of the present study are to investigate the effect of wall suction and injection on the global modes of the boundary layer. Many researchers have studied the effect of wall suction and injection on the boundary layer stability in the past using the local stability approach. To the best of our knowledge in all previous works, the local stability approach has been applied for the stability analysis.

2 Problem Formulation

The flow of incompressible fluid has been considered on a porous materials plate to allow the transpiration of mass trough the wall in the normal direction to the flow. The base flow is non-parallel due to the viscous effect of the fluid. The displacement thickness ($\delta*$) at the inflow boundary of the computational domain is considered as a length scale to compute the Reynolds number. The suction and injection velocities are normalized with the uniform inlet velocity. The suction and injection velocity have been normalized with the uniform inlet velocity (U_∞). The three different magnitudes of 0.01, 0.025, and 0.050 at Re = 226 have been considered for the present study. The governing stability equations have been derived in the rectangular coordinates for the small disturbances using standard procedure. The normal modes form of the disturbances is considered for the stability analysis. The spanwise direction (Z) is homogeneous, and thus, spanwise wavenumber (β) has zero magnitudes. The two-dimensional normal mode form of the disturbances is considered.

$$\mathrm{Re} = \frac{U_\infty \delta^*}{\nu} \tag{1}$$

The instantaneous flow quantities can be presented as the sum of basic and small disturbances as,

$$\overline{U} = U_b + u_p, \overline{V} = V_b + v_p, \overline{P} = P_b + p_p \tag{2}$$

The disturbance amplitudes are the functions of flow direction (x) and normal wall direction (y).

$$u_p(x, y, t) = \hat{u}_p(x, y)\mathrm{e}^{-i\omega t}, v_p(x, y, t) = \hat{v}_p(x, y)\mathrm{e}^{-i\omega t}, p(x, y, t) = \hat{p}(x, y)\mathrm{e}^{-i\omega t} \tag{3}$$

The linearized Navier–Stokes equations for the instability analysis are as follows.

$$\frac{\partial u_p}{\partial t} + U_b\frac{\partial u_p}{\partial x} + u_p\frac{\partial U_b}{\partial x} + V_b\frac{\partial u_p}{\partial y} + v_p\frac{\partial U_b}{\partial y} + \frac{\partial p_p}{\partial x} - \frac{1}{\mathrm{Re}}\left[\nabla^2 u_p\right] = 0 \tag{4}$$

$$\frac{\partial v_p}{\partial t} + U_b\frac{\partial v_p}{\partial x} + u_p\frac{\partial V_b}{\partial x} + V_b\frac{\partial v_p}{\partial y} + v_p\frac{\partial V_b}{\partial y} + \frac{\partial p_p}{\partial y} - \frac{1}{\mathrm{Re}}\left[\nabla^2 v_p\right] = 0 \tag{5}$$

$$\frac{\partial u_p}{\partial x} + \frac{\partial v_p}{\partial y} = 0 \tag{6}$$

$$\text{where, } \nabla^2 = \frac{\partial^2}{\partial x^2} + \frac{\partial^2}{\partial y^2} \tag{7}$$

The following different boundary conditions are considered for the solution of linearized Navier–Stokes equations. No-slip and no-penetration conditions have been considered at the plate wall for velocity disturbances.

$$u_p(x, 0) = 0, \quad v_p(x, 0) = 0, \tag{8}$$

In the far-field, far away from the wall, velocity and pressure disturbances exponentially reduce and vanish to zero.

$$u_p(x, \infty) = 0, \quad v_p(x, \infty) = 0, \quad p_p(x, \infty) = 0 \tag{9}$$

The global stability analysis required boundary conditions in the flow direction at the inlet and outlet. In the present work, we considered that small disturbances are evolving within the flow domain only. Thus, homogeneous Dirichlet conditions have been considered for the disturbances at the inflow boundary [10].

$$u_p(0,\ y) = 0, \quad v_p(0,\ y) = 0, \tag{10}$$

At the outflow boundary, physically, it is difficult to predict how disturbances will leave the domain. Artificial numerical conditions in the form of an extrapolated type conditions have been considered for the velocity disturbances [11].

$$u_{p_{n-2}}(x_n - x_{n-1}) - u_{p_{n-1}}(x_n - x_{n-1}) + u_{p_n}(x_{n-1} - x_{n-2}) = 0 \tag{11}$$

$$v_{n-2}(x_n - x_{n-1}) - v_{p_{n-1}}(x_n - x_{n-1}) + v_{p_n}(x_{n-1} - x_{n-2}) = 0 \tag{12}$$

where x_{n-2}, x_{n-1}, and x_n are the most exterior grid points toward the downstream. No physically any conditions exist at a wall for the pressure perturbations. However, compatibility conditions are collocated at the wall from the governing stability equations itself.

$$\frac{\partial p_p}{\partial x} = \frac{1}{\text{Re}} \frac{\partial^2 u_p}{\partial y^2} \tag{13}$$

$$\frac{\partial p_p}{\partial y} = \frac{1}{\text{Re}} \frac{\partial^2 v_p}{\partial y^2} \tag{14}$$

The above-discussed boundary conditions and discretization of the linearized Navier–Stokes equations together form a general eigenvalues problem. Arnold's iterative algorithm has been used to compute a few selected eigenmodes and eigenfunctions to study the evolution of small disturbances. Readers are requested to refer [9] for a detailed description of the discretization and solution of the eigenvalues problem.

3 Base flow Solution

The steady two-dimensional Navier–Stokes equations are solved numerically using finite volume code, ANSYS Fluent. The second-order upwind scheme has been used for the spatial discretization of the N-S equations. The numerical solution of the base flow has been checked for grid convergence. The steady two-dimensional Navier–Stokes equations are solved numerically using finite volume code, ANSYS Fluent. The second-order upwind scheme has been used for the spatial discretization of the N-S equations. The numerical solution of the base flow has been checked for grid convergence.

$$U_b \frac{\partial U_b}{\partial x} + V_b \frac{\partial U_b}{\partial y} = \left(-\frac{\partial P_b}{\partial x} \right) + \frac{1}{\mathrm{Re}} \left(\frac{\partial^2 U_b}{\partial x^2} + \frac{\partial^2 U_b}{\partial y^2} \right) \tag{15}$$

$$U_b \frac{\partial V_b}{\partial x} + V_b \frac{\partial V_b}{\partial y} = \left(-\frac{\partial P_b}{\partial y} \right) + \frac{1}{\mathrm{Re}} \left(\frac{\partial^2 V_b}{\partial x^2} + \frac{\partial^2 V_b}{\partial y^2} \right) \tag{16}$$

$$\frac{\partial U_b}{\partial x} + \frac{\partial V_b}{\partial y} = 0 \tag{17}$$

The following boundary conditions have been considered to close the above problem.

Inflow boundary: $U_b(0, y) = 1; \quad V_b(0, y) = 0;$
Solid wall: $U_b(x, 0) = 0; \quad V_b(x, 0) = \pm V_w$

Free-stream: $U_b(x, \infty) = 1; \quad V_b(x, \infty) = 0;$
Outflow: $\frac{\partial U_b}{\partial x} = 0; \frac{\partial V_b}{\partial x} = 0; \quad P_b = 0$

where V_w is the wall suction or injection, the positive V_w has an injection effect, and negative V_w has a suction effect. Figures 1 and 2 show the effect of wall suction and injection on the streamwise velocity (U) and its first derivative (dU/dy) at streamwise location $x = 1.00$ m from the leading edge of the boundary layer. The base velocity profile U moves toward the wall under the effect of suction and away from the wall under the effect of the injection. The magnitude of the first derivative (dU/dy) increases with the increased suction rate and decreases with the increased injection rate. Thus, the characteristics of the basic state of the velocity profile are directly affected by wall suction and injection. It is having a significant effect on the stability characteristics of the boundary layer.

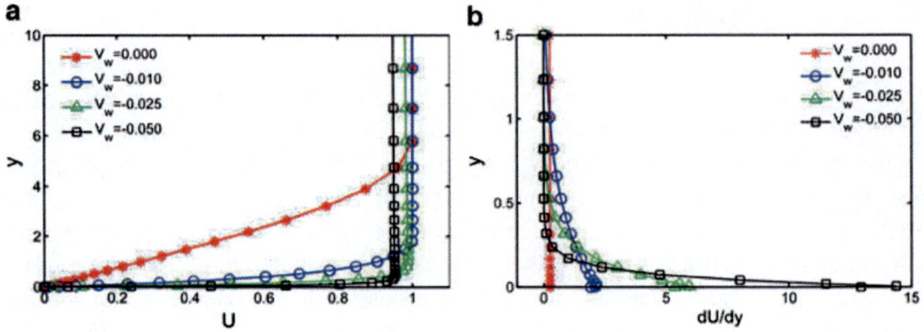

Fig. 1 Variation of **a** U and **b** dU/dy for different suction velocity (V_w) at streamwise location $x =$ 1.00 m for Re = 226

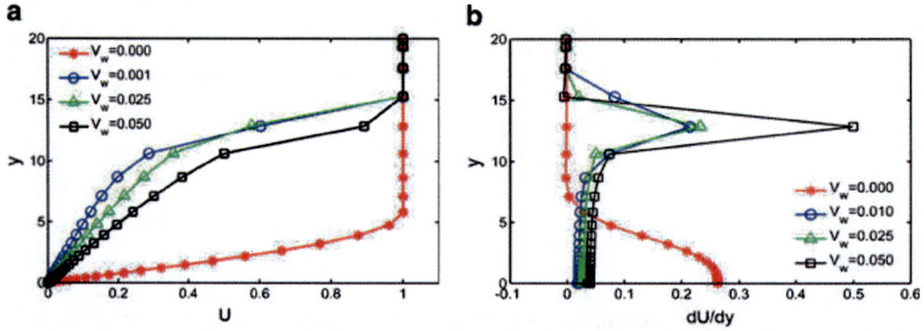

Fig. 2 Variation of **a** U and **b** dU/dy for different injection velocity (V_w) at streamwise location $x =$ 1.00 m for Re = 226

4 Results and Discussions

In the present stability analysis of the boundary layer, we considered Reynolds number of 226 based on the displacement thickness along with the suction and injection rates of 0.01, 0.025, and 0.050 (normalized with inlet velocity). The computational domain of size 550 and 20 have been considered. In the flow direction, 241 and wall-normal direction 61 collocation points have been considered. The base flow and eigenvalues problem solution have been tested for the grid convergence. Sponging with appropriate strength and length has been applied near the outlet region to avoid any non-physical solution.

Figure 3a shows the comparison of the eigenspectra for the different suction velocities at Re = 226. Spectra with $V_w = 0$ is for a boundary layer without any suction. It has been found from the spectra that the increase in suction reduces the magnitudes of ω_i, which makes flow more stable. In the present computations, the imaginary part of the global modes is negative. Thus, flow is globally stable. The pattern of the discrete part of the spectra has been found different from the suction effect. Two distinct branches exist with the suction effect. The distribution of the

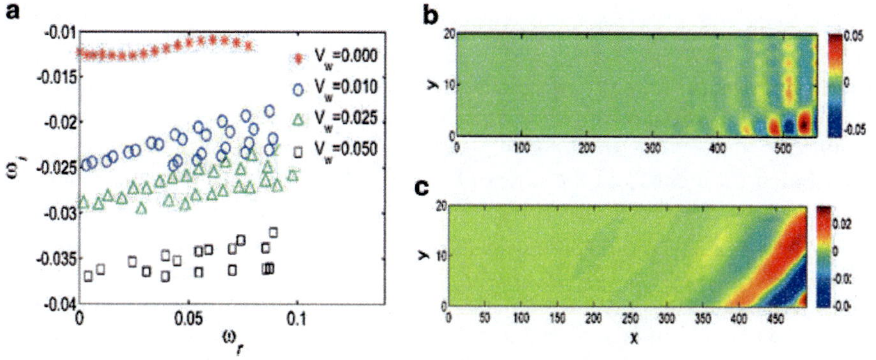

Fig. 3 **a** Comparison of the spectra at different suction velocity. **b** Contour plots of the streamwise velocity disturbances (u_p) for eigenmode, $\omega = 0.0607 - 0.0109i$ and $V_w = 0$. **c** Contour plot of velocity disturbance (u_p) for the eigenmode for $\omega = 0.0668 - 0.0245i$ and $V_w = -0.025$. The associated Reynolds number is considered as 226.

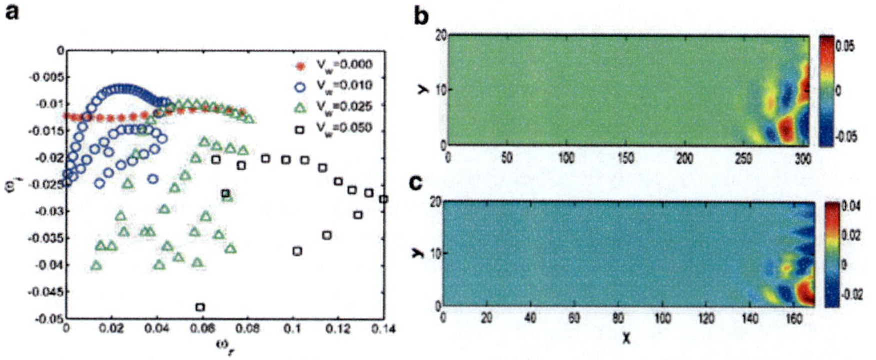

Fig. 4 **a** Comparison of the spectra at different injection velocity. **b** Contour plots of the streamwise velocity disturbance (u_p) for eigenmode, $\omega = 0.0232 - 0.0072i$ and $V_w = 0.010$. **c** Contour plot of u_p velocity disturbances of the eigenmode at $\omega = 0.0532 - 0.0101i$ and $V_w = 0.025$. The Reynolds number considered is 226

frequency even changed and a cluster of two frequencies have appeared with the suction. We have also plotted the 2D spatial structure of the temporal eigenmodes nearly $\omega \cong 0.06$. Figure 3b, c presents contour plots of u disturbance velocity with $V_w = 0$ and $V_w = 0.025$. The comparison clearly shows that the largest values of disturbance amplitudes have been reduced for $V_w = 0.025$. The elongated structure of the disturbances is found with the effect of suction. Wave-like behavior of the disturbance amplitude has been observed with and without suction. Thus, the study of temporal and spatial eigenmodes computed with ($V_w \neq 0$) and without ($V_w = 0$) the effect of suction prove that the effect of suction is to stabilize the boundary layer.

Figure 4a presents a comparison of the eigenspectra for different injection velocity at Re = 226. As shown, for $V_w = 0.01$, the temporal growth of the disturbances has been found, and a further increase in injection velocity reduces the temporal growth. However, $\omega_i < 0$ found for all the injection velocities considered here. Thus, flow is globally stable. The structure of spectra and distribution of the frequency has been found different from the effect of the injection. Even it is observed that the discrete part of the spectra has shifted toward the right side as injection velocity has increased. Similar to the previous case, more than one branch is found in the spectra also. Figure 4b, c present the 2-D spatial structure of the eigenmodes corresponding to the least stable temporal eigenmodes (largest values of $\omega_{i)}$ for injection velocity of 0.010 and 0.025. It is observed that the magnitudes of the disturbances are larger to increase injection velocity. The wave-like nature of the disturbances is found in the streamwise direction; however, the structure is not elongated in the streamwise direction like an under the effect of suction. One important and consistent observation found in the variation of the dU/dy at the wall for suction and injection both and its significant effect has been reflected in the stability of the boundary layer. The magnitude of the dU/dy increases with the increased suction.

5 Conclusions

Bi-global stability analysis of the laminar flat-plate boundary layer has been performed to study the effect of wall suction and injection on the temporal and spatial stability properties of the laminar boundary layer at Re = 226. The global temporal modes have been computed at different wall suction and injection rate, and spatial structure of the least stable temporal eigenmodes has been extracted. It has been found that at the considered value of Re and different wall suction/injection rate, the computed eigenmodes are found stable, and thus, flow is temporally stable. However, the growth of the small disturbance amplitudes has been found in the streamwise direction for the least stable eigenmodes, which indicates that the flow is spatially unstable. The increased wall suction rate from 0.0 to 0.050 has moved the profile of U velocity toward the wall and increased the magnitudes of the dU/dy at the wall. The effect of the modified base flow has been found directly on the temporal and spatial growth of the small disturbances. The magnitudes of the temporal growth $(\omega_{i)}$ and spatial growth of the disturbance amplitudes have been found reduced with the increased suction rate. However, the increased rate of injection initially moves the U velocity profile away from the wall, then as injection increases profile moves toward the wall. However, the curvature of the profile is not the same as that of suction. A similar trend has been found for the dU/dy at the wall—this modified base flow due to the wall injection reflected the boundary layer's stability characteristics. The growth of the disturbances initially increases with the increased injection rate and then reduces. Thus, overall suction has stabilized, and injection has a destabilizing effect on the laminar boundary layer.

References

1. Reynolds GA, Saric WS (1986) Experiments on the stability of the flat-plate boundary layer with suction. AIAA J 24:202–207
2. Aydin O, Kaya A (2005) Laminar boundary layer flow over a horizontal permeable flat-plate. App Math Comput 161:229–240
3. Saric W, Reed L (1986) Effect of suction and weak mass injection on boundary-layer transition. AIAA J 03:383–389
4. Watanabe T (1987) Effect of uniform suction or injection on the magnetohydrodynamic boundary layer flow along with a flat plate with a pressure gradient. Acta Mech 73:33–44
5. Lingwood RJ (1997) On the effects of suction and injection on the absolute instability of the rotating-disk boundary layer. Phys Fluids 9:1317
6. Fransson JHM, Alfredsson PH (2002) On the disturbance growth in an asymptotic suction boundary layer. J Fluid Mech 482:51–90
7. Wang S, Lei J, Zhen H, Lee C (2014) Numerical investigation of wall cooling and suction effect on supersonic flat-plate boundary layer transition using large eddy simulation. Adv Mech Eng 7(2):493194
8. Huang A, Wu X (2017) 2017 A local scattering approach for the effects of abrupt changes on boundary-layer instability and transition: a finite Reynolds number formulation for isolated distortions. J Fluid Mech 822:444–483
9. Hinvi LA, Monwanou AV, Chbi Orou JB (2018) Linear stability analysis of fluid flow between two parallel porous stationary plates with small suction and injection. https://arxiv.org/abs/1412.0909v1
10. Theofilis V (2003) Advances in global linear instability analysis of non-parallel and three-dimensional flows. Prog Aerosp Sci 39:249–331
11. Bhoraniya R, Narayanan V (2019) Global stability analysis of the spatially developing boundary layer: effect of the streamwise pressure gradient. Fluid Dyn 53(6):84–98

Experimental Determination of Drag on Cylindrical Objects with Superhydrophobic Coating

Adharsh Shankaran and K. R. Sreenivas

Abstract The experiments involve dropping of a cylindrical specimen from a height above the water surface. The specimen is a 3D printed hollow cylinder with length to diameter ratio 10 as shown in Fig. 1. The outer surface is modified by engraving channels of radius 0.5 mm along the longitudinal axis, with breakers at an interval of 20 mm to decrease depletion of air at high shear flow. The specimen is then spray coated with superhydrophobic substance. The specimen is loaded at bottom with conical brass head to maintain metacentric height above centre of gravity and easy entrance into water. Specimen is released from a height of 20 cm from the surface of water. The maximum Reynolds number achieved is ~$1.9857 * 10^4$. Drag reduction of ~8.189% is achieved for grooved, and ~3.67% is achieved for plane cylinder.

Keywords Terminal velocity · Superhydrophobic surface · Skin friction · Drag coefficient

1 Introduction

Reducing drag of objects due to skin friction at the fluid–solid interface is highly researched area with earliest known research in 1931 [1]. For a ship sailing in water, 70–80% of total resistance is because of surface roughness [2]. As there are limitations in both application and manufacturing smooth surface, other options are explored such as introducing a fluid of lower viscosity(mostly gas) between the solid and viscous fluid. Injecting gas/air between the solid–fluid boundary [3], superhydrophobic coating the solid surface [4] are some of the methods employed to decrease drag by achieving slip condition between solid–fluid interface. Injecting gas requires

A. Shankaran (✉)
Department of Mechanical Enigineering, Ramaiah Institute of Technology, Bangalore 560054, India
e-mail: chiadharsh@gmail.com

K. R. Sreenivas
Engineering Mechanics Unit (EMU), Jawaharlal Nehru Centre For Advanced Scientific Research, Jakkur P.O., Bangalore 560064, India

L. Venkatakrishnan et al. (eds.), *Proceedings of 16th Asian Congress of Fluid Mechanics*, Lecture Notes in Mechanical Engineering,
https://doi.org/10.1007/978-981-15-5183-3_55

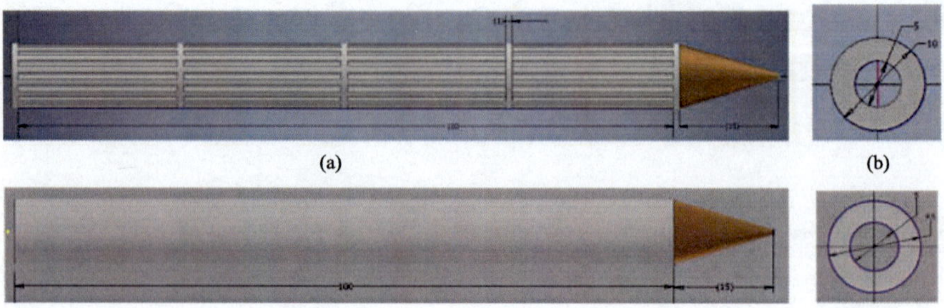

Fig. 1 **a** Grooved specimen with conical brass head front view **b** Grooved specimen top view **c** Plane specimen with conical brass head front view **d** Plane specimen top view

continuous supply of gas which is both energy-consuming and space-consuming especially in submarine applications. Superhydrophobicity can be achieved by forming the microlevel roughness and coating with hydrophobic substance on the surface [5] (Fig. 1).

Experiments conducted on superhydrophobic-coated microchannels of various widths showed reduction in pressure drop of ~40% [6] for Re < 1000, ~20% [7] in laminar range, 20–30% [8]. Drag reduction when a rotating disc was superhydrophobic coated was ~30% in laminar and 15% in turbulent flow [9]. For external flow over a flat plate drag reduction of ~50% in laminar and lower levels after transitioning to turbulence was reported due to superhydrophobic surface [5]. Similar trend was seen in [10] where ~30% drag reduction in transition regime ($10^5 < Re < 10^6$) and increased drag in turbulent regime ($10^6 < Re < 10^7$). In another study, drag reduction of sphere due to hydrophobic coating was accounted by measuring terminal velocity and showed 5 to 15% decrease in drag [11].

In the current study, long hollow cylindrical specimen with total length to diameter(l/d) ratio 11.5:1 is studied. Since [10] reported high drag due to the depletion of air at high shear flows, surface modifications are done to decrease depletion of air even at Reynolds Number of ~3.5 * 10^5. The high l/d ratio specimen is chosen to understand drag reduction for applications like submarines, underwater and aerial torpedoes. Change in drag for similar specimen with and without superhydrophobic coating is accounted by measuring terminal velocity of specimen in water. The specimen is released in air at a particular height from water, reaches water surface at a velocity which may be lower or higher than terminal velocity of body in water, then accelerates or decelerates to reach terminal velocity. The forces acting on a body are weight of body(F_g), buoyant force(F_b) and drag force(F_d). As the body travels through the fluid, it reaches a state where $\mathbf{F_g = F_b + F_d}$. The body no longer accelerates and thus has reached terminal velocity.

2 Experimental Set-up

The hollow cylindrical specimen is 3D printed using polylactic acid(PLA) material and has l:d ratio of 10:1 with inner diameter 5 mm and thickness of 2.5 mm. The specimen surface has geometrical constrains such as semi-circular channels with radius 0.5 mm running through out channel with breakers at an interval of 20 mm. From here on, superhydrophobic coated is termed coated and when no coating is given, it is termed uncoated. The specimen is coated with commercially available superhydrophobic spray(Rust-Oleum 274,232 NeverWet Liquid Repelling Superhydrophobic Anti-wet Spray Kit) in two steps. A base coating is given and dried for an hour, and then a top coating is provided and left for drying for around 6 h. An iron ring is attached to one end for electromagnetic dropping and a conical brass head of length 1.5 cm is attached to other end so that specimen moves through a vertical path without wobbling. The central hole is blocked with wax on the face where ring is attached to avoid filling the hole with water. The total l/d ratio of specimen comes to 11.5. A representative figure can be seen in Fig. 2(a) and (b). Figure 2(c), (e) shows uncoated specimens, and Fig. 2(d), (f) shows coated specimens, where entrapped air can be clearly seen in coated specimen.

Figure 3 shows contact angle between water coloured with food dye and hydrophilic and superhydrophobic surface. Water tends to bead-up when the surface is non-wetting.

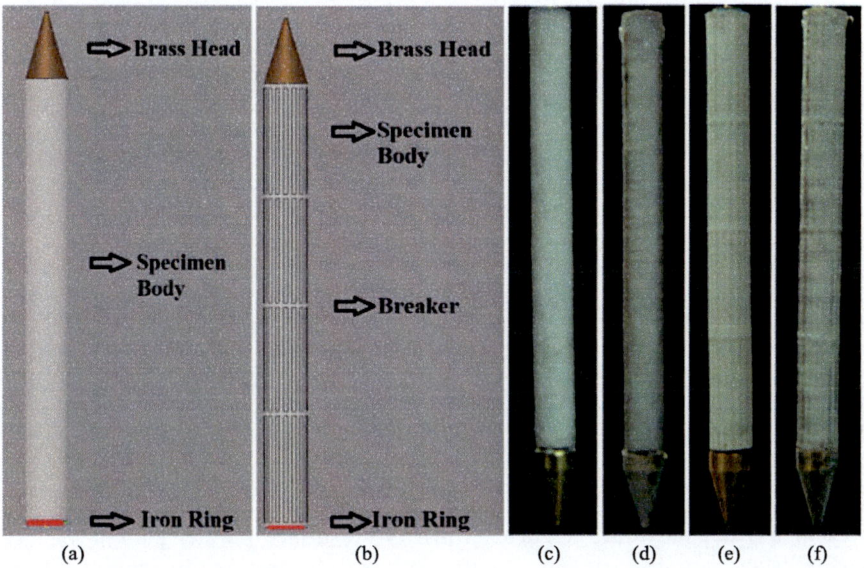

Fig. 2 **a** Representative figure of plane specimen indicating different parts **b** Representative figure of grooved specimen indicating different parts **c** Uncoated plane specimen in water **d** Coated plane specimen in water **e** Uncoated grooved specimen in water **f** Coated grooved specimen in water

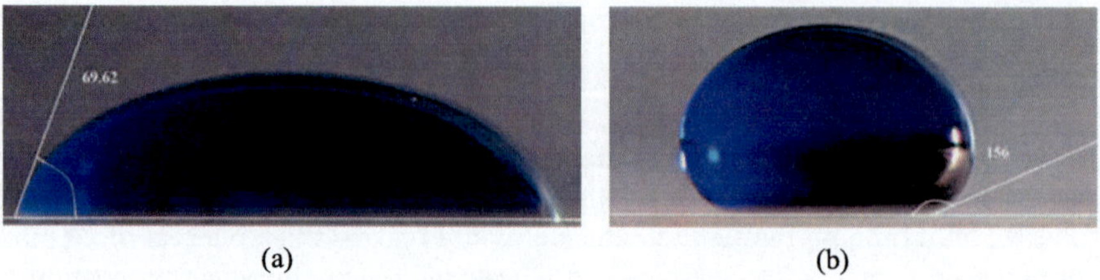

Fig. 3 Contact angle of water mixed with aludine dye on a surface **a** Without superhydrophobic coating **b** With superhydrophobic coating with angles shown in degrees

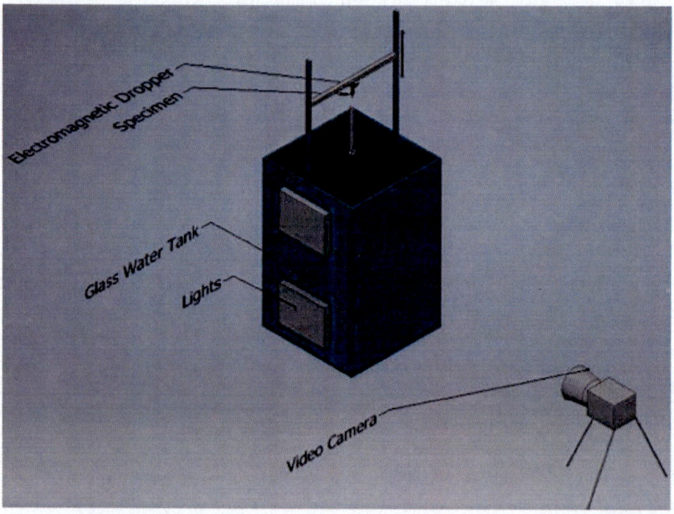

Fig. 4 Schematic representation of experimental set-up

Experimental set-up consists of a water tank of dimensions 0.5 m * 0.5 m * 0.8 m. A 12 V electromagnetic dropper is used to drop the specimen from required height. The arrangement is shown in Fig. 4. The specimen falling through the tank is imaged using Motion Pro X3 Camera at 200 frames per second(fps).

3 Results

Specimen is dropped from a height of 0.2 m from the surface of water. Plots of velocity versus distance are shown in Fig. 5. It can be clearly seen that the terminal velocity of coated plane specimen is ~1.76 m/s, and uncoated plane specimen is ~1.7 m/s. Once grooves are made, the terminal velocity further reduces to ~1.62 m/s, and once this is coated, terminal velocity raises to ~1.706 m/s. Figure 6 shows variation of drag

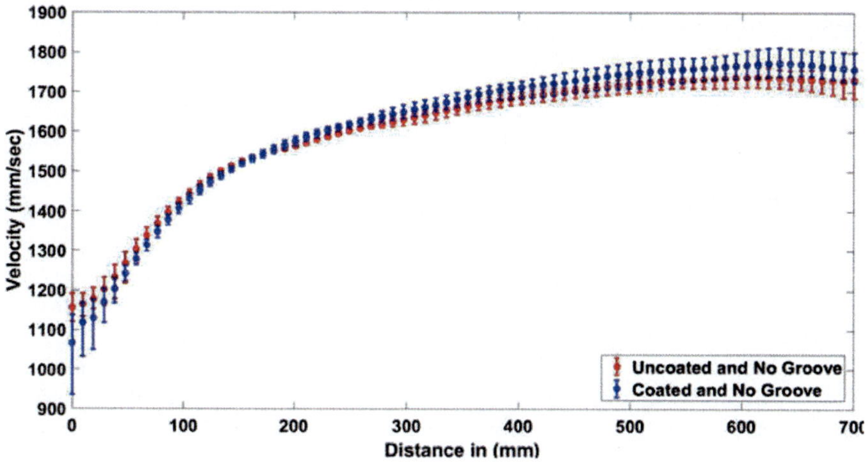

(a)

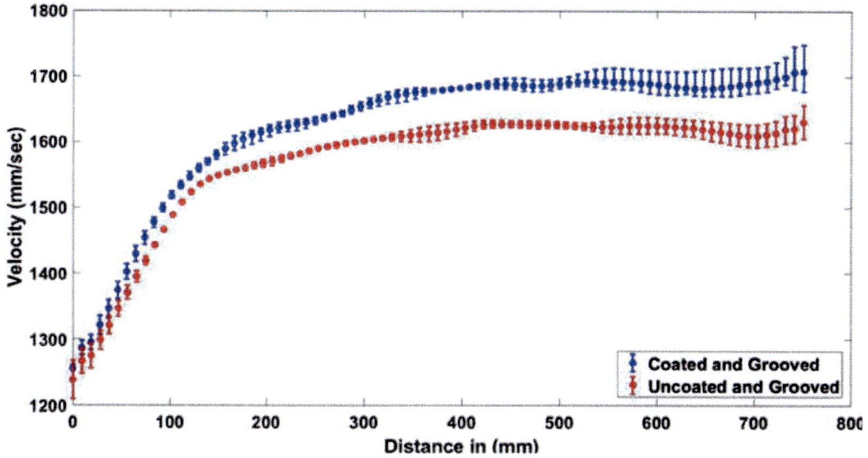

(b)

Fig. 5 Plot of velocity versus distance. **a** Plane specimen **b** Grooved specimen

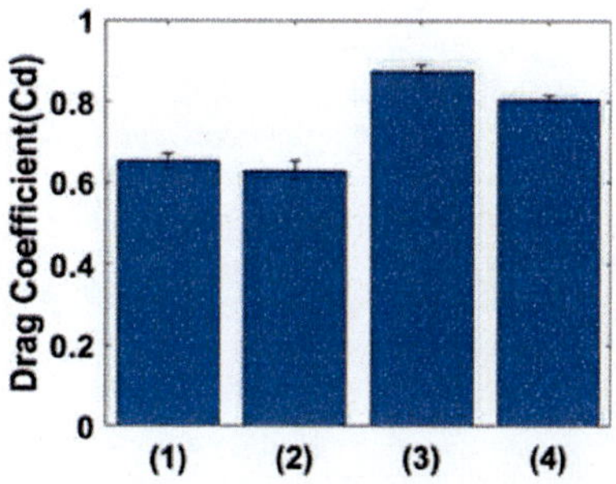

Fig. 6 Plot showing drag coefficient(Cd) for mean velocity vs type of surface **1** No groove and no coating **2** no groove and coated **3** Grooved and no coating **4** Grooved and coated

coefficient with different surfaces. Coating the plane specimen resulted in ~3.67% decrease in drag, whereas coating of grooved specimen resulted in ~8.189% decrease in drag.

4 Conclusions

Current experiments show higher drag reduction for grooved specimen, but the drag coefficient value itself is higher than no grooves and coated. However, the results obtained are encouraging as further reduction in drag coefficient of grooved specimen can be achieved by optimization of groove geometry, which would also increase the air layer stability as done in [12] for liquid infused surfaces to prevent failure. Further, experiments have to be done to comment on the retention of air layer and drag on the body with time.

Acknowledgements I would like to thank JNCASR for providing me opportunity as Summer Research Fellow and all the facilities to conduct experiment. I would also like to thank all my colleagues for their active support.

References

1. Forrest F, Grierson GA (1931) Friction losses in cast iron pipe carrying paper stock. Paper Trade J 92(22):39–41
2. Yunqing G, Tao L, Jiegang M, Zhengzan S, Peijian Z (2017) Analysis of drag reduction methods and mechanisms of turbulent. Appl Bionics Biomech
3. Ceccio SL (2010) Friction drag reduction of external flows with bubble and gas injection. Ann Rev Fluid Mech 42:183–203
4. Rothstein JP (2010) Slip on superhydrophobic surfaces. Ann Rev Fluid Mech 42:89–109
5. Henoch C, Krupenkin T, Kolodner P, Taylor J, Hodes M, Lyons A, Peguero C, Breuer K (2006) Turbulent drag reduction using superhydrophobic surfaces. In: 3rd AIAA flow control conference, p 3192
6. Ou J, Perot B, Rothstein JP (2004) Laminar drag reduction in microchannels using ultrahydrophobic surfaces. Phys Fluids 16(12):4635–4643
7. Hao PF, Wong C, Yao ZH, Zhu KQ (2009) Laminar drag reduction in hydrophobic microchannels. Chem Eng Technol: Ind Chem-Plant Equip-Process Eng-Biotechnol 32(6):912–918
8. Choi CH, Ulmanella U, Kim J, Ho CM, Kim CJ (2006) Effective slip and friction reduction in nanograted superhydrophobic microchannels. Phys Fluids 18(8):087105
9. Moaven K, Rad M, Taeibi-Rahni M (2013) Experimental investigation of viscous drag reduction of superhydrophobic nano-coating in laminar and turbulent flows. Exp Thermal Fluid Sci 51:239–243
10. Aljallis E, Sarshar MA, Datla R, Sikka V, Jones A, Choi CH (2013) Experimental study of skin friction drag reduction on superhydrophobic flat plates in high Reynolds number boundary layer flow. Phys Fluids 25(2):025103

11. McHale G, Shirtcliffe NJ, Evans CR, Newton MI (2009) Terminal velocity and drag reduction measurements on superhydrophobic spheres. Appl Phys Lett 94(6):064104
12. Wexler JS, Jacobi I, Stone HA (2015) Shear-driven failure of liquid-infused surfaces. Phys Rev Lett 114(16):168301

Modeling of Flow in a Weir and Pool Fishway With Orifices for Optimizing the Fishway Design

W. M. L. K. Abeyratne, P. S. Weerasinghe, and S. B. Weerakoon

Abstract Fishways that facilitate fish to migrate upstream by passing the obstructions such as dams built across waterways are given increased attention with respect to their efficiency and cost. A computational fluid dynamics (CFD) model, ANSYS CFX 17.2 software was used to investigate the three-dimensional flow in a weir and pool fishway. The model results were verified using published data and measurements obtained by a laboratory experiment. The model was then applied to compute the water depth, velocity, turbulent kinetic energy and energy dissipation rate distributions of a weir, and pool type fishway under different slopes. Economic design of steeper weir and pool type fishways with high fishway efficiency is feasible by controlling the depth at the downstream of the fishway.

Keywords Computational fluid dynamics · Numerical modeling · Weir and pool fishways · Fishway efficiency

1 Introduction

Fish migratory routes have been badly affected by anthropogenic obstructions in rivers throughout the world. Fishways or fish ladders are hydraulic structures constructed to allow fish to migrate upstream bypassing obstructions such as dams, weirs and barrages. There are several types of engineered fishways distinguished by the arrangement of baffles or weirs in the channel. These are mainly classified as vertical slot, weir and pool, denil, and culvert fishways. Nature-like fishways are constructed using natural or soft materials like boulders, gravel, and log and are designed to mimic the hydraulic, bed, and bank conditions of natural stream [4]. Different hydraulic and physical characteristics make these fishways suitable for some fish species and not for others. An effective fishway attracts fish, allow them to enter, pass through and exit safely with minimum time, and spending least amount of energy [3].

W. M. L. K. Abeyratne (✉) · P. S. Weerasinghe · S. B. Weerakoon
Faculty of Engineering, University of Peradeniya, Kandy, Sri Lanka
e-mail: abeyratnelka@eng.pdn.ac.lk

L. Venkatakrishnan et al. (eds.), *Proceedings of 16th Asian Congress of Fluid Mechanics*, Lecture Notes in Mechanical Engineering,
https://doi.org/10.1007/978-981-15-5183-3_56

The weir and pool type engineered fishway is a type of fishway for species having strong swimming ability. It is composed of consecutive baffles (weirs) with or without orifice and notch. Orifices can be provided in the weirs for lower depth swimming fish and to accommodate fish movement during dry seasons. The weir and pool type of fishways are simple to construct, and it is the least expensive compared to other types [3]. There are two main types of flow over these weir fishways depending on depth of flow for a given slope and pool length, i.e., plunging flow and streaming flow. With the increasing discharge, the flow regime in weir and pool fishway varies from plunging to streaming flow. According to the experimental data in weir and pool fishways, both orifice and notch are active for a narrow range of discharge values. Also, the measured velocity values have indicated that orifice may lead to critically high velocities which are not suitable for several fish types. The highest turbulence has been observed in the layer between two main layers "Flow over weir" and "Flow through orifice," because of the relative intensive momentum transfer between the two layers [7].

Higher the energy dissipation rate, harder it is for fish to swim. Fish normally tend to avoid strongly turbulent regions because these regions reduce their critical speed (maximum speed that they can sustain in a current) and increases their energy expenditure [2]. In the weir and pool fishways, notches are provided in the weirs for the passage of middle and surface swimmer fish. The bottom staggered orifices are provided for the bottom swimmer fish and to allow fish passage during the dry seasons [6].

Fishway efficiency is affected by flow characteristics such as, turbulent kinetic energy, average volumetric energy dissipation rate, and maximum velocity. Lower the values of these parameters higher the efficiency of fishway. The focus of this study is to investigate the effect of fishway slope and downstream condition on its hydraulic efficiency.

2 Computational Model

2.1 Governing Equations of Flow

The commercial software ANSYS-CFX 17.2 was used for the numerical simulation and it solves the continuity and momentum equations:

$$\frac{\partial \rho}{\partial t} + \frac{\partial \rho u_j}{\partial x_j} = 0 \tag{1}$$

$$\frac{\partial \rho u_i}{\partial t} + \frac{\partial \rho u_j u_i}{\partial x_j} = -\frac{\partial p}{\partial x_i} + \frac{\partial \left\{ (\mu + \mu_t)\left(\frac{\partial u_i}{\partial x_j} + \frac{\partial u_j}{\partial x_i} \right) - \frac{2}{3}\rho k \delta_{ij} \right\}}{\partial x_j} + (\rho - \rho_a) g_i \tag{2}$$

Here, i or j denotes 1, 2 or 3, x_1, x_2, x_3, denotes the longitudinal (x), vertical (y) and transversal (z) directions, respectively, u_1, u_2, u_3 denotes the average velocities in x, y, z directions, respectively, δ_{ij} is the kronecker delta, g is gravitational force, k is turbulent kinetic energy, μ is the molecular viscosity of the fluid, μ_t is the turbulent viscosity of fluid, p is the static pressure, ρ and ρ_a are the densities of water and air respectively. The standard k–ε model, where ε is the turbulent kinetic energy dissipation rate is used to calculate turbulent viscosity. k–ε model is based on eddy viscosity concept so that,

$$\mu_{\text{eff}} = \mu + \mu_t \tag{3}$$

where μ_{eff} is the effective viscosity accounting from turbulence. The k–ε model assumes the turbulence viscosity which is linked to the turbulent kinetic energy and dissipation by Eq. (4), where C_μ is a constant.

$$\mu_t = C_\mu \rho \frac{k^2}{\varepsilon} \tag{4}$$

2.2 Verification of the Model Setup

The model setup was verified by applying it to a reported study on Vianney-Legendre fishway in Quebec, Canada [5]. The geometry of the turning pool of Vianney-Legendre fishway (Fig. 1a) was setup in SOLIDWORKS 2016, and fluid domain was extracted using ANSYS-CFX 17.2. Unstructured tetrahedral elements of 0.11 m size were used to mesh the fluid domain. Average water depth of the pool 13 is 2.34 m.

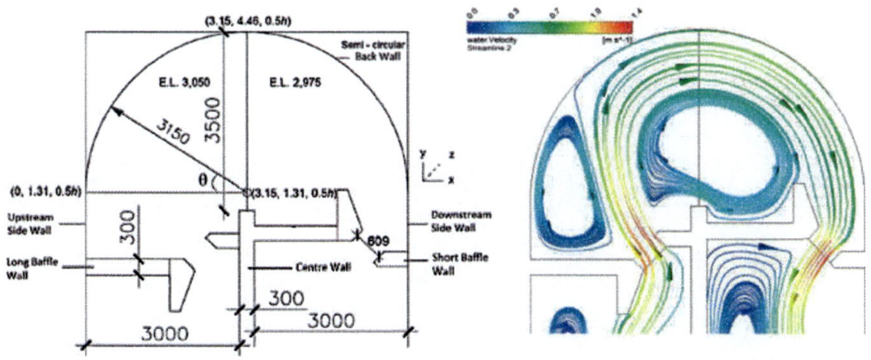

(a) (b)

Fig. 1 **a** Geometry of the turning pool, **b** streamlines on an xy plane at $z = 1.74$ m

Computation was carried out under the same boundary conditions in [5]. Flow rate of 1.63 m^3/s was applied at the inlet boundary, and the outlet boundary was specified as a pressure boundary. Top surface was set as an open boundary. No-slip condition with roughness height of 2 mm was applied at the walls. Simulation was run under steady state, and the convergence criterion for each equation is 0.0001.

Grid independency of results was verified, and results were compared with those reported in [5]. The maximum velocity (V_m) at the slot and the maximum turbulent kinetic energy (k_m) at planes $z = 0.13$ h, $z = 0.5$ h and $z = 0.8$ h, where h is the water depth and z is measured upwards from pool floor (Fig. 1), were obtained. The model results compare well with those reported in [5] as shown in Table 1. The velocity field in x–y plane at an elevation (z) of 1.74 m above the pool floor were compared using velocity vectors as in Fig. 2.

Table 1 Comparison of results

Plane	Maximum velocities at the slot V_m (m/s)		Maximum turbulent kinetic energy (k_m) (m^2/s^2)	
	Marriner et al. [5]	Model results	Marriner et al. [5]	Model results
$z = 0.13$ h	1.35—1.4	1.407	0.07	0.062
$z = 0.5$ h		1.409	0.1	0.086
$z = 0.8$ h		1.379	0.119	0.107

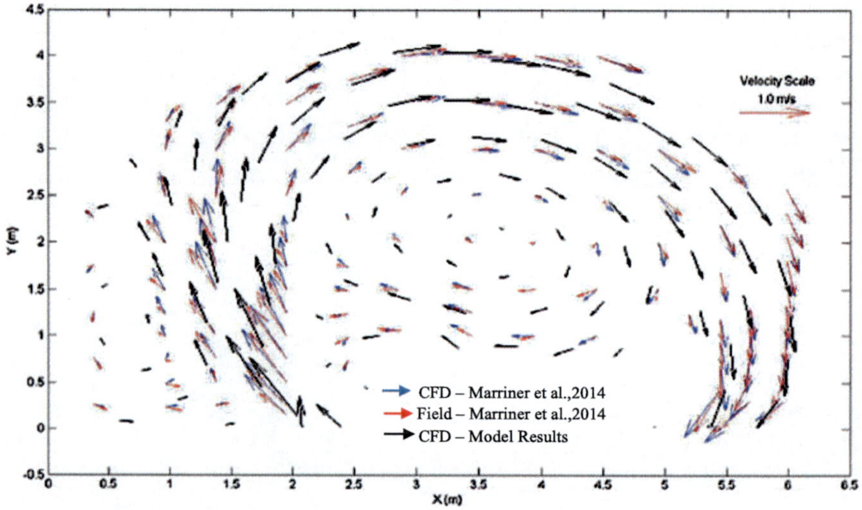

Fig. 2 Velocity field diagram comparison at 1.74 m above pool floor level

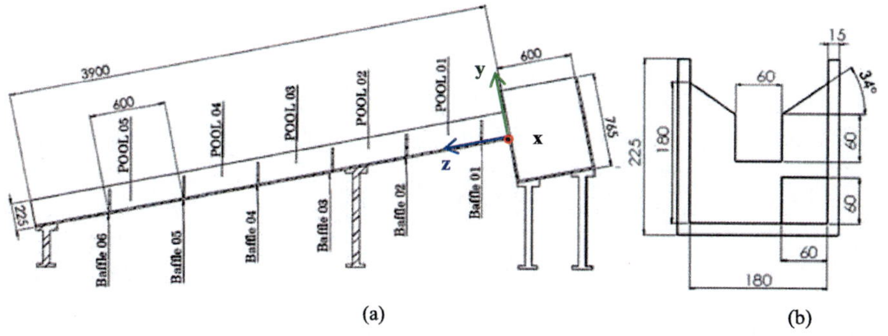

Fig. 3 **a** Experimental setup, **b** dimensions of the baffle (all dimensions are in millimeters)

Table 2 Details of laboratory test cases

Test case	I-1	I-2	II-1	II-2
Slope	11%	13%	11%	13%
Downstream depth	160 mm	200 mm	Free fall	Free fall

3 Model Application

3.1 Experimental Setup

The experimental setup (Fig. 3a) in the laboratory was built to 1:5 geometric scale of the fishway of Moragolla Hydropower Plant (MHP) with a slope of 11%. The slope of the experimental setup is adjustable. Flow was re-circulated, and the downstream depth can be controlled. Experimental runs were carried out for four cases given in Table 2. The discharge of 5.35 l/s was used in the experiment based on Froude Similarity (Fr = 0.33). Velocity measurements were carried out by an electromagnetic flow meter (accuracy to ±0.5%) and water depths along the middle plane of the fishway was measured using a depth gauge (accuracy ±0.1 mm).

3.2 Computation

For the application of the ANSYS-CFX model to simulate the four different cases in Table 2, Hex-dominant mesh of size 0.01 m was used to mesh the fluid domain. Flow rate of 5.35 l/s was set at the inlet boundary condition and outlet was considered as a pressure boundary. Walls were considered as smooth and no-slip wall conditions were applied. The top surface of the fluid domain was set as an open boundary and is a pressure boundary. The volume of fluid (VOF) method solves set of equations through the domain while maintaining a record of the two phases in each computational cell

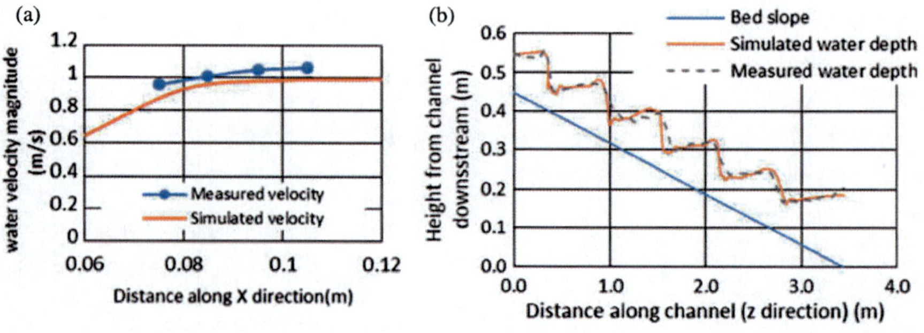

Fig. 4 **a** Velocity comparison in *y*–*z* plane at 10 cm above channel at the downstream of baffle No. 02 in Case I-1, **b** comparison of experimental and simulated water surface profile in Case I-2

[5]. Computations were run under steady-state conditions as a standard free surface model, and the convergence criterion used for each equation is 0.00001. Four mesh sizes were tested to ascertain the grid independency of the results.

The water depth profile at middle vertical plane parallel to *y*–*z* were compared in all the four cases. The velocity comparison of flow over the weir in Case I-1 is in Fig. 4a. The water depth comparison in the Case I-2 is in Fig. 4b. According to the comparisons, good agreement appears between experimental and CFD model results. The small deviations could be due to experimental errors—velocity at an exact point cannot be measured by the electromagnetic flow meters.

Table 3 gives the maximum velocity (U_m), the maximum turbulent kinetic energy (TKE_m), and average energy dissipation rate (ε) for the four different cases derived from the model. These are parameters used to compare the efficiency of the fishways. Results show that the U_m, TKE_m, and ε increases with the increasing slope. Thus, lower slopes are favorable for fish passage. However, this increase can be minimized by increasing the downstream water depth and thus to maintain high fishway efficiency at higher slopes. Hence, the designs can be economized by increasing the slope and controlling the downstream water depth so that the U_m does not exceed the burst speed of ascending fish.

Table 3 Effect of downstream control and slope on fishway efficiency parameters

Case	I-1	I-2	II-1	II-2
U_m (m/s)	1.227	1.592	1.96	2.00
TKE_m (m^2/s^2)	0.118	0.239	0.207	0.262
ε (m^2/s^3)	2.246	3.046	3.054	3.984

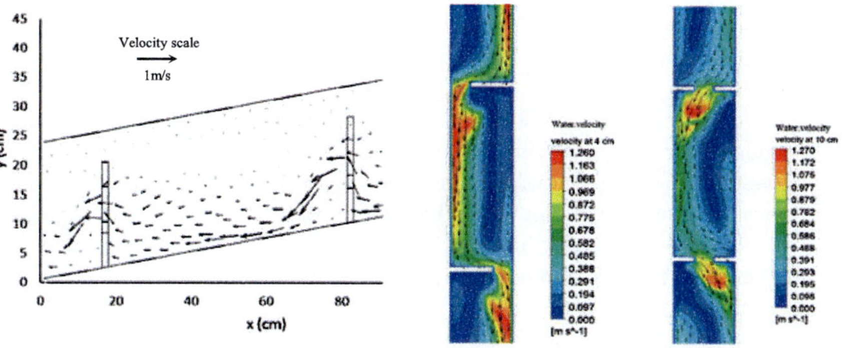

(a) (b) (c)

Fig. 5 Velocity fields in test case I-1 **a** plunging flow regime, **b** velocity field 1—through the orifice in the pool 4 at 4 cm above the pool floor, **c** velocity field 2—in pool 4 at 10 cm above the pool floor

Table 4 Predicted maximum velocities of the prototype

Slope	V_m in model (m/s)	V_m in prototype (m/s)
11% (Case I-1)	1.227	2.74
13% (Case I-2)	1.592	3.56

3.3 *Flow Pattern and Velocity Field*

Based on the lowest values of U_m, TKE_m, ε for each case, the optimum Case I-1 was selected for further analysis. The flow in all simulations were in the plunging flow regime of weir and pool fishway (Fig. 5a). There are two distinct flow patterns at the lower and upper water levels in the fishway pool. Submerged jet flow through the orifice travels from one side of the pool and moves to the next orifice on the opposite side of the pool (Fig. 5b). Whereas, the flow over the weir (nappe) from the weir deflects towardf the side of the pool where the orifice is located in the downstream weir plate. Therefore, the alternating orifices cause weir flow to form recirculation areas for the fishes to swim.

The calculated U_m for the prototype based on Froude similarity is shown in Table 4. Burst speed range of target river water fish in the proposed fishway in the MHP is 4–7 m/s [1]. The obtained maximum velocities are below this range. Therefore, the fishway of slope 13% was found to be acceptable.

4 Conclusions

1. The computed velocity fields show the occurrence of two distinct flow patterns at lower and higher depth levels in the pools of weir and pool type fishways

with bottom orifices. Moreover, there exists a region of low velocity required for ascending fish to rest although high velocities occur at the orifice and the nappe.
2. The model is a useful tool for selecting the optimum slope for an economical construction of orifice and weir type fishway while satisfying the flow conditions conducive for the target fish to ascend through it.
3. Economic design of steeper weir and pool type fishways with high fishway efficiency is feasible by controlling the depth at the downstream of the fishway. For the proposed fishway of the MHP, where the burst speed of target fish is in the range of 4–7 m/s, the fishway slope of 13% is found to be acceptable with downstream depth control.

References

1. Das MK, Hassan MA (2008) Statutes of Fish Migration and passes with special reference to India (Central Inland Fisheries Research Institute, Bulletin no. 156)
2. Enders E, Boisclair D, Roy A (2005) A model of total swimming costs in turbulent flow for juvenile Atlantic salmon (Salmo salar). Can J Fish Aquat Sci 62(5):1079–1089
3. Katopodis C (1992) Introduction to fishway design. Freshwater Institute, Central and Arctic Region, Department of Fisheries and Oceans, 501 University Crescent, Winnipeg, Manitoba, Canada, p. 3–6
4. Katopodis C, Kells J, Acharya M (2001) Nature-like and conventional fishways: alternative concepts? Can Water Resour J 26(2):211–232
5. Marriner BA, Baki ABM, Zhu DZ, Thiem JD, Cooke SJ, Katapodis C (2014) Field and numerical assessment of turnical pool hydraulics of a vertical slot fishway. Ecol Eng 63:88–101
6. Nippon Koei, Fichtner (2017) Fish Ladder, Moragolla Hydropower Project. Ministry of Power and Renewable Energy, Ceylon Electricity Board, Democratic Socialist Republic of Sri Lanka
7. Yagci O (2010) Hydraulic aspects of pool-weir fishways as ecologically friendly water structure. Ecol Eng 36(1):36–46

Airflow and Particle Transport to the Terminal Bronchioles During Heliox Breathing

Suvash C. Saha and Mohammad S. Islam

Abstract The understanding of the airflow and particle transport to the deeper airways is important for targeted drug that transport to the terminal bronchioles. The available literature reports only a small portion of the pharmaceutical drug can reach to the targeted position of the lower airways. This study aims to develop a numerical framework to transport the drug particle to the lower airways of a whole lung model. A digital 17-generation lung model is used to perform the airflow and particle transport study. ANSYS Fluent 19.2 solver is used for the numerical calculations. Discrete phase method and species transport model are used for the particle transport and heliox gas mixture, respectively. The numerical result illustrates higher near-wall velocity magnitude for the air-breathing than heliox breathing. The species transport model shows lower deposition at the upper airways than the air-breathing. The findings of the present study would improve the understanding of heliox therapy, which potentially improves the targeted drug delivery efficiency to the lower airways.

Keywords Helium-oxygen · Aerosol particle · Deposition · Drug delivery

1 Introduction

Breathing gas-exchange in obstructive and narrowed airways of the human lung is complex. Different respiratory diseases like asthma, chronic obstructive pulmonary disease (https://lungfoundation.com.au/health-professionals/clinical-resources/copd/copd-the-statistics/) occur the different morphological change of the lung airways, including reduced lung volume, and decrease of the thoracic cavity. The reduced lung volume and narrowed airway occur shortness of breath and chest tightness. The human lung also experiences reduced volume with aging as the airway muscles, and tissues lose their ability to keep the bronchioles open [23]. Acute asthma and COPD patient experience higher airflow resistance within the bifurcating airways

S. C. Saha (✉) · M. S. Islam
School of Mechanical and Mechatronic Engineering, University of Technology Sydney, Sydney, NSW 2007, Australia
e-mail: suvash.saha@uts.edu.au

L. Venkatakrishnan et al. (eds.), *Proceedings of 16th Asian Congress of Fluid Mechanics*, Lecture Notes in Mechanical Engineering,
https://doi.org/10.1007/978-981-15-5183-3_57

[24]. Helium-oxygen (Heliox) gas mixture has been used for decades to reduce the airflow resistance within the obstructive airways. The heliox gas mixture improves the gas-change efficiency [24] and improves the mechanical ventilation of the narrowed airways of acute asthma and bronchiolitis [14]. Heliox gas mixture is less dense than air, and the lower density of heliox mixture influences the gas-exchange by decreasing the airway resistance [18]. The less dense heliox mixture creates less turbulent intensity at the larynx and tracheal section of the lung and reduces the pressure gradient at the extra-thoracic region. The use of helium gas for the treatment of acute asthma and obstructive airways is first reported by Barach [3]. The use of heliox gas mixture for the treatment of respiratory diseases becomes popular in later 1970s when the death rate of asthma patient rise [20]. The use of heliox inhalation therapy improves airway ventilation [5] and decreases airway acidosis [13]. A wide range of clinical measurement has been conducted for the acute asthma patient, and the heliox inhalation therapy is found more efficient for heliox-based nebulization than air-driven nebulization [2, 15, 16, 21]. Apart from the all positive findings, some clinical measurement is found less or no beneficial for the heliox-driven nebulization for adult asthma patient [4, 6, 19]. However, their study is performed for a specific subject of disease and not applicable for all asthma patient. Almost, all of the heliox-based study only consider fluid flow in the extra-thoracic airways. The first particle-based inhalation therapy reports heliox mixture that is more efficient than air [1]. Later, nebulized aerosol therapy for obstructive child airway demonstrates that heliox improves the efficiency of the radio-aerosol transport to the distal alveoli [17]. Recently, a numerical study for CT-based upper airway model reports that heliox inhalation therapy is better than air inhalation [7]. The numerical study considered only the first three generations of the lung, and the understanding of the particle transport to the lower airways during heliox breathing is still unknown. The heliox inhalation therapy for a whole lung model is essential for better understanding of the heliox inhalation. This study aims to develop an advanced modeling framework for heliox breathing for a large-scale airway model.

2 Numerical Methods

The present large-scale digital reference model developed by Schmidt et al. [22], which is derived from high-resolution computed tomography imaging of an vitro preparation. The large-scale 17 generations model consists of 1453 bronchi and exhibits highly asymmetric branching pattern. The airway branching pattern is based on Horsfield order. ANSYS meshing module is used to construct the unstructured tetrahedral mesh for the bifurcating airways. A fine inflation layer mesh is employed near the wall for complex flow fields. Dense tetrahedral elements are generated at the bifurcating airways. A proper mesh refinement test has been performed, and the final mesh contains about 34 million of the computational cells. The detail of the mesh information and grid refinement can be found in the author's other study

[8–10]. ANSYS 19 solver is used for the overall calculations. Euler-Lagrangian-based discrete phase model [11] (DPM), and species transport model is considered to calculate the microparticle transport and deposition in the upper airways. The air is considered as the primary phase, and the particle is the secondary phase in DPM. The equation for conservation of mass can be written as follows:

$$\frac{\partial \rho}{\partial t} + \nabla.(\rho \vec{v}) = S_{\mathrm{m}} \tag{1}$$

The source S_{m} is the mass added to the continuous phase from the dispersed second phase and any user-defined sources. The conservation of momentum in an inertial (non-accelerating) reference frame is described by;

$$\frac{\partial}{\partial t}(\rho \vec{v}) + \nabla.(\rho \vec{v}\vec{v}) = -\nabla p + \nabla.\left(\mu\left[\left(\nabla \vec{v} + \nabla \vec{v}^{T}\right) - \frac{2}{3}\nabla.\vec{v}I\right]\right) + \rho \vec{g} + \vec{F} \tag{2}$$

where p is fluid static pressure, $\rho \vec{g}$ is body force due to gravity, and $\vec{F}$ is body force due to external (particle-fluid interaction) force. The stress tensor is associated with the molecular viscosity (μ) and the unit tensor (I).

The chosen k-ω model is based on the Wilcox [25], which incorporates modification of compressibility, low-Reynolds number effects and shear flow spreading. This is also based on model transport equations for turbulence kinetic energy (k) and the specific dissipation rate (ω).

Turbulence kinetic energy and specific dissipation rates are obtained from the following transport equations:

$$\frac{\partial}{\partial t}(\rho k) + \frac{\partial}{\partial x_i}(\rho k u_i) = \frac{\partial}{\partial x_j}\left(\Gamma_k \frac{\partial k}{\partial x_j}\right) + G_k - Y_k + S_k \tag{3}$$

And

$$\frac{\partial}{\partial t}(\rho \omega) + \frac{\partial}{\partial x_i}(\rho \omega u_i) = \frac{\partial}{\partial x_j}\left(\Gamma_\omega \frac{\partial \omega}{\partial x_j}\right) + G_\omega - Y_\omega + S_\omega \tag{4}$$

where G_k is the turbulence kinetic energy due to mean velocity gradients, G_ω is the generation of ω. Γ_k and Γ_ω are the effective diffusivity of k and ω, respectively. Y_k and Y_ω represent the dissipation of k and ω due to turbulence, respectively. S_k and S_ω are source terms.

The effective diffusivities for the k-ω model are given by

$$\Gamma_k = \mu + \frac{\mu_t}{\sigma_k} \tag{5}$$

$$\Gamma_\omega = \mu + \frac{\mu_t}{\sigma_\omega} \tag{6}$$

where σ_k and σ_ω are turbulent Prandtl numbers, and μ_t is turbulent viscosity defined by

$$\mu_t = \alpha^* \frac{\rho k}{\omega} \tag{7}$$

Low Reynolds correction is used for the k-ω option. The coefficient α^*, damps turbulent viscosity, causing a low-Reynolds number correction, which is given by

$$\alpha^* = \alpha^*_\infty \left(\frac{\alpha^*_0 + \frac{\mathrm{Re}_t}{R_k}}{1 + \frac{\mathrm{Re}_t}{R_k}} \right) \tag{8}$$

The species transport equation for ith species

$$\frac{\partial}{\partial t}(\rho Y_i) + \nabla.(\rho \vec{v} Y_i) = -\nabla.\vec{J}_i + R_i + S_i \tag{9}$$

where the net rate of production of species i is R_i and S_i is the rate of creation by addition from the dispersed phase plus any user-defined sources. An equation of this form is solved for $n - 1$ species where n is the total number of fluid phase chemical species present in the system. Since the mass fraction of the species must sum to unity, the nth mass fraction is determined as one minus the sum of the $n - 1$ solved mass fractions. To minimize numerical error, the nth species is selected as that species with the overall largest mass fraction.

The mass diffusion for the turbulent case is defined as;

$$\vec{J}_i = -\left(\rho D_{i,m} + \frac{\mu_t}{SC_t} \right) \nabla Y_i - D_{T,i} \frac{\nabla T}{T} \tag{10}$$

where SC_t is the turbulent Schmidt number.

The overall numerical investigation is performed for different physical activity conditions, and different diameter microparticles are considered. The aerosol particle density is considered as 1100 kg/m^3. For the heliox mixture model, 80% helium and 20% oxygen are used. The velocity inlet and pressure outlet boundary conditions are used for the present study [12]. A total of 15,000 particles are injected from the surface of the tracheal inlet. The stationary wall and no-slip shear condition are used. The wall DPM boundary condition is used as trap, and heat flux thermal condition is used at the wall. All of the particles are injected at once.

3 Results

The present numerical study analyzed the airflow and particle transport to the terminal bronchioles of a large-scale 17-generation model. The overall investigation is performed for heliox and air-breathing. The numerical simulation is performed for 15 and 60 l/min flow rates. The airflow and pressure distribution are analyzed during heliox and air-breathing. The overall investigation is carried out for one-way inhalation, and monodisperse particles are considered for the present study.

Airflow contours at the tracheal region and the different lobes of the lung are investigated for heliox and air breathing. Figure 1 shows the velocity contours at different selected planes of the lung at 15 l/min flow rate. The velocity contour at different selected planes on five different lobes is found different for heliox and air-breathing. At the tracheal region, the velocity contour shows a fully developed flow contour for heliox and air-breathing. The velocity contour illustrates a lower velocity magnitude near the tracheal wall for heliox breathing than air-breathing. The density of the heliox mixture (0.5 g/l) is significantly lower than the air density (1.25 g/l). The lower density of the heliox mixture creates lower resistance for the airflow than air and reduces the turbulent fluctuation at the tracheal wall. The lower turbulent intensity and lower velocity magnitude near the airway wall for heliox breathing eventually reduce the deposition efficiency at the upper airways. The velocity contours at the RUL and RML lobes are found significantly different for heliox and air-breathing. For air-breathing case, the vortex is generated at the RUL and RML at 15 l/min flow rate. The higher turbulent fluctuation, pressure-driven force, and complex curvature of the bifurcating airways influence the flow pattern for highly dense air than less dense heliox mixture. The velocity contour at RUL and RML lobes also shows lower velocity magnitude near the airway wall for heliox breathing than air. The velocity contours at the LUL and RUL show a similar velocity distribution at 15 l/min flow rate. The overall velocity contour for the LUL and LLL for heliox breathing is found different than air-breathing.

The pressure contour for the 17-generation model at different flow rates is investigated. Figure 2 shows the pressure contour comparison for heliox and air-breathing. Figures 2a, b show the pressure contour for heliox and air-breathing, respectively. The overall pressure magnitude shows higher pressure at the upper airways for air-breathing than heliox breathing. The higher pressure at the upper airways indicates higher airway resistance for air-breathing. The maximum calculated pressure for air-breathing is slightly higher than heliox breathing. Figures 2c, d show the pressure contour for heliox and air-breathing at 60 l/min flow rate, respectively. The maximum pressure for air-breathing is found significantly higher than air-breathing at 60 l/min flow rate. Figure 2e shows the pressure drop at RLL for heliox and air-breathing.

The aerosol particle transport to the terminal airways of the 17-generations model has been investigated. Figure 3 shows the deposition pattern of 5-μm diameter aerosol particle for air and heliox breathing. Figure 3a shows the deposition pattern for air-breathing, and Fig. 3b shows the deposition pattern for heliox breathing. The overall deposition pattern shows more aerosol particle deposition for air-breathing than heliox breathing. The general deposition scenario of 5-μm diameter particle reports that a higher number of particles are deposited at the upper airways for air-breathing than heliox breathing. The higher density of the air influences the overall deposition pattern during air breathing. The calculated velocity contour in Fig. 1 also supports the deposition pattern of Fig. 3. For air-breathing, the velocity magnitude

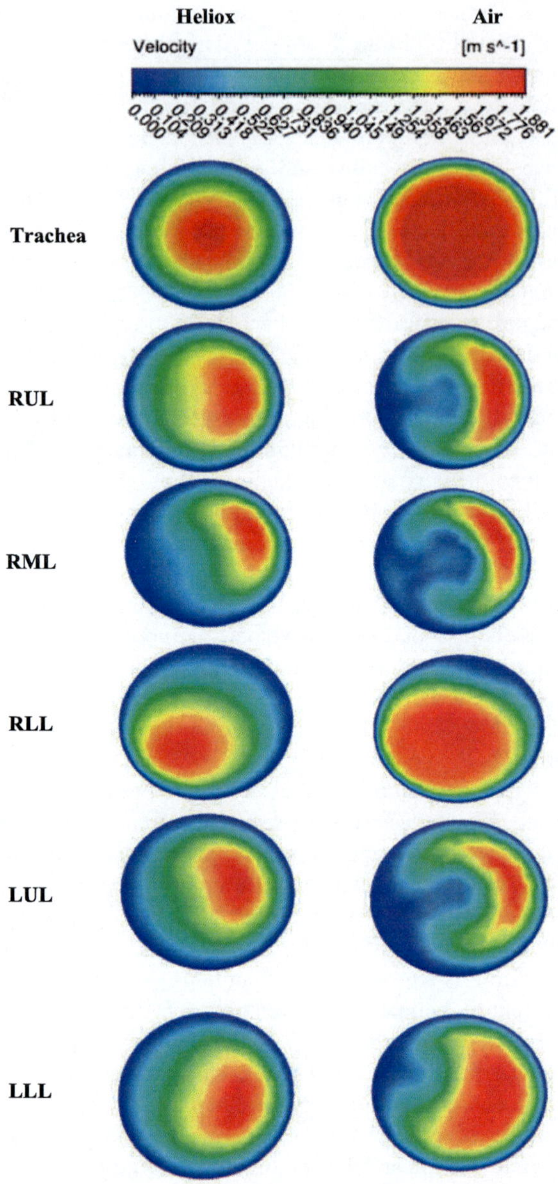

Fig. 1 Velocity contour at the selected planes of the 17-generation model for heliox and air-breathing (RUL = right upper lobe, RML = right middle lobe, LLL = right lower lobe, LUL = left upper lobe and LLL = left lower lobe) at 15 l/min flow rate

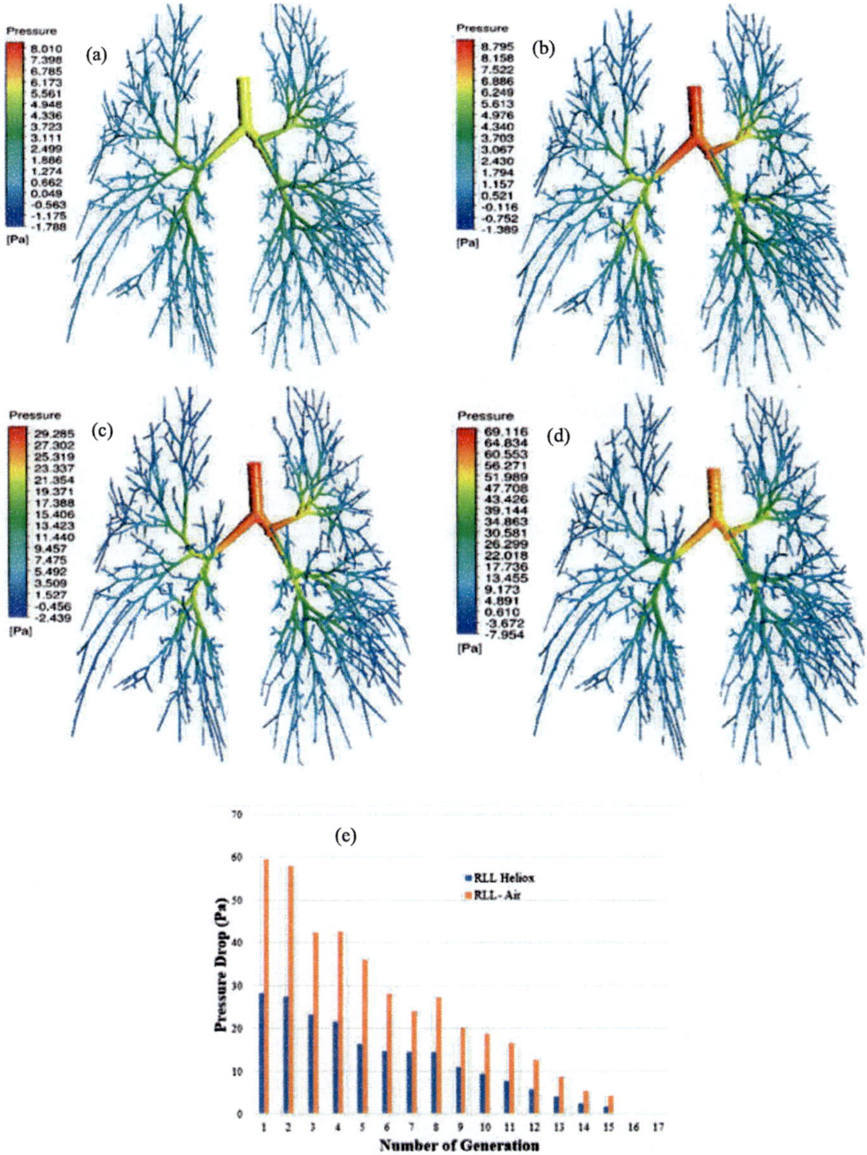

Fig. 2 The pressure distribution for 17-generations model at different flow rates, **a** heliox-15 l/min, **b** air-15 l/min, **c** heliox-60 l/min, **d** air-60 l/min and **e** pressure drop at 60 l/min

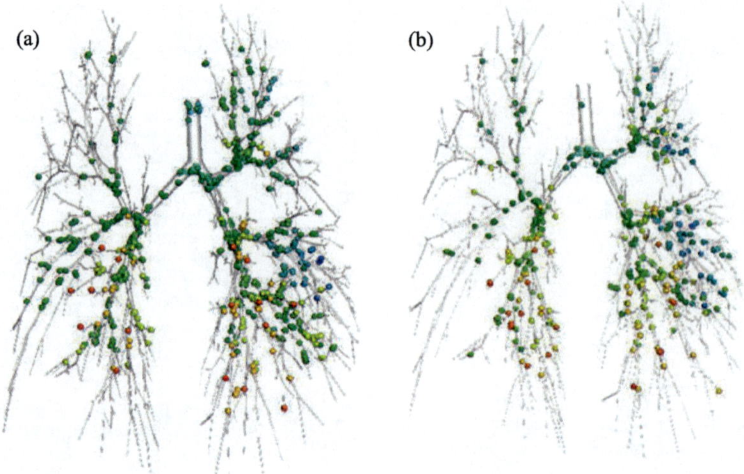

Fig. 3 Aerosol particle deposition pattern in a 17-generation lung model, **a** 60 l/min flow rate-Air and **b** 60 l/min flow rate-Heliox

near the airway wall is found higher than heliox breathing, and the higher near-wall velocity magnitude influences the deposition pattern.

4 Conclusions

This study investigates aerosol particle transport and deposition to the terminal bronchioles for heliox breathing. The airflow and particle transport comparison is performed for different flow rates. Air and helium-oxygen breathing are considered for the present study. The numerical study calculated deposition pattern for different flow rates. The major conclusions from the present study are listed below:

- The velocity magnitude near the airway wall for heliox breathing is lower than the air-breathing.
- The calculated pressure at the upper airways for air-breathing is higher than the heliox breathing at 15 l/min flow rate. At 60 l/min flow rate, the maximum pressure for air-breathing is found 2.36 times higher than heliox breathing.
- The overall deposition is found higher for air-breathing than heliox breathing.

The findings of the present study, along with more case-specific study, will increase the knowledge of the field. A comprehensive patient-specific mixture model will be considered in the future.

Acknowledgements The authors would like to acknowledge the computing facility of UTS Tech Lab. The authors also would like to thank Dr Tevfik Gemci for providing the geometry data.

References

1. Anderson M, Svartengren M, Bylin G, Philipson K, Camner P (1993) Deposition in asthmatics of particles inhaled in air or in helium-oxygen. Am J Respir Crit Care Med 147:524–528
2. Bag R, Bandi V, Fromm RE Jr, Guntupalli KK (2002) The effect of heliox-driven bronchodilator aerosol therapy on pulmonary function tests in patients with asthma. J Asthma 39:659–665
3. Barach AL (1935) The use of helium in the treatment of asthma and obstructive lesions in the larynx and trachea. Ann Intern Med 9:739–765
4. Dorfman TA, Shipley ER, Burton JH, Jones P, Mette SA (2000) Inhaled heliox does not benefit ED patients with moderate to severe asthma. Am J Emerg Med 18:495–497
5. Gluck EH, Onorato DJ, Castriotta R (1990) Helium-oxygen mixtures in intubated patients with status asthmaticus and respiratory acidosis. Chest 98:693–698
6. Henderson SO, Acharya P, Kilaghbian T, Perez J, Korn CS, Chan LS (1999) Use of heliox-driven nebulizer therapy in the treatment of acute asthma. Ann Emerg Med 33:141–146. https://lungfoundation.com.au/health-professionals/clinical-resources/copd/copd-the-statistics/
7. Islam MS, Saha SC, Young PM (2018a) Aerosol particle transport and deposition in a CT-based lung airway for helium-oxygen mixture. 21st Australasian Fluid Mechanics Conference, Adelaide, Australia
8. Islam MS, Saha SC, Gemci T, Yang IA, Sauret E, Gu YT (2018b) Polydisperse microparticle-transport and deposition to the terminal bronchioles in a heterogeneous vasculature tree. Sci Rep 8:16387
9. Islam MS, Saha SC, Sauret E, Gemci T, Gu Y (2017a) Pulmonary aerosol transport and deposition analysis in upper 17 generations of the human respiratory tract. J Aerosol Sci 108:29–43
10. Islam MS, Saha SC, Sauret E, Gemci T, Yang, IA, Gu Y (2017b) Ultrafine particle transport and deposition in a large scale 17-generation lung model. J Biomech
11. Islam MS, Saha SC, Sauret E, Gu Y, Molla MM (2017c) Numerical investigation of diesel exhaust particle transport and deposition in the CT-scan based lung airway. In AIP Conference Proceedings 1851(1):020092 https://doi.org/10.1063/1.4984721
12. Islam MS, Saha SC, Sauret E, Gu Y, Ristovski Z (2015) Numerical investigation of aerosol particle transport and deposition in realistic lung airway. In: Proceedings of the International Conference on Computational Methods, vol 2. ScienTech Publisher, United States of America, pp 1–9
13. Kass JE, Castriotta RJ (1995) Heliox therapy in acute severe asthma. Chest 107:757–760
14. Kim IK, Corcoran T (2009) Recent developments in heliox therapy for asthma and bronchiolitis. Clin Pediatr Emerg Med 10:68–74
15. Kress JP, Noth I, Gehlbach BK, Barman N, Pohlman AS, Miller A, Morgan S, Hall JB (2002) The utility of albuterol nebulized with heliox during acute asthma exacerbations. Am J Respir Crit Care Med 165:1317–1321
16. Lee DL, Hsu CW, Lee H, Chang HW, Huang YCT (2005) Beneficial effects of albuterol therapy driven by heliox versus by oxygen in severe asthma exacerbation. Acad Emerg Med 12:820–827
17. Piva JP, Barreto SSM, Zelmanovitz F, Amantéa S, Cox P (2002) Heliox versus oxygen for nebulized aerosol therapy in children with lower airway obstruction. Pediatr Crit Care Med 3:6–10
18. Reuben A, Harris A (2004) Heliox for asthma in the emergency department: a review of the literature. Emerg Med J 21:131–135
19. Rivera ML, Kim TY, Stewart GM, Minasyan L, Brown L (2006) Albuterol nebulized in heliox in the initial ED treatment of pediatric asthma: a blinded, randomized controlled trial. Am J Emerg Medicine 24:38–42
20. Rodrigo G, Rodrigo C, Pollack C, Travers A (2001) Helium-oxygen mixture for nonintubated acute asthma patients. Cochrane Database Syst Rev, CD002884-CD002884

21. Sattonnet P, Plaisance P, Lecourt L, Vicaut E, Adnet F, Goldstein P, Chollet C, Marx J, Ecollan P, Lambert Y (2004) The efficacy of helium-oxygen mixture (65%–35%) in acute asthma exacerbations. Emerg Med Australasia 16:A71
22. Schmidt A, Zidowitz S, Kriete A, Denhard T, Krass S, Peitgen H-O (2004) A digital reference model of the human bronchial tree. Comput Med Imaging Graph 28:203–211
23. Skloot GS (2017) The effects of aging on lung structure and function. Clin Geriatr Med 33:447–457
24. Valli G, Paoletti P, Savi D, Martolini D, Palange P (2007) Clinical use of Heliox in asthma and COPD. Monaldi Arch Chest Dis 67
25. Wilcox DC (1998) Turbulence modeling for CFD. DCW industries La Canada, CA

Proper Orthogonal Decomposition of Turbulent Flow in a Straight Square Duct

Hamid Hassan Khan, Syed Fahad Anwer, Nadeem Hasan, and Sanjeev Sanghi

Abstract The self-sustaining turbulent flow in a straight square duct is simulated using direct numerical simulation (DNS). The underlying flow dynamics in a square duct is numerically investigated by using the three-dimensional proper orthogonal decomposition (POD) technique. The method of snapshot POD is an efficient tool to educe coherent structures from the fluctuating component of the DNS database at frictional Reynolds number of 300 (based on duct width and frictional velocity). The coherent structures are manifested through the spatial analysis of POD modes. The spatial POD modes depict the first two most energetic modes as streamwise-independent rolls (or non-propagating modes). The third and fourth most energetic mode are observed as streamwise-dependent structure (or propagating modes). The higher POD modes demonstrate the loop-like structure near the wall region and detached structures in the wake region of the square duct.

Keywords Proper orthogonal decomposition · Turbulent flow · Square duct

1 Introduction

Application of duct flow is ubiquitous in oil and gas pipelines, heat exchangers, turbomachinery, and nuclear engineering. The duct with square cross-sectional geometry often shares similar characteristics to a pipe and channel flow. While the turbulent flow in a square duct is distinguishable due to the presence of Prandtl's second kind secondary flow. The eight-vortex secondary flow in a fully turbulent flow is referred to the mean cross-flow plane orthogonal to the streamwise direction reported by

H. H. Khan (✉) · S. Sanghi
Department of Applied Mechanics, Indian Institute of Technology Delhi, Hauz Khas, New Delhi 110016, India
e-mail: hamid.hasankhan@gmail.com

S. F. Anwer · N. Hasan
Department of Mechanical Engineering, ZHCET, Aligarh Muslim University, Aligarh, Uttar Pradesh, India

L. Venkatakrishnan et al. (eds.), *Proceedings of 16th Asian Congress of Fluid Mechanics*, Lecture Notes in Mechanical Engineering,
https://doi.org/10.1007/978-981-15-5183-3_58

Gavrilakis [1], Huser and Biringen [2]. The two popular philosophies for investigating turbulence are the statistical theory and the coherent structures of turbulence. The classical statistical theory of turbulence mostly contributes to the current understanding of turbulence, but the different directional motions of the turbulent flow are obfuscated in this theory. Since the last few decades, the study of coherent structures (Robinson [3]) for turbulence research is growing. The visualization of coherent structures with instantaneous flow field is unable to capture or buries the flow pattern in the random flow field. Hence, the POD techniques make a better choice for identification of coherent structures. Proper orthogonal decomposition (POD) is a powerful technique to study the most energetic contributions of structures and to understand the dynamics of turbulence. Bakewell and Lumley [4] first used POD to study near-wall coherent structures. Proper orthogonal decomposition is used to decompose the flow fields into an orthonormal set of functions. Each function or modes are associated with different spatial structures and dynamics. The method of snapshots first proposed by Sirovich [5] is used in the present study. Webber et al. [6], Alfonsi and Primavera [7] had shown that for turbulent channel flow, there are two kinds of coherent structures. Later, Duggleby et al. [8] and Duggleby et al. [9] extended the use of POD technique for turbulent pipe flow to investigate the flow dynamics and identify the coherent structures. Reichert et al. [10] and Matin et al. [11] applied POD analysis in a square duct and reported eigenfunctions in a two-dimensional plane.

Three-dimensional study of the flow physics has become a necessity due to the inherent three dimensionality of flow in a square duct. POD has become a potent tool for analysis of a flow field to extract coherent structures for different modes. In the present work, the motivation behind the application of POD on the turbulent flow in a square duct is to study the three-dimensional spatial dynamics of coherent structures. The extended/full version of this work has been published in Khan et al. [12].

2 Methodology

The square duct geometry of width h is displayed in Fig. 1. The dimensionless incompressible Navier-Stokes equations for a streamwise periodic square duct (Huser

Fig. 1 Schematic diagram of a square duct

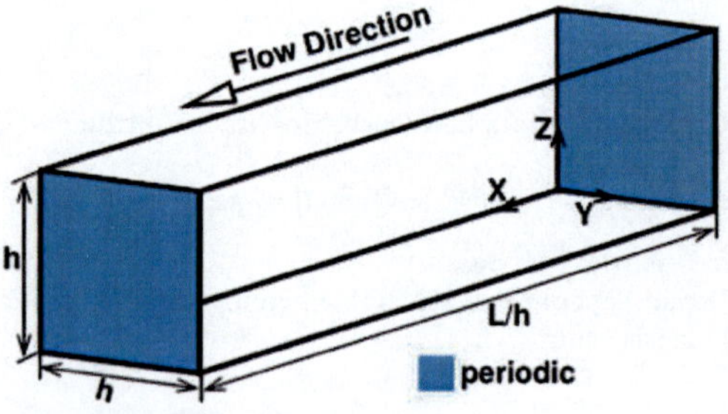

and Biringen [2]) are shown in Eqs. (1) and (2);

$$\nabla \cdot U = 0 \tag{1}$$

$$\frac{\partial U}{\partial t} + (U \cdot \nabla)U = -\nabla P + \frac{1}{\mathrm{Re}_\tau}\left(\nabla^2 U\right) + 4i \tag{2}$$

The governing equations are normalized by frictional velocity (u_τ) and width of the duct (h). The Reynolds number is defined as $\mathrm{Re}_\tau = u_\tau\ h/\nu$, where ν is kinematic viscosity. The flow is periodic in streamwise direction, maintained with the constant pressure gradient. No-slip boundary conditions are imposed on four solid walls. The numerical methods employed to generate DNS data and the technique used for POD analysis of the generated DNS data are explained in the following section.

2.1 Numerical Details

The incompressible Navier-Stokes equations are discretized on a collocated grid using finite difference method. The discretized equations are fourth-order accurate in space. For second-order accuracy in time, implicit Crank-Nicolson scheme is used for the viscous term. Pressure correction Poisson equation (PCPE) is solved using the predictor–corrector scheme similar to simplified marker and cell (SMAC) method. Subsequently, PCPE is solved by parallel GMRES solver. The space resolution and time-step are kept fine enough to capture all turbulence scale. To capture the wall effect, non-uniform grids are employed using the hyperbolic-tangent function in transverse and span-wise directions, whereas the streamwise grid is uniform.

The friction Reynolds number of 300 is used to study fully turbulent flow in a square duct. The length of the square duct in streamwise direction is considered as $L_x = 4\pi h$ and in wall normal direction $L_y = L_z = 1$. The grid points (N_x, N_y, N_z) are 384, 128, 128, respectively; where N_x, N_y and N_z are numbers of grid points in x, y and z-direction. The present numerical procedure was verified with a reference DNS data from Gavrilakis [1]. The mean flow is obtained only for validation of DNS data. The averaging is done through time integration of streamwise average data. The present DNS data was first compared with the mean streamwise velocity profile of reference results, as shown in Fig. 2a, b. The Prandtl second kind eight-vortex secondary flow with streamwise velocity contours are shown in Fig. 2c. The mean velocity profile and secondary flow in Fig. 2 show good agreement with the published results of Gavrilakis [1].

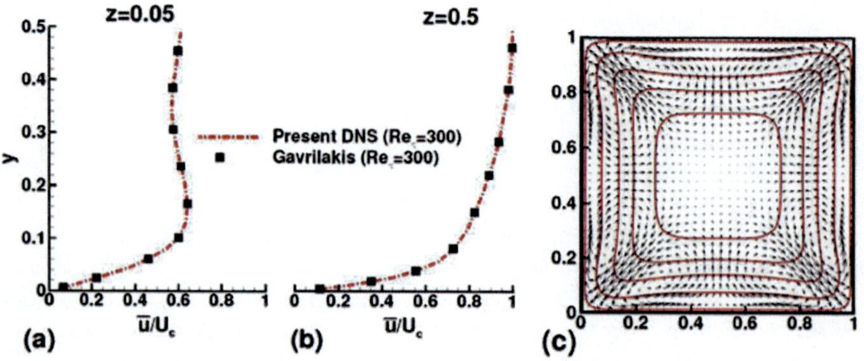

Fig. 2 **a, b** Comparison of present DNS results of mean streamwise velocity profile with reference DNS data of Gavrilakis [1]; **c** Mean secondary flow and streamwise flow contours

2.2 *POD Analysis*

The sampled instantaneous velocity field $U(\mathbf{x}; t)$ obtained from DNS is considered as a snapshot; where $\mathbf{x} = (x; y; z)$ is the position vector in the square duct domain. To study turbulent structures, fluctuating velocity $u(\mathrm{x}; t)$ is used for the rest of the POD analysis. The POD steps are illustrated in the literature [5]. The general procedure starts with the use of fluctuating velocity field (DNS data) to construct a cross-correlation matrix. The cross-correlation matrix eigenfunctions represent the spatial POD modes, while the corresponding eigenvalues indicate energy contributions. The POD modes are arranged in the decreasing order of the energy content. The most energetic modes containing 90% (percentage depends on the flow physics being investigated) of energy are retained to study the flow structure. The obtained POD modes are projected back on to the fluctuating velocity field to acquire the POD coefficients (temporal evolution of the modes). Thus, from the snapshot POD methodology proposed by Sirovich [5], the equations which are inherently used for computation are: (i) Correlation tensor, (ii) eigenproblem, (iii) eigenfunctions or POD mode, and (iv) temporal coefficient. The fidelity check of the POD analysis is done by comparing the reconstructed velocity field and the original velocity field from DNS data, as shown in Fig. 3.

3 Results

Three-dimensional POD analysis is studied using the fluctuating velocity field. The results of POD depend on the number of snapshots and sampling time. The snap independence study gives the total number of snapshots to be used for a particular case. The optimum number of snapshots is accepted for the analysis when the magnitude of computed eigenvalue remains almost the same with further increase in snapshots.

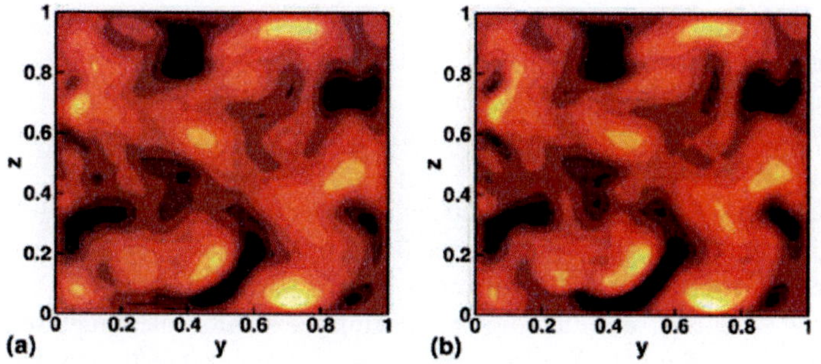

Fig. 3 Comparison of mid-cross-plane of the square duct. **a** reconstructed (90% energy contribution) and **b** the DNS fluctuating streamwise velocity

Therefore, the snap independence study illustrates that the distribution of energy among modes are constant with increase in snapshots. The number of statistically independent snapshots acquired for turbulent flow in the square duct for the present work is 13,000.

The contribution of (90%) energy in POD modes is shown in Fig. 4a. The largescale most energetic structures are attributed by lower POD modes. While the higher POD modes feature small-scale less energetic structures. The first few modes of energy spectra (Fig. 4a) depict higher energy with a substantial drop. The remaining modes contain less energy with lower decay in energy content. The large contribution of lower modes ascribes to most of the flow dynamics. The magnitude of most energetic eigenvalue corresponds to the eigenvalue depicted from 2D POD analysis by Reichert et al. [10] for fully turbulent flow at $Re_\tau = 600$. The first 25 POD modes contain 20% of total energy. The most energetic mode has 1.25% energy contribution

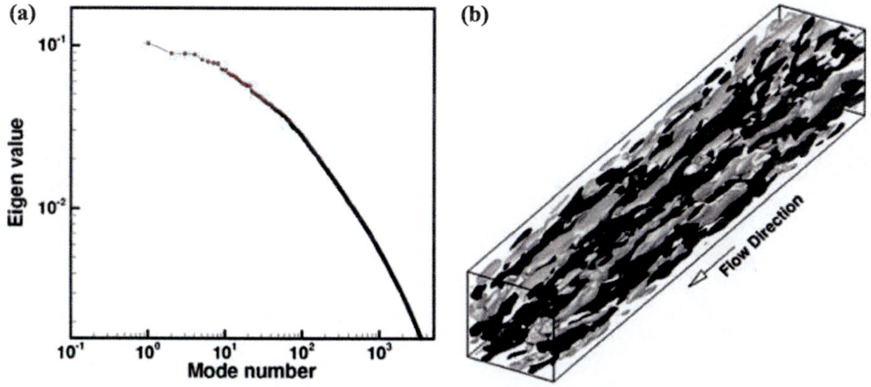

Fig. 4 **a** Energy (eigenvalue) spectra for 5003 POD modes containing 90% of energy, **b** Iso-surfaces of streamwise velocity fluctuation reconstructed from first 5003 POD modes containing 90% of energy (the light surface is positive and the dark surface is negative)

which is reported first time for fully turbulent flow ($Re_{\tau} = 300$) in the square duct and the percentage contribution is found similar to pipe flow reported by Duggleby et al. [8]. The cumulative energy distribution reports 5003 modes are required to represent 90% of total energy (not shown). This illustrates that a large number of structures represent the flow physics. The iso-surfaces of streamwise velocity reconstructed using 5003 modes, which contributes 90% of total energy content are shown in Fig. 4b. It is observed that most of the flow structures are closer to the walls and aligned toward the streamwise direction.

The flow structures are filtered by neglecting the small-scale structure and using only the most energetic POD mode. The POD mode represents the spatial stationary flow structures over time. The first four most energetic POD modes represented by iso-surface of streamwise velocity are shown in Fig. 5; and the streamwise velocity contours on the enlarged cross-plane are also depicted. The structures in the first two POD modes are streamwise-independent, whereas the third and fourth POD modes are streamwise-dependent. Mode 1 and Mode 2 (shown in Fig. 5a, b) are similar to roll modes reported by Weber et al. [6] in a channel flow. Also, the two roll modes

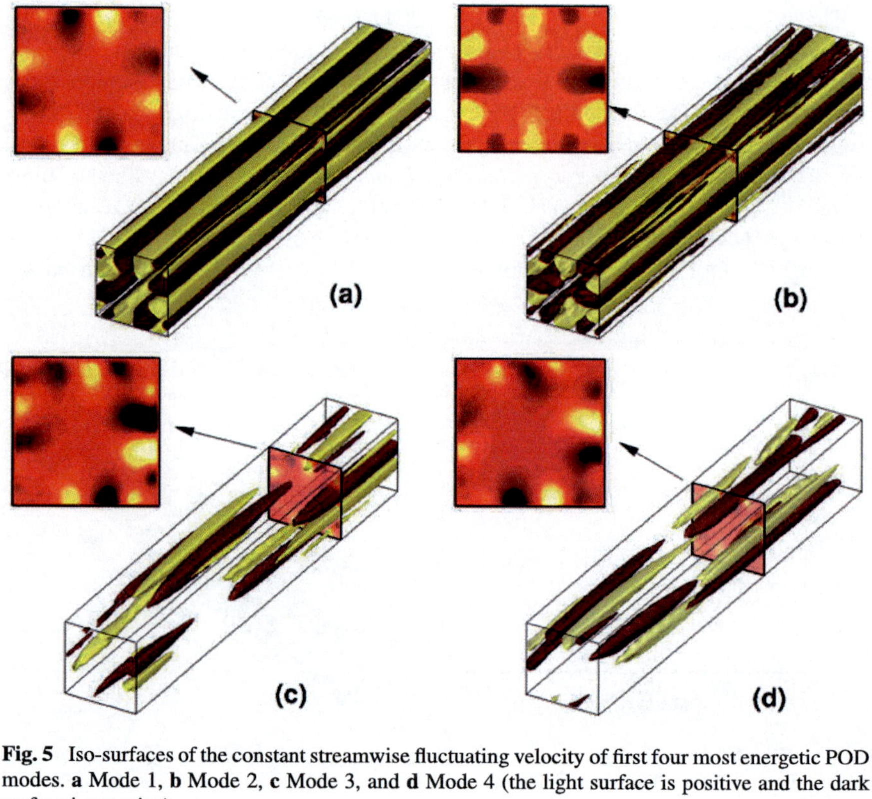

Fig. 5 Iso-surfaces of the constant streamwise fluctuating velocity of first four most energetic POD modes. **a** Mode 1, **b** Mode 2, **c** Mode 3, and **d** Mode 4 (the light surface is positive and the dark surface is negative)

(opposite in sign) are closer to each wall in a square duct, and rolls of the smaller radii are depicted near the corners of the duct. The smaller radii rolls observed for Mode 1 at the corners of the duct show the overlap of the boundary layer at the corner of the duct. Unlike, in pipe flow [8, 9], the streamwise rolls have constant radii.

These streamwise roll modes do not propagate in the flow and are termed as non-propagating modes. The non-propagating modes ensure the existence of low and high-speed streaks. The average low-speed streaks spacing calculated as 105 (approx.) wall units is pertinent with typical streak spacing in a self-sustaining turbulence square duct reported in previous studies. The third most energetic mode (Mode 3) displayed in Fig. 5c is a first streamwise-dependent flow structure. These structures in Fig. 5c, d are visualized as quasi-streamwise rolls elongated in the flow direction. The streamwise-dependent modes are addressed as propagating modes in the previous study [6], on account of its wave-like behavior. The time-independent spatial structures observed for the first two most energetic mode (roll modes) are almost similar to pipe [8, 9] and channel [6, 7] flow structures, while the corner effect in square duct shows additional streamwise rolls with smaller radii. The spatial structures of propagating modes (streamwise-dependent modes) in a square duct are visualized as a quasi-streamwise structure similar to pipe [8, 9] flow, unlike bean-shaped structures in channel flow [7].

The iso-surface of the streamwise velocity representing higher POD modes is shown in Fig. 6. The Mode 50 (Fig. 6a) and Mode 500 (Fig. 6b) illustrate how the structures are evolving both in the axial and cross-flow directions. The streamwise velocity on the enlarged cross-plane in Fig. 6a, b indicates that the structures represent loop-like connections extended from the wall to the outer layer. The nested loop-like connected structures are found to be larger for lower mode (Fig. 6a) as compared to the higher mode in Fig. 6b–d. The highest POD modes 1000 and 1500 display the random small-scale structures of the turbulent flow. These small-scale structures (enlarged cross-plane) represent multiple nested loop-like structures in the wall region and detached structures in the wake region.

4 Conclusions

We conducted DNS of the turbulent flow through a straight square duct at frictional Reynolds number of 300. Three-dimensional POD technique is applied on turbulent flow DNS data to extract 3D coherent structures in the square duct. The powerful POD technique categorizes the three-dimensional coherent structures in lower and higher POD modes based on the energetic contribution. The lower POD modes (or most energetic) structures presented by the first two POD modes depict streamwise-independent rolls (or non-propagating modes). The two dominant counter-rotating streamwise-independent rolls larger in radii are observed adjacent to each wall. While streamwise-independent rolls of smaller radii are observed on account of boundary layer overlap at the four corners of the square duct. The non-propagating modes ensure the existence of low and high-speed streaks, and it helps in momentum

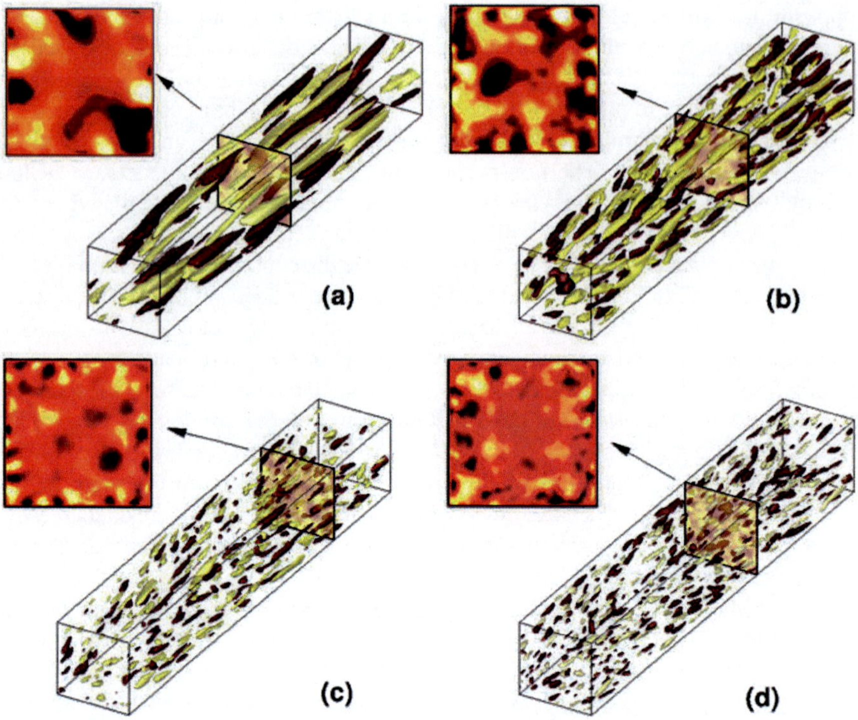

Fig. 6 Iso-surfaces of the constant streamwise fluctuating velocity of higher POD modes. **a** Mode 50, **b** Mode 500, **c** Mode 1000 and **d** Mode 1500 (the light surface is positive and the dark surface is negative)

transfer between the wall and outer region. The third and fourth POD modes illustrate streamwise-dependent structures aligned in the flow direction, suggesting similar to wave-like behavior. The higher POD modes demonstrate the presence of the loop-like connected structures extending from the wall region to the wake region. The small-scale structure representing the highest POD mode illustrates the multiple nested loop-like structures near the wall region and detached structures in the wake region. The loop-like structures conjecture the presence of hairpin structures.

Acknowledgements The authors thank IIT Delhi HPC facility for computational resources.

References

1. Gavrilakis S (1992) Numerical simulation of low-Reynolds-number turbulent flow through a straight square duct. J Fluid Mech 244:101–129
2. Huser A, Biringen S (1993) Direct numerical simulation of turbulent flow in a square duct. J Fluid Mech 257:65–95
3. Robinson SK (1991) Coherent motions in the turbulent boundary layer. Annu Rev Fluid Mech 23(1):601–639
4. Bakewell HP Jr, Lumley JL (1967) Viscous sublayer and adjacent wall region in turbulent pipe flow. Phys Fluids 10(9):1880–1889
5. Sirovich L (1987) Turbulence and the dynamics of coherent structures-I. Coherent structures. Q Appl Math 45(3):561–571
6. Webber GA, Handler RA, Sirovich L (1997) The Karhunen-Loeve decomposition of minimal channel flow. Phys Fluids 9(4):1054–1066
7. Alfonsi G, Primavera L (2006) The structure of turbulent boundary layers in the wall region of plane channel flow. Proc R Soc A: Mathematical, Physical and Engineering Sciences 463(2078):593–612
8. Duggleby A, Ball KS, Schwaenen M (2008) Structure and dynamics of low Reynolds number turbulent pipe flow. Philos Trans R Soc A: Mathematical, Physical and Engineering Sciences 367(1888):473–488
9. Duggleby A, Ball KS, Paul MR, Fischer PF (2007) Dynamical eigenfunction decomposition of turbulent pipe flow. J Turbul 8:N43
10. Reichert RS, Hatay FF, Biringen S, Huser A (1994) Proper orthogonal decomposition applied to turbulent flow in a square duct. Phys Fluids 6(9):3086–3092
11. Matin R, Hellström LH, Hernández-García A, Mathiesen J, Smits AJ (2018) Coherent structures in turbulent square duct flow. Int J Heat Fluid Flow 74:144–153
12. Khan HH, Anwer SF, Hasan N, Sanghi S (2020) Dynamics of coherent structures in turbulent square duct flow. Phys Fluids 32(4):045106

Influence of Buoyancy on Flow Past a Circular Cylinder Near a Moving Wall

Sartaj Tanweer, Anupam Dewan, and Sanjeev Sanghi

Abstract Influence of buoyancy on vortex shedding for flow past a circular cylinder placed near a moving wall is studied for different values of the gap-ratio. It is introduced by heating or cooling the cylinder. A finite-volume-based tool OpenFOAM is used for the computations. Influence of buoyancy is investigated by observing the mean velocity in the gap-region between the cylinder and the moving wall, and vorticity on the moving wall. A positive buoyancy (Ri > 0) deflects the flow downward which causes an increase in the velocity in the gap-region while a negative buoyancy (Ri < 0) decreases it. For Ri > 0, the baroclinic production of vorticity is negative in the gap-region while for Ri < 0, it is positive. Therefore, for Ri > 0, vorticity on the moving wall increases which causes the strength of the lower shear-layer of the cylinder to decrease while for Ri < 0, the strength of the shear-layer increases. At high values of *G/D*, vortex shedding occurs for Ri > 0 while at low values of *G/D*, vortex shedding occurs for Ri < 0. Interesting results have been observed for *G/D* = 0.3, where vortex shedding occurs for both Ri > 0 and Ri < 0.

Keywords Circular cylinder · Moving wall · Vortex shedding · Buoyancy · Baroclinic production

1 Introduction

Flow past a circular cylinder is an area of significant engineering interest. It is encountered in various practical applications like heat exchangers, hot wire anemometers, seabed pipelines, etc. [1]. At low values of Reynolds number $\mathrm{Re} = {}^{UD}/_{\nu}$, the flow around a circular cylinder is steady, and beyond a critical Re, approximately equal to 50, the flow becomes unsteady and phenomenon of vortex shedding in the form of von Karman street is observed. The onset of vortex shedding in the wake region influences the flow field and heat transfer characteristics significantly. Flow in the

S. Tanweer · A. Dewan (✉) · S. Sanghi
Department of Applied Mechanics, Indian Institute of Technology Delhi, Hauz Khas, New Delhi 110016, India
e-mail: adewan@am.iitd.ac.in

L. Venkatakrishnan et al. (eds.), *Proceedings of 16th Asian Congress of Fluid Mechanics*, Lecture Notes in Mechanical Engineering,
https://doi.org/10.1007/978-981-15-5183-3_59

wake region is considerably influenced by nearby objects, such as, a wall and also by the presence of external forces, such as, buoyancy. Buoyancy effect is measured by a non-dimensional parameter called Richardson number (Ri), which is the ratio of the buoyancy force to the inertial force. When a circular cylinder is placed in a free-stream, the buoyancy tries to trigger the vortex shedding [2]. On the other hand, when cylinder is placed near a stationary wall, the buoyancy tries to suppress the vortex shedding [3]. This dual nature of buoyancy is quite interesting and needs to be investigated in detail. When a circular cylinder is placed near a stationary wall, the boundary-layer formed on the stationary wall interacts with the shear-layer formed on the cylinder surface and influences the flow. The influence of buoyancy cannot be observed properly due to the presence of the wall shear-layer. In order to clearly study the influence of buoyancy on flow, the stationary wall is replaced with a moving wall in the present paper. In the case of a moving wall, no conventional boundary-layer is formed and the influence of buoyancy can be observed clearly.

Some studies have been reported on investigation of flow past a circular cylinder near a moving wall. Nishino et al. [4] experimentally studied flow at two values of Re (0.4×10^5 and 1×10^5) and observed that for the gap-ratio (*G/D*) of 0.4, the vortex shedding became intermittent and a suppression of the vortex shedding occurred for $G/D \leq 0.3$. Huang and Sung [5] examined the vortex shedding from a circular cylinder and observed that the critical value of *G/D* for the suppression of vortex shedding decreased with an increase in Re and for Re > 500, it was independent of Re. They also proposed a method in terms of the streamwise maximum value of the mean velocity in the gap to determine the critical value of *G/D*. Yoon et al. [6] examined the vortex shedding and heat transfer characteristics of a circular cylinder near a moving wall. They observed that with a decrease in the value of *G/D*, the flow stabilized and wake was suppressed. Yoon et al. [7] numerically studied flow and heat transfer characteristics for a circular cylinder placed near a moving wall for $G/D = 0.1$–4. They reported that the pattern of vortex shedding changed from two rows to a single row at $G/D = 0.5$ and a suppression of the vortex shedding occurred for small values of *G/D*. Zhen and Wu [8] examined the process of vortex shedding using the Rayleigh's and Fjortoft's theorems. They examined the streamwise velocity in the gap and observed two inflection points for an unsteady flow while for the steady flow, no inflection point was observed. Bimbato et al. [9] examined flow at a high value of Re and reported that a suppression of the vortex shedding occurred due to the Venturi effect.

Rao et al. [10] performed a linear stability analysis to calculate the critical value of Re for the onset of an unsteady flow. They reported that the critical Re first decreased and then increased with an increase in the gap-ratio. They further reported that for $G/D \leq 0.25$, the flow changed from a 2D steady flow to a 3D steady flow and for $G/D \geq 0.25$, it changed from a 2D steady flow to a 2D unsteady flow. Li et al. [11] studied vortex shedding from a cylinder using the immersed interface method and calculated the critical gap-ratio for the onset of vortex shedding. They reported that it decreased with an increase in Re. Recently, Jiang et al. [12] performed direct numerical simulation (DNS) for two- and three-dimensional flows to examine the instability in a wake. They investigated the role of mean flow rate around the cylinder, shear-layer formed on the moving wall, and the confinement of the wake on the instability of flow. To the best of our knowledge, no study has been carried out to examine the influence of buoyancy on a flow past a circular cylinder placed near a moving wall. In the present paper, flow past a circular cylinder placed near a moving wall is numerically investigated for $0.1 \leq G/D \leq 0.5$ and $-1 \leq \mathrm{Ri} \leq 1$. The problem statement and formulation are provided in Sect. 2. Results and discussion are presented in Sect. 3 followed by conclusions in Sect. 4.

2 Methodology

A circular cylinder of diameter D was placed near a moving wall at a distance of $9D$ (L_u) from the inlet and $40D$ (L_d) from the outlet (Fig. 1). The distance between the moving wall and the cylinder is denoted as G. A slip boundary was placed above the cylinder at a distance of $20D$. The cylinder was maintained at a constant temperature of T_h with the free-stream temperature T_∞. For a positive buoyancy (Ri > 0), $T_h > T_\infty$ and for a negative buoyancy (Ri < 0), $T_h < T_\infty$. The direction of buoyancy was downward (in the negative y direction). The Boussinesq approximation was used for solving the governing equations.

For a two-dimensional, incompressible, laminar flow, the non-dimensional governing equations may be written as

1. Continuity:

$$\frac{\partial u}{\partial x} + \frac{\partial v}{\partial y} = 0 \tag{1}$$

2. Momentum:

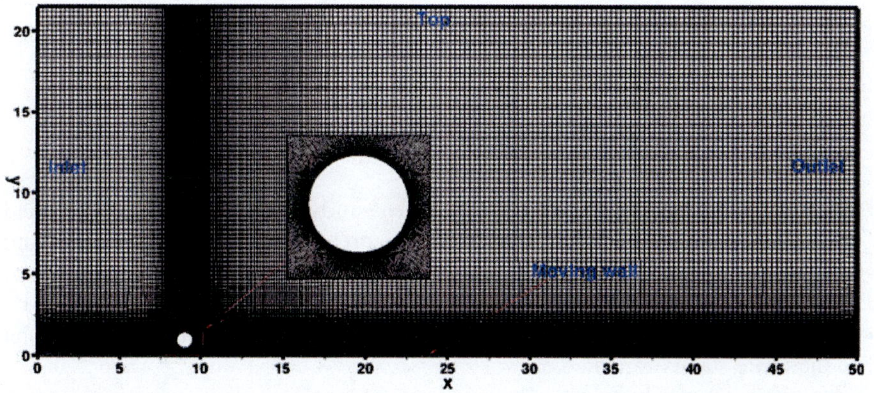

Fig. 1 Computational domain with non-uniform grids. Grids are fine near the cylinder (zoomed view)

$$\frac{\partial u}{\partial t}+\frac{\partial (uu)}{\partial x}+\frac{\partial (uv)}{\partial y}=-\frac{\partial p}{\partial x}+\frac{1}{\mathrm{Re}}\left(\frac{\partial^2 u}{\partial x^2}+\frac{\partial^2 u}{\partial y^2}\right) \tag{2}$$

$$\frac{\partial v}{\partial t}+\frac{\partial (uv)}{\partial x}+\frac{\partial (vv)}{\partial y}=-\frac{\partial p}{\partial y}+\frac{1}{\mathrm{Re}}\left(\frac{\partial^2 v}{\partial x^2}+\frac{\partial^2 v}{\partial y^2}\right)+\mathrm{Ri}\theta \tag{3}$$

3. Energy:

$$\frac{\partial \theta}{\partial t}+\frac{\partial (u\theta)}{\partial x}+\frac{\partial (v\theta)}{\partial y}=\frac{1}{\mathrm{Re}\,\mathrm{Pr}}\left(\frac{\partial^2 \theta}{\partial x^2}+\frac{\partial^2 \theta}{\partial y^2}\right) \tag{4}$$

Boundary conditions:

- Inlet: uniform flow. $u = 1$, $v = 0$ and $\theta = 0$.
- Moving wall: $u = 1$, $v = 0$ and $\theta = 0$.
- Cylinder: no slip. $u = 0$, $v = 0$ and $\theta = 1$.
- Outlet: advective. $\frac{\partial \Phi}{\partial t} + U\frac{\partial \Phi}{\partial x} = 0$ where Φ is u, v, θ.
- Top: slip. $\frac{\partial u}{\partial y} = v = 0$ and $\frac{\partial \theta}{\partial y} = 0$.

The governing equations were numerically solved using a finite-volume-based open-source code OpenFOAM-4.1. The time-derivative terms were discretized by using the second-order backward scheme. For the convection terms, the fourth-order cubic scheme and for the diffusion terms, the second-order central difference scheme were used. A linear interpolation of the cell center values was used for obtaining values at the face centers. The PISO algorithm was used for the pressure–velocity coupling. The solution was assumed to be converged for residual less than 1×10^{-6}.

A non-uniform orthogonal grid with a finer mesh near the cylinder and the moving wall was used. The minimum grid size was kept equal to 0.003D. For choosing an optimal grid, three different sets of grid sizes 273×161, 353×211, and 433×261 were tested and based on the obtained results the grid size of 353×211 was chosen for the computations. The time step (Δt) was chosen such that Courant number was less than 1 in the entire computational domain for all the simulations.

3 Results

In order to validate the code used in the present study, simulations were performed for flow past a square cylinder placed in a free-stream for various values of Re and for flow past a square cylinder placed near a stationary wall for various values of Ri and the present results were compared with the reported results [3, 13]. It can be concluded from Figs. 2 and 3 that the present code provides quite accurate results and therefore stands validated.

In this section, the influence of buoyancy on the vortex shedding process is studied in detail for $G/D = 0.1$, 0.3, and 0.5. The presence of a moving wall influences the flow in two ways: first due to the presence of the wall flow in the gap-region between

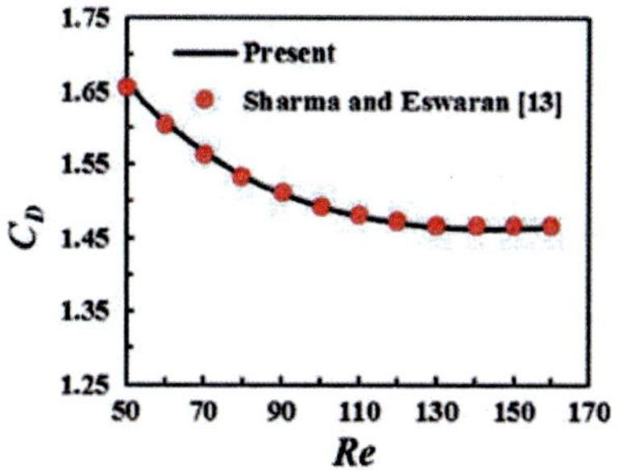

Fig. 2 Drag coefficient for unbounded flow past a square cylinder

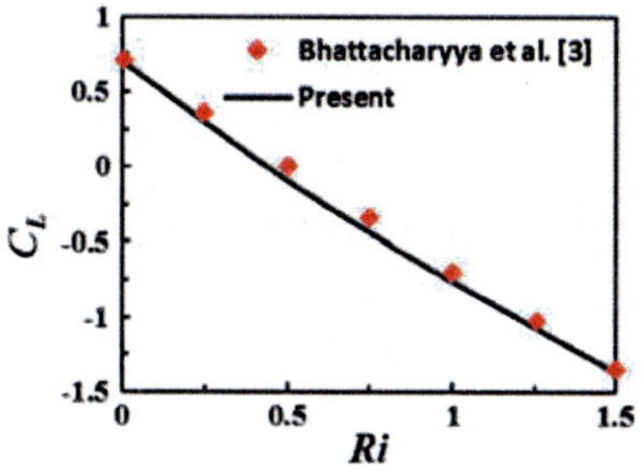

Fig. 3 Lift coefficient for flow past a square cylinder near a wall at Re = 125

the cylinder and the wall changes (it may increase or decrease depending on the gap-ratio and Re); second a boundary-layer is formed on the wall due to a relative velocity between the wall and fluid in the vicinity of the wall. This boundary-layer is quite weak in comparison to the boundary-layer formed in the case of a stationary wall; however, it has a strong influence on the flow. Therefore, the role of buoyancy has been investigated in terms of two parameters: one is the mean velocity in the gap-region between the cylinder and moving wall and the other is the vorticity on the moving wall in the gap-region.

An increase in velocity in the gap-region assists the vortex shedding process by increasing the strength of the lower shear-layer of the cylinder. The boundary-layer formed on the wall interacts with the shear-layer formed near the lower surface of the cylinder. This interaction may assist the vortex shedding process or suppress it. If the vorticity on the moving wall is of the same sign as the vorticity in the lower shear-layer of the cylinder, then the strength of the lower shear-layer increases and the vortex shedding process prevails. On the other hand, if the vorticity on the moving wall is of the opposite sign, then it decreases the strength of the lower shear-layer which may cause the suppression of the vortex shedding.

Figure 4a presents the mean velocity in the gap-region for non-buoyant and buoyant flows. The flow is steady at $Ri = 0$ for the considered case. A positive buoyancy ($Ri > 0$) increases the flow in the gap-region while a negative buoyancy ($Ri < 0$) decreases it. The reason behind this behavior is that in the case of $Ri > 0$, the incoming streamlines are deflected downwards (Fig. 5a) which cause an increase in velocity in the gap-region and for $Ri < 0$, streamlines are deflected upwards (Fig. 5b)

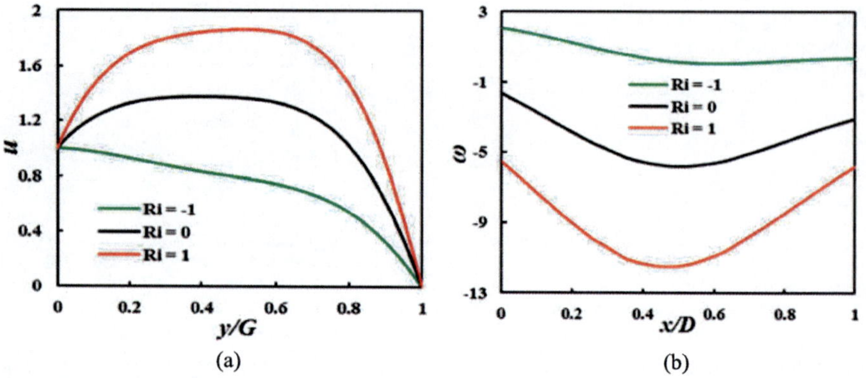

Fig. 4 **a** Mean velocity in the gap-region; **b** vorticity on the moving wall for $G/D = 0.5$ at Re = 70

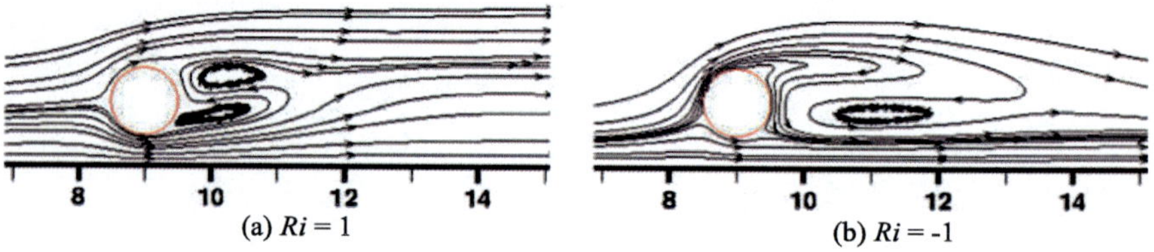

Fig. 5 Time-averaged streamlines for $G/D = 0.5$ at Re = 70

and hence the flow rate in the gap-region decreases. Therefore, it can be concluded that in terms of the mean velocity in the gap-region, a positive buoyancy assists the vortex shedding process while a negative buoyancy tries to suppress it.

Another important factor is the vorticity on the moving wall, which is negative (opposite to the shear-layer vorticity) for Ri > 0 (Fig. 4b) and it decreases the strength of the lower shear-layer formed near the cylinder surface due to its interaction with the wall boundary-layer. For Ri < 0, the vorticity is positive (Fig. 4b) which increases the strength of the lower shear-layer. The change in the vorticity due to the buoyancy is investigated by plotting the baroclinic production of vorticity (the vorticity production due to buoyancy) in Fig. 6. The baroclinic production of vorticity is negative for Ri > 0 and it is positive for Ri < 0. Therefore, in terms of the vorticity on the moving wall, a positive buoyancy tries to suppress the vortex shedding while a negative buoyancy assists the vortex shedding process. For buoyant flows, these two factors (mean velocity and vorticity) are opposite to each other and the vortex shedding process depends on which factor is dominant.

At $G/D = 0.5$ (a high value), a positive buoyancy starts the vortex shedding process (Fig. 7b) while for Ri < 0 no vortex shedding is observed (Fig. 7c). Therefore, we can say that at high values of G/D, mean velocity in the gap-region is a dominant factor and it triggers the vortex shedding process. At $G/D = 0.1$ (a low value), the flow is steady at the considered value of Re for non-buoyant flows (Fig. 7d). For Ri

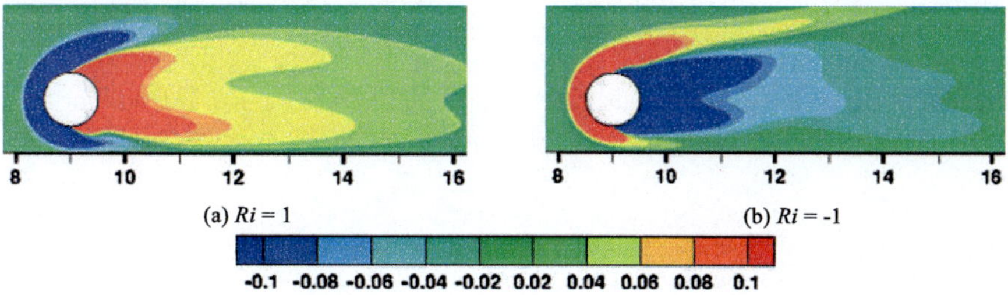

Fig. 6 Time-averaged baroclinic production of vorticity $\mathrm{Ri} \times \frac{\partial \theta}{\partial x}$ for $G/D = 0.5$ at $\mathrm{Re} = 70$

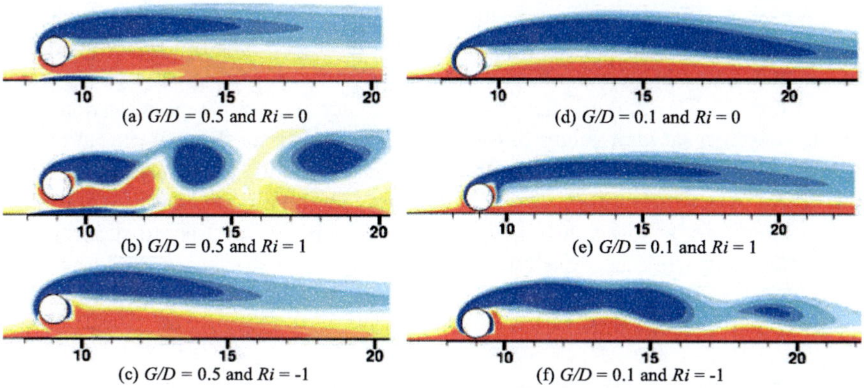

Fig. 7 Instantaneous vorticity contours for $G/D = 0.5$ at $\mathrm{Re} = 70$ and $G/D = 0.1$ at $\mathrm{Re} = 120$ (red: positive and blue: negative)

> 0, no vortex shedding is observed while for Ri < 0, an onset of vortex shedding occurs. Therefore, it can be stated that at low values of *G/D*, vorticity on the moving wall is a dominant factor which increases the strength of the lower shear-layer and causes an onset of the vortex shedding.

An interesting result is obtained for $G/D = 0.3$. At $\mathrm{Re} = 105$, the flow is steady for a non-buoyant case and the lift coefficient (C_{L}) is constant with time (Fig. 8). In

Fig. 8 Instantaneous lift coefficients for $G/D = 0.3$ at $\mathrm{Re} = 105$

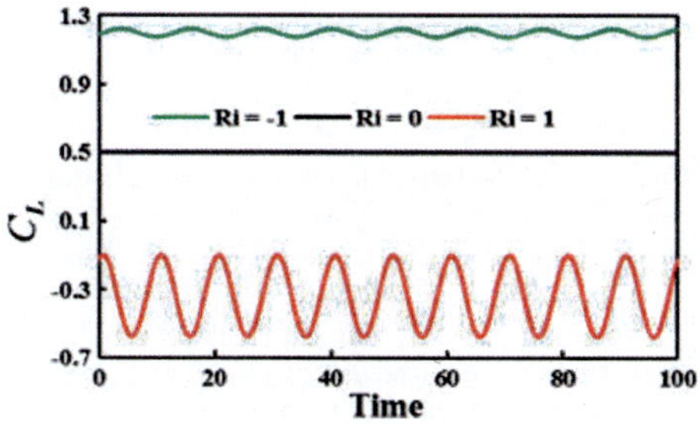

this case, an onset of the vortex shedding occurs for both Ri > 0 and Ri < 0 which can be demonstrated by the time-varying lift coefficient (Fig. 8). For Ri > 0, the vortex shedding occurs due to an increase in the mean velocity in the gap-region. For Ri < 0, the vortex shedding occurs due to a decrease in the vorticity on the moving wall.

4 Conclusions

The influence of buoyancy on the vortex shedding process for the flow past a circular cylinder placed near a moving wall has been studied for different values of the gap-ratio for $-1 \leq \mathrm{Ri} \leq 1$. For Ri > 0, the time-averaged velocity in the gap increases which assists the vortex shedding process while the magnitude of negative vorticity on the wall increases which tries to suppress the vortex shedding. At high values of the gap-ratio ($G/D = 0.5$), the time-averaged velocity in the gap-region is a dominant factor. Therefore, at high values of *G/D*, a positive buoyancy (Ri > 0) tries to trigger the vortex shedding while the negative buoyancy (Ri < 0) tries to suppress it. At low values of the gap-ratio ($G/D = 0.1$), vorticity on the moving wall is a dominant factor. Therefore, at low values of *G/D*, a positive buoyancy tries to suppress the vortex shedding while negative buoyancy tries to trigger the vortex shedding. At a moderate value of the gap-ratio ($G/D = 0.3$), the vortex shedding occurs for both positive and negative values of Ri.

Acknowledgements The first author would express his thanks to Council of Scientific and Industrial Research, New Delhi, India for providing financial support through the award of Junior Research Fellowship [Ref. 09/086(1244)/2015-EMR-I].

References

1. Kieft R, Rindt CCM, Van Steenhoven AA (2007) Near-wake effects of a heat input on the vortex-shedding mechanism. Int J Heat Fluid Flow 28(5):938–947
2. Biswas G, Sarkar S (2009) Effect of thermal buoyancy on vortex shedding past a circular cylinder in cross-flow at low Reynolds numbers. Int J Heat Mass Transf 52(7–8):1897–1912
3. Bhattacharyya S, Maiti DK, Dhinakaran S (2006) Influence of buoyancy on vortex shedding and heat transfer from a square cylinder in proximity to a wall. Numerical Heat Transfer, Part a: Applications 50(6):585–606
4. Nishino T, Roberts GT, Zhang X (2007) Vortex shedding from a circular cylinder near a moving ground. Phys Fluids 19(2):025–103
5. Huang WX, Sung HJ (2007) Vortex shedding from a circular cylinder near a moving wall. J Fluids Struct 23(7):1064–1076
6. Yoon HS, Lee JB, Chun HH (2007) A numerical study on the fluid flow and heat transfer around a circular cylinder near a moving wall. Int J Heat Mass Transf 50(17–18):3507–3520
7. Yoon HS, Lee JB, Seo JH, Park HS (2010) Characteristics for flow and heat transfer around a circular cylinder near a moving wall in wide range of low Reynolds number. Int J Heat Mass Transf 53(23–24):5111–5120

8. Zhen H, Wu F (2012) Lattice Boltzmann simulation of flow past a circular cylinder near a moving wall. Int J Numer Meth Fluids 69(11):1753–1761
9. Bimbato AM, Pereira LAA, Hirata M.H (2013) Suppression of vortex shedding on a bluff body. J Wind Eng Ind Aerodyn 121: 16–28
10. Rao A, Thompson MC, Leweke T, Hourigan K (2013) The flow past a circular cylinder translating at different heights above a wall. J Fluids Struct 41:9–21
11. Li Z, Jaiman RK, Khoo BC (2016) An immersed interface method for flow past circular cylinder in the vicinity of a plane moving wall. Int J Numer Meth Fluids 81(10):611–639
12. Jiang H, Cheng L, Draper S, An H (2017) Two-and three-dimensional instabilities in the wake of a circular cylinder near a moving wall. J Fluid Mech 812:435–462
13. Sharma A, Eswaran V (2004) Heat and fluid flow across a square cylinder in the two-dimensional laminar flow regime. Numerical Heat Transfer, Part a: Applications 45(3):247–269